Genetic Resources of Rice in India

Past and Present

S. D. Sharma

Formerly Principal Scientist and Head
Genetic Resources Division
Central Rice Research Institute
Cuttack (Orissa)

Today & Tomorrow's Printers and Publishers
4436/7, Ansari Road, Daryaganj, New Delhi - 110 002 (India)
Ph : 23242621; 23241021; Fax 23242621; E-mail: ttpp@vsnl.net

ISBN 81-7019-457-1 (India)
ISBN 1-55528-314-4 (USA)

Published by
Today and Tomorrow's Printers and Publishers
4436/7, Ansari Road, Daryaganj, New Delhi - 110 002 (India)
Ph : 23242621; 23241021; Fax 23242621; E-mail: ttpp@vsnl.net

FOREWORD

Rice is the most important cereal of India and is cultivated under a wide range of growing conditions, such as below the sea level farming in Kuttanad in Kerala to high altitude farming in the Himalayas. Because of its adaptation to such variable agro-ecosystems, rice will play an even greater role in the future in our National Food Security System, since we are likely to face adverse alterations in temperature, precipitation and sea level in the coming decades as a result of climate change. Fortunately, India is rich in genetic variability in rice as brought out clearly by the different authors in this timely book. There are rice varieties which can rise above flood level and there are others which can grow under moisture stress and soil salinity.

It is our responsibility to conserve this national treasure for posterity. The conservation continuum starts with *in situ* on-farm conservation by tribal and rural families and extends upto *ex situ* preservation under permafrost conditions as in Svalbard in Norway and Chang La in Ladakh. Regions like the Koraput region of Orissa, which were once rich in genetic diversity in rice, have experienced considerable genetic erosion during the last few decades. Habitat destruction and monoculture play an important role in causing a severe shrinkage in genetic diversity. Steps like the development of Koraput as a globally important agricultural heritage site and the institution of the Genome Saviour Award for recognizing the invaluable contributions of tribal families to genetic resources conservation and enhancement, are likely to reverse the trend in the field of genetic erosion.

Recombinant DNA technology has opened up uncommon opportunities to create novel genetic combinations. For example, the Indian Council of Agricultural Research has initiated a programme for producing a C_4 rice. Photosynthesis takes two different forms – C_3 and C_4 – and rice is a C_3 plant. In C_3 leaves, the CO_2 diffuses into each leaf cell separately where it is captured by rubisco to be turned into sugars. Rubisco, however, occasionally binds to oxygen instead of CO_2. In C_4 leaves, on the other hand, concentric rings of cell "pump" CO_2 into specialized, rubisco-containing cells. By increasing the concentration of CO_2 around the enzyme, this system reduces rubisco's error rate, according to Julian Hibberd of the University of Cambridge and a member of an international C_4 rice project, funded by the Bill and Melinda Gates Foundation.

We owe a deep sense of gratitude to Dr S D Sharma for compiling this important book, in which different chapters have been contributed by eminent rice scientists and geneticists. This is the first time that all the available information on rice genetic resources have been brought together in a reader-friendly manner. The book brings out clearly the diversity of germplasm in rice in India and the interaction between cultural diversity and biodiversity. The book will help to launch an era of biohappiness in rice growing and consuming areas, based on conservation, sustainable use and equitable sharing of benefits. The book will also help breeders to design and develop new breeding material essential for launching a climate resilient rice farming system. The book is thus an appropriate contribution to the International Year of Biodiversity (2010).

M. S. Swaminathan

PROF M. S. SWAMINATHAN
Member of Parliament (Rajya Sabha)
Chairman, M S Swaminathan Research Foundation
Third Cross Street, Taramani Institutional Area
Chennai - 600 113 (India)
Tel: +91 44 2254 2790 / 2254 1229; Fax: +91 44 2254 1319
Email: swami@mssrf.res.in / msswami@vsnl.net

Preface

India lies within the large belt of primary centre of genetic diversity for Asian cultivated rice *Oryza sativa.* Up to 1920s, the genetic diversity of rice available in India was well maintained by its farmers in their farms. Though around 450 varieties of rice were bred during 1920 to 1960, most of these were pure line selections from the local varieties and a few were derivatives of crosses among these pure line selections. As the spread of these improved ones was very limited during those years, there was no danger of loss of genetic diversity of rice in the country. However, the situation changed dramatically with the advent of high yielding rice varieties in mid-sixties of the twentieth century. Vigorous campaign by agricultural extension agencies as well as popularity of these new high yielding rice varieties (HYV) led the farmers to give up cultivation of their traditional ones in favour of these high yielding ones. The age-old traditional rice varieties are, however, the results of unconscious selection by the farmers for millennia; hence conserving this genetic treasure for future crop improvement in the country is quite essential.

Efforts to collect, evaluate and utilize genetic diversity of rice was started in this country with the establishment of the first agricultural research station for improvement of rice in pre-partition India at Dacca (now in Bangladesh) in 1911. This was soon followed by the establishment of a Paddy Breeding Station at Coimbatore in 1912. As more and more rice research stations were established in subsequent years, collection, conservation, evaluation and utilization of the genetic resources of rice gathered momentum. In fact, the first ever project on collection of genetic diversity of rice was launched by the Central Rice Research Institute, Cuttack in 1950s in south Orissa. The second project to collect and conserve the genetic diversity of rice in India was the collection of traditional rice varieties of northeast India by the Indian Agricultural Research Institute, New Delhi soon after the introduction of high yielding rice varieties in the country. The international awareness and endeavour to collect and conserve the genetic diversity of rice came up with the spread of high yielding rice varieties and associated shrinkage of area under traditional ones in 1960s.

To promote conservation of genetic resources of crops world over, the International Board of Plant Genetic Resources (IBPGR) was established by the Consultative Group for International Agricultural Research (CGIAR) at the headquarters of FAO at Rome in 1974. Soon thereafter, the Indian Council of Agricultural Research (ICAR) established the National Bureau of Plant Genetic Resources (NBPGR) at New Delhi in 1976. The Bureau with a network of its substations in the country initiated intensive efforts to collect, conserve and document genetic resources of all crops of the country. Rice being an important crop in the country with its own ecotypic diversity and genetic system received due attention and the NBPGR and the Central Rice Research Institute (CRRI) in collaboration and cooperation of state agricultural universities collected much of the genetic diversity of the crop during 1975 to 1980. The last major drive to explore and collect genetic resources of rice from all over the country was implemented through the National Agricultural Technology Project (NATP) for which NBPGR was the lead centre and in which all the national agricultural institutes and state agricultural universities cooperated. Thus the efforts made so far with regard to collection and conservation of genetic resources of rice have been quite satisfactory though much more needs to be done especially for its characterisation, evaluation, documentation and utilisation.

The editor felt the need for a book that provides comprehensive information on the genetic resources of rice: information on the past efforts to collect this genetic diversity, information on what has been collected from where and when, information on its characterisation, evaluation and documentation and information on what has been utilized and why and how. Not all this information has been documented earlier and whatever has been documented is scattered in annual reports of research stations (which are often out of reach of young scientists now), scientific journals and chapters of books written with different purposes. He felt that it is high time that a book devoted exclusively to genetic resources of rice – their exploration, collection, rejuvenation, characterization, evaluation, documentation, conservation and utilization — should be made available to our new generation of scientists who have completed their education on agricultural sciences but have not yet involved themselves with rice genetic resources.

The book starts with an overview of the situation of the country with regard to genetic resources of rice and then discusses what has been done at various national institutes which are mandated to work on rice. A separate chapter has been devoted to wild rices as their ecotypes differ and they have different types of genetic barriers. A chapter on biotechnology in so far as it provides new tools to help characterize, evaluate or incorporate genes especially from distant species into the cultivated species has also been included. Similarly, a separate chapter on legal and regulatory aspects about the movement of genetic resources and intellectual property rights associated with the use of the material has also been included. Some of the state agricultural universities (earlier their agricultural departments) have played significant roles with regard to collection, conservation, evaluation or utilization of rice germplasm in the past and some others have taken up the work with greater interest in recent years; hence their contributions have also been documented. Many non-government organizations have now come up to take up the task of conserving traditional rice varieties with participation of the farming communities and also with value-addition to the landraces. The book has, therefore, two chapters on the work being done by two such organisations: (a) the M.S. Swaminathan Research Foundation (MSSRF), Chennai and the Participatory Rural Development Foundation (PRDF), Gorakhpur, Uttar Pradesh. The MSSRF has been doing a commendable job with regard to conservation of genetic resources of rice in the Jeypore tract of Orissa that has been considered, based on the morphological and genetic diversity in its rice varieties, a secondary centre of origin of cultivated rice and the PRDF has been collecting, cataloguing and conserving rice germplasm of eastern Uttar Pradesh where there are archaeological evidences of rice cultivation since Neolithic period. Lastly, a chapter on the genetic diversity of rice as gleaned from ancient literature of India has been added. Thus the book has attempted to provide a comprehensive picture of all that has been done so far with regard to rice germplasm in India and also indicates the directions in which the work is expected to advance in the coming years.

The editor hopes that this book will be quite useful not only for students, trainees and young rice scientists who start their career on management of genetic resources of rice in India but also to scientists who help its characterization, evaluation and utilization. Besides, the book will be a valuable record of the work done during the last one hundred years on rice genetic resources in this country. Incidentally, the book celebrates completion of one hundred years of rice research in the country and the year 2010 as the International Year of Biodiversity.

The book has nineteen chapters and covers the information on genetic resources of rice in the entire country ever since the work started in 1911 up to the present time. This could be possible due to active cooperation by fifty eight rice scientists working in various national institutes and state agricultural universities where rice research is an important component. The editor expresses his gratitude to all the contributors for their contribution and cooperation.

The contribution of Prof M S Swaminathan in the field of rice genetic resources in India is indelible. He could visualise the loss of genetic diversity of rice in the country when the semi-dwarf rice varieties had hardly made a mark in the country and initiated and led a project on exploration and collection of rice germplasm from northeast India which was then least explored. He has played a prominent role in the evaluation and utlisation of this crop in India. He was responsible for the establishment of the National Bureau of Plant Genetic Resources which has been playing a major role in the conservation of genetic resources of all crops including rice. The editor is grateful to Prof M S Swaminathan for writing a foreword for this book.

S.D. Sharma

Contributors

J.P. Aditya, Vivekananda Parvatiya Krishi Anusandhan Sansthan, Almora, Uttarakhand.

P.K. Agrawal, Vivekananda Parvatiya Krishi Anusandhan Sansthan, Almora, Uttarakhand.

T. Ahmed, Chief Scientist, Regional Agricultural Research Station, Assam Agricultural University, Titabar–785 630, Assam.

Sharik Ali, Senior Research Fellow, Crop Improvement Division, Central Soil Salinity Research Institute, Karnal, Haryana-132001.

S. Anitha, C/o Dept. of Botany (Genetics), Calicut University, Calicut, Kerala.

L. Behera, Central Rice Research Institute, Cuttack-753006, Orissa.

J.C. Bhatt, Vivekananda Parvatiya Krishi Anusandhan Sansthan, Almora, Uttarakhand.

M. Borthakur, Nagaland University, Medziphema, Nagaland.

U. Chaitanya, Directorate of Rice Research, Hyderabad-500030, Andhra Pradesh.

R. C. Chaudhary, Chairman, Participatory Rural Development Foundation (PRDF), Gorakhpur, India. Email: Ram.Chaudhary@gmail.com, PRDF2008@gmail.com

Susanta Sekhar Chaudhury, Senior Scientist, Biju Pattanaik Medicinal Plants Garden & Research Centre, M.S. Swaminathan Research Foundation, Makaput, Jeypore-764002, Koraput, Orissa.

S.K. Chetia, Senior Scientist, Regional Agricultural Research Station, Assam Agricultural University, Titabar–785 630, Assam.

G.R. Das, Retd Chief Scientist, Regional Agricultural Research Station, Assam Agricultural University, Titabar–785 630, Assam.

T. P. Devi, ICAR Research Complex for NEH Region, Barapani, Meghalaya.

S.R. Dhua, Central Rice Research Institute, Cuttack-753006, Orissa.

N. Dikshit, Senior Scientist, Opp PDKV Campus, Akola-444104, Maharashtra. Email: dikshitn@yahoo.com

J.L. Dwivedi, Professor, Plant Breeding & Sr. Rice Breeder, Crop Research Station, Masodha, P.O. Dabhasemer-224133, NDUAT, Faizabad, Uttar Pradesh.

R.K. Gautam, Senior Scientist (Rice Breeding), Crop Improvement Division, Central

Soil Salinity Research Institute, Karnal, Haryana-132001.

D.K. Hore, Former Officer-in-Charge, NBPGR Regional Station, Umroi Road, NEH Complex (ICAR), Barapani, Shillong-793103, Meghalaya.

P. Khate, ICAR Research Complex for NEH Region, Barapani, Meghalaya.

P. Kumar, Vivekananda Parvatiya Krishi Anusandhan Sansthan, Almora, Uttarakhand.

Arjun Lal, Division of Germplasm Exchange, National Bureau of Plant Genetic Resources, New Delhi-110012.

Leena Kumary S., Professor, Kerala Agricultural University, Rice Research Station, Moncompu Thekkekkara-688503, Alleppey Dt, Kerala.

B. Mishra, Vice Chancellor, Sher-e-Kashmir University of Agricultural Sciences and Technology, Chatha–180009, Jammu, J&K.

S. B. Mishra, Senior Research Fellow, Participatory Rural Development Foundation (PRDF), Gorakhpur, India

Smita Mishra, Network and Training Coordinator, M.S. Swaminathan Research Foundation, 3rd Cross Street, Taramani Institutional Area, Taramani, Chennai 600113. Email: smita@mssrf.res.in

N.K. Motiramani, Indira Gandhi Krishi Vishwavidyalaya, Raipur-492006, Chhattisgarh.

R.L. Pandey, Indira Gandhi Krishi Vishwavidyalaya, Raipur-492006, Chhattisgarh.

D.R. Pani, Scientist (SS) & Officer-in-Charge, NBPGR Base Centre, CRRI Campus, Cuttack-753006. Email-drpani@rediffmail.com

B.C. Patra, Central Rice Research Institute, Cuttack-753006, Orissa

A. Pattanayak, ICAR Research Complex for NEH Region, Barapani, Meghalaya.

K. Pavithran, Professor (Retd), Calicut University, Calicut-673653. Present Address: L.K. Gardens, Pallikal P.O., Malappuram-673653, Kerala.

D. J. Pophaly, Indira Gandhi Krishi Vishwavidyalaya, Raipur- 492006, Chhattisgarh.

J.C. Rana, Principal Scientist, NBPGR Regional Station, Phagli, Shimla-171004, Himachal Pradesh. Email-ranajc2003@yahoo.com

S. Robin, Professor, Department of Rice, Centre for Plant Breeding and Genetics, TNAU, Coimbatore–641003.

G. R. Sahu, Indira Gandhi Krishi Vishwavidyalaya, Raipur-492006, Chhattisgarh.

A. K. Sarawgi, Indira Gandhi Krishi Vishwavidyalaya, Raipur-492006, Chhattisgarh.

B. K. Sarma, Ex-Head, Division of Plant Breeding, ICAR Research Complex for NEH Region, Barapani, Meghalaya.

S. Saxena, Principal Scientist (Policy & IPR), ICAR, Krishi Bhawan, New Delhi-110001. Email: ssaxena.icar@nic.in

A.D. Sharma, Technical Officer, Germplasm Conservation Division, NBPGR, Pusa Campus, New Delhi-110012.

R.K. Sharma, Vivekananda Parvatiya Krishi Anusandhan Sansthan, Almora, Uttarakhand, India

S.D. Sharma, Ex-Head, Division of Genetic Resources, Central Rice Research Institute, Cuttack-753006, Orissa. Present address: Neelachakra Enclave, N3/57, IRC Village, Bhubaneswar-751015, Orissa.

S.K. Sharma, Director, NBPGR, Pusa Campus, New Delhi-110012.

N. Shobha Rani, Principal Scientist & Head, Crop Improvement Section, Directorate of Rice Research, Rajendra Nagar, Hyderabad-500030, Andhra Pradesh.

A.K. Singh, Programme Leader (Rice), Genetics Division, Indian Agricultural Research Institute, New Delhi-11012.

J.K. Singh, ICAR Research Complex for NEH Region, Barapani, Meghalaya.

R.K. Singh, Rice Breeder for Salt Affected Areas, Plant Breeding, Genetics and Biotechnology Division, International Rice Research Institute, Los Banos, Manila, Philippines.

V.P. Singh, Ex-Principal Scientist and Programme Leader` (Rice), Division of Genetics IARI, New Delhi. Present address: E-9, Sector 40, NOIDA-201 301, UP.

Kalyani Srinivasan, Germplasm Conservation Division, NBPGR, Pusa Campus, New Delhi-110012.

M.N. Shrivastava, Indira Gandhi Krishi Vishwavidyalaya, Raipur-492006, Chhattisgarh.

J. Stanley, Vivekananda Parvatiya Krishi Anusandhan Sansthan, Almora, Uttarakhand.

L.V. Subba Rao Diretorate of Rice Research, Hyderabad-500030, Andhra Pradesh.

M. Subramanian, Director of Research, TNAU, Coimbatore-641003. Present address: Plot 9, VOCC (Krishnapuram Colony), 7th Cross Street, Chokkanathapurm, Madurai – 625 014, Tamil Nadu

K. Thiyagarajan, Director, Centre for Plant Breeding and Genetics, TNAU, Coimbatore–641003.

Vandana Tyagi, Division of Germplasm Exchange, National Bureau of Plant Genetic Resources, New Delhi-110012.

G.S. Varaprasad, Diretorate of Rice Research, Hyderabad-500030, Andhra Pradesh.

Vinod Verma, Senior Research Fellow, Crop Improvement Division, Central Soil Salinity Research Institute, Karnal, Haryana-132001.

B.C. Viraktamath, Directorate of Rice Research, Hyderabad-500030, Andhra Pradesh.

CONTENTS

Genetic Resources of Rice in India : 1-45
Editor : S. D. Sharma
Today & Tomorrow's Printers and Publishers, New Delhi - 110 002 (India)

Chapter 1

Genetic Resources of Rice in India: An Overview

S.D. Sharma

Historical Background

Archaeologists have evidences that wild rice was being harvested and eaten in Proto-Neolithic period at Chopani Mando (ca 8080 BC), Koldihwa (ca 6500 BC), Mahagara (5440 BC) and Lahuradewa (ca 5300 BC), all located in the middle Ganga valley (Singh, 2008). By the end of the third millennium BC, rice was being cultivated over a large geographical area stretching from Pirak (Balochistan in Pakistan), Gufkral (Kashmir), Lothal (Gujarat), Ahar (Rajasthan) and Chirand (Bihar) (Mehra, 2005).

The Harappan civilization (2900-1900 BC) was spread over Punjab, Sind and Nothwestern Frontier Province of Pakistan and over the Punjab, Haryana, Rajasthan and Gujarat states of India. However, rice does not seem to be the main crop of these people evidently because rice was not a suitable crop for this agro-climatic region.

The early Aryans (1500–1000 BC) spread themselves from Afghanistan to Punjab, Haryana and northern part of Rajasthan; they were basically pastoral people and cattle was their main wealth. Ratnakar (2006) gives us an idea of their way of life from the *Rig Veda*, their earliest text.

> *The crop most frequently mentioned is* yava *or barley, a cereal that withstands drought, can adapt to high and low altitudes and has a short growing season. The term* odana *may possibly have meant a variety of rice as it was later known and occurs rarely;* vrihi *or rice as it was later known does not occur in the Rig Veda.*
>
> *Prayers for wealth and fertility abound in the Rig Veda, there being prayers for many sons and for large flocks of animals*

> *but not for the control or ownership of extensive fields. A tribal chief was called* gopa *or* gopati, *protector of cattle. Gifts made by patrons to their family priests comprised cattle, slaves, chariots and horses, rather than 'sacks of grains' or 'grants of lands'.*

References to rice, however, abound in the post-Vedic literature known as *Brahmanas, Aranyakas, Samhitas* and *Upanishads*. Rice became a 'respectable' crop for them in this period as it was used in their rituals. They started recognizing varietal differences in rice as we find references to *Ashudhanya, Hayana, Krishnavrihi, Mahavrihi, Shuklavrihi* besides *Nivara* (wild rice) (Mehra, 2005).

By 600 BC, Aryans had moved eastward and formed Mahajanapadas (large settlements) in the middle Ganga valley; rice now became an important crop for them. There are frequent references to different rice varieties such as *Sali, Kalmasali, Gandhasali, Raktasali*, hill paddy, etc in Jain and Buddhist literature.

The first treatise on agriculture *Krishi-Parashara* was written around 400-100 BC by Parashara. He classified the rice varieties into two groups: those directly sown and those that are transplanted. We get a better picture of the varietal diversity of rice in the medical treatises *Charaka Samhita* and *Sushruta Samhita* of the early Christian era. These texts describe many rice varieties and mention their medicinal properties (Nene, 2005).

Some of the names of the rice varieties mentioned by Charaka are *Raktashali* (red), *Mahashali* (large and fragrant), *Kalama* (thick stem), *Shakunahrita* (curved), *Turnaka* (quick maturity), *Deerghashuka* (long awned), *Panduka* (yellowish), *Langula* (tall?), *Sugandhika* (fragrant), *Lohawal* (red), *Shariva* (pointed?), *Pramodaka* (fragrant), *Patanga* (resembling grasshopper/ locust?) and *Tapaniya* (golden or maturing in hot weather), *Shavak* (barley-like), *Hayana* (golden), *Panshu* (dusty), *Vapya* (oblong), and *Naishadaka* (from Nishadha country i.e. modern Kumaon hills), *Shastika* (maturing in 60 days).

Sushruta mentioned many new varieties of rice such as *Lohitaka* (red husk), *Kardamaka* (growing in slimy soil), *Panduka* (yellowish), *Sugandhaka* (fragrant), *Shakunahrita* (curved), *Pushpandaka* (resembling ovary of flower), *Pundarika* (white), *Mahashali* (large and fragrant), *Shitabhiruka* (cold sensitive), *Rodhrapushpaka* (red), *Dirghashuka* (long awned), *Kanchanaka* (golden husk), *Mahishamastaka* (resembling buffalo head), *Hayanaka* (golden), *Dushika* (rice resembling pencil), and *Mahadushika* (?), *Pramodaka* (fragrant), *Mahashastika* (large seeded, maturing in 60 days?), *Kedaraka* (from mountains), *Pitaka* (yellow grain), *Krishnavrihi* (black), *Jatumukha* (freckled), *Nandimukha* (shape like bullock face), *Lavakshaka* (curved grain?), *Twaritaka* (early rice),

Kukkutandaka (oval), *Paravataka* (small, oval) and *Patal* (red).

Around 800 AD, another treatise on agriculture, *Kashyapiya-krishisukti,* was written by one Kashyapa. Kashyapa recognizes three groups of rice: Shali, Kalama and Vrihi and mentions a number of varieties in each group. He also mentions that the different varieties of rice need different types of soils and climates.

Another *Materia Medica*, probably the last one written in the ancient tradition in the series, was by Bhavamisra in the 16th century and is known as *Bhavaprakasa Nighantu.* As Nene (2005) puts it

> *Sri Bhavamisra essentially uses* vrihi *for 4-month and* shali *for 6-month rices. Most of the names are the same as we find in the texts of Charaka and Susruta referred earlier. However, Sri Bhavamisra also states that the number of rice varieties was too large and thus could not be included in the chapter of his treatise.*

We get a view of the varietal diversity of rice in *Ain-e-Akbari* written by Abul Fazal Alami ca 1590. He describes a number of rice varieties that were procured for the royal kitchen and provides the prices of these rice varieties. The varieties mentioned by him are *Sukhdas, Dewzira, Jinjin, Dakah, Dunaprasad, Mushkin, Sada, Samjira, Sathi, Shakarchini* and *Zirhi.* Of these, *Mushkin* was the costliest; *Sada* and *Sukhdas* were next to *Mushkin*. Sathi was the cheapest of them all. It is noteworthy that he does not mention Basmati in this list. The first reference to Basmati is found in *Heer Ranjha* of Waris Shah of Punjab (Nene, 2005).

In the nineteenth century, many British plant explorers came to India to survey and explore this country for its plant diversity. They prepared plant specimens and deposited them in herbaria at Shibpore (Kolkata) or at Saharanpur[1] in India or carried the specimens to England and deposited them at Kew. Roxburgh (1832) has written *Flora Indica* and Hooker (1897) *Flora of British India.* The specimens of *Oryza sativa* available at these herbaria provide the first real view of the diversity in the cultivated rice of that period. Based on the information available up to his time, George Watt (1891), in his *Dictionary of Economic Products of India,* provides detailed information on rice varieties of India and estimates that the total number of rice varieties of Bengal and India. Writing about his own experience (in third person), Watt (1891:529) says:

---

[1] The herbarium of Saharanpur was later shifted to Dehradun and is now in the Forest Research Institute, Dehradun

— he had the opportunity of examining a very extensive series of specimens of rice numbering about 4,000. These were the rices of Bengal alone and he therefore was induced to affirm that if a similarly exhaustive collection had been made in all the other provinces of India, the total number of named cultivated rice might have been found to be little short of 10,000.

The above brief survey indicates that the knowledge about the varietal diversity of rice in India was progressively increasing since the Vedic period up to the end of the nineteenth century but there is no evidence that this knowledge was put to any use for the improvement of rice varieties except that a few seed multiplication and distribution centres were established towards the fag end of the nineteenth century. Serious efforts for the improvement of Indian agriculture with a scientific approach started in the twentieth century only after the country was afflicted with a series of famines in the second half of the nineteenth century.

In the historical perspective, it is interesting to note that the British Government of India made sincere and consistent efforts to introduce Carolina rice of USA into India. This rice was introduced into Carolina state of USA in the seventeenth century presumably from Madagascar and became very successful there. There was a profitable trade between USA and Europe with regard to this rice in the eighteenth and nineteenth centuries and hence the British Government tried to replicate that success story in India. Repeated efforts were made to grow this variety in India in the nineteenth century. According to Watt (1891:638), the difficulty was "not so much to grow the rice and produce a large yield but failure to induce the people to grow it"

The First Phase (1911 to 1928)

The British Government took over the administration of India from the East India Company in 1858. During the next forty years, India suffered through a series of droughts and famines. After each major famine, the Government of India appointed a Commission to go into the causes of famines and suggest remedial measures to ward off such famines in future and to take necessary measures to mitigate the suffering of the people. After the Great Deccan Famine of 1876-78, Lord Litton, the then Governor General of India, appointed a Commission under the presidency of Richard Strachey to go into the whole question of famine and relief. The Commission suggested, along with several other measures, creation of irrigation works (to save the crop from drought as much as possible) and railways (to facilitate transport of food to target areas during famines). It also recommended the setting up of Departments of Agriculture in provinces and also at the centre. Accordingly, a Department of Revenue and Agriculture was set up at the centre and Departments of Agriculture were established in some of the provinces. The famines in 1896-97 and again in 1899-1900 were very severe; they were spread over a very large area and there was heavy loss of life. Lord Curzon, the then Governor-General of India, appointed a Famine Commission under the presidency of Sir Anthony McDonnell. The Commission, in its report submitted in 1901, recommended establishment of an agricultural research institute for the country, colleges of agriculture for providing scientific education on agriculture and departments of agriculture in all the provinces. The Government accepted most of the recommendations of the Commission and, accordingly, an Imperial Agricultural Research Institute was established at Pusa in Darbhanga district (then in Bengal, now of Bihar). Agricultural Colleges were established (or reorganized) at Sabour, Kanpur, Poona, Nagpur and Coimbatore.[2]

The Imperial Agricultural Research Institute (IARI) established at Pusa in 1905 started work on wheat and did some work on rice as well but it was soon realized that Pusa does not provide the ideal location for rice research. Therefore two rice research stations were established to work mainly on rice: an Agricultural Research Station at Dacca (now in Bangladesh) in 1911 and a Paddy Breeding Station at Coimbatore (now in Tamil Nadu) in 1913. Later, a Rice Research Station was established at Raipur for the Central Provinces and Berar (now in Chhattisgarh).

These stations were then under the Departments of Agriculture of their respective provinces and were headed by Dr J.M. Hector (Dacca), Mr M.R. Parnell (Coimbatore) and Dr R.J.D. Graham (Raipur). Dr Hector was the first to start research work on rice in India and made a very good beginning. Dr Hector worked there for 17 years from 1911 to 1928 (Ramiah 1953:51). At Coimbatore, the first Economic Botanist (later designated as Paddy Specialist), Mr F.R. Parnell, was appointed in 1912. He worked there up to 1924[3] and was succeeded by Dr K. Ramiah who had worked under Dr Parnell. In 1937, Dr Ramiah was succeeded by Mr C.R. Srinvasan. Nagpur was then the capital of the Central Provinces and Berar[4] and the Agricultural Station of the province was at Raipur. Dr R.J.D.

---

[2] The other two were established at Lyallpur in Punjab (now in Pakistan) and Mandalay in Myanmar (then a part of British India).

[3] After leaving India, he became Cotton Breeder in South Africa under British Cotton Growing Corporation.

[4] Before the re-organisation of Indian states in 1956, Central Provinces and Berar comprised of Chhattisgarh, Vidarbha region of Maharashtra and the Narmada valley region of the present Madhya Pradesh.

Graham was the first Economic Botanist of the province stationed at Nagpur. He looked after the Station by visiting Raipur frequently. After his retirement in 1915, the work was looked after by a Deputy Director of Agriculture of the province until his successor Mr D.N. Mahta was appointed in 1924. Mr B.B. Dave was appointed as the Rice Research Officer in 1932; he shifted his headquarter to Raipur.

It was soon realized that there are many types of soils and climates even within a province and the agricultural experiments carried out in one agro-climatic zone of the province do not hold good in another zone of the same province. As a result, Rice Research Stations were established in different provinces and in different agro-climatic zones of each province of the country. Table 1.1 provides the list of rice research stations established during 1911 to 1929 in the country.

Table 1.1. Rice Research Stations (or Agricultural Research Stations with rice research as a component) established in India during 1911 up to 1929.

Year	Location of Rice Research Stations/ Agricultural Research Stations	Established in the then province (or princely state) of	Now located in the state of	For the region/ area
1903 (1905?)	Raipur	Central Provinces and and Berar	Chhattisgarh	Chhattisgarh
1905	Palur	Madras Presidency	Tamil nadu	For deep water rice
1909	Samalkota	Madras Presidency	Andhra Pradesh	For East Godavari District
1911	Dacca	Bengal	Bangladesh	For Bengal
1913	Coimbatore	Madras Presidency	Tamil Nadu	For South India
1913	Anakapalli	Madras Presidency	Andhra Pradesh	For rainfed tracts of Vishakhapatnam
1913	Karimganj	Assam Province	Assam	For Barak valley
1917	Karjat	Bombay Presidency	Maharashtra	For Konkan region
1917	Nagenhalli	Mysore State	Karnataka	For Mysore state
1921	Mugad	Bombay Presidency	Karnataka	For Malnad tract
1921	Aduthurai	Madras Presidency	Tamil Nadu	For Cauvery delta
1923	Titabar	Assam Province	Assam	For Upper Assam
1922	Bankura	Bengal Presidency	West Bengal	For less rainfall areas of western part of Bengal
1923	Pratap Model Farm	Jammu & Kashmir State	Jammu & Kashmir State	For high altitude and colder climate
1924	Titabar	Assam	Assam	For Assam valley
1924	Kanpur (Shifted to Nagina	United Provinces	Uttar Pradesh	For central Gangetic plain

	in 1932)			
1925	Maruteru	Madras Presidency	Andhra Pradesh	For Gadavari delta area
1926	Kala Shah Kaku	Punjab Province	Pakistan	For Punjab
1927	Pattambi	Madras Presidency	Kerala	For Malabar coast
1928	Himayat Sagar	Hyderabad State	Andhra Pradesh	For Hyderabad state
1929	Nagrota Bagwan	Punjab	Himachal Pradesh	For Kangra valley

The period between 1911 and 1929 laid the foundation of rice research in this country. The rice workers of this period collected the local varieties of paddy of their respective regions to study morphology of the rice plant, its floral biology, physiology, traditional agronomic practices, response to manuring (there were no chemical fertilizers in those days), its productivity, insect pests and diseases, etc. They standardized the nomenclature of plant parts and agricultural practices for experimentation and system of documentation. They collected, purified (as many of them were often admixtures of varieties when collected from farmers), maintained and evaluated these traditional rice varieties. They also started hybridization work for the study of inheritance of many of its traits and initiated breeding rice varieties by pure line selection. And, above all, they prepared the next generation of rice researchers for the country. Table 1.2 lists the names of rice workers of that period in India and periods of their scientific contributions.

Table 1.2. Rice scientists who worked on rice genetic resources in India during 1911-1941.

States	Scientists (Periods of their research activity)
Assam	S.K. Mitra (1924-1938) P.M. Ganguly (1932-1938)
Bengal	G.P. Hector (1913-28) G. Sarangapani (1928-1934) M. Hedayatullah (1933-41) H.K. Nandi (1936-41)
Bihar	M. Alam (1932-40)
Chhattisgarh	R.J.D. Graham (1911-13) D.N. Mahta (1931) B.B. Dave (1937-40) S.C. Roy (1927)
Maharashtra	R.K. Bhide (1922-27) S.G. Bhale Rao (1926-30) B.S. Kadam (1932-43)

	G.G. Patil (1933-41)
Orissa	P. Parija (1938-43) P.D. Dixit (1941)
Uttar Pradesh	R.L. Sethi (1930-40) T.R. Mehta
Tamil Nadu	F.R. Parnell (1917-22) K. Ramiah (1924-45) C.R. Srinivasan (1926-41) N. Parthasarathy (1927-42) S. Ramanujam (1935-41)

Source: Ramiah (1953)

Publication of journals on agricultural sciences was also started in India in this period. Publication of *Agricultural Journal of India* was started in 1906, *Memoirs of Department of Agriculture (Botanical Series)* in 1911, *Madras Agricultural Journal* in 1913 and *Indian Journal of Agricultural Sciences* in 1930. These journals provided platform for publishing findings on rice research done during that period.

Prior to 1901, Departments of Agriculture were established in a few provinces as per the report of the Famine Commission of 1861. The United Provinces (now called Uttar Pradesh) was the first province to create a Department of Agriculture in 1875. After the Famine Commission Report of 1901, Departments of Agriculture were established in other provinces as well. In 1915, the Provincial Agriculture Departments were revamped and their activities were enlarged.

> *Before this time, tentative but rather unsuccessful, efforts had been made to establish Departments of Agriculture in the various Provinces. These failed partly because of the inadequacy of the information and statistics, partly because there was no expert scientific staff and partly because attention was mainly concentrated on the introduction of exotic plants* (Anstey, 1977:165).

The agricultural research in India got a set-back due to the First World War (1914-1918) but continued thereafter. According to 1921 administrative reforms, agriculture became a provincial subject and hence all the rice research stations of the provinces henceforward came under the administrative control of the Departments of Agriculture of the provincial governments. The appointments of Directors of Agriculture in those days used to be from the ICS cadre only.

Classification of rice varieties

The rice research stations collected the traditionally grown rice varieties from their localities, characterized them and tried to group them according to a logical system. Traditionally, the farmers of India group the varieties according to the cropping season that these varieties are adapted to. For example, the farmers of Bengal classify their rice varieties into *aus* (early), *aman* (late) and *boro* (summer). The farmers of Assam and Orissa also follow a similar system though their nomenclatures differ. The farmers of Tamil Nadu classify the rice varieties into *kuruvai, samba* and *thaladi*. The earliest effort to classify the rice varieties was done by Graham (1913). He tried to classify the rice varieties of Central Provinces and Berar on the basis of anthocyanin pigmentation of leaf sheath, spikelet and kernel characters and vegetative plant parts. Duration of rice varieties was also considered as an important character for classification. His style of classifying rice varieties was followed by many botanists in subsequent years.

As a result of varietal trials conducted over a number of years at Dacca, six *aman* varieties were recommended to the cultivators in 1922. The first rice variety that was recommended by the Paddy Breeding Station at Coimbatore was GEB 24[5]. According to Ramiah (1937:233),

> *GEB 24 was the first successful strain and was a selection from a bulk crop of Konamani grown at the central farm in 1913-14. Evidences are accumulating that it must have arisen first as a mutation. This strain is a short duration* samba *variety, very popular in this Presidency on account of its high quality of table rice. —— It is grown all through the Province and has entirely replaced the local varieties in the central tracts. It is the chief variety of rice in the adjoining Mysore State.*

Raipur (then in Central Provinces and Berar, now in Chhattisgarh) had bred eleven varieties of rice during 1913 to 1923 (Shrivastava and Kumar, 2005). Table 1.3 gives the list of rice varieties bred in that early phase of rice breeding in India.

Dr Hector was the first British scientist to start work on rice in India in 1911and he was also the last British rice scientist who left this country in 1928. Ever since then, rice research in this country has been in the hands of the Indian agricultural scientists. Evidently, the Agricultural Colleges established

---

[5] GEB stands for 'Government Economic Botanist'. Evidently, no system of designating a rice variety was developed when this variety was named.

in India in the first decade of the twentieth century provided the required manpower to take up this role.

Table 1.3. Rice varieties bred during 1911 to 1929.

Rice Research Station	Varieties
Dacca (Bangladesh)	Six varieties
Habiganj (Bangladesh)	Three varieties: Indrasail, Georgesail, Latisail
Coimbatore (Tamil Nadu)	Eight varieties: GEB 24, CO 1, CO 2, CO 3, CO 4, CO 5, CO 6, CO 7
Aduthurai (Tamil Nadu)	Seven varieties: ADT 1 to ADT 7
Raipur (Chhattisgarh)	Nine varieties: EB-17, Parewa No. 143, Bhondu No. 11, G 13, Bhata Gurmatia, Bhondu No. 10, Parewa No. 22, Gurmatia No. 11, Dilbuxa No. 35
Pratap Model Farm (Jammu & Kashmir)	Four varieties

The Second Phase II (1929 to 1949)

The Government of India appointed a Royal Commission of Agriculture in 1926 headed by Lord Linlithgow to examine the condition of agriculture and rural economy in India and to recommend measures to improve the condition. The Commission recommended, among other things, to set up a council to supervise the agricultural research in this country; advise and guide the provincial governments with regard to agricultural research and provide financial help to initiate such research. The Government of India accepted this recommendation of the Commission and established the Imperial Council of Agricultural Research (ICAR). (Many other recommendations of the Commission were shelved in view of the Great Depression that followed). The setting up of the ICAR was an important and fruitful decision of the Government. It helped setting up of many new rice research stations especially in those agro-climatic zones where rice research stations had not been set up till that time (Table 1.4). Thus in the next decade (1930 to 1939), with the monitoring, guidance and financial support of the ICAR, rice research stations could be established in almost all the agro-climatic regions of India. Rice Research Stations continued to be set up even during the war period up to 1942.

Table 1.4. Rice Research Stations established in India during 1930 to 1947.

Year	Rice Research Stations	Province where it was established	State where it is situated now
1930	Rudroor	Hyderabad State	Andhra Pradesh
1932	Chinsurah	Bengal	West Bengal
1932	Berhampur	Madras Presidency	Orissa
1932	Cuttack	Bihar and Orissa	Orissa
1932	Kanke (Ranchi)	Bihar and Orissa	Jharkhand
1932	Sabour (Bhagalpur)	Bihar and Orissa	Bihar
1932	Warangal	Hyderabad State	Andhra Pradesh
1934	Habiganj	Bengal	Bangladesh
1935	Pattukottai (Thanjavur)	Madras Presidency	Tamil Nadu
1937	Buchireddipalem	Madras Province	Andhra Pradesh
1937	Ambasamudram	Madras Province	Tamil Nadu
1937	Kayankulam	Madras Province?	Kerala
1938	Jeypore	Orissa	Orissa
1940	Moncompu	Madras Province	Kerala
1941	Igatpuri	Bombay Province	Maharashtra
1941			
1942	Tirurkuppam	Madras Province	Tamil Nadu
1942	Mangalore	Madras Province	Karnataka
1942	Rudrur	Madras Province	Andhra Pradesh

Source: Ramiah (1953), Ghose *et al.* (1960)

Following the style of Graham (1913), Sethi and Saxena ((1930) classified traditional rice varieties of United Provinces, Mitra and Ganguly (1932) of Surma valley, Hector and Sarangapani (1934) of Bengal, Kashiram and Saravayya (1934) of Orissa and Bihar, Alam (1935) of Bihar and Kashiram and Ekbote (1936) of Punjab and western part of United Provinces. Similar exercise was also done for the rice varieties of Sind Province and lower Burma that were then parts of British India. All these authors used the morphology of plant characters, distribution of anthocyanin pigment in various plant parts and size and shape of the spikelets for their classification. Additional characters considered were presence or absence of awn, size of glumes, kernel colour, etc. But as Ramiah (1953:12) has put it,

> *A large number of varieties grown in the Madras Presidency has been collected and grown at the Paddy Breeding Station, Coimbatore since 1915, which collection at present amounts to 2000. But the workers there realized that it is very difficult to devise a satisfactory and accurate system of classifying these rices. Owing to the large amount of intergrading of most of the characters, it was felt that a very broad classification even if possible was hardly worth the trouble. They never attempted any classification and were content with a description of the types as they believed that it was easier to identify a variety by eye when one is dealing with them intensively than by reference to key tables.*

In pre-independent India, the period between 1929 and 1939 was the most productive period in so far as collection, maintenance, evaluation and utilization of rice germplasm are concerned. The rice research stations collected the traditional rice varieties of their respective areas, conducted varietal trials to assess their yield potential, made pure line selections in the popular/promising varieties and recommended the better ones for general cultivation. They also evaluated the varieties for tolerance to various physiological stress situations such as drought, flood and cold and against diseases and insect pests. Table 1.5 provides the number of rice varieties bred and recommended for general cultivation in the country. Comprehensive lists of rice varieties bred before the advent of high yielding rice varieties have been provided by Ramiah (1953), Ghose *et al.* (1960) and Richharia and Govindaswami (1990). It may, however, be mentioned again that there were no chemical fertilizers and no pesticides in those days. There was no formal authority to 'release' a variety as in the modern sense of this term and no seed corporations to multiply and distribute the seeds. There were, however, some seed multiplication-cum-demonstration farms under the Departments of Agriculture in every province where seeds of recommended types were multiplied and supplied to the farmers of their locality.

Table 1.5. Number of improved rice varieties bred in India in the pre-HYV period.

State	No. of improved varieties bred by		Total number of varieties bred
	Pure line selection	Hybridization	
Assam	23	6	29
Andhra Pradesh	52	7	59
Bihar	15	2	17
Bombay	44	1	45
Hyderabad	17	—	17
Madhya Pradesh	9	9	18
Madras	103	10	113
Mysore	17	8	25
Orissa	31	3	34
Punjab	8	—	8
Travancore & Cochin	14	—	14
Uttar Pradesh	22	2	24
West Bengal	42	3	45
Total	**394**	**51**	**445**

Source: Gangadharan (1985)

The breeding procedure was, in most cases, pure line selection of the existing popular varieties. Hybridization was followed occasionally and, whenever adopted, both the parents were indigenous varieties only. The varieties bred and recommended by a rice research station were generally designated as the selection numbers of that station. For example, varieties bred and recommended by the Rice Research Station at Aduthurai were designated as Adt 1, Adt 2, Adt 3, etc. The second style of naming a pure line selection was to add the selection number to the local variety from which the selection was made, e.g. Basmati-370, Jhona-349, Ambemohar-159, etc. A third style was to name a pure line after its collection number such as T 141, T 812, etc.

The impact of this simple pure line selection with a few seed multiplication and demonstration farms cannot be underestimated as, in those years, the farmers were often growing an admixture of seeds of rice varieties. As Ramiah (1937:189) has put it:

> *During the 1933-34 season, the total rice area under the improved strains in the* [Madras] *province has been estimated*

> *to be at least eight lakhs of acres* [3.2 lakh ha]. *Taking the average acre yield at a very modest figure of 1,500 lb per acre* [1705 kg/ha], *an improvement of ten per cent in yield by the growing of strains would amount to 150 lb* [68 kg] *of grain which in money value at the present price is equivalent to Rs. 4. Thus the value of the improvement over the whole area comes to 32 lakhs of rupees. Improved seed enough to sow about 2½ lakhs of acres* [1.14 lakh ha] *has been distributed during the year 1934-35.*

As the local rice varieties collected by various rice research stations was increasing in number and each station was adopting its own style of presenting the data, it was felt that the nomenclature of plant parts and their description should be standardized. This work was entrusted to a Special Committee with Dr K. Ramiah as its convener. The report of the Committee was published in the *Indian Journal of Agricultural Sciences* for general adoption (Hutchinson and Ramiah, 1938).

The other aspects of genetics of rice also received their due attention. A series of papers on genes governing various morphological characters and anthocyanin pigmentation were published and linkage groups were proposed. Cytological studies of cultivated and wild rices were initiated, crosses between cultivated and wild rices were made and their cytological behaviour was reported. Haploids, autotriploids and autotetraploids of *O. sativa* were also reported and their cytology was studied. Mutations (spontaneous and induced) were reported. Physiological studies on sterility, flowering duration, plant height, tillering, lodging, shattering of grains, flood tolerance, drought tolerance, flood resistance and soil salinity were studied. Many wild species of Oryza and the African cultivated species (*O. glaberrima*) were introduced for their cytological and phylogenetic studies. All these works of this period have been summarized by Ramiah (1953) in his monograph *Rice Breeding and Genetics.*

The decade of 1940s was traumatic for the country. During this period, India had to suffer from the consequences of the Second World War (1939-1946), political turmoil due to Quit India movement launched by Mahatma Gandhi, the Famine of Bengal and pre-partition communal riots followed by large scale migration of people across the borders during partition. These incidences diverted the attention as well as funds of the Government that could have been otherwise available for agricultural research and development. India had become a rice deficit country even before the World War II but was managing by imports from Burma (which was then under British occupation). Occupation of Burma by Japan during World War II stopped that source. As all the major rice growing countries of the world (these are the countries of east, southeast and south Asia) had suffered from the devastation of World War II, there was hardly any country in the world from where rice could be imported. India had no option but to increase its own rice production on an urgent basis. In fact, all the South and Southeast Asian countries faced the same challenge.

When one looks back, the research activities on rice during the decade of 1940s were on a very low key. The very few academic activities of the War period were standardization of gene symbols for rice by Kadam and Ramiah (1943) and preparation of a monograph on rice *Rice Genetics and Breeding* by Ramiah (1953)[6]. The formation of Indian Society of Genetics and Plant Breeding and the publication of the *Indian Journal of Genetics and Plant Breeding* in 1941 were also noteworthy as the journal has been publishing important papers on genetic resources of rice ever since it was started.

India became independent in 1947. Even before attaining Independence, the (then provisional) national Government of India decided to set up a national rice research institute for the country. The Central Rice Research Institute was established at Cuttack (Orissa) in 1946 and Dr K. Ramiah became its founder Director.

During 1935 and 1940, some Chinese rice varieties were introduced and were found promising (Seetharaman *et al.* 1971). In mid-1940s, these were made available to other provinces. Some of these turned out to be quite successful. For example, Ch 1039 became a widely grown rice variety of Kashmir for quite some time. Another variety, Ch 988, was popular in Himachal Pradesh. A list of successful Chinese introductions is provided in Table 1.6. Some of these have been found to be good donors and some have been used in breeding rice varieties. For example, Ch 13 and Ch 45 are resistant to helminthosporiose; Ch 4 has been used as a parent in breeding rice variety NSJ 98 in Uttar Pradesh and Ch 1039 in breeding VL Dhan 221 for Uttarakhand. Some Russian varieties were also introduced; one of them (R 13073) was cultivated at high altitudes in Kashmir and two of them (Russ 2652, Russ 3264) were used as parents in breeding rice varieties (Kashi, Rohela, Varuna) in Uttar Pradesh.

---

[6] The monograph *Rice Breeding and Genetics* was completed by K. Ramiah in 1944 during British rule (in war days) but the book could be published only in 1953 (after Independence) by the Indian Council of Agricultural Research, New Delhi.

Table 1.6. Introduction of Chinese varieties showing wider adaptability.

Chinese Introductions	Found adapted in the state of
Ch 4, Ch 10	Uttar Pradesh, West Bengal
Ch 45	Andhra Pradesh, Bihar, Karnataka, Kashmir, Orissa, Punjab, Uttar Pradesh, West Bengal
Ch 10, Ch 41, Ch 42, Ch 43, Ch 45, Ch 62, Ch 1007, Ch 1039, Ch 1040	Bihar
Ch 41, Ch 43, Ch 62, Ch 63	Maharashtra
Ch 45, Ch 47, Ch 988,	Punjab

Source: Richharia and Govindaswami (1990)

By 1960, there were 69 rice research stations in the country and these rice research stations had bred and recommended 438 improved varieties for general cultivation (Parthasarathy, 1972). According to Gangadharan (1985), however, there were 82 rice research stations in the country by 1950 and these stations had bred 448 rice varieties by then (Table 1.5). The main objectives of breeding rice varieties in this period were (a) high yield, (b) response to heavy manuring, (c) earliness, (d) tolerance to deep water and flood, (e) resistance to salinity and alkalinity of soil, (f) tolerance to drought, (g) resistance to lodging, (h) non-shedding of grain, (i) dormancy, (j) control of wild rice, and (k) resistance to pests and diseases (Ghose, Ghatge and Subrahamanyam, 1960).

The Third Phase (1950 to 1964)

After Independence, the immediate challenge before the country was to increase rice production. Since there was hardly any scope to increase area under rice crop, the only other alternative was to increase its productivity.

The method of rice cultivation in Japan was advanced in comparison with that of India and the productivity of this crop in that country was much higher than in India. Impressed with the Japanese method of rice cultivation, a country-wide campaign was started in 1953-54 to popularize Japanese method of rice cultivation.

> *In all, an area of 1.32 million acres was brought under the Japanese method. This resulted in an estimated increased production of 0.745 million tons of rice, or approximately 1,300 lb of additional production per acre* (Ghose *et al.* 1960:275).

But more than the Japanese method of cultivation, the Japanese rice varieties had played a major role in boosting rice production in Japan. Japanese scientists had developed rice varieties that were responsive to application of chemical fertilizers. The *japonica* varieties were thermo-sensitive so when these were introduced into India, they became lean and lanky, produced very small panicles with a few grains and yield was very poor. The other possible option was to cross the *japonica* types with *indica* types and transfer the fertilizer-responsive traits of the *japonica* types into *indica* genetic background.

Japonica-indica hybridisation (1952–1964)

The rice situation was not unique for India. The whole of South and Southeast Asia was facing similar problems after the World War II. To solve the problems of all these countries, the Food and Agricultural Organization (FAO) set up an International Rice Commission in 1949. The Commission started a scheme for maintenance of world genetic stocks of rice at five centres, one each in India, Indonesia, Pakistan, Japan and USA. The Central Rice Research Institute, Cuttack was vested with the responsibility of maintaining the *indica* varieties of South and Southeast Asia. During 1956-57, out of 795 registered FAO germplasm, 721 accessions were received at CRRI for multiplication and maintenance. Since its inception till late 1977, Director, CRRI continued to be in charge of overall supervision of the world genetic stock of the FAO scheme

The International Rice Commission in its first meeting held at Rangoon in 1950 decided to launch an international programme to cross the prominent local rice varieties of south and southeast Asian countries with the *japonica* types that have proved responsive to chemical fertilizers and to select from the progeny fertilizer responsive *indica* types for increasing the productivity of rice varieties of south and southeast Asian countries (Parthasarathy, 1972). The CRRI was chosen as the venue where the *japonica* types were assembled and the trainees from all these countries came with their own popular varieties. A total of 192 *indica* varieties were crossed with *japonica* types and 710 *japonica-indica* F_1 hybrids were produced.[7] The trainees were trained and sent back with F_2 seeds to their respective countries to make further selections. Convinced with this plan of IRC, the ICAR started a parallel programme in which all the Indian states were involved. Both the programmes were terminated in 1956. To ensure that the programme ends up with expected results in the country, the ICAR launched the second phase of this scheme in 1960 and created eleven additional posts of rice breeders (then called Assistant Botanists) and posted them in different rice growing states (where *japonica-indica* hybridization was going on) to liaison the

[7] The *japonica* parents used were: *Aikoku, Asahi, Fukoku, Norin 1, Norin 6, Norin 8, Norin 17, Norin 18, Norin 20, Norin 23, Norin 36, Oshu No. 2, Rikuu 132, Shenei, Taichu 65, Taihoku 6 and Zuiho*

work between CRRI and state centres. This programme was abruptly ended in 1966 with the advent of semi-dwarf rice varieties.

The final outcome of the *japonica-indica* hybridization project was only a single rice hybrid derivative (Adt 27) in India, two (Malinja and Mahsuri) in Malaysia, one (Circna) in Australia and one (Norelon Strain 340) in the Philippines. Mahsuri, a cross between Myang Ebos 80 x Taichu 65 and developed in Malaysia was later introduced in Tamil Nadu. From there, it spread to Andhra Pradesh and Bihar states of India. The success of the programme was therefore very limited if not a failure.

What went wrong with so ambitious a project that was recommended by an august body like the International Rice Commission? Was it launched without doing much "homework"? It was argued that inter-racial hybrids between *japonica* and *indica* are often partially sterile and there is hybrid breakdown in subsequent generations resulting ultimately with parental types only. Ghose, Ghatge and Subrahamnyan (1960) have provided data on the sterility pattern of the F_1 and F_2 populations. The cytologists and geneticists came out with their explanations for such a situation (see reviews by Oka, 1964; Samppath 1964; Shastry 1964) but the breeders viewed the situation differently. Jennings (1966), after conducting a well-planned research on this aspect at IRRI, concluded:

> *A thorough review of literature, the data reported above and field experience suggest that the sterility accompanying* indica x japonica *hybridizations neither imposes direct adverse consequences nor acts as a barrier to orderly achievement of the breeders objectives. The data from the F_2 studies discussed earlier failed to give evidence that F_2 sterility causes or is associated with abnormal gene frequencies, modified segregation ratios, lower heterozygote fertility, reduction in expected numbers of recombinants of paired characters, or reduced recombinant fertility in comparison to segregates showing parental characters. Neither does the writer give much credence to other theoretically proposed consequences of hybrid sterility such as: difficulty in transference of traits from one subspecies to the other, non-random zygote elimination, or substantial numbers of unthrifty plants in F_2.*

In fact, in later years, rice breeders of India have successfully bred *japonica* x *indica* hybrids and have combined the desirable traits of both the subspecies. For example, Jagriti is a cross between Rikuu 132 (*japonica*) and R4 a variety bred at Raipur (Shrivastava and Kumar, 2004). CR 2002 bred at CRRI is a hybrid derivative of Norin 6 and GEB 24. Utkal Prabha is a hybrid derivative of Waikoku (*japonica*) and CR 1014 which again is a hybrid between T 90 (an *indica* variety from Orissa) and Urang Urangan (a *javanica*). It was bred at CRRI and released by the State Variety Release Committee of Orissa. Annada is a hybrid derivative of the cross between MTU 15 and Waikoku (*japonica*) released by the Central Variety Release Committee (CVRC).

The causes of failure in breeding *japonica-indica* hybrids have been attributed to (a) choice of wrong *japonica* parents, (b) lack of skilled breeders during that period, and (c) lack of proper monitoring and coordination of the programme (Gangadharan, 1985). As Chang (1972) and Rao and Nagaraju (1974) have pointed, had *japonica* types of warmer areas (such as Taiwan) been taken as parents in this programme, there was a greater chance of success in achieving the results. The unfortunate part of the whole project was that the country lost more than a decade when it was suffering from food deficit.

The Central Rice Research Institute had also been testing some of the Japanese varieties under Indian conditions but their performance was quite unsatisfactory except that a few of them were successful in the hills of Himachal Pradesh and Kashmir. For example, Shenei was introduced in Kashmir valley and Norin 12, Norin 18 and Asahi in Himachal Pradesh. The failure of the *japonicas* as well as *japonica-indica* hybrids led CRRI to try some introductions from Taiwan. Taiwan grows *indicas* as well as *japonicas*, the latter introduced and acclimatized there during the Japanese rule. Some of the *japonica* types of Taiwan (called *ponlai* types) proved quite successful in some of the states. For example, Tainan 3 and Kaohsiung 18 were successful in Kerala and Taichung 65 in Karnataka (Gangadharan, 1985). VL 8, a secondary selection from Kaohsiung 22, was found suitable in areas 1000-1500 m above mean sea level in Uttarakhand. K 84, a mutant of Taichung 65, was suitable for high altitude areas of Kashmir.

Fourth Phase (1965 onwards)

Establishment of IRRI

The International Rice Research Institute (IRRI), Manila was established jointly by the Ford Foundation and Rockefeller Foundation at Los Banos in the Philippines in 1960. It hired distinguished scientists from many countries such as Japan, Taiwan, the Philippines, Indonesia, India, Sri Lanka and USA. In course of time, this Institute became a prime mover with regard to exploration, collection, characterization, evaluation, utilization and distribution of rice germplasm world over. In addition, it became an important hub for training young scientists on various disciplines and sub-disciplines of rice. In course of time, IRRI built a truly international treasure of rice genetic resources and the credit for this goes to Dr Te-Tzu Chang who was in charge of germplasm at this Institute from its initial years for thirty years.

Taiwan was under Japanese occupation from 1895 to 1945. Before the Japanese occupation, Taiwan was growing *indica* types but, during Japanese occupation, Japanese scientists introduced *japonica* types into Taiwan and selected types adapted to Taiwanese condition. The Japanese Government of Taiwan encouraged the cultivation of such *japonica* types (called *ponlai* types in Taiwan) to meet the rice requirement of Japan. After the end of Japanese rule, Taiwanese rice scientists restarted their work on the native *indica* rice varieties. At the District Agricultural Improvement Station, Taichung, they crossed a semi-dwarf *indica* rice variety (Dee-Geo-Woo-Gen)[8] with Tsai-Yuen-Chung (a tall *indica* variety) and developed a dwarf variety which they named Taichung (Native) 1. They released this semi-dwarf variety in 1956 for general cultivation in Taiwan. When IRRI was established and Dr T.T. Chang joined IRRI as a geneticist, he brought this variety from Taiwan to IRRI. It was confirmed at IRRI that this variety is responsive to higher doses of chemical fertilizers and has the potential to yield very high compared to tall *indicas*.

Taichung (Native) 1 and IR 8

Taiwanese rice varieties such as Tainan 3, Taichung 65, Kaohsiung 18, Kaohsiung 22, Dee-Geo-Woo-Gen, Taichung (Native) 1, etc were introduced into India in 1964 and were evaluated at CRRI for their response to application of

---

[8]. According to Dalrymple (1986), "it is not known precisely when and where semi-dwarf rices first made their appearance but it was probably in Fujian province, China or Taiwan. — The 1871, local records of Taiwan (*Taiwan Fan-Chih*) mention a variety called Woo-gen that was introduced from Fujian Province. In 1906, D-G-W-G and Woo-gen were recorded by the Taiwan Agricultural Experiment Station." According to Huang, Chang and Chang (1972), the semi-dwarf rice variety Dee-geo-woo-gen had already become a leading variety of China as back as 1939. According to Lu (1979), in 1956, a semi-dwarf plant was found in a farmer's field in Kwangtung Province in the field of a tall rice variety, *Nan-te 16*. It seems to have been a spontaneous mutation. It was released as *Ai-chiao-nan-te*, the first semi-dwarf rice variety in mainland China. In 1957, Kwangtung Academy of Agricultural Sciences used *Ai-tsai-chan*, a local semi-dwarf variety from Kwangsi Province as a female parent to breed new semi-dwarf rice varieties *Kwang-chang-ai* and *Chen-chu-ci* and released them in 1959 and 1962 respectively. By 1967, these two rice varieties were being grown in 160 m ha in southern China. While exploring and collecting landraces of rice in Northeast India, a number of semi-dwarf landraces of rice were collected by S.D. Sharma and his team (Sharma *et al*., 1971; Sharma, 1982). Two semi-dwarf rice varieties named Khonorullo and Ngoba were released by Rice Research Station, Shillong (Meghalaya) in 1965 just when Taichung (Native) 1 was being introduced in India. It appears that occurrence of spontaneous mutation from tall to semi-dwarf stature is not a rare phenomenon in rice and has occurred throughout the region stretching from Assam (India) to Kwangtung and Fujian (China). The word "Dee-geo" means dwarf and "woo-gen" means brown apiculus (see Lu, 1979).

chemical fertilizers, non-lodging nature and grain yield.[9] After conducting trials at many locations and getting convinced about its potential, it was released in India in 1966. The International Rice Research Institute (IRRI), Manila crossed Dee-Geo-Woo-Gen with Peta (a tall *indica* variety of the Philippines) and developed IR 8. IR 8 was introduced into India and released for general cultivation in 1966.

The plant type of the two introduced rice varieties, Taichung (Native) 1 and IR 8, was different from that of the other Indian rice varieties. Chandler (1972) has explained this plant type as follows:

> *For purpose of clarification, I wish to define "improved tropical plant type." It is a plant of short stature that, under good growing conditions, has a total height of 90 to 110 cm. The culms are not only short but also relatively thick and sturdy so that they do not break or bend at high fertility levels or when heavy rains or moderately strong winds occur. Its leaves are erect, rather short, and not too wide. It has an inherent heavy tillering capacity. Each of these characters has a beneficial effect on the grain yield (Tanaka, Kuwano and Yamaguchi, 1966; International Rice Research Institute, 1968, p 17-45; Tanaka et al., 1969). The short sturdy straw prevents lodging. The short erect leaves permit greater penetration of sunlight and thus increase the efficiency of photosynthesis. The heavy tillering capacity aids in producing more panicles per unit area of land, allowing a stand of rice to compensate for missing hills or, in direct seeded rice, for any thinly sown area.*

It is noteworthy that ever since the introduction of Chinese varieties in pre-Independence era, India has been looking for the exotic germplasm for the improvement of its rice varieties; first Chinese types, then Russian types, then *japonica* types from Japan, then *ponlai* types of Taiwan, then the hybrids between *japonica* and *indica* and lastly the semi-dwarf plant types from Taiwan. Either the genetic diversity of rice available in India could not meet country's requirement or the genetic diversity of rice in the country was not fully explored and collected or properly evaluated and utilized. For example, the semi-dwarf rice varieties of northeast India, Khonorullo and Ngoba, were the result of collection, evaluation and release by the Rice Research Station, Shillong in 1965. Many such semi-

---

[9] Huang, Chang and Chang (1972) mention "A group of 15 *ponlai* and four semi-dwarf native varieties sent to a University of Missouri team in Calcutta during 1960 resulted in the release of two *ponlais* in West Bengal under the names Kalimpong I and Kalimpong II in 1965. Taichung and several *ponlais* were entered in the cooperative variety trials sponsored by FAO and the International Rice Commission —." They cite N. Parthasarathy (unpublished) as their source.

dwarf traditional rice varieties were collected during Assam Rice Collection (Sharma *et al.*, 1971; Sharma, 1982).

Breeding for semi-dwarf plant type

In the early phase of the introduction and adoption of semi-dwarf rice varieties, the idea was advanced that growing a few high yielding varieties throughout the country would solve many of the problems associated with seed production, quality control and marketing. But not all Indian rice scientists subscribed to this view. Reviewing the rice breeding in India, Gangadharan (1985:77) has aptly put it:

> *But rice being cultivated during the monsoon, when no other cereal could be grown in heavy rainfall areas, normally faced problems of adaptation to specific ecology and the newly introduced semi-dwarf types were unsuitable in a variety of stress situations such as water-logging, salinity, drought, low solar radiation due to clouded atmospheric condition, etc. When these semi-dwarf varieties were cultivated under high fertility conditions, they were found susceptible to most of the pests and diseases of rice. Continuous and intensive cultivation of these semi-dwarfs caused disease and pest epidemics which gave premonitions of famine or ruins in Bihar, Andhra Pradesh, Kerala, Indonesia and elsewhere. — In such a situation, it is unwise to advocate varieties of identical genetic constitution over wide tracts. Genetic diversity is still the best insurance against disease and pest epidemics as illustrated by the Indonesian and Kerala catastrophes.*

The semi-dwarf plant type as idealized by Taichung (Native) 1 and IR 8 became successful in those ecological conditions where there was ample sunshine (as during the summer season), assured irrigation along with good drainage and chemical fertilizers were applied to realize their fertilizer-responsive capabilities. As these varieties are photoperiod insensitive, two crops could be taken in the same field in a year and thus boost the production per unit area in a year.

> *But high-yielding semidwarf varieties, structured on a few pairs of dwarfing gene or with narrow genetic variability, under high fertilizer inputs, were susceptible to almost all the pests and diseases of rice. Continuous and intensive cultivation of HYV has aggravated the severity of major pests as well as enlarged minor pests into major ones*" (Gangadharan, 1985:85).

For all these reasons, India had to start another cycle of rice breeding with the same objectives as mentioned earlier (enumerated by Ghose, Ghatge and Subrahmanyam, 1960) with added emphasis on resistance to pests and diseases. In fact, India had to start from the phase of exploration and collection of genetic resources of rice for reasons discussed below.

Exploration and Collection of Rice Germplasm

When the rice research stations were established in various agroclimatic zones of India, the stations no doubt collected the traditional rice varieties of their localities either to study their variability and evaluate them for their performance under normal as well as stress situations or to study genetics of their traits. The first ever project on exploration and collection of rice germplasm was the project on exploration and collection of traditional rice varieties from the southern districts of Orissa presumably because Ramiah and Ghose (1951) considered this tract as the secondary centre of origin of cultivated rices. This project sponsored by the ICAR and executed by the Central Rice Research Institute (CRRI) during 1955-59 is known as Jeypore Botanical Survey[10]. Under this project, Govindaswami and Krishnamurty (1959) collected 1753 traditional varieties of rice. This material is generally known as 'JBS' material and has been studied for its morphological variation and evolutionary significance (Oka and Chang, 1962) but rarely for its possible utilization in rice breeding mainly because rice breeders of that period had high hopes in *japonica-indica* hybridization programme that was going on with full vigour during that period. Subsequently, advent of the high yielding rice varieties eclipsed its importance as well as its significance. The objective of this project was, however, "botanical survey" as its name implied. The loss of genetic diversity of crop plants and urgency to explore, collect and conserve the genetic diversity of crop plants was realized in the 1960s and 1970s only when HYVs made inroads into the farmers' fields and started replacing the traditional types. This led to setting up of International Board of Plant Genetic Resources (IBPGR) in 1974.

Dr R.H. Richharia, then Director CRRI, initiated efforts to collect rice genetic resources from Manipur State of northeast India and, with this purpose, deputed one of his staff (A.C. Sharma) at Wangbal (Manipur). He collected about 900 accessions during 1965 to 1969.

The Indian Agricultural Research Institute came out with a project to

---

[10] The southern Orissa now comprising of Rayagada, Koraput, Nabarangpur and Malkangiri districts was then known as Jeypore tract.

explore and collect rice genetic resources of rice from northeastern region of India as it was least explored till then. The project was sponsored by the International Rice Research Institute (IRRI) and financially supported by the United States Department of Agriculture (USDA) with a PL-480 grant. Dr M.S. Swaminathan (then Director, IARI) and Dr S.V.S. Shastry (then Project Coordinator, All India Co-ordinated Rice Improvement Project i.e. (AICRIP) were the Principal Investigators of this project. During the project, the whole of Arunachal Pradesh, Nagaland, Manipur, Meghalaya and Tripura were explored; Mizoram could not be explored because of political disturbances there; from Assam only Mikir and North Cachar Hills (a hill district), North Lakhimpur (a district of upper Assam), Guwahati (a district of middle Assam) and Goalpara (a district from lower Assam) districts were explored. The project was operative for a period of five years (1967-1972) and a total of 6630 samples were collected from the region. These collections are collectively known as ARC (Assam Rice Collections). A catalog was prepared and submitted to ICAR for its publication.[11] A large number of accessions of this collection were screened by the All India Coordinated Rice Improvement Project for insect pests and diseases and many donors were identified (Shastry *et al.*, 1971). Genetic studies were also carried out at CRRI. First, a preliminary and, later, a summary report of the project were published by S.D. Sharma who was the team leader of this exploration and collection (Sharma, Vellanki, Hakim and Singh, 1971; Sharma, 1982). As per the project proposal, a duplicate set of the material was supplied to IRRI. Unlike CRRI in the case of JBS material, AICRIP was liberal in supplying the ARC material to other institutes for its evaluation and utilization. As a result, the material was well evaluated, donors were identified at national and international levels and many of the donors have been utilized in breeding. A pure line selection from one of the accessions has also been released as a variety (Suvarnamodan) in Kerala.

After his retirement from CRRI as Director, Dr R.H. Richharia made an extensive and intensive exploration and collection of rice germplasm from the Chhattisgarh state and rice growing regions of the present state of Madhya Pradesh. These collections were assembled at Raipur and were maintained and characterized by him. Incidentally, his is the single largest collection of rice germplasm under any one mission in the country so far. After he relinquished this project, the Indira Gandhi Agricultural University has continued to enlarge, maintain and evaluate this collection. A detailed report about this project is presented in Chapter 14 (Rice Germplasm of Chhattisgarh: Collection, Evaluation, Utilization and Conservation).

---

[11] This catalog, however, could not be published by the publication Division of ICAR. The manuscript was ultimately not traceable in the Publication Division. A duplicate copy of the catalog could not be kept as photocopy system was not available in those days.

Organizational changes

While these efforts to collect rice germplasm were going on in India, some important organizational changes were taking place in the country with regard to agricultural research and education. When Indian states were reorganized in 1956, each state designated one of its rice research stations as the main rice research station and the other ones as its substations. The first agricultural university was established in the country at Pantnagar (then in Uttar Pradesh, now in Uttarakhand) in 1960. Since then, the process progressed rapidly to cover all the states of the country. All agricultural research activities of the states have come under the control of the State Agricultural Universities (SAU). Under the National Agricultural Research Project (NARP) in 1979, the agricultural research in the states was re-organized again. The main agricultural research station of an agroclimatic zone in a state was designated as the Zonal Agricultural Research Station with emphasis on the major crop of that zone but was mandated to work on other crops that were also grown in that zone. Accordingly, one of the Zonal Agricultural Research Stations in a state was recognized as the main station for rice research but working on other crops also that fit with the rice-based cropping system.

The Indian Council of Agricultural Research established the All India Coordinated Rice Improvement Project (AICRIP) at Hyderabad in 1965 to coordinate the rice research activities in the country. The AICRIP made a concerted effort to coordinate the rice research activities of the State Agricultural Universities (SAUs) to avoid duplication of efforts in the country. Besides, some of the research stations of states were specifically financed to conduct AICRIP trials. In 1975, AICRIP was designated as Directorate of Rice Research (DRR). The AICRIP/DRR has been playing a prominent role in coordinating rice research in the country. In addition, the Indian Council of Agricultural Research established new Institutes such as ICAR Research Complex for the Northeastern Hill Region (Umiam), Central Soil Salinity Research Institute (CSSRI, Karnal), Vivekanad Parvatiya Krishi Anusandhan Sansthan (VPKAS, Almora) to meet the regional and/or problem-specific needs of specific areas. All these institutes have rice as a component of their research activities.

The development of high yielding rice and wheat brought green revolution at a cost. The farmers who had been cultivating the traditional varieties and thus maintaining the genetic diversity of these crops were giving up their traditional varieties in preference to these HYVs. As a result, there was loss of genetic diversity of crops particularly of rice and wheat. This was not limited to India only; it was happening in other countries as well. The Consultative Group of International Agricultural Research (CGIAR) and Food and Agricultural

Organization (FAO) decided to establish an international board with headquarters in the FAO at Rome. Accordingly, the International Board of Plant Genetic Resources (IBPGR) was established in 1974 at the headquarters of FAO in Rome. The IBPGR has supported and monitored field collections, seed storage facilities and training field collectors mainly through international agricultural research centers but also directly through the national organizations. The IBPGR and IRRI jointly sponsored a Rice Advisory Committee that prepared a standard set of descriptors and descriptor states for uniformity in recording data by rice scientists so as to facilitate transportability of data among rice scientists and rice institutes world over (IBPGR-IRRI Rice Advisory Committee, 1980).

The Indian Council of Agricultural Research established the National Bureau of Plant Genetic Resources (NBPGR) in 1976 to look after all aspects of genetic resources of all crops in the country. In fact, the idea had its humble beginning much earlier as Plant Introduction Section within the Division of Genetics (then called Botany Division) of the Indian Agricultural Research Institute, thanks to the foresight of Dr B.P. Pal when he was the Director of the Institute. Later, it became a separate Division of IARI before it became a separate Bureau. The Bureau (NBPGR) has seven regional stations (Srinagar, Shimla, Bhowali, Jodhpur, Akola, Hyderabad, Thrissur and Shillong) and two base centres (Cuttack and Ranchi) that cover the entire country. The Bureau soon took up its responsibilities and, in association with the ICAR crop research institutes, started a series of explorations and collections from all parts of the country. There were many explorations jointly organized by NBPGR and CRRI with the logistic support from its Regional Stations and Base Centres. In addition, the CRRI had its own explorations to make its collection of rice germplasm as comprehensive as possible.

Intensive exploration and collection

In 1975, CRRI collected about 1038 collections from rice growing states of the country. During 1978-80, CRRI collected 6,349 collections from about 100 districts of 14 states of the country in collaboration with State Agricultural Universities, State Departments of Agriculture and ICAR centres. During 1990-2000, CRRI and NBPGR jointly collected about 2,400 accessions of both cultivated and wild rices. Between 1985 and 2002, a total 2639 accessions of rice germplasms including their wild relatives have been collected (Hore, 2005). Vivekanad Parvatiya Krishi Anusandhan Sansthan (VPKAS, Almora) explored hilly region of Uttar Pradesh for rice germplasm and collected 1,247 primitive cultivars.

The National Agricultural Technology Project (NATP), a World Bank-aided project, was implemented by the Indian Council of Agricultural Research and the Department of Agriculture and Co-operation (DAC) during 1999 to 2005. One of its components was a mission mode programme of collection of germplasm of crop plants and their wild relatives from the entire country and their characterization and conservation. This programme was executed by the NBPGR. With regard to collection of landraces of rice, it was, in a way, a mopping-up operation of rice germplasm as extensive collections of rice germplasm had already been made from all over the country before this programme.

Before the establishment of NBPGR, the Central Rice Research Institute was maintaining about 6500 accessions of rice germplasm which included 1344 Jeypore collections, 904 Manipur collections and 970 FAO World Genetic Stocks. In addition, various Rice Research Stations in the country were maintaining 24,553 accessions which included 3135 exotic collections as well (Seetharaman *et al.*, 1972) (Table 1.7). The NBPGR has developed long term storage facility for germplasm of crop plants. Besides, medium term storage facilities have come up at major rice germplasm depositories in the country. The availability of long term storage facility at NBPGR has helped the various central and state rice research centres to deposit a duplicate set of their germplasm at NBPGR for safe-keeping. Whereas the NBPGR holds the national collection of rice germplasm, CRRI has the responsibility of seed rejuvenation, seed multiplication and characterization. The responsibility for evaluation of rice germplasm for abiotic and biotic stresses is the responsibility of the DRR. After the NATP project, the total number of accessions now being maintained at various rice research centres (excluding NBPGR) in the country is 1,22,432. The NBPGR holds 74,857 accessions of which some are direct collections by the Bureau staff and the rest has been deposited by other rice research centres of the country. Subramanian *et al.* (2007) have provided a comprehensive list of collections made by various agencies in the country.

Table 1.7. Collections maintained at some of the major Rice Research Stations in India in 1972.

State / Institute	Research Station (No. of collections)
CRRI	General (3282 approx), Jeypore (1344), Manipur (904), FAO Catalogue (970)
Andhra Pradesh	Maruteru (721), Rajendranagar (1076), Others (326)
Assam	Karimgnj (1915), Titabar (605), Raha (125)
Bihar	Sabour (828), Pusa (800), Dholi (465)
Chhattisgarh	Raipur (850)
Gujarat	Nawagam (632)
Jammu & Kashmir	Khudwani (426)

Himachal Pradesh	Nagrota Bagwan (100)
Karnataka	Mandya (1850)
Kerala	Pattambi (810), Mannuthy (436), Kayankulam (124), Moncompu (120), Kottarakkara (60)
Madhya Pradesh	Rewa (500), Waraseoni (400)
Maharashtra	Karjat (1119), Panwel (201), Sakoli (106)
Orissa	Bhubaneswar (491), Berhampur (250), Jeypore (61)
Punjab	Ludhiana (450), Kapurthala (125)
Rajasthan	Banswara (50)
Tamil Nadu	Coimbatore (2306), Aduthurai (507)
Uttarakhand	Garampani (256)
Uttar Pradesh	Faizabad (900), Gograghat (400)
West Bengal	Chinsurah (3500), Bankura (1150), Kalimpong (426), Hathwara (86)

Source: Seetharaman, Sharma and Shastry (1972)

NB: The figures in brackets include local collections, collections from other states of India and exotic collections.

Exploration and Collection of Rice Germplasm in India by other Countries

Because of the richness of genetic diversity of cultivated and wild rices in India, many foreign organizations have visited this country to explore and collect the genetic resources of rice. In a comprehensive paper on genetic resources of rice in India, Dr Mangla Rai (1998) states:

> *Many joint explorations with other countries have also been undertaken. Japanese exploration teams led by H. Kihara in the early 1960s and T. Watabe in the late 1960s and early 1970s made systematic collections in Western Uttar Pradesh, Bihar, Andhra Pradesh and parts of Maharashtra. H.I. Oka travelled extensively on collection trips in different parts of India. French team from IRAT and ORSTOM collaborated with ICAR for collection, particularly of* O. nivara, O. rufipogon *and* O. officinalis *from Goa, Karnataka, Maharashtra, and Gujarat in 1986. During 1987 to 1989, ICAR and IRRI scientists undertook intensive collection for wild rice in South India and West Bengal.*

Introduction of exotic germplasm into India

In the initial years (1911-1935), the traditional rice varieties of the country were collected, characterized and evaluated. The varieties released during this phase were mostly pure line selections of the traditional rice varieties only. Some varieties were, however, released after hybridization between these local varieties followed by selection. In the second phase, some Russian and Chinese varieties were introduced and, after their agronomic evaluation, were recommended mainly for hilly tracts where cooler temperature prevails.[12] When the International Rice Commission (IRC) recognized CRRI as a centre for the maintenance of world genetic stocks of rice varieties, especially of the *indica* types, many rice varieties of south and southeast Asian countries were introduced for their maintenance at CRRI, this provided an opportunity to Indian rice scientists to test and recommend a few of them for general cultivation. Again, when CRRI was recognized as the main centre for the *japonica-indica* hybridization, many *japonica* types were introduced into India. The participants of the southeast and south Asian countries came with their own rice varieties for hybridization. This provided a further opportunity to Indian rice scientists to grow and study the rice varieties of other countries. Some of the Japanese types when tried in hilly regions were found suitable for direct introduction. Among the *japonica-indica* hybrids bred in this period, Mahsuri bred in Malaysia was introduced into India and became very popular.

Before Japanese occupation of Taiwan in 1895, the island was growing *indica* rices only. The Japanese Government of Taiwan introduced many *japonica* varieties from Japan, made selections for better yielding *japonica* types and encouraged cultivation of *japonica* types which were called *ponlai* types in Taiwan. After the World War II and liberation of the island, work on *indica* rices was also started. One of the *indica* varieties bred and recommended for release there was Taichung (Native) 1. Many of these *ponlai* types along with the Taichung (Native) 1 were introduced into India in early 1960s. Taichung (Native) 1 turned out to be highly susceptible to bacterial blight and hence was withdrawn. However, it convinced the rice scientists of the country about the potential of its plant type.

Dee-Geo-Woo-Gen, Taichung (Native) 1 and other semi-dwarf rice varieties were introduced from Taiwan to IRRI (Philippines) and IR8 was bred there. When IR8 was introduced into India, it proved to be a successful introduction because of its resistance to bacterial blight and superior yield. Later many IRRI bred varieties were released in this country directly after their trials in the country.

---

[12] According to R. Seetharaman, the Chinese and Russian varieties were introduced during 1935 to 1940 (See Seetharaman *et al.*, 1971).

Characterisation, Rejuvenation and Seed Multiplication

Characterisation of rice germplasm has been an ongoing exercise at CRRI. During the Silver Jubilee celebration of CRRI in 1971, a Catalogue of rice germplasm was published. The catalogue of entire collection of rice germplasm of Assam Rice Collection was prepared and submitted for its publication to the Publication Division of ICAR in 1976. Unfortunately, this could never get published. Instead, the manuscript was lost by the Publication Division of ICAR. The Indira Gandhi Krishi Vishwavidyalaya (IGKV) has brought out a catalogue of its rice collections (See Chapter 14: Rice Germplasm of Chhattisgarh: Collection, Evaluation, Utilization and Conservation).

The National Bureau of Plant Genetic Resources has now developed long term storage facility for the germplasm of all crops including that of rice. Medium term storage facility is now available at CRRI, DRR and IGKV which maintain sizeable collections of rice germplasm in the country. In 1987, it was decided that all curators of rice germplasm in the country should transfer their rice germplasm to NBPGR for its long term storage. Accordingly, most of the rice germplasm of the country has now moved to NBPGR for its long term storage. CRRI has been recognized as the National Active Collection Centre for Rice with the responsibility of rejuvenation of collections as and when necessary, their characterization, seed multiplication, medium term storage and seed supply to various rice research centres.

Evaluation

After the establishment of Central Rice Research Institute (CRRI) in 1946 and before the start of All India Coordinated Rice Improvement Project (AICRIP) in 1965, CRRI was the main centre where rice collections were evaluated in the country. In the early years, no standard procedures for screening rice germplasm against pests and diseases were available. It was done under natural infection only by planting the material on different dates and at several locations to ensure infection for screening. Subsequently, the promising ones were tested under laboratory conditions. As Sridhar (1996) puts it

> *Long before the establishment of the coordinated rice improvement programme in India, both brown spot and blast resistant varieties identified at the institutional testing at CRRI were sent for evaluation for their resistance in multi-location trials in India. The resistant varieties were tested in 46 centres all over India consecutively for three years during 1955-58. This was significant for the formation of multi-location trials*

for evaluation of resistant donors and breeders' material during the semi-dwarf era about a decade later.

Even in those early years when techniques for screening rice material for various pests and diseases were yet to be developed and perfected, a number of donors could be identified. Under the initiative of the FAO/IRC, an international cooperation for testing rice varieties for blast was initiated and thus the Uniform Blast Nursery technique was developed. Table 1.8 provides a list of donors identified for various pests and diseases by CRRI during that period.

Table 1.8. Sources of resistance to pests and diseases identified before HYV era.

Resistant/Tolerant to	Varieties
Blast	CO 4, BJ 1, S 67, SM 6, SM 9, CH 48, AC 1613, Aichi-Asahi, Seluz 44, Tetep, Tadukan, Zenith
Helminthosporose	CH 13, CH 45, T 141, CO 20, BAM 10, AC 1351, AC 2041, AC 2045, AC 2559, T 498-2A
Bacterial Leaf Blight	Lacrose x Zenith-Nira, Wase Aikoku, Early prolific, BJ 1
Stem Rot	Basmati 370, Bara 62
Stem Borers	TKM 6, CH 67, W 1263, CB 1, CB 2, BU 3
Gall Midge	PTB 10, PTB 18, PTB 21, PT 27, BU 3, AC 1368, Leaung 152, Peykeo E53, Peykeo P129

Source: Seetharaman, Sharma and Shastry (1972)

After its establishment, the AICRIP started coordinated programmes to test the breeding lines developed at the central and state research stations under different agroclimatic regions at three successive stages all over the country and called them (a) initial evaluation trial (IET), (b) preliminary varietal trial (PVT), and (c) uniform varietal trial (PVT). AICRIP (later named DRR) screens germplasm and breeding lines for their tolerance and/or resistance to abiotic and biotic stresses and also for quality parameters. In every state, AICRIP/DRR has its cooperating centres that participate in these coordinated trials. These evaluations have helped identify a number of rice varieties that were bred in the pre-HYV era as resistant to various abiotic and biotic stresses. During 1993 to 1998, DRR evaluated a large number of collections, especially IGKV collections, and identified many donors for various desirable traits (See Chapter 14: Rice Germplasm of Chhattisgarh: Collection, Evaluation, Utilization and Conservation).

A network programme on characterization and evaluation of rice germplasm called Evaluation of Rice Germplasm for Biotic Stresses was implemented during 1993 to 1998 with Directorate of Rice Research as its coordinating centre. In this programme, more than 13,000 accessions (comprising mostly of Raipur accessions) were characterized for 21 morphological characters and screened for three major diseases (blast, bacterial leaf blight and sheath blight) and three major insect pests (brown plant hopper, gall midge and stem borer) at two lead centres (NBPGR and IGKV) and at 12 hot spots. A number of donors were identified that can be utilized in the rice breeding programme.

The International Rice Research Institute, established in 1960, could standardize screening procedures for resistance to various pests and diseases, identify races of pathogens and biotypes of insect pests and study genetics of the resistance. Indian rice scientists actively participated in such programmes and made valuable contributions.

In 1975, IRRI launched its International Rice Testing Programme (IRTP) initially with the financial support from the UNDP. It was later named as International Network of Genetic Evaluation in Rice (INGER). Under this programme, IRRI makes the world's elite breeding lines and genetic resources available to all rice scientists for direct use and/or for crosses within their breeding programmes. It provides an opportunity to rice scientists to help identify genetic resources with resistance to major biotic stresses and tolerance for abiotic stresses. This programme has promoted international cooperation and interaction among rice scientists of the world. On an average, about a thousand genetically diverse materials contributed by NARES[13] and IARC[14] scientists are evaluated yearly under different rice ecosystems and for biotic and abiotic stresses. More than 600 experiment stations in 80 countries participate in these annual evaluations.

Utilisation

As mentioned already, Taichung (Native) 1 or IR 8 could not fit in all the ecological situations of the country. Besides, they were not resistant to the spectrum of pests and diseases prevalent in India. Convinced about the better plant architecture of these varieties, the Indian rice breeders started breeding new rice varieties with plant type of the semidwarf varieties. As the varieties bred during 1920s and 1930s had the adaptability to various ecosystems of India, these semi-dwarf varieties were crossed with those improved varieties. In other words,

---

[13] NARES: National Research and Extension Systems

[14] IARC: Indian Agricultural Research Centres

the varieties bred during 1920s to 1940s reappeared with semidwarf plant type.

In the early years (1960s) of HYV breeding, Taichung (Native) 1 and IR 8 were directly used as parents but with passage of time varieties already bred or breeding lines already developed were used as parents. However, the narrow genetic base of Dee-Geo-Woo-Gen persists in all these varieties and makes them vulnerable to new biotypes of insects or races of diseases. In fact, the vulnerability is much more in the sense that most often the breeders have used the semidwarf varieties as the female parent and thus have caused a narrow base for the cytoplasm as well. By 2000, ICAR had released 632 rice varieties for all the ecosystems present in India (Prasada Rao, 2004) (Table 1.9). Recently a comprehensive list of high yielding rice varieties bred and released by the ICAR since 1964 has been brought out by DRR (Shobha Rani *et al.,* 2008).

Table 1.9. Number of rice varieties released for different ecosystems in India during 1965 to 2000.

S. No.	Ecosystem	No. of varieties released
1	Irrigated	314
2	Upland	84
3	Lowland	123
4	Semi-deep water	30
5	Deep water	14
6	Saline alkaline	15
7	High altitude	33
8	Scented fine grained	19
	Total	**632**

Source: DRR (2000)

New Tools and Technologies

Hybrid rice

The phenomenon of heterosis in rice was first reported by Jones (1926) for its number of culms and yield. Later, Kadam *et al.* (1937), Sethi (1937) and Ramiah and Ramaswamy (1941) also observed heterosis in rice. Success of developing commercial hybrid in maize, pearl millet and sorghum further helped in exploring the possibility of developing commercial hybrid varieties in rice (Swaminathan *et al.,* 1972). For developing successful commercial rice hybrids,

CMS lines are quite useful.

The first male sterile plant with abortive pollen was found by Li Bi-Hu in a population of common wild rice (*Oryza sativa* L. f. *spontanea*) on Hainan Island in China (Lin and Yuan 1980). Since it was found in a wild population and had aborted pollens, this cytoplasm was given the name Wild Abortive (WA). Utilizing this cytoplasm three CMS lines (Er-Chui-Nan 1A, Zhen-Shan 97A and V20A) were developed in China and shared with IRRI. From IRRI, these CMS lines were received at IARI, New Delhi during *kharif* 1981 (courtesy Dr. M.S. Swaminathan). Later, many other sources of cytoplasmic male sterility have been identified but the WA source has been the most stable and widely used source of CMS system.

In 1989, the Indian Council of Agricultural Research started a Mission Mode Project on basic as well as applied aspect of hybrid rice under the leadership of Dr E.A. Siddiq. The financial support was received from UNDP and technical support from FAO. Later, Mahyco Research Foundation also supported the project financially. DRR became the coordinating centre of the project; ICAR institutes and some State Agricultural Universities participated in the project (Table 1.10). Later, many private sector seed companies also started work on hybrid rice. In the process, thirty-six CMS lines, their maintainers and restorers have been identified by different centres (DRR 1999-2000; Sarial and Singh 2000). Among the CMS lines, IR58025A, IR62829A, Pusa 6A, PMS3A and PMS10A have proved very promising.

Table 1.10. Hybrid rice research network in India.

Objective	Centres responsible
Basmati hybrids	IARI, New Delhi.
Hybrids for shallow lowland	CRRI, Cuttack
Long duration rice for coastal region	Rice Research Station, Maruteru
Two line hybrids	DRR, Hyderabad TNAU, Coimbatore GBPUAT, Pantnagar
Development of interspecific hybrids (*indica*/tropical *japonica*)	DRR, Hyderabad TNAU, Coimbatore GBPUAT, Pantnagar
Region specific rice hybrids	TNAU, CoimbatoreGBPUAT, Pantnagar, and Respective State Agricultural Universities

The Indian Agricultural Research Institute, New Delhi has made a remarkable contribution in development of hybrid rice technology with special reference to Basmati quality rice hybrids. A systematic programme for development of Basmati quality CMS and restorer lines was initiated at IARI in late eighties. For this, Pusa Basmati 1 formed the foundation. By transferring the WA cytoplasm to the nuclear background of Pusa Basmati 1, a CMS line (Pusa 3A) with typical Basmati quality was developed. In addition to Pusa 3A, a number of other Basmati quality CMS lines (Pusa 6A, Pusa7A, Pusa 8A, Pusa 9A, Pusa 10A and Pusa 11A) were developed. These CMS and restorer lines are being used in development of quality rice hybrids. One of the superfine grain aromatic rice lines (Pusa 167-120-3-2) developed by IARI was used by IRRI for conversion (IR48483A/Pusa167) into a CMS line by transferring WA cytoplasm which led to development of IR 58025A, the most widely used CMS in the hybrid rice programme not only in India but world over. In India, more than two dozen rice hybrids under commercial cultivation, are based on IR 58025A.

So far, thirty-seven rice hybrids of different growth duration, plant height, grain type and quality have been developed. In the early group, Pusa Rice Hybrid 10 (PRH 10) maturing in 110 days from seed to seed with superfine aromatic grain has become most popular and the farmers have realized a yield up to 10 tons/ha.

In situ conservation

The cultivated rice (*O. sativa*) does not occur in nature in wild state except as an occasional escape. Therefore, on-farm conservation could be the only possible way for its *in situ* conservation. Until 1920, there was hardly any improved rice variety grown by the farmers in India. And even though a fairly large number of rice varieties were bred up to 1960, the area covered under these varieties was not significant. The varietal diversity of rice in India was, therefore, conserved *in situ* by rice farmers without any deliberate attempt for it. With the advent of high yielding rice varieties in 1960s and their adoption by more and more farmers, especially due to vigorous campaigning by the agricultural extension workers in favour of these varieties, the genotypic diversity of rice started disappearing at a very fast rate. For example, when CRRI explored the Jeypore tract of Orissa (i.e. southern districts of Orissa, inhabited by tribal people) in 1950s, 1745 landarces were collected. When a similar survey was conducted by the same Institute after fifty years, hardly 350 landraces could be collected.

On-farm conservation has its own merits that cannot be achieved by *ex situ* conservation. In on-site conservation, population of a landrace is always exposed to ever-changing edaphic, climatic and biotic factors and the population of a landrace goes on readjusting itself to the changed conditions. There is constant gene flow within the plants of this population and also with the neighbouring populations of the same species. There is also flow of genes from the populations of the related wild species that may be growing around the landrace. This process helps creating recombinants and thus enriching genetic variation in a crop species. In *ex situ* conservation, this process of microevolution is arrested.

Anil Gupta (1992) has discussed the various aspects of biodiversity conservation especially in the diversity-rich economically poor communities and says:

> *Several modern technologies, including DNA fingerprinting, recombinant DNA, tissue culture, etc have generated false hope that genetic diversity can be preserved entirely by laboratory methods. I do not share this optimism. In my view, biodiversity is not a static concept. People living in a bio-rich and economically poor environment keep on making selections of one kind or the other to enrich the given diversity; therefore, the need for institutional mechanisms that safeguard the ability of these people to continue the process of enriching biodiversity.*

In *in-situ* conservation, the population of a landrace is often much larger than that of the number of seeds kept in a seed store. Since the number of accessions maintained in an experiment station is often large and the space available in the field is limited for seed rejuvenation and multiplication, only a very small population of an accession is grown. This limits the plant population raised for rejuvenation and often results in random genetic drift. Besides, the accessions that are collected from diverse ecotypes are grown under uniform field conditions of the experiment station year after year; there is also the danger of genetic shift.

There are several aspects of on-farm conservation that have not yet been worked out. As Bellon, Pham and Jackson (1997:266) remark:

> *In spite of increasing interest in on-farm conservation, which is even addressed in the Convention of Biological Diversity, there is still limited knowledge about what this approach means, and even less understanding of its various social, economic, cultural and genetic aspects.* ———
>
> *Many questions remain to be answered about the viability of on-farm conservation for genetic conservation. In what way do varieties change over time? Do they conserve varieties or do they conserve traits such as aroma, plant architecture or disease resistance, for example? What is the importance of seed exchange among farmers for enriching their germplasm? Why do some farmers continue to grow their local varieties and yet*

others have abandoned them in favour of improved varieties? What is the degree of outcrossing between varieties in farmers' fields, and so on.

There has not been much effort for on-farm conservation of genetic diversity of rice in India as it involves not only biological but also social, cultural, ethical, economic and even administrative logistics. In India, any effort for on-farm conservation can be easily frustrated by an administrator (a minister, a district officer) who may not be able to appreciate the significance of on-farm conservation. There are, however, pockets in India which deserve on-farm conservation. The Jeypore tract of Orissa, Palakkad area of Kerala and Apatani valley of Arunachal Pradesh are some good candidates for on-farm conservation.

The *in situ* conservation of wild rices faces a different kind of situation. There are two species of wild rices (*O. officinalis* and *O. granulata*) that grow in forest shades. There *in situ* conservation is, therefore, linked with the conservation of forests. The other two species (*O. rufipogon* and *O. nivara*) grow in open habitat in the edges of ponds and in the seasonal ditches on the roadsides. These habitats are disturbed and very much vulnerable to increasing human encroachment. *In situ* conservation of these species is therefore challenging.

Wide hybridisation

The genus Oryza has 22 wild and two cultivated species. The wild species are distributed all over the tropics. Some of them are diploid (2n=24) whereas others are tetraploids (2n=48). The cultivated rices and some of the wild species are diploid and have the genomic constitution AA whereas others have genomes CC, BBCC, CCDD, EE, FF, GG, HHJJ. The wild species harbour genes for resistance to many insect pests and diseases as well as for tolerance to many types of physiological stresses. Some of these genes have already been successfully transferred from the wild species to the cultivated rices. For example, the genes for resistance to grassy stunt virus have been transferred from *O. nivara* to the cultivated rice; similarly the genes for resistance to brown planthopper have been transferred from *O. officinalis* into *O. sativa*. The cytoplasmic male sterility has been transferred from *O. sativa* f. *spontanea* to the cultivated rice to develop hybrid rice. The cultivated rice has been crossed with many of the wild species and hybrids have been produced for their possible use in rice breeding. This has been discussed in detail in Chapter 9 (Exploration, Collection, Conservation and Utilization of Wild Rices for the Improvement of Cultivated Rice).

Biotechnological tools

Biotechnological tools can help enriching genetic resources by using two approaches: molecular markers and gene transformation. With the advancement of science in molecular markers, highly saturated linkage map is available in many crops including rice. The whole genome of rice is already sequenced and is available in the public domain. This has made precise transfer of genes resistant against biotic and abiotic stresses from wild relatives into cultivated varieties possible (Brar *et al.,* 1996; Brar and Khush, 1997; Jena and Khush, 1984; Khush *et al.,* 1990; Multani *et al.,* 1994; Multani *et al.,* 2003). This approach also helped in reducing the genetic drag while transferring such genes from wild relatives into the cultivated one. Molecular markers also makes it possible to pyramid two or more genes into a genetic background which is otherwise not easily possible using conventional techniques. Many laboratories in India have transferred genes for resistance against bacterial blight, blast and tungro and are being evaluated for their agronomic worth besides being used as donors and genetic stocks. In early nineties, many Indian laboratories started transferring genes for resistance against blast and bacterial blight into cultivated ones. Some of them are Central Rice Research Institute (Cuttack), Punjab Agricultural University (Ludhiana) and University of Agricultural Science (Bangalore). All these centres initiated the marker assisted selection (MAS) work in collaboration with the International Rice Research Institute, Philippines. Many genes for blast resistance (*Pi* 1, *Pi* 2, *Pi* 4, *Pi* $k^{(h)}$, and *Pi* 9) and for bacterial blight resistance (*xa*-5, *Xa*-13 and *Xa*-21) have successfully been transferred into rice cultivars (Hittalmani *et al.,* 2000; Joseph *et al.,* 2004; Singh *et al.,* 2001; Sundaram *et al.,* 2008) in India. Two rice cultivars (bacterial blight resistant Pusa Basmati 1 from IARI and blast resistant Samba Mahsuri from DRR, Hyderabad) have been released by the CVRC for general cultivation in India. All those breeding materials are valuable genetic stocks for breeding as well as MAS work.

The other approach to transfer an alien gene is by using one of the methods of plant transformation. Two most acceptable plant transformation methods are: *Agrobacterium*–mediated gene transfer and biolistic method (particle bombardment). The advantage of plant transformation is that a gene from any source can be transferred into plants, may it be from plants, insects-pests, microbes, animals or human beings. This opens up scope to practically transfer any gene from any source into any plant system. In India genes have been transferred into rice plants for disease resistance (Patkar and Chatoo. 2006), insect resistance (Rao *et al.,* 2009), and other traits including marker genes (Agrawal *et al.,* 2005a). Agrawal *et al.* (2005a) reported a modified biolistic system for multiple gene transfer in rice and for the development of marker-free transgenic plants. Rice transformation work is being carried out at many institutes in India including CRRI

(Cuttack), DRR (Hyderabad), PAU (Ludhiana), IARI (New Delhi), TNAU (Coimbatore) and VPKAS (Almora). Many transgenic plants with resistance against diseases and insect-pests have been reported from these centers. These transgenic plants are not only under agronomic evaluation but can also be useful as genetic stocks for breeding and basic work.

Future Prospects

The farmers of India had been holding a fabulous genetic wealth of rice even up to the middle of the twentieth century. During the last sixty years, the situation has changed dramatically. The farmers have dropped many of their traditional rice varieties and have gone for the high yielding varieties because of the development of high yielding rice varieties by the rice breeders and vigorous campaigning by the agricultural extension workers. Fortunately, a large part of this variability has been already collected and conserved in the germplasm repositories of the country. The varieties bred and breeding lines developed by rice scientists during the last sixty years now form new sources of elite germplasm and are now used in rice breeding. The National Bureau of Plant Genetic Resources has been conserving these elite lines also in the national genebank.

In India, exploration and collection of rice germplasm have moved faster than their characterisation and evaluation and the latter have moved faster than their utilisation. In other words, a large number of landraces have been conserved but have not been characterised or evaluated and a large number of donors have been identified but only some of them have been utilised in breeding. This is neither unexpected nor undesirable. The donors identified but not utilised so far act as 'reserve' against the new biotypes and races and many unforeseen situations that might arise in future. It would be perilous to consider or treat the unevaluated or unutilised germplasm accessions unworthy of conservation. The environment is ever changing, the spectrum of insect pests and diseases are changing too. Even human preferences and usages are changing. And it is impossible to predict the direction of these changes. The genetic diversity conserved in our repositories is a guarantee against any food insecurity in future. It is safer to conserve this variation created by generations of mankind and nature for our unforeseen use in future.

The rice germplasm now available in the national genebank has many duplicates. In the past, there has been a lot of exchange of rice germplasm among various rice research stations of the country and, in the process, each station has assigned its own accession number to a collection received from other stations. Besides, each station has assigned its own unique prefix to its accession number to maintain the station's identity. This has resulted in the multiplicity of prefixes and duplicates of material resulting now in unnecessary wastage of space in the fields for its rejuvenation and screening, wastage of manpower and material in its evaluation and handling, wastage of space and electricity in the cold storage and wastage of funds. This could have been avoided by developing a national accession number system; the process can be arrested even now. It is presumed that when all the collections are received at the NBPGR for their safe-keeping and in the process NBPGR assigns its own accession numbers, the NBPGR accession numbers will be used as the National Accession Numbers.

Some of the challenges that the breeders of the next generation may face are shortage of land for growing rice vis-à-vis growing population, shortage of water for irrigation, shortage of fertilizers and pressure to increase production of rice per unit area (due to ever increasing human population and need to feed them). There may be increasing international pressure to reduce wetland rice cultivation to check methane emission. Global warming will create many new problems. It is feared that coastal rice lands will be inundated under the rising sea and the rivers emerging from Himalayas may dry up in summer. The spectrum of diseases and pests will, no doubt, change as it has changed in the past. The breeders then may have to look for different kinds of genes in germplasm than what we have looked for so far.

The efforts to explore, collect, conserve, characterize, evaluate, document and utilize rice germplasm have been done by a large number of scientists belonging to many institutes and research stations of the country during the past four or five generations. In this endeavour, the Indian Council of Agricultural Research has played a prominent role and has kept the National Bureau of Plant Genetic Resources, the Central Rice Research Institute and the Directorate of Rice Research in the forefront. Even the role of the state agricultural universities (and state departments of agriculture as their predecessors) has been quite commendable. But the story is not as simple as said. It has many facets and many dimensions, many successes and many failures from which the new generations of rice scientists of the country should learn. The following chapters of this book are meant to address the efforts made by the different national institutes, state agricultural universities (including the departments of agriculture that were their predecessors) and some non-government organisations and narrate the successes achieved and failures encountered.

Acknowledgement:

The author thankfully acknowledges the inputs provided by Dr V. P. Singh (IARI), Dr P. K. Agrawal (VPKAS), Dr B.C. Patra (CRRI) and Dr D. R. Pani (NBPGR, Cutack) in preparing this manuscript

References:

Agrawal PK, Twyman R, Christou P, Kohli A. 2005a. Efficient multiple gene transfer and expression in rice using minimal expression cassettes. *Molecular Breeding* 16(3):247-260.

Alam M. 1935. *Annual Reports of the Rice Research Station, Sabour for 1934-35, 1935-36, 1937-38, 1938-39 and 1939-40.* Dept of Agri, Govt of Bihar and Orissa, Patna.

Anstey Vera. 1977. *The Economic Development of India.* Arno Press (A New York Times Co), New York .

Bellon MR, Pham J-L, Jackson MT. 1997. Genetic conservation: A role for rice farmers. Pages 263-289 in Maxted N, Ford-Lloyd BV, Hawkes JG (eds) *Plant Genetic Conservation: The In-situ Approach.* Chapman Hall, London.

Brar DS, Dalmacio R, Elloran R, Aggrawal R, Angeles R, Khush GS. 1996. Gene transfer and molecular characterization of introgression from wild Oryza species into rice. Pages 477-485 in Khush GS (ed) *Rice Genetics III.* International Rice Research Institute, Manila, Philippines.

Brar DS, Khush GS. 1997. Alien introgression in rice. *Plant Mol Biol* 35:35–47.

Chandler RF. 1972. The impact of improved plant tropical plant type on rice yields in South and Southeast Asia. Pages 77-84 in IRRI (ed) *Rice Breeding.* International Rice Research Institute, Los Banos, Philippines.

Chang TT. 1972. See Discussion on Parthasarathy (1972).

Christou P, Ford TL, Kofron M. 1991. Production of transgenic rice plants from agronomically important *indica* and *japonica* rice varieties via electric discharge particle acceleration of exogeneous DNA into immature zygotic embryos. *BioTechnol* 5:957-96.

Dalrymple Dana G. 1986. *Development and Spread of High-Yielding Rice Varieties in Developing Countries.* Agency for International Development, Washington, DC.

Directorate of Rice Research. 2000. *Development and Use of Hybrid Rice Technology: Progress Report 1999-2000.* Directorate of Rice Research (ICAR), Rajendranagar, Hyderabad 500 030.

Gangadharan G. 1985. Breeding. Pages 73-109 in Padmanabhan, SY (ed) *Rice Research in India.* Indian Council of Agricultural Research, New Delhi.

Ghose RLM, Ghatge MB, Subrahmanyan V. 1960. *Rice in India.* Indian Council of Agricultural Research, New Delhi.

Govindaswami S, Krishnamurty A. 1959. Genetic variability among cultivated rices of Jeypore tract and its utility in rice breeding. *Rice News Teller* 7(4/4):12-15.

Graham RJD. 1913. Preliminary note on the classification of rice in the Central Provinces. *Mem Dept Agri India, Bot Ser* 6:209-229.

Gupta Anil. 1992. Biodiversity, poverty and intellectual property rights of third world peasants: A case for renegotiating global understanding. Pages 1-5 in Swaminathan MS and Jana S (eds) *Biodiversity: Implications for Global Food Security.* Macmillan India Ltd, Madras.

Hector GO, Sarangapani SG. 1934. Varietal characters and classification of rices. *Indian J Agric Sci* 4:1-80.

Hittalmani S, Parco A, Mew TV, Ziegler RS, Huang N. 2000. Fine mapping and DNA marker-assisted pyramiding of the three major genes for blast resistance in rice. *Theor Appl Genet* 100:1121-1128.

Hooker JD. 1897. *Flora of British India.* Vol. VII. Reeve & Co Ltd.

Hore DK. 2005. Rice diversity collection, conservation and management in Northeastern India. Genetic Resources and Crop Evolution 52(8):1129-1140.

Huang CH, Chang WL, Chang TT. 1972. Ponlai varieties and Taichung Native 1. Pages 31-46 in IRRI (ed) *Rice Breeding*. International Rice Research Institute, Los Banos, Philippines.

Hutchinson JB, Ramiah K. 1938. Description of crop plant characters and their ranges of variation. II. Variability in rice. *Indian J Agri Sci* 8:592-616.

IBPGR-IRRI Rice Advisory Committee. 1980. *Descriptors for Rice* (*O. sativa* L). IRRI, Los Banos, Philippines.

International Rice Research Institute. 1968. *Annual Report 1968. Rice Breeding*. International Rice Research Institute, Los Banos, Philippines. 402 pp

Jena KK, Khush GS. 1984. Embryo rescue of interspecific hybrids and its scope in rice improvement. *Rice Genet Newsl* 1:133–134.

Jennings PR. 1966. *Evaluation of Partial Sterility in* Indica x Japonica *Rice Hybrids.* Technical Bulletin No. 5. International Rice Research Institute, Manila.

Jones JW. 1926. Hybrid vigour in rice. *J Am Soc Agron* 18:423-428.

Joseph M, Gopalakrishnan S, Sharma S, Singh RK, Singh VP, Singh AK, Singh NK, Mohapatra T. 2004. Combining bacterial blight resistance and basmati quality characteristics by phenotypic and molecular marker assisted selection in rice. *Mol Breed* 13:377-387.

Kadam BS, Patil GG, Patankar VK. 1937 Hetrosis in rice. *Indian J Agric Sci* 7:118-126.

Kadam BS, Ramiah K. 1943. Symbolization of genes in rice. *Indian J Genet & Plant Breed* 3:125-132.

Kashiram and Ekbote RB. 1936. Classification of autumn rices of the Punjab and Western United Provinces. *Indian J Agri Sci* 6: 930-7.

Kashiram, Saravayya CV. 1934. Classification of rices of Bihar and Orissa. *Indian J Agri Sci* 4: 4-537.

Khush GS, Bacalangco E, Ogawa T. 1990. A new gene for resistance to bacterial blight from *O. longistaminata. Rice Genetic Newsl* 7:121-122.

Lin Shih-Cheng, Yuan Loung-Ping. 1980 Hybrid rice breeding in China. Pages 35-52 in IRRI (ed) *Innovative Approaches to Rice Breeding* (Selected papers from the 1979 International Rice Research Conference, 1979). International Rice Research Institute, Manila.

Lu Yung-ken. 1979. Semi-dwarf gene sources of rice in China. *IRRN* 4:2.

Lu Yung-ken. 1979. The origin of semi-dwarf genes. *IRRN* 4:2.

Mehra KL. 2005. Rice in Indian history and culture. Pages 1-31 in Sharma SD and Nayak BC (eds) *Rice in Indian Perspective*. Today and Tomorrow's Printers and Publishers, New Delhi.

Mitra SK, Ganguly PM. 1932. Studies in Surma valley rices and their classification. *Indian J Agri Sci* 2: 271-9.

Multani DS, Jena KK, Brar DS, Delos Reyes BC, Angeles ER and Khush GS. 1994. Development of monosomic alien addition lines and introgression of genes from *Oryza australiensis* Domin to cultivated rice *O. sativa* L. *Theor Appl Genet* 88:102-109.

Multani DS, Khush GS, delos Reyes BG, Brar DS. 2003. Alien genes introgression and development of monosomic alien addition lines from *Oryza latifolia* Desv. to rice, *Oryza sativa* L. *Theor Appl Genet* 107:395–405.

Muralidharan K, Laha GS, Krishnaveni D, Reddy CS, Reddy JN, Sridhar R. 2004. Field performance of genotypes generated by molecular-assisted breeding to bacterial leaf blight in India. *Rice Genet Newsl* 21:29-31.

Nanda JS, Virmani SS. 2000. Hybrid rice. Pages 23-52 in Nanda JS (ed) *Rice Breeding and Genetics: Research Priorities and Challenges.* Oxford and IBH Publishing Co, New Delhi.

Nene YL. 2005. Rice research in South Asia through ages. *Asian Agri-History* 9(2):85–106.

Oka HI, Chang WT. 1962. Rice varieties intermediate between wild and cultivated forms and the origin of the *japonica* type. *Bot Bull Acad Sin* 3:109-131.

Oka HI. 1964. Considerations on the genetic basis of intervarietal sterility in *O. sativa.* Pages 159-174 in IRRI (ed) *Rice Genetics and Cytogenetics.* Elsevier, Amsterdam.

Parthasarathy N. 1972. Rice breeding in tropical Asia up to 1960. Pages 5-29 in IRRI (ed) *Rice Breeding.* International Rice Research Institute, Los Banos, Philippines.

Patkar RN, Chatoo BB. 2006. Transgenic indica rice expressing ns-LPT like proteins shows enhanced resistance to both fungal and bacterial pathogens, *Molecular Breeding* 17:159-171.

Prasad Rao U. 2004. Role of Directorate of Rice Research in genetic improvement of rice varieties of India. Pages 101-140 in Sharma SD and Prasada Rao U (eds) *Genetic Improvement of Rice Varieties of India.* Today and Tomorrow Printers and Publishers, New Delhi.

Rai M. 1998. Rice germplasm evaluation and enhancement in India: Issues, status, options and future plan of action. Pages 83-91 in Rutger JN, Robinson JF, Dilday RH (eds) *Proceedings of the International Symposium on Rice Germplasm Evaluation and Enhancement.* Stuttgart, Arkansas, Aug 30 to Sept 2, 1998. Dale Bumpers National Rice Research Center, USDA-ARS and Rice Research and Extension Center, Division of Agriculture, University of Arkansas, USA.

Ramiah K. 1937. *Rice in Madras: A Popular Handbook.* Government Press, Madras.

Ramiah K. 1953. *Rice Breeding and Genetics.* Indian Council of Agricultural Research, New Delhi.

Ramiah K, Ramaswami K. 1941 Hybrid vigour in rice (*Oryza sativa* L). *Indian J Genet & Pl Breed* 1:4-12.

Rao MJBK, Nagaraju M. 1974. The *japonica-indica* rice hybridisation in retrospective and prospective. *IRC Newsletter* 23(1):41.

Rao MVR, Behera KS, Baisakh N, Datta SK, Rao GJN. 2009. Transgenic *indica* rice cultivar 'Swarna' expressing a potato chymotrypsin inhibitor *pin2* gene show enhanced levels of resistance to yellow stem borer. *Plant Cell Tiss Organ Cult* 99: 277- 285.

Ratnakar, Sherren. 2006. Agro-pastoralism and the migrations of the Indo-Iranians. Pages 157-192 in Thapar, Romila *et al.* (eds) *India: Historical Beginnings and the Concept of Aryan*. National Book Trust, New Delhi.

Richharia RH, Govindaswami S. 1990. *Rices of India.* 2nd Edition. Academy of Development Sciences, Karjat Taluk, Kashele-410201 (Maharashtra).

Roxburgh W. 1832. *Flora Indica. Vol. II.* W. Thacker & Co, Calcutta and Parbury Allen & Co, London. (See pages 200-206 for Oryza)

Sampath S. 1964. The significance of hybrid sterility in rice. Pages 175-186 in IRRI (ed) *Rice Genetics and Cytogenetics.* Elsevier Publ Co, Amsterdam.

Sarial AK, Singh VP. 2000. Identification of restorers and maintainers for developing Basmati and non-Basmati hybrid in rice. *Plant Breeding* 119:243-247.

Seetharaman R, Sharma SD, Shastry SVS. 1971. Germplasm conservation and use in India. Pages 187-200 in IRRI (eds) *Rice Breeding.* International Rice Research Institute, Los Banos, Philippines.

Sethi RL, Saxena BP. 1930. Classification and study of characters of cultivated rices in the United Provinces. *Mem Dept Agri India, Bot Ser* 18:149-210.

Sethi RL. 1937. *Annual Report of Rice Research Station, Nagina.* Dept of Agriculture, United Provinces, India.

Sharma SD, Vellanki JMR, Hakim KL, Singh RK. 1971. Primitive and current cultivars of rice in Assam: A rich source of valuable genes. *Curr Sci* 40:126-128.

Sharma SD. 1982. Collection and evaluation of rice germplasm from northeast India. *IBPGR Plant Genetic Resources Newsletter* 50:62-69.

Shastry SVS, Sharma SD, John VT, Krishnaiah K. 1971. New sources of resistance to pests and diseases in the Assam rice collections. *International Rice Commission Newsletter* 20(3):1-16

Shastry SVS. 1964. Is sterility genic in *japonica-indica* rice hybrids? Pages 154-157 in IRRI (ed) *Rice Genetics and Cytogenetics*. Elsevier Publ Co, Amseterdam.

Shobharani N *et al.* 2008. *High Yielding Rice Varieties of India*. Directorate of Rice Research, Hyderabad. 188 pp.

Shrivastava MN, Kumar A. 2005. Genetic improvement of rice varieties in Chhattisgarh and Madhya Pradesh. Pages 435-475 in Sharma SD and Prasada Rao (eds) *Genetic Improvement of Rice Varieties of India.* Today and Tomorrow Printers and Publishers, New Delhi.

Singh P. 2008. Origin of agriculture in the middle Ganga plain. Pages 3-18 in Lallanji Gopal and Srivastava VC (eds) *History of Agriculture in India (up to C 1200 AD)*. Concept Publishing Co, New Delhi.

Singh S, Sidhu JS, Huang N, Vikal Y, Li Z, Brar DS, Dhaliwal HS, Khush GS. 2001. Pyramiding three bacterial blight resistance genes (*xa5, xa13 and Xa21*) using marker-assisted selection into *indica* rice cultivar PR106. *Theor Appl Genet* 102:1011-1015.

Singh VP, Satya P, Joseph M. 2005. Hybrid rice: Methodology, seed production and present status in India. Pages 489-502 in Sharma, SD and Nayak BC (eds) *Rice in Indian Perspective*. Today and Tomorrow Printers and Publishers, New Delhi.

Sridhar R. 1996. *Fifty Years of Research at CRRI (1946-1995).* Central Rice Research Institute, Cuttack.

Subramanian M, Nilakanta Pillai KN, Thirumeni S, Kannaiyan S. 2007. *Biodiversity of Rice and its Conservation.* National Biodiversity Authority, Chennai, Tamil Nadu. 92 pp.

Sundaram RM, Vishnupriya MR, Biradar SK, Laha GS, Reddy GA, Rani NS, Sarma NP, Sonti RV. 2008. Marker-assisted introgression of blight resistance in bacterial blight resistance in Samba Mahsuri, an elite *indica* rice variety. *Euphytica* 160:411-422.

Swaminathan MS, Jana S.1992. Introduction. Pages 1-5 in *Biodiversity: Implications for Global Food Security.* Macmillan India Ltd, Madras.

Swaminathan MS, Siddiq EA, Sharma SD. 1972. Outlook for hybrid rice in India. Pages 609-613 in IRRI (ed) *Rice Breeding.* International Rice Research Institute, Manila, Philippines.

Tanaka A, Kawano K, Yamaguchi J. 1966. *Photosynthesis, Respiration and Plant Type of the Tropical Rice Plant.* IRRI Bulletin No 7. Rice Breeding. International Rice Research Institute, Los Banos, Philippines. 46 pp.

Tanaka T, Matsushima S, Kojyo S, Nitta H. 1969. Analysis of yield-determining process and its application to yield-prediction to and culture improvement of lowland rice. XC. On the relation between the plant type of rice plant community and the light-curve of carbon assimilation. *Proc Crop Sci Soc* (Japan) 38:287-293.

Watt G. 1891. *Dictionary of Economic Products of India. Vol.V.* Reprint Edition, Periodical Experts, Delhi.

Genetic Resources of Rice in India : 47-77
Editor : S. D. Sharma
Today & Tomorrow's Printers and Publishers, New Delhi - 110 002 (India)

Chapter 2

Rice Germplasm at Central Rice Research Institute: Exploration, Collection, Conservation and Utilisation

B.C. Patra and S.R. Dhua

The cultivated rice (*Oryza sativa* L.) is said to have originated in the east or southeast or south Asia. India forms a major part of this region and is traditionally rich in the genetic diversity of cultivated rice including that of its wild progenitors. Rice research in India started when the country was facing widespread drought and famine. This was further geared up when millions of people died or suffered due to Bengal famine in 1943 which was caused due to the brown spot disease of rice in an epidemic form. Since then, systematic research in developing biotic (disease/insect) and abiotic stress (drought/submergence/salinity) resistant/tolerant varieties have been undertaken through pure line selection from a vast range of landraces available in the country. A comprehensive account of rice germplasm collection through exploration or acquisition, its conservation, characterization, evaluation, documentation and utilization has been given in this chapter.

During the first quarter of the 20th century and more specifically in the years between 1910 and 1920, F.R Parnell, G.P. Hector and R.J.D. Graham initiated collection of traditional rice varieties from Madras Presidency, Bengal and Central Provinces respectively. The first Agricultural Research Station devoted to rice research was established in Dacca (now Dhaka in Bangladesh) in 1911 and the Paddy Breeding Station was established at Coimbatore in 1913. Subsequently, rice research stations were established at Karimganj (Assam) in 1913, and Titabar (Assam) in 1923, Pattambi (Kerala) and Nagina (Uttar Pradesh) in 1927, Chinsurah (West Bengal) in 1932, Sabour (Bihar) in 1932, Habiganj (now in Bangladesh) in 1934, and at many more places in different agro-climatic zones of the country (see Tables 1.1 and 1.4). These rice research stations collected traditional rice varieties from their respective regions/localities. They also practised pure line selection to identify higher yield potential lines and recommended them for general cultivation.

Germplasm Acquisition

The Central Rice Research Institute was established at Cuttack in the year 1946 with K. Ramiah as its first Director. He brought with him a nucleus set of about 2,000 accessions of rice germplasm from Coimbatore, which was being maintained at the Paddy Breeding Station. This became the starting point of the building up National Germplasm Collection for rice at CRRI. Subsequently, many explorations, collection, introduction and acquisition through exchange activities have helped enrich this Gene Pool (now Gene Bank) (Appendix 2.1).

Germplasm Collection through Exploration

In 1955, when Dr. N. Parthasarathy was the Director, the CRRI undertook its first planned exploration and collection mission of rice germplasm in the Jeypore tract (now Koraput district of Orissa). The collection programme continued for 5 years (1955-59) by a team of scientists led by Dr. S. Govindaswami, supported by a scheme sanctioned by the ICAR. This mission was popularly known as Jeypore Botanical Survey (JBS) and was the first of its kind, ever organized in the world to collect rice germplasm (Chang, 1989).The team explored about 27,000 sq km and collected a total of 1,745 cultivated rice and 150 wild rice accessions (Govindaswami and Krishnamurty, 1959).

Later when Dr. R.H. Richharia became Director of CRRI, he initiated the exploration and collection of rice germplasm from Manipur and Nagaland during 1965-69. The exploration and collection work was done by Mr A.C. Sharma who was posted in Manipur for this work. Mr Sharma was a Senior Research Associate of the Institute but since he hailed from Manipur, he knew the area and the people very well. This mission was spread over a period of five years and a total of 874 accessions were collected.

A PL-480 project on collection of rice germplasm was operative during 1967-72 with Dr M.S. Swaminathan (then Director, IARI) and Dr S.V.S. Shastry (then project Co-ordinator, AICRIP, Hyderabad). In this programme, during a period of five years, Dr S.D. Sharma and his associates collected a total of 6,630 accessions from all the districts of Arunachal Pradesh, Nagaland, Manipur, Tripura and Meghalaya states and from North Lakhimpur, Guwahati and Goalpara districts of Assam. This mission is popularly known as Assam Rice Collection (ARC). A total of 6,630 collections were made in this project from the stipulated area. During 1970-79, a special programme was undertaken to collect rice germplasm from all the rice growing districts of Madhya Pradesh by Dr. R.H. Richharia after he left CRRI in 1969. He explored 42 districts and collected a total of 19,226 accessions. During that time, Chhattisgarh was a part of Madhya Pradesh, hence his collections were from the present states of Chhattisgarh as well as Madhya Pradesh.

In 1975, under the leadership of Dr J.K. Roy, a comprehensive exploration and collection programme was drawn for the whole country especially for the traditional rice growing areas of Karnataka, Maharashtra, Madhya Pradesh, Uttar Pradesh, Bihar, West Bengal and Orissa covering 30 districts of 7 states. This programme popularly known as National Collection from States (NCS) resulted in a collection of 1,038 accessions. Similarly, another collaborative programme was in operation at CRRI with the agricultural universities, state departments of agriculture and ICAR centres for collection of rice germplasm all over the country during 1978-80. The materials were collected from over 100 districts of 14 States and a total of 6,349 accessions were collected. From 1990-2000, in collaboration with NBPGR, the scientists of CRRI collected about 2,400 accessions of both cultivated and wild rice germplasm. With the introduction of a mission mode NATP project on plant biodiversity (by the NBPGR as its lead centre), a new thrust was made on the exploration and collection of rice germplasm with special reference to wild rice. Table 2.1 provides a comprehensive list of explorations and collections of rice germplasm mainly in the country since 1910 to 2008.

Table 2.1. Major rice germplasm explorations and collections by CRRI.

Year	Collectors/Institute	Region	No. of accessions
1946	K. Ramiah (CRRI)	Paddy Breeding Station, Coimbatore	2000
1955-59	S. Govindaswami, A. Krishnamurty and N.K. Chyaupatnaik (CRRI)	Jeypore Botanical Survey (JBS)	1895
1965	A.C. Sharma	Manipur	874
1967-72	S.D. Sharma (IARI)	Assam Rice Collection (ARC)	6630
1970-79	R.H. Richharia (IGKVV)	Chhattisgarh & Madhya Pradesh (42 dist)	19226
1975	J.K. Roy (CRRI)	National Collection from States (30 dist; 7 states)	1038
1978-80	J.K. Roy (CRRI) & other SAUs	100 dist; 14 states	6349
1984-89	S.D. Sharma *et al.* (CRRI)	9 states	3697
1991	B.C.Patra & S. Duhoon (CRRI & NBPGR)	Western UP & Haryana (Basmati)	88

1991	S.S.C. Patnaik & J.B. Tomar (CRRI & NBPGR)	Eastern Uttar Pradesh (wild rice)	40
1992	B.C. Patra & J.B. Tomar (CRRI & NBPGR)	Chhotanagpur region (upland)	80
1994-98	B.C. Patra *et al.* (CRRI)	Orissa	921
1999-03	S.R. Dhua *et al.* (CRRI)	Eastern India (wild rice; NATP)	483
1999-00	S.K Pareek *et al.* (NBPGR & CRRI)	Rescue mission to super cyclone devastated areas of Orissa	857
2000	B.C. Marandi & K. Pande (CRRI)	*Bao* (deep water) rice from Assam	126
2000-02	B.C. Patra *et al.* (CRRI)	*Boro* rice from Eastern India	208
2001	B.C. Marndi (CRRI)	*Aman* rice from West Bengal	69
2003	B.C. Patra & B.C. Marandi (CRRI)	Medicinal rice from Chhattisgarh	71
2004	B.C. Patra & B.C. Marandi (CRRI)	Saline tolerant rice from Kerala	51
2005	B.C. Patra & B.C. Marandi (CRRI)	Traditional rice from Maharashtra & Sikkim	113
2006	B.C. Patra & B.C. Marandi (CRRI)	Traditional rice from Gujarat & Tripura	109
2007	B.C. Patra	Traditional rice including aromatic short-grained Kalanamak types from eastern U.P.	36
2008	B.C. Patra	Basmati rices from Punjab	60

There are four distinct nomenclaturally valid wild species of rice in India; they are *O. nivara, O. rufipogon, O. granulata* and *O. officinalis*. Three other species like *O. indandamanica, O. malampuzhaensis* and *O. jeyporensis* described earlier from Indian peninsula have later been merged with *O. granulata, O. officinalis and O. sativa* respectively. Apart from the species of genus *Oryza*, several allied weedy species of other genera like *Porteresia coarctata, Leersia hexandra, Hygroyza aristata* have also shown importance in having different abiotic stress tolerant genes and form the tertiary genepool.

The importance of wild rices was felt when grassy stunt virus resistant genes were identified from among the collections of *O. nivara* made from India. Since then efforts are on to collect wild rices from different regions of the country by CRRI. During 1984-89, a total of 3,697 accessions of cultivated rice and 204 accessions of wild species namely *O. nivara, O. rufipogon,* and *O. officinalis* were collected from northern Orissa, South Bihar, West Bengal, Sikkim, Mizoram, Maharashtra, Karnataka, Goa and Gujarat in collaboration with the NBPGR. Exploration and collection of *Porteresia coarctata* from one of the biodiversity rich mangrove forests of Bhitarkanika in three different seasons have shown tremendous variability (Patra & Dhua, 1996) (Table 2.2).

All activities associated with the exploration, collection, maintenance, characterization and supply of rice genetic resources were a part of the activities of the scientists of Botany Division later named as Genetics Division of the Institute. In 1984, this division was renamed as Plant Breeding and Genetics Division and a separate division named as Genetic Resources Division was created to look after all such activities on rice germplasm. But this Division was again merged with the newly created Division of Crop Improvement in the year 2001. Evaluation of rice germplasm was done in the Divisions of Physiology, Plant Pathology and Entomology. CRRI is recognized as the National Active Collection Centre (now National Active Germplasm Site) by the NBPGR.

Table 2.2. Collection of wild/weedy species of rice by CRRI.

Year	Species	No. of collections	Area/State
1984-89	*O. nivara, O. rufipogon, O. officinalis*	204	Orissa, West Bengal, Maharashtra, Karnataka, Goa, Gujarat
1994-95	*Porteresia coarctata*	9	Bhitarkanika (Orissa)
1999-2003*	*O. nivara*	161	Orissa
	O. rufipogon	151	
	O. sativa f. *spontanea*	6	
1999-2003	*O. nivara*	78	West Bengal
	O. rufipogon	84	
	O. sativa f. *spontanea*	1	

* Under NATP

Exploration and collection for specific traits or for specific ecosystems were also executed and collections were made from specific areas. The major ones are listed below:

Medicinal rice: Seventy two (72) accessions of rice germplasm were collected from Bastar region of Chhattisgarh in which some medicinal rices namely Gudmatia, Bhejari, Danwar, Baisur, Gathuwan, etc were also reported.

Saline tolerant rice: 51 accessions of saline tolerant rices mostly from Pokkali region (the GI) of Kerala state were collected.

Speciality rices: Attempts are on to collect aromatic rice, soft rice, wine rice, glutinous or waxy rice, colour rice (brown, black, red), beaten rice, pop rice, organic rice, nutritional rice, etc.

Basmati rice: Eighty eight accessions of long slender basmati rice germplasm were collected from eight districts of western Uttar Pradesh and 6 districts of Haryana.

Aromatic short grained rice: Sixty seven accessions of scented rice "Kalanamak" germplasm were collected from eastern U.P.

Bao rice: About 126 accessions of Bao rice germplasm were collected from Assam and Meghalaya.

Aman rice: A set of 69 accessions of Aman rice germplasm were collected from West Bengal.

Wild Rice: About 495 accessions of wild rice germplasm (*O. nivara, O. rufipogon*, and *Porteresia coarctata*) were collected in 12 exploration trips from Orissa and West Bengal under National Agricultural Technology Project.

Rescue mission: Five hundred and fifty five accessions of landraces and primitive cultivars of rice were collected through a rescue mission from super-cyclone devastated areas of Orissa.

Boro rice: A total of 208 accessions of Boro rice germplasm were collected from Assam, North Bihar, North Bengal and eastern Uttar Pradesh (Table 2.3).

Germplasm Conservation

Until recent past, conservation of rice germplasm was synonymous with repeated rejuvenation in the field. This process of maintenance subjected the germplasm to a threat of losing their identity because of random and non-random processes due to sampling. Also loss due to unforeseen natural calamity of the type of super-cyclone followed by flood that affected Orissa during the year 1999 devastating the entire germplasm grown in the field or unprecedented heavy early monsoon rain which caused water stagnation and severe flood in CRRI farm can not be ruled out in so far as *ex situ* conservation is concerned.

Realizing the danger of such genetic erosion, the effort of developing a

Table 2.3. Detailed account of explorations made in eastern India for *boro* rice germplasm.

Year	State/Region	Period	Longitude & Latitude	No. of Districts	Name of the districts	No. of collections made	Important varieties
2000	Assam	24th May to 6th June	24° to 28° E & 90° to 96° N	12	Kamrup, Marigaon, Nagoan, Golaghat, Jorhat, Sonitpur, Darrang, Nalbari, Barpeta, Karimganj Hailakandi & Cachar	92	Aijung, Boga, Jagili, Boni, Kala joha, Boro joha, Kala boro, Baismuthi, Phulbasang, Khaiya boro, etc.
2001	North Bihar	14th to 23rd May	24° to 26° E and 84° to 88° N	14	Katihar, Purnea, Saharsa, Supaul, Madhubani, Darbhanga, Samastipur, Vaishali, Muzaffarpur, Begusarai, Patna, Khagaria, Rohtas & Bhojpur	40	Gautam, Sarjoo 52, Sita & PR 3
2001	North Bengal	3rd to 13th June	24° to 27° E and 88° to 90° N	7	Murshidabad, Malda, Dakshin Dinajpur, Uttar Dinajpur, Darjeeling, Jalpaiguri & Cooch Bihar	25	Parhi, Nayanmuni, Boro dhan, China boro, Parimal, Naldugi, Paijam, BR 10, IR 36 & IR 50
2002	Eastern UP	21st May to 8th June	25° to 28° E and 82° to 85° N	12	Balia, Ghazipur, Sant Rabi Das Nagar, Varanasi, Basti, Sant Kabir Nagar, Siddharth Nagar, Gorakhpur, Maharajganj, Deoria, Mau & Kushi Nagar	24	Kala boro, Bonglai, Kacheri, China boro, Sarjoo 52, Gautam & Krishna Hansa

medium term cold storage system for rice germplasm was initiated at CRRI in 1984. However, though a Gene Bank system was made operative in 1988, it faltered and failed on several occasions due to some reason or other and it could never be made dependable for effective germplasm conservation as it was not able to keep the desired temperature and humidity. Meanwhile, during 1986, it was decided by Dr. R.S. Paroda, the then Director, NBPGR and Dr S.D. Sharma, the then Head, Genetic Resources Division, CRRI to conserve all the germplasm of CRRI at the National Gene Bank of NBPGR. Since then, a total of 30,741 accessions have been deposited in the long-term storage of the NBPGR (Table 2.4).

Table 2.4. CRRI rice germplasm deposited at NBPGR, New Delhi for long term storage till date.

Year of storage	No. of accessions
1986	6,763
1988	3,863
1989	1,129
1990	2,523
1992	2,058
1998	3,328
1999	2,473
2003	1,391
2004	1,023
2005	2,249
2006	2,336
2007	1,605
2008	2,387
2009	2,491
Total	**35619**

Germplasm Evaluation and Utilization

In the past, the scientists involved with crop improvement programme at different research stations undertook the evaluation of germplasm and were choosing the varieties through pure line selection. Some examples are Latisail released in West Bengal in 1948 through selection from a bulk collection of

Bangladesh, FR 13A released in West Bengal in 1953 from Orissa collections, FR 43B and SR 26B released in 1953 in Orissa from local germplasm, NC 1281 released in West Bengal in 1960 through selection from materials collected from lowlying areas of 24 Parganas district of West Bengal, CB 1 released in 1952 from *boro* germplasm collected in Hooghly district of West Bengal, CB 2 released in 1956 from a collection of Midnapore district of West Bengal and Adday released in Sikkim in 1955 through selection from a local germplasm of Darjeeling. However, in the absence of a National Numbering System and proper conservation of such evaluated materials, the information generated often suffered from deficient identification and thus utilization of such information remained inadequate. Now, the total number of germplasm accessions maintained by various research stations in the country might exceed 80,000 (Table 2.5).

Table 2.5. Rice germplasm holdings at major Rice Research Stations in India.

State	Institute, Location	No. of accessions
Andhra Pradesh	APRRI, Maruteru,	17,335
	DRR, Hyderabad	10,219
	NBPGR Regional Station, Hyderabad	258
Assam	RARS (AAU), Titabar	3,299
Bihar	ARI (RAU), Patna	2,020
	RAU, Pusa	812
	Bihar Agri. College (RAU), Sabour	355
Chhattisgarh	IGAU, Raipur	20,374
Delhi	NBPGR, New Delhi	74,857
Gujarat	RRS (GAU), Nawagaon	933
Haryana	CSSRI, Karnal	960
Himachal Pradesh	CSKHPKV, Malan	496
Jammu & Kashmir	RRS (SKUAT), R.S. Pura	426
Jharkhand	BAU, Ranchi	51
Karnataka	RRS (UAS), Mandya	11,850
Kerala	ARS(KAU), Pattambi	570
	ARS(KAU), Moncompu	150
	NBPGR Regional Station, Trichur	2,873

Maharashtra	ARS (KKV), Karjat	1,119
Manipur	RRS, Wangbal	240
Meghalaya	ICAR Research Complex, Barapani	53
	NBPGR Regional Station, Shillong	52,600
Orissa	CRRI, Cuttack	30,741
	NBPGR Base Centre, Cuttack	2,248
	OUAT, Bhubaneswar	840
Punjab	PAU, Kapurthala	1,000
Rajasthan	ARS, Banswara	2,570
Tamil Nadu	TNAU, Coimbatore	2,210
	TNRRI, Aduthurai	4,092
Uttar Pradesh	NDUAT, Faizabad	1,548
	BHU, Varanasi	200
Uttaranchal	GBPUAT, Pantnagar	1,112
	NBPGR Regional Station, Bhowali	367
	VPKAS, Almora	1,577
West Bengal	RRS, Chinsurah	5,012
	RRS, Bankura	980

NB: Updated from S.S. Malik and S.D. Sharma (2005)

Thanks to the Indo-USAID collaborative project, a cold module was gifted to CRRI by NBPGR. After its prompt installation, the facility became operative since 1998 with a controlled temperature of 4 ± 2^0 C & 33±5 % RH and found to be rather dependable. The Gene Bank facility thus created is meant for medium term storage (MTS) and the seeds could be kept viable for 6-8 years. At CRRI about 12,000 accessions of rice germplasm are conserved as active collection and about 30,000 accessions of rice germplasm are deposited in National Gene Bank as base collection. However, the space inside this module is inadequate to house all the active collections of rice. It is now desirable to have another module of a bigger size for accommodating all the germplasm at the National Active Germplasm Site.

Now the stage is set for streamlining and systematizing the conservation of rice germplasm in the country. In this endeavor, the first step would be to allot a National Accession Number to all these collections which has already been initiated by the NBPGR, New Delhi. This has to be followed by seed increase in sufficient quantity at the National Active Collection site for further deposit in the long-term storage and also in the medium term storage of the active collection in a phase-wise manner. At the same time they need to be characterized and documented. Further evaluation and utilization has to be followed up for the purpose of crop improvement. However, the possible area/source of resistance and tolerance for some important biotic and abiotic traits have been identified over the years (Table 2.6).

Table 2.6. Sources of resistance/tolerance for some important traits.

Traits	**Possible source**
Resistance against different pests and diseases	Landraces of insect- or disease-specific endemic areas.
Drought tolerance and recovery at vegetative stage	Dry seeded upland germplasm and landraces grown in the hilly region under shifting (*jhum*) cultivation.
Cold tolerance at vegetative stage	Landraces of *boro* cultivation
Low light intensity	Landraces of traditional *sali, asra* and *bao* cultivation areas of Assam and traditional *aman* varieties of Bengal and Orissa.
Flash flood	Coastal areas of Orissa and West Bengal. Also *sali* and *asra* germplasm of the Bramhaputra valley of Assam.
Deep water situation	*Bao* germplasm of Assam and landraces from deep-water areas of Eastern UP, Bihar and West Bengal.
Earliness	*Ahu* germplasm of Assam and aus group of materials from West Bengal.
Aromatic rice	*Joha* varieties of Assam, Kalanamak group of varieties of UP, small grain scented varieties of Sikkim, West Bengal and Orissa.
Non-aromatic fine grain	*Lahi* group of materials from Assam, Balami grain type of Orissa and short slender grain types of West Bengal.

More than 15,000 accessions of rice germplasm have so far been evaluated for both biotic and abiotic stress situations at CRRI during last one decade. Several donors have been identified and they are being utilized for crop improvement program.

Rice plants, many a time, are exposed to stress prone environments that adversely affect the crop leading to low productivity. Among the more often experienced stresses, drought brings maximum damage to the rice physiology and causes setback to its production potentiality. Various USDA statistical analyses identified drought as the single most yield-reducing factor. Drought at vegetative stage causes irreparable loss of canopy whereas at flowering stage, water deficit hampers anthesis and seed setting leading to higher spikelet sterility and lower yields. Keeping this in view about 2100 rice germplasm consisting of uplands, medium lands, lowlands and 97 accessions of wild rice (*O. nivara*) were screened against vegetative stage drought under field condition. Dry seeding was done in the first week of February and the crop was grown with adequate soil moisture for 30 days with sprinkler irrigation system. On 31 days after sowing the irrigation was suspended until the susceptible check showed permanent wilting and then reirrigated for plant recovery. The soil moisture content during drought period decreased from 34% to 3% of the soil profile of 45 cm. Data on leaf rolling and recovery were recorded using SES of IRRI. The lines with early leaf rolling after suspension of sprinkler irrigation showed higher score for drought tolerance (7-9); lines with delayed leaf-rolling recovered faster after re-irrigation. Two accessions of *O. nivara* collected from West Bengal having Identity No as (SRD01-17, AC-100374, IC-330470) and (BCPW-30, AC-100476, IC-330611) were found to be tolerant to drought. They are being processed for registration at NBPGR. While characterizing these accessions for their agro-morphological traits as per IRRI-IPGRI descriptor, it was found that both are morphologically same except for a few characters like ligule, collar, internode and auricle colour, culm and flag leaf angle, culm strength and panicle exsertion. Days to 50% flowering also differ from each other, being 107 days for AC-100374 and 118 days for AC-100476. It bears significance because the habitat of this species is usually swampy and marshy places along the roadside seasonal ditches.

The genetic pool of submergence tolerant rice cultivars is very small. For screening against submergence tolerance, the seeds were direct seeded in specially designed field cemented tanks (40 x 8 m) with known submergence tolerant (e.g. FR 13A) and susceptible (e.g. IR 42) cultivars as checks. Twenty-one-day-old seedlings were submerged for 12 days under 80 cm of water. Survival count was taken visually on 10th day of de-submergence. The characteristics of the floodwater in terms of light transmission (%) were measured at 1200 h and water temperature and oxygen concentration were determined at 0600 h and 1630 h. Light intensity at 60 cm water depth varied between 37.6 and 41.2 % of the incident irradiance above the floodwater. The oxygen concentration at the same water depth was 4.4-7.4 mg L^{-1} at 0600 h and 6.2-11.6 mg L^{-1} at 1630 h. The temperature did not vary greatly, being 29.1-33.3°C throughout the period of experiment. After thorough screening of 11000 germplasm lines, few landraces have been identified as tolerant to complete submergence. They are Khoda, Khadara, Kusuma, Matiaburush, Gangasiuli, Atirang and Kalaputia.

Direct seeding is common in rainfed lowlands. In eastern India sometimes-early rain causes water stagnation in the field just after sowing which results in poor crop establishment. We have identified two cultivars namely Panikekoa and T 1471 from our huge genetic stock. They establish very well under water in anaerobic condition. This work has immense value for irrigated condition also in reducing the cost of cultivation.

Agro-morphological characterization of rice germplasm was another important project. More than 6712 accessions of cultivated rice have been characterized with the help of descriptors developed by IRRI-IPGRI. Every year the characterized data of 45 morpho-agronomic characters are documented and published as a spiral bound copy. The wild species of cultivated rice represent an untapped reservoir of genetic variability for agronomic traits. They harbour significantly higher genetic diversity than does the cultivated species. Therefore, an attempt was made under NATP to collect, evaluate and conserve the wild rice germplasm from eastern India where it is found abundantly. About 483 accessions of wild rice species namely *Oryza nivara* and *O. rufipogon* were collected and multiplied for seed increase, then characterized, evaluated and conserved. The wild rices show a strong dormancy, so to facilitate quick germination and raise seedlings the seeds were dehusked and put in the germination paper in petri dishes. Once the seeds sprouted, these were planted in small earthen pots containing sterilized soil. Preliminary observations on percentage of germination, days to 2nd leaf emergence, days to 3rd leaf emergence, 15 days seedling height and leaf sheath pigmentation were recorded. Then the seedlings from the earthen pots were transferred to the experimental field after a period of 30-35 days. These accessions comprising of *O. nivara* and *O. rufipogon* were grown in the field with two replications following Complete Randomized Block Design (CRBD) and evaluated for various traits as presented in Table 2.7 and Table 2.8.

These wild rice accessions were evaluated against bacterial leaf blight (BLB) pathotypes and some were found resistant. Among the *O. nivara,* the resistant accessions are IC-309815, 282947, 282948, 330651, 330652, 330653, 330656, 330657, 330583, 330585, 330589, 330595, 330598, 330612, 330616, 330617 and IC-330621. In the case of *Oryza rufipogon,* the resistant accessions are IC-330669, 330670, 330597, 330599, 330601, 330623 and IC-330625.

Table 2.7. Range of variation observed in *O. nivara* collections maintained at CRRI.

Character (unit)	Lowest value	Observed in Accession No.	Highest value	Observed in Accession No.
Days to 2nd and 3rd leaf emergence (days)	Minimum	AC 100023	Maximum	AC 100128
Height of 15 days old seedlings (cm)	Lowest	AC 100385	Highest	AC 100228
Leaf length (cm)	11.4	AC 100353	72.68	AC 100406
Leaf width (cm)	0.56	AC 100400	1.7	AC 100404
Ligule length (cm)	0.59	AC 100313	3.32	AC 100422
Culm length ((cm)	159.5	AC 100452	31.22	AC 100353
Days to 50% flowering (days)	94	AC 100024	168	AC 100044
No. of ear-bearing tillers (no.)	5	AC 100091	147	AC 100387
Panicle length (cm)	10.7	AC 100353	35.42	AC 100414
Awn length (cm)	1.15	AC 100150	13.35	AC 100418
L/B ratio of spikelet	2.31	AC 100423	4.45	AC 100452
100-grain weight (g)	1.04	AC 100450	2.76	Ac 100023

Table 2.8. Range of variation observed in *O. rufipogon* collections maintained at CRRI.

Character (unit)	Lowest value	Observed in Accession No.	Highest value	Observed in Accession No.
Days to 2nd leaf emergence (days)	minimum	AC 100055	maximum	AC 100138
Days to 3rd leaf emergence (days)	minimum	AC 100201	maximum	AC 100147
Height of 15 days old seedlings	lowest	AC 100018	highest	AC 100062
Leaf length (cm)	13.04	AC 100151	69.85	AC 100035
Leaf width (cm)	0.58	AC 100159	1.52	AC 100443
Ligule length (cm)	0.55	AC 100151	1.38	AC 100189
Culm length ((cm)	46.13	AC 10039	176.8	AC 100173
Days to 50% flowering (days)	172	AC 100259	172	AC 100310

No. of ear-bearing tillers (no.)	5.5	AC 100168	101.6	AC 100173
Panicle length (cm)	12.62	AC 100156	37.93	AC 100173
Awn length (cm)	1.98	AC 100151	11.24	AC 100453
L/B ration of spikelet	2.12	AC 100152	4.64	AC 100401
100-grain weight (g)	0.86	AC 100160	2.64	Ac 100265

Donors Identified for Different Traits

Screening and testing rice genetic resources for their tolerance to physiological stresses and resistance to insect pests and diseases are an on-going affair at the Central Rice Research Institute. During all these years, the Institute has identified a number of donors. The important ones are listed below:

i) *Bacterial blight:* AC-33523 (Tarical), AC-33557 (Dulla karma), AC-33562 (Kangpui), AC-3094 (TKM-6), AC-8368 (BJ-1), AC-26903 (DV-85), Chinsurah boro-II, Somera mangga, Wase Aikoku 3, Malagkit, Sung son

ii) *Blast:* AC-55 (CH-55) AC-8368 (BJ-1), AC-8369 (S-67), SM-6, SM-8, SM-9, CP-6, AC-293 (AKP-8), AC-294 (AKP-9), AC-360 (PTB-10), AC-26904 (Tetep), Fukunishike, CO-4, CO-29 Tadukan, Zenith, Carreon.

iii) *Tungro Virus:* ARC-7125, ARC-7149, DW-8, AC-368 (PTB-18), AC- 5079 (Kataribhog), Bhagirathi, Boitalpakhia, AC-34558 (Nalini) AC-17933 (Kamod-153), AC-34650 (Usha), AC-273 (ADT-20), AC-297 (ASD-1), AC-304 (CO-1), AC-315 (CO-12), AC-351 (PTB-1), AC-360 (PTB-10), AC-3094 (TKM-6), AC-8396 (CB-1), etc.

iv) *False smut:* AC-26570 (ADT-33), AC- 40119 (PTB-23), AC-40124 (PTB-26), AC-3070 (PTB-32), etc.

v) *Sheath rot:* AC- 26904 (Tetep), Ram tulasi

vi) *Brown spot:* Kataktara, Bhutmuri, BAM 10, SR-26B, CH-45, CO-20

vii) *Stem borer:* AC-3094 (TKM-6), AC-392 (SLO-12), AC-267 (ADT-14), AC-8396 (CB-1), AC-344 (MTU-15), AC-20006 (JBS-1638), Tepa-1, etc.

viii) *Gall midge:* AC-35 (Ningar small), AC-39 (CNAB white rice), AC-210 (Bhadas-79), AC-391 (Bikiri sannam), ARC-5984 (Suto syamara), ARC-10660, ARC-12508 (Khauji), ARC-12586 (Vale matse), ARC-12588 (Amamma matse), ARC-12670 (Nien sah), ARC-13166 (Jaksa), ARC-13210 (Yangbelok), ARC-14915 (Maich dol), ARC-14967 (Galong), AC- 352 (PTB-2), AC-362 (PTB-12), AC-371 (PTB-21), PTB-18, PTB-24, AC-

26704 (Phalguna) and Leaung-152.

ix) *Brown plant hopper:* ARC-6650 (Gomiribora), AC-34969 (Baidyaraj), AC-34993 (Ghusuri), AC-34997 (Jhupjhupa), AC-35014 (Naldhan), AC-371 (PTB-21), AC-40634 (PTB-33), AC-30300 (MR-1523), AC-35181 (Salkathi), AC-35184 (Dhoba numberi), AC-35228 (Jalakanthi), AC-35066 (Banspati), AC-35070 (Panidubi), AC-35108 (China bali), AC-17912 (Ganga sagar), AC-20363 (Kalachudi), Tarapith, Haldiganthi, etc.

x) *Leaf folder:* AC-33849 (Bundei), AC-35034 (Harisankar), AC-33831 (Sunakathi), AC-33832 (Surjana), Juli, AC-35338 (Saruchinamali), etc.

xi) *Yellow stem borer:* AC-33515, AC-33526, AC-33538, AC-33563, ARC-5984, AC-30300 (MR-1523) and AC-30349 (Aganni).

xii) *Nematode:* AC-26594 (TKM-6), AC-40083 (MTU-17), AC-467 (Lalnakanda-41), Hasma, Bahagia, AC-40509 (Manoharsali), Amla, AC-17134 (Sathia), AC-22899 (Anang), AC-23652 (Kalakeri), Kanyakaprashant, etc.

xiii) *Drought:* AC-254, AC-263, AC-304, AC-511, AC-2298, AC-3035, AC-3111, AC-3577, AC-9066, AC-9387, ARC-7063, AC-45 (CH-45), AC-40083 (MTU-17), W-691, AC-467 (Lalnakanda-41), AC-35207 (Dular), AC-37077 (Dhan gora), AC-37127 (Black gora), AC-37291 (Kalakeri), AC-8205 (Surjamukhi), AC-34440 (Salumpikit), AC- 34256 (Kabiraj Sal), AC- 34296 (Bombay murgi), AC-34992 (Salkiann), AC-35021 (Kalondani) AC-35038 (Godhi akhi), AC-35046 (Naditikar), AC-35059, (Phutkibari), AC-35060 (Bhuska), AC-35143 (Baihunda), AC-35452 (Karama) AC-100374 (*Oryza nivara),* AC-100476 (*Oryza nivara),* etc.

xiv) *Submergence:* AC-24682 (FR-13A), AC-35741 (Telgri), AC-35323 (Chaula pakhia), AC-35675 (Biesik), AC-36107 (SL-276), AC-36470 (Khoda), Khadara, Kalaputia, AC-26670 (Janki), AC-40844 (Manasarovar), Sarumuli, AC-40916 (Jalamagna), AC-40604 (Jaladhi-1), Kanawar, etc.

xv) *Salinity:* AC-2405 (SR-26B), AC-8532 (Pokkali), Pateni-2, AC-41360 (Nonabokra), AC-35255 (Rahaspanjar), Canning-7, Ravana, etc

Rejuvenation of Cultivated Rice Germplasm

When a variety is collected either through exploration or through correspondence/exchange, it is given an Accession No. in the CRRI Germplasm Register which runs serially from 1 onwards. After recording the passport characters like place of collection, farmers' name and other preliminary plant characteristics the seeds are kept in the cold storage until the cropping season starts in the month of June. On the onset of monsoon, the *kharif* season starts and newly acquired seeds are taken out along with other rice germplasm which are to be regenerated.

To avoid weed and other agronomic management problems, we usually go for seed bed sowing followed by transplanting. The seed beds are of 1 meter width and 20-25 meters long; in each bed 120-130 lines are drawn and in each line an accession of germplasm is sown in dry condition. By the time rain comes, the seeds are properly conditioned in the soil with temperature and start sprouting. To minimize mistakes, human errors or to know if any line has not germinated we give a purple rice variety as a marker in every 20 line interval; also at the edge of each line a pinch of the same purple marker is sown to keep the line distinct from the other when the plants grow up with the competition of weeds.

When the seedlings become 3-4 weeks old, they become ready for uprooting and transplanting. We prepare 40-cents (40 m x 40 m) plots with thorough ploughing, followed by irrigation, then leave them for 6-7 days so that whatever rice or other weed seeds have fallen earlier from the previous season will germinate. Then we fill the plot with water from a pump up to 30-40 cm and leave it overnight. Next day with a tractor drawn rotavator, the experimental field is puddled and all the undesired sprouted seedlings perish. A basal dose of fertilizer is applied before leveling so that the chemical fertilizer is uniformly mingled in the soil. Then through a leveler, drawn by the bullocks, the plot is leveled manually.

Meanwhile in the nursery seed beds, for every variety a bamboo peg of about a meter length is put with an aluminium label or waxed label so that it becomes easy to transport and arrange in the puddled field. Further, care is taken to see that the label remains intact until harvesting stage. Then sufficient water is given in the channels between the seed beds so that it becomes easy to uproot the seedlings and wash the roots making them clean from soil. Every variety is uprooted and tied with the bamboo peg with label and kept in a tray. Then we arrange them in the puddled field serially for line transplanting. We make 10 ranks in each plot giving about 1 meter gap between the ranks so that it becomes easy to monitor the plants as well as for wind flow to the culm base of the varieties. In each rank 50-60 varieties are put. Each variety is grown in three lines of 25-30 plants, with a spacing of 20 cm between plants and 30 cm between lines. A blank line or gap between varieties is left as a measure to check cross pollination. After every 20 varieties we give one line of purple variety as a number counting mark. The agro-morphological observations are recorded at vegetative and reproductive stages. The post harvest grain characteristics are taken in the laboratory. A single plant from the middle row is selected and the panicles are harvested in a paper packet for next season's sowing, to assure genetic purity. Soon after panicle initiation stage, rogueing is done to discard the off types. Depending upon the maturity duration the varieties are harvested when the seeds become physiologically matured.

Seed Supply

Supply of rice germplasm to different researchers within the country as and when they need it is another important aspect of the Institute. In the past, CRRI has supplied rice germplasm even to the foreign agencies but in the present IPR regime, the supply is now restricted to the researchers within the country only. Further, an SMTA (Standard Material Transfer Agreement) has been introduced before supplying any germplasm to any researcher. In 2000, when the RRS, Chinsurah lost its entire genetic stock due to heavy flood, CRRI supplied about 815 germplasm lines to start their research programme. Likewise, HPAU, Palampur also lost its germplasm over the years. Fortunately, the University had deposited a set with CRRI which was restored to HPAU. Apart from this, every year hundreds of germplasm accessions are being supplied to different researchers in the country. A detailed account of germplasm supply is provided in Table 2.9.

Table 2.9. Number of accessions of rice germplasm supplied by CRRI during 1962 to 1973.

Year	Supplied to		No. of accessions
	Within India	Foreign country	
1973		9 countries	1339
1972	748	185 (18 countries)	
1976	899	145 (10 countries)	1044
1975	1224	243	1467
1977	500	2885 (ARC to IRRI) + 50	3435
1966	18 States	9 countries	751
1967			731
1949-51			596
1951-52			128
1953-54			340
1954-55		11 countries	331
1955-56			627 (including 6 wild species)

1956-57		13 countries	246
1957-58		11 countries	239
1958-60			1793
1961	660	36 (9 countries)	696
1962		1630 (14 countries, 659 to IRRI))	1630

Research on Origin of Cultivated Rice at CRRI

Biswal (1998, 1990) made an extensive collection of *O. nivara* from different states of India and studied variation for ten morphological characters. She subjected 68 accessions of the *O. sativa* complex comprising 4 accessions of *O. rufipogon* collected from India, 51 accessions of *O. nivara* collected from different parts of the country and 13 accessions of *O. sativa* (*japonica*:2, *aus*:3, *aman*:3 and *javanica*:5) to multivariate analysis following modified Mahalnobis D^2 analysis. On this basis, she could divide these collections into 9 clusters of which *O. rufipogon* was placed in one cluster, *O. nivara* in 5 clusters, and *O. sativa* in 3 clusters with overlaps. The five clusters of *O. nivara* did not group accessions on any geographical basis. However, when she intercrossed accessions of *O. nivara*, she observed that the pollen fertility of their F_1 hybrids ranged from 4 to 91%. When the accessions of *O. nivara* were arranged according to pollen fertility, she found that the wider the spatial separation of two parental accessions, the higher their pollen sterility. She concluded that populations of *O. nivara*, which grow in small pools, have undergone genetic differentiation before the evolution of cultivated forms.

Biswal (1988) crossed Indian *O. nivara* with 'seihr' (southeastern Indian hill rices) types of *O. sativa* and observed sterility in their F_1 hybrids as in *nivara* x *nivara* F_1 hybrids. She interpreted that 'seihr' types behaved as *O. nivara* and hence could be the direct progeny of *O. nivara*. Tripathy (1994) made 'seihr' x 'seihr' crosses, which like *nivara* x *nivara* hybrids also proved sterile. Crosses between *O. nivara* and *japonica* and between 'seihr' and *japonica* likewise behaved as *nivara* x *nivara* or as 'seihr' x 'seihr' crosses (Biswal, 1988; Tripathy, 1994). Modern 'seihr' types have retained many of the primitive characters of *O. nivara*. Sharma *et al.* (2000) have therefore concluded that southeastern Indian hill rices form a basic group that originated directly from *O. nivara*. Sharma (2003) also suggests that three basic ecotypes appear to have originated directly from *O. nivara* in three different regions of Asia: the *seihr* types in the northeastern part of the Deccan plateau and the central Ganga valley, the 'hrmsea' (hill region of mainland southeast Asia) types in the hilly region of mainland southeast Asia and *japonica*-like type in the southwestern part of mainland China. The *aus* type could have evolved directly from the 'seihr' type in eastern India, *japonica* from *japonica*-like type in southwestern China and *bulu* from the 'hrmsea' type in Indonesia. This implies that cultivation of rice started in uplands and was carried to lowlands later.

The *aman* ecotype evolved from the *aus* ecotype as a result of introgression of *rufipogon* genes into the *aus* ecotype in the lower Gangetic valley. According to Ramiah and Ghose (1951) and Chang (1976), the deepwater types of rice cultivars are the product of introgression of *rufipogon* characters into *O. sativa*.

Vaughan (1989) reported a new species of *Oryza* from Sri Lanka. It is a diploid species closely related to *O. eichingeri* of Sri Lanka and *O. officinalis* of south and south East Asia. The genome of *O. rhizomatis* was determined by crossing with other known genomes of *Oryza* species and studying the pairing of chromosomes as the meitotic stage in the F_1 hybrids. Dhua (1994) worked on the genome analysis of *O. rhizomatis* and found that it has DD genome.

The spikelets of *O. rhizomatis* are characterized by a wash of purple pigmentation. The size of spikelets is the largest in *O. australiensis* followed by that of *O. punctata, O. grandiglumis* and *O. alta*. The smallest size of spikelets is seen in *O. minuta* and O. *eichingeri*.

Registration of Rice Germplasm

After thorough screening of rice genetic resources, a few landraces have been identified as tolerant to complete submergence. They are Khoda, Khadara, Kusuma, Matiaburush, Gangasiuli, Atirang and Kalaputia. Out of these, Khoda has been registered (INGR No. 04001 dt 31May 2004). This germplasm is awnless and distinctly different from international tolerant check FR 13A. After screening large number of germplasm lines for anaerobic seeding tolerance, we have identified T-1471 (Kodiyan) and Panikekoa as tolerant to anaerobic seeding. Out of these, T-1471 (Kodiyan) has been registered (INGR No. 05001 dt 11 August 2005) with the Germplasm Registration Committee of NBPGR, New Delhi. Four more accessions for submergence and one for drought tolerance have been registered in 2008.

References:

Biswal J. 1988. Genetic differentiation in *Oryza nivara*. PhD thesis, Utkal University, Bhubaneswar, Orissa (India).

Biswal J. 1990. Genetic diversity of *O. nivara* and its allied taxa. *Oryza* 27(2):145-152.

Chang TT. 1989. Domestication and spread of the cultivated rices. Pages 408-417 in Harris DR and Hillman GC (eds) *Foraging and Farming: The Evaluation of Plant Exploration.* Unwin Hyman, London.

Dhua SR. 1994. Genome analysis of *Oryza rhizomatis* Vaughan. PhD thesis, Visva Bharati, Sriniketan (India).

Govindaswami S, Krishnamurty A. 1959. Genetic variability among cultivated rices of Jeypore tract and its utility in rice breeding. *Rice News Teller* 7:12-15.

Patra BC, Dhua SR. 1996. Genetic diversity in *Porteresia coarctata* collected from Bhitarkanika mangrove forest, Orissa. *J Econ Tax Bot* 20(1):241-244.

Sharma SD. 2003. Species of genus *Oryza* and their inter-relationships. Pages 73-111 in Nanda JS and Sharma, SD (eds) *Monograph on Genus Oryza.* Science Publishers, Enfield, New Hampshire, USA & Oxford & IBH, New Delhi.

Sharma SD, Tripathy S, Biswal J. 2000. Origin of *O. sativa* and its ecotypes. Pages 349-369 in Nanda JS (ed) *Rice Breeding and Genetics: Research Priorities and Challenges.* Science Publishers, Enfield, New Hampshire, USA & Oxford & IBH, New Delhi.

Tripathy S. 1994. Ecogenetic differentiation in *Oryza sativa* L. PhD thesis, Utkal University, Bhubaneswar, Orissa (India).

Vaughan DA.1989.The genome Oryza L.: Current status of taxonomy. *IRRI Research Paper Series* 138:1-21

Appendix 2.1. Detailed account of rice germplasm acquisitions at CRRI.

Year	Acquired from	Source	Accessions	Remarks
1964-1983	D. P. Bhattacharya	Plant Physiology Division, CRRI, Cuttack	AC 8601-8672	Drought tolerant lines
	PB & G Division	CRRI, Cuttack	AC 8701-8857	Fixed lines of *japonica* x *indica* hybrids
1969 to 1983	S. Govindaswami & A.K. Ghosh	CRRI, Cuttack	AC 8901-8932	Scented types
1960-1964	PB & G Division	CRRI, Cuttack	AC 8951-8969	Improved cultures
1972 - 1983	R. K. Bhattacharya	Central Soil Salinity Research Institute, Substation, Canning, Canning town, 24 Parganas, West	AC 8001-8573	Salt tolerant lines

		Bengal		
1973	S.D. Sharma	IARI Regional Station, Hyderabad	AC 16785-16849	
1974	R.K. Arora	Plant Introduction Division, IARI, New Delhi	AC 16850-16881	Collections from Meghalaya
1974	T.A. Thomas	Plant Introduction Division, IARI, New Delhi	AC 16882	Collection Madhya Pradesh
1974		Plant Introduction Division, IARI, New Delhi	AC16883-16887	Collections from Uttar Pradesh
1983	A. Panda	CRRI, Cuttack	AC 16888-16960	
13-19 October, 1976	D.P. Srivastava	CRRI, Cuttack	NCS 1-152 (AC 17001-17152)	Collections from Ranchi and Chainbasa districts of Jharkhand
November,1976	D. Choudhury	CRRI, Cuttack	NCS 153-212 (AC17153-17212)	Collections from Gonda, Bahraich, Azamgarh, Mirzapur, Gorakhpur and Faizabad districts of UP
18-24 October, 1976	R.N. De &S. Biswal	CRRI, Cuttack & Rice Specialist, Chinsurah	NCS 213 –335 (AC17213-17335)	Collected from Bankura, Purulia and Jalpaiguri districts of WB
1976	J.K. Roy	CRRI, Cuttack	NCS 336-401 (AC17336-17401)	Sambalpur, Dhenkanal, Sundargarh and Bolangir districts of Orissa
1976	M. Nagaraju	CRRI, Cuttack	NCS 402–413 (AC17402-17413)	Collections from Andhra Pradesh
13-31 October, 1976	D.P. Ghorai	CRRI, Cuttack	NCS 414-620 (AC17414-17620)	Collections from Rewa and Raipur districts of MP

1976	S.N. Ratho	CRRI, Cuttack	NCS 621-784 (AC17621-17784)	Collections from Karnataka
1976	P.J. Jachuck	CRRI, Cuttack	NCS 785-1038 (AC17785-17999	Collections from Pune, Satara, Gangabarda, Ajra, Ratnagiri, Nipanni, Patan, Koyna valley of Maharashtra
1976	S.D. Sharma	IARI, Regional Station, Hyderabad	ARC18001-18623 (AC18001-18623)	ARC types
Feb.1977	Jt. Director of Agriculture, Sikkim		AC 18701-18788	Collections from Sikkim
23rd May 1978	Rajendra Agril. University	Agril. Institute, Patna	NCS 1039-1126 (AC19039-19126)	Collections of Bihar
16.8.1977	K.D. Sharma	Head, HPAU, Palampur	NCS 1127-1453 (AC19127-19453)	Collections of Himachal Pradesh
1976-78	N.D. Desai, Main Rice Research Station	GAU, Nawagam Gujarat	NCS 1454-1594 (AC19454-19594)	Collected from rice growing areas of Gujarat
1977	J.K. Roy from AICRIP	NDUAT, Faizabad	NCS 1595-1610 (AC19595-19610)	Deep-water types of Ghaghraghat, Uttar Pradesh
4-6 October, 1978	N.P.S. Varde and R.C. Mandal	ICAR Research Complex, Goa (CPCRI), Margao	NCS 1611-1646 (AC19611-19646)	Collections of Goa
1978-79	Rice breeder	Regional Research Station, University of Agril. Sciences, VC Farm, Mandya	NCS 1647-1750 (AC19647-19750)	Collections of Karnataka
1978-79	S.M. Ahmed	Asst.Eco. Botanist, University of Udaipur, Agril. Res. Station, Banswara, Rajasthan	NCS 1751-1954 (AC19751-19954)	Collections from four districts of Rajasthan
1978-79	Associate Professor	RRS, Moncompu, Dist Alleppey, Kerala	NCS 1955-1979 (AC19955-19979)	Collected from Kerala
			NCS 1980-1995 (AC19980-19995)	Received from Gujarat

1978-79	Ashutosh Roy	Rice specialist, Chiplima, Sambalpur (Orissa)	NCS 1996-2185 (AC19996-19999 and AC22000-22185)	Collections from Orissa
1979	Mansoor Kazim	NBPGR, New Delhi	NCS 2186-2340 (AC22186-22340)	Collected by Agril. Research Station, Vadgaon (Maval), Dist Pune, M.S.
1978-79	D.M. Maurya	Professor & Head, Plant Breeding and Genetics, NDUAT, Faizabad	NCS 2341-2563 (AC22341-22563)	Collections from U.P.
	A.T. Roy	Rice Specialist, Chiplima, Sambalpur (Orissa)	NCS 2564-2670 (AC22564-22670)	Orissa
	J.L. Dwivedi	Asstt. Rice Breeder, Rice Research Station, Titabar, Assam	NCS 2671-2689 (AC22671-22689)	Assam
1978-79	Associate Professor (B)	Rice Research Station, Pattambi, Kerala	(AC22690-22813 and AC 24785-24900)	Trichur district of Kerala
1979-80	M.S. Patel, Rice Scientist	Main Rice Research Station, Gujarat Agril. University, Nawagam	NCS 2814-2873 (AC22814-22873)	Kheda district, Gujarat
1979	Plant Germplasm Quarantine Center, USDA	BARC, East, Beltsville, Maryland, USA	AC 16001-16008	Blast differentials
1979	NBPGR	N.I. Vavilov All Union Institute of Plant Industry, 44, Harzen Street, Leningrad, USSR	AC 16009-16015 (did not germinate)	
1979	IRRI	Philippines	AC 16016-16025	
1980	K. Maruyama	National Hokiriku Agri. Expt. Sta., 1-2-1, Indane, Joetsu-Shi, Niigata-Ken, Japan	AC 16026-16073	Leading varieties of Japan
1980	Chief of Plant	All Union Institute	AC16074-16076	

	Industry	of Plant Industry, 44, Harzen street, Leningrad, USSR		
1980	M.N. Koppar	NBPGR, New Delhi	AC-16077-16190	Landraces from Himachal Pradesh
1980	T.A. Thomas	NBPGR, New Delhi	AC 16091-16196	Collections from Sikkim
1980	T.A. Thomas	NBPGR, New Delhi	AC 16197-16779	Collections from Nigeria, West Africa
1980	T.A. Thomas	NBPGR, New Delhi	AC16780-16784	
1978 -1980	R.M. Singh, Junior Rice Breeder	Department of Genetics and Plant Breeding, BHU, Varanasi	NCS 2889-2960 (AC22889-22960)	Collections from U.P.
1980	S. A Dadlani	Sr. Scientist (Seed Exchange), NBPGR	AC 22961-23043	Collected by Rice Research Station, Kapurthala, Punjab
1980		Raipur, MP	AC23044-23112	Collections from M.P.
1980	A. N. Asthana, Sr. Scientist (Plant Breeding)	ICAR Research Complex for NEH Region, Shillong	AC23113-23414	Collections from NE hill region
1980	S. R. Das, Asst. Economic Botanist	Department of Plant Breeding and Genetics, College of Agriculture, Bhubaneswar	AC 23415-23916	Collections from Orissa
1980	R. C. Chaudhary Chief Scientist (Rice)	Agril. Res. Institute (Rajendra Agril. University), Mithapur, Patna	AC 23917-24006	Collections from Bihar
1980	A. T. Roy	Chiplima, Sambalpur, Orissa	AC 24007-24066	Orissa
1980	J. K. Roy	CRRI, Cuttack	AC 24067-24152	Coastal areas of Orissa
1980	G. J. N. Rao	CRRI, Cuttack	AC 24153-24212	Choudwar area of Cuttack,

				Orissa
1980	S. C. Prasad Sr. Rice Breeder	Agril. College, Kanke, Ranchi	AC 24213-24513	Chhotanagpur, Bihar
1980-81	H.K. Nagar, Jr. Rice Breeder	Dept. Of Genetics & Plant Breeding, NDUAT, Faizabad	AC 24514-24554	Collections from U.P.
1980-81		CRRI, Cuttack	AC 24555-24678	Collected from Chilka and Basudevpur area (coastal saline areas) of Orissa
1980-81	J. K. Roy through AICRIP		AC 24679-24685	Alkaline resistant materials from U.P
1980-81			AC 24686-24697	Collections of Saline resistant materials from Kerala
1980-81			AC 24698-24700	Collections of Saline resistant materials from Orissa
1980-81			AC 24701-24707	Tamilnadu
1980-81		NBPGR	AC 24708-24720	Monkompu, Kerala
1981	R.K. Arora	NBPGR, New Delhi	AC 24721-24784	Collected from Sikkim
			AC 24785-24900	Collected from Trichur Dist, Kerala
1981	J. K. Roy	CRRI, Cuttack	AC 24901-24960	Nayagarh, Daspalla & Narsinghpur areas of Puri district, Orissa
1981	N.K.C. Patnaik	CRRI, Cuttack	AC 24961-24964	24-Parganas district, West Bengal

1981	A. Panda	CRRI, Cuttack		Balasore district, Orissa
1981	J. K. Roy	CRRI, Cuttack	AC 24973-24985	Nirgundi of Cuttack district, Orissa
1981	R. K. Arora	NBPGR, New Delhi	AC 24986–25019	Ratnagiri and Pune districts of Maharashtra
1981	M. N. Koppar	NBPGR, New Delhi	AC 25020-25037	Champaran district of Bihar
1981	M. N. Koppar	NBPGR, New Delhi	AC 25499-25506	Collections from U.P, H.P, Punjab and Haryana
1981	K. P. S. Chandel	NBPGR, New Delhi	AC 25507-25510	Collections from J&K
1981		NBPGR, New Delhi	AC 25511-25518	
1981		NBPGR, New Delhi	AC 25519-25522	
1981		NBPGR, New Delhi	AC 25523	
1982	R. K. Arora	NBPGR, New Delhi	AC 25038-25055	Darjeeling district of West Bengal
1982	J. K. Roy	CRRI, Cuttack	AC 25056-25102	Daspalla, Nayagarh, Kantilo, Bolagarha of Puri district in Orissa
1982	R. H. Richharia, M. N. Srivastava Rice Breeder	Zonal Agril. Res. Station, Raipur (MP)	AC 25103-25264	Collections from M.P.
1982		NBPGR, New Delhi	AC 25265-25454	Collections from Kalahandi, Koraput, Bolangir, Sambalpur districts of Orissa
1982		NBPGR, New Delhi	AC 25455-25486	??
1982	J.K. Roy	CRRI, Cuttack	AC 25487-25497	Paradip area of

				Cuttack, Orissa
1982	H. K. Pande	Director, CRRI, Cuttack	AC 25498	From a village 5 km from Bhopal, M.P.
1983		NBPGR, New Delhi	AC 26403-26407	Received from Afghanistan
1983-84	NBPGR scientists	NBPGR, New Delhi	AC 26408-26416	??
1984	N.D. Desai	Res. Scientist (Rice) Main Rice Research Station, Nawagaon, Gujarat	AC 26417-26431	Collections from Gujarat
1984	V.H. Patil	Rice Specialist, Agricultural Research Station, Karjat	AC 26432-26462	Raigad, Maharashtra
1984	Director	Regional Rice Research Station, Kapurthala, Punjab	AC 26463-26480	Punjab
1984	P. C. Katoch, Rice Breeder	HPKV, Palampur, HP	AC 26481-26490	H.P.
1984	R. S. Tripathy, Paddy Specialist	Agril. Res. Station, Dubad Road, Banswara, Rajasthan	AC 26491-26499	Rajasthan
1984	Siva Subramaniyan Professor	Tamilnadu Agril. University, Coimbatore	AC 26500-26626	Tamilnadu
1984		Agril. Res. Station, Vadgaon (Maval), Pune	AC 26627-26628	Maharashtra
1984	Mr. Singh	Ph.D. student of Ranchi	AC 26629-26630	Manipur
1984	Sr. Scientist	Pl. Breeding & Genetics, JNKVV, ZARS (Rice zone) Raipur	AC 26631-26661	M.P.
1984	R. C. Chaudhary, Chief Scientist (Rice)	Rajendra Agril. University, Bihar, Agril. Res. Instt. Mithapur, Patna	AC 26662-26694	Bihar

1984	Assoc. Professor	Kerala Agril. University, Rice Res. Station, Kayamkulam	AC 26695	Kerala
1984	M. Gopal Krishna Rao, Plant Scientist (Paddy)	Regional Res. Station, VC Farm, Mandya	AC 26996-26720	Karnataka
1984	Assoc. Professor	Rice Res. Station, Cochin, Kerala	AC 26721-26724	Kerala
1984	Chief Scientist	RRS, Titabar,	AC 26725-26729	Assam
1984	Sr. Scientist	Pl. Breeding & Genetics, JNKVV, Jabalpur	AC 26730-26732	M.P.
1984	H.K. Gardoo Scientist (Rice)	Rice Research Station, Khudwani, Anantnag, J&K	AC 26733-26740	J&K
1984	Sr. Scientist	Pl. Breeding & Genetics, JNKVV, Jabalpur	AC 26730-26732	MP
1984	Professor (Pl. Br.)	Rice Res. Station, Moncompu, Kerala Agril. University	AC 26761-26765	Kerala
1984	S. Sardana	ICAR Res. Complex for NEH Region, Tripura Centre, Lembichera	AC 26766-26769	Tripura
1984	S.C. Pradhan	Technical Asst., CRRI, Cuttack	AC 26770-26777	Amiyajhar, Cuttack, Orissa
1984	P.N. Sreedharan	CRRI, Cuttack	AC 26778-26782	Paradip area, Orissa
1984	P.N. Sreedharan	CRRI, Cuttack	AC 26783-26786	Rajnagar, Coastal Saline area, Orissa
1984	Sr. Rice Breeder	Birsa Agril. University, Kanke, Ranchi	AC 26787-26821	Bihar
1984	P. P. Khanna	Head, Division of Germplasm Conservation, NBPGR, New Delhi	AC 26884-26897	Lalitpur, Jhansi, Banda districts of U.P.

1984	S. N. Ratho	CRRI, Cuttack		Orissa
1984	S. Devadath	Plant Pathology Divn., CRRI, Cuttack	AC 26898-26908	Orissa
1984	R. Sreedhar	Plant Pathology Division, CRRI	AC 26909-26920	Orissa
1984	J. S. Nanda, Prof. of Genetics	G.B. Pant University of Ag & Tech., Pantnagar, U.P.	AC 26921-26931	U.P.
1984	D. M. Maurya, Prof. & Head	Genetics & Plant Breeding, NDUAT, Faizabad, UP	AC 26932-26945	U.P.
1984	S. D. Sharma	Head, Division of Genetic Resources, CRRI, Cuttack	AC 26946-26950	Collections from Sundargarh, Orissa
1984	Assoc. Director	Regional Agril. Res. Station, Pattambi	AC 26652-26970	Kerala
2000	B.N. Singh	West African Rice Development Board (WARDA)	96 accessions	Exotic origin
2001	OIC	RRURRS, Hazaribagh	742 accessions	Upland rice germplasm
2001	NBPGR	New Delhi through ARBN	90 accessions	*Oryza sativa* and *O. glaberrima* crossbred introgression lines from IRRI
2001	NBPGR	New Delhi through ARBN	56 accessions	*O. glaberrima*
2002	NBPGR	New Delhi	673 accessions	WCG lines from IRRI
2002	NBPGR	New Delhi	48 accessions	NPT lines from IRRI
2002	OIC	NBPGR Regional Station, Ranchi	67 accessions	Upland germplasm
2004	Directorate of Rice Research (DRR)	Hyderabad	1061 accessions	Indian germplasm

2004	OIC	NBPGR Regional Station, Ranchi	200 germplasm	Upland
2004	NBPGR	New Delhi	56 accessions of Pokkali collections	Coastal saline tolerant rice germplasm from Kerala
2005	Directorate of Rice Research (DRR)	Hyderabad	1224 accessions	Indian germplasm
2005	NBPGR	New Delhi	2262 accessions	Indian germplasm
2006	Directorate of Rice Research (DRR)	Hyderabad	1224 accessions	Indian germplasm
2006	NBPGR	New Delhi	2005 accessions	Indian germplasm
2007	NBPGR	New Delhi	2556 accessions	Indian germplasm
2008	NBPGR	New Delhi	1911 accessions	Indian germplasm

Genetic Resources of Rice in India : 79-133
Editor : S. D. Sharma
Today & Tomorrow's Printers and Publishers, New Delhi - 110 002 (India)

Chapter 3

Role of Directorate of Rice Research in Evaluation, Utilization and Conservation of Rice Germplam in India

N. Shobha Rani, G.S. Varaprasad, L.V. Subba Rao, U. Chaitanya and B.C. Viraktamath

New horizons were opened for rice development when the country ushered in the era of plant type based high yielding varieties. The promotion of the first semi-dwarf variety Taichung (Native) 1 in India with high yield potential coupled with wide adaptability albeit some major weaknesses raised both hopes and fears. Hopes came from the fact that the barrier to high yields was at long last broken in *indica* rice, while disease problems such as bacterial leaf blight and rice tungro virus, triggered apprehensions with regard to risks involved in the cultivation of this strain. The urgency of intensifying research efforts leading to the development of locally adapted varieties was recognized immediately which resulted in the initiation of the All-India Coordinated Rice Improvement Project (AICRIP) in 1965. The All-India Coordinated Rice Improvement Project is one of a number of Coordinated Crop Improvement Projects initiated by the Indian Council of Agricultural Research (ICAR) with the objective of accelerating the research and the application of research findings to increase production to meet the demands of the ever increasing population. The task of AICRIP was to speedily develop a unified national grid of co-operative experimentation on all aspects of rice production. More than one hundred rice research stations and personnel involved therein were brought together to achieve convergence and synergy both in terms of research strategies and their implementation towards bringing in India's rice revolution. The anticipated rice research outputs demanded a closer linkage with the International Rice Research Institute on one hand and with the national institutes and agricultural universities on the other, thus making AICRIP an outstanding example of inter-disciplinary, inter-institutional and international collaboration. The fruits of these collaborative efforts led to the 'green revolution,' a transition period from despair to an era of prosperity and self sufficiency in the country.

The focus of research under the disciplines of Breeding, Agronomy, Pathology and Entomology was on the problems related to increasing production and other aspects. Research under AICRIP which is now known as the Directorate of Rice Research (DRR) concentrated on the introduction and breeding of new plant type based varieties that provided the basis for substantial increase in yield potential which were non-lodging, fertilizer-responsive, with good tillering ability and erect dark green leaves. Involving the major rice research centres of the country in programme planning, free exchange of germplasm and breeding materials and conduct of multilocation trials resulted in rapid generation of data which proved invaluable in releasing the appropriate varieties for cultivation. Concomitantly, development of suitable management practices to be used with the new varieties and evaluation of selections and varieties/landraces for their reaction to insects and diseases was undertaken (Anonymous, 1973). Although the rice germplasm related activities in terms of collection, evaluation and conservation *per se* were not originally included in the objectives of the All India Coordinated Rice Improvement Project, the facility of multilocation evaluation that is ingrained in its structure and functioning, facilitated in the assessment of several rice germplasm collections in the earlier period in search of new donors for biotic and abiotic stresses. However, in the last two decades under various externally funded projects special emphasis was laid by DRR on accelerating the evaluation activities which contributed significantly towards sustainable management of rice germplasm, as increasing population and rapid technological strides are putting tremendous pressure on this valuable genetic resource.

Collection of Germplasm

Collection of rice germplasm from Northeast Hills (Assam Rice Collection)

Between 1967 and 1971, the Indian Agricultural Research Institute and the All-India Coordinated Rice Improvement Project, with funds from PL-480, conducted an extensive survey and collected the rice germplasm from northeast india. The need to preserve the genetic diversity of this region was urgent since high yielding varieties were soon expected to replace local types (Seetharaman *et al.*, 1972). The region in which the survey and collection was made extended approximately from 22°N to 30°N and covered the states of Assam, Meghalaya, Nagaland and the Union Territories of Manipur, Tripura and the North East Frontier Agency (NEFA). Elevation ranged from 150 to 3,500 meters. In the past, ethnic groups from many countries immigrated to the region and probably brought diverse plant materials with them. Poor communication and tribal rivalries restricted the exchange of seeds thus maintaining the distinctiveness of the varieties. Hilly regions grow varieties with cold tolerance, an attribute of the *japonica* ecotype which is not common to the plains. The hilly terrain itself provided a great diversity of climatic conditions within relatively short distances. The survey was conducted by direct collection from the farmers' fields in the plains and indirectly with the assistance of local revenue and agricultural officials from the inaccessible hilly regions. The latter system was used to avoid violation of local traditions. The collection which includes 6,730 varieties can be claimed to be exhaustive in the locations covered (Table 3.1).

Table 3.1. Details of rice collections made from different areas of northeast India, 1967-1971.

State	District	No. of Collections
Assam (Plains)	North Lakhimpur	658
	Sibsagar	15
	Kamrup	231
	Goalpara	27
	Reasearch Station (Titabar)	182
Assam (Hills)	Mikir Hills	518
	North Cachar Hills	75
NEFA*	Kameng	138
	Subansiri	383
	Siang	330
	Luhit	374
	Tirap	302
Meghalaya	Garo Hills	808
	Khasi and Jaintia Hills	548
	Research Station (Upper Shillong)	43
Nagaland	Tuensang	270
	Mokokchung	349
	Kohima	230
Manipur	Manipur East	172
	Manipur West	61
	Manipur North	70
	Manipur South	109
	Manipur Central	586
Tripura	Tripura North	109
	Tripura South	137
	Total	**6730**

NB: Arunachal Pradesh was earlier known as North East Frontier Agency (NEFA).

Evaluation of Germplasm

Evaluation of Assam Rice Collection

Systematic assessment of collection from Northeast India resulted not only in the identification of new donors with resistance/tolerance to insect pests and/ or diseases but also enabled recognition of existence of specific zonal differentiation of these characters (Table 3.2) (Hakim *et al.,* 1986).

Table 3.2. Localized distribution of traits in Assam rice collection.

Character	Distribution zone	Entries
Short statured types	Around North Cachar hills district of Assam and Arunachal Pradesh	ARC Nos: 5824, 5829, 5923, 5929, 5981, 5981A, 6638, 6579, 11229
Drought-tolerant	Garo hills, Kamrup district, Lakhimpur district	ARC Nos: 6113, 6114, 6138, 7046, 7090, 11095
Gall midge resistance	Mikir hills (Assam), Khasi and Jaintia hills, Garo hills (Meghalaya)	ARC Nos: 5838, 5984, 6158, 7308, 10395, 11210
Green leafhopper resistance	Kamrup district (Assam) and Garo hills (Meghalaya)	ARC Nos: 7128, 7412, 10154, 10220, 10230
Brown planthopper resistance	Mikir and North Cachar hills (Assam), Subansiri (Arunachal Pradesh), Khasi and Jaintia hills (Meghalaya)	ARC Nos: 5830, 6611, 6650, 10666, 14766
Blast resistance	Khasi, Jaintia (Meghalaya) and Mikir hills (Assam)	ARC Nos: 5805, 6231, 6241, 7105, 7124, 10998.
Rice tungro virus resistance	Lakhimpur, Goalpara (Assam), Khasi and Jaintia hills (Meghalaya), Luhit region (Arunachal Pradesh)	ARC Nos: 5943, 13766, 13788, 14031, 14058

For the first time with the constitution of the screening trials on a multilocation basis, the ARC germplasm was screened for blast at Anakapally, Nellore and Ponnampet; bacterial leaf blight (BLB) at AICRIP, Hyderabad, CRRI and Maruteru; rice tungro virus (RTV) at AICRIP, Hyderabad; brown spot at Pattambi; stem borer at Maruteru, CRRI and Warangal; gall midge at Warangal and for leafhoppers at AICRIP, Hyderabad under field and green house conditions both during *kharif* and *rabi* seasons (Shastry *et al.,* 1971). Several new sources of resistance to these pests and diseases were identified. The distribution of resistant varieties followed a pattern with most of the BLB resistant varieties originating from NEFA, while leafhopper and stemborer resistant varieties originated from Garo Hills. Varietal diversity was far better preserved in hilly regions than in the plains. The collection also exhibited wide range of variation for plant height, plant habit, tillering potential, leaf characteristics, grain size and flowering duration (Sheshu *et al.,* 1984). While the dwarf *indica* breeding programmes have been relying mainly on the single source of semi-dwarf gene, it has been considered desirable to search for new sources of dwarfism in order to achieve a more effective combination of traits particularly related to resistance to diseases, pests and the cooking characteristics. The pursuit of this objective led to the identification of alternative source of dwarfism such as ARC 5981A (early), ARC 12805 (mid early) and ARC 5929 (mid duration) with promising yield attributes (Table 3.3) and those with resistance to insect pests and diseases (Tables 3.4 and 3.5).

Table 3.3. Donors from ARC identified for short stature as an alternate source of dwarfism.

	Dwarf Cultivars	
Duration group /ARC No.	**Name**	**Source**
Early		
5824	Theki	NC hills
5829	K. Chikari-34	RRS, Shillong
5981A	Haldiram	Mikir hills
Medium		
5929	Khagari Biron	NC hills
6576	Bukongsei	NC hills
6600	Tinfooti	NC hills
10608	Batdong	K & J hills
11229	Ngoba	RRS, Shillong
12897	Akughu Khoghi	Kohima
Mid-early		
5817	Sabit	RRS, Shillong
5940	Maizumgajaw	NC hills
6638	Sencher	NC hills
12805	Khawchaku	Mokokchung
Late		
12814	Muzulo	Mokokchung

Table 3.4. Donors from ARC identified as resistant/tolerant to insect pests.

ARC No.	Name	Source	Days to maturity	Glume color	Kernel color	Grain type
Gall midge						
6087	Garu-chakana bora	RRS, Titabar	150	BF	W	M
6136	Kharika joha	RRS, Titabar	155	BF	R	C
6140	Boka chakua	RRS, Titabar	150	BF	DW	C
6158	Khanimunshi	RRS, Titabar	140	S	W	F
6202	Kula joha	N. Lakhimpur	145	B	W	F
6221	Anapachi dhan	N. Lakhimpur	140	S	W	F
6232	Kula sali	N. Lakhimpur	150	BF	W	C
6234	Ranga sali	N. Lakhimpur	150	BF	W	C
6238	Jiton-sali	N. Lakhimpur	150	G	W	C
6557	Haldhar sali	M & NC Hills	150	S	W	C
6605A	Bora dhan	M & NC Hills	155	BR	W	M
6631	Taraboli	M & NC Hills	145	DS	W	F
6210	Ahu bora	M & NC Hills	150	S	W	C
7077	Morong	Garo Hills	150	BF	W	C
7138	Rangakesor	Garo Hills	155	BF	DW	C
7213	Boni sali	Garo Hills	150	BR	R	C
7308	Hathi sali	Garo Hills	155	BF	W	C
7316	Malshira	Garo Hills	145	BR	W	C
10395	Lynter heh	K & J Hills	140	BF	W	M
10494	Kba nar	K & J Hills	140	S	W	C
10534	Mikotchi gisim	K & J Hills	155	S	W	C
10654	Suamara	K & J Hills	150	S	W	F
10817	Bungoi	K & J Hills	140	S	W	F
10963	Sunkuchi	Kamrup	160	G	W	C
11210	Hatimal sali	Kamrup	160	S	W	C
11307	Badshabhog	Kamrup	150	G	W	F

Stem borer						
6033	Bar joha	RRS, Titabar	160	S	W	M
6045	Chakua dhan	RRS, Titabar	155	DS	W	C
6049	Chanasar	RRS, Titabar	150	S	W	F
7316	Malshira	Garo Hills	145	BR	W	C
10217	Bangdam	Garo Hills	140	DS	DW	C
10346	Dut bilni	Garo Hills	-	BR	W	M
10379	Agali	Garo Hills	115	BF	R	M
10386	Agalia	Garo Hills	120	BF	R	M
10528	Malbik	K & J Hills	145	S	W	M
11261	Dudh-kalam	Kamrup	145	S	R	M
11313	Kekua bao	Kamrup	150	B	R	C
11332	Kekowa bao	Kamrup	140	BR	R	C
GLH						
7128	Bangal sali	Garo Hills	155	DS	R	M
10154	Kanchi sali	Garo Hills	145	G	W	C
10220	Minil joha	Garo Hills	150	BR	W	F
7252	Pangapasi	Garo Hills	155	S	W	F
7412	Malbog	Garo Hills	145	DS	W	C
11307	Badshabhog	Kamrup	150	G	W	F

BF: brown furrow; S: Straw; B: Black; G: Gold; BR: Brown; DS: Dirty straw; W: White; R: Red; DW: Dirty white; M: Medium; C: Coarse; F: Fine

Table 3.5. Donors from ARC identified as resistant/tolerant to diseases.

ARC No.	Name	Source	Days to maturity	Glume color	Kernel color	Grain type
Rice blast						
5805	Horo sali	N. Lakhimpur	150	BR	W	C
6225	Khamti sali	N. Lakhimpur	145	BF	W	F
6231	Lalit ahu	N. Lakhimpur	140	S	R	F
6620	Mohanbhog	M & NC Hills	150	S	W	C

5987	Satuki	M & NC Hills	150	BF	W	M
7105	Sill sali	Garo Hills	160	S	W	C
7124	Gioal bini	Garo Hills	150	G	W	M
10345	Kalagasi	Garo Hills	150	S	W	C
10686	Kba Lieh	K & J Hills	115	S	W	M
10709	Kba Nonglwai	K & J Hills	115	BF	R	C
10998	Sanjul	Kamrup	150	S	W	F
11215	Khakhan Boka	Kamrup	145	S	W	C
Bacterial leaf blast						
6114	Kishor Bardhan	RRS, Titabar	160	S	W	C
11524	Kamti sali	Siang	130	-	-	M
11527	Pumpu	Siang	130	BF	W	M
11611	Newly	Siang	135	-	-	M
11654A	Amkha	Siang	135	BF	W	C
11956	Ampang	Tirap	135	-	-	M
12163	Sugamoni	Subansiri	145	BF	W	C
12164	Gumuri	Subansiri	145	BR	W	M
12181	Kumol	Subansiri	145	BR	W	C
12414	Epyu	Mokokchung	140	DS	R	C
Rice tungro virus						
5846	Badal sali	Sibsagar	155	BR	W	M
5905A	Peohali	M & NC Hills	140	BR	W	M
5981	Haldhiram (Dwarf)	M & NC Hills	130	S	W	M
5988	Betguti sali	M & NC Hills	160	S	W	C
6044	Santi boka	RRS, Titabar	160	S	W	F
6064	Ronga bordhana	RRS, Titabar	160	R	W	C
6094	Bhaholi joha	RRS, Titabar	140	S	W	M

Evaluation of other germplasm

Over 6,181 germplasm accessions from different sources such as IRRI, Philippines (5158), JBS Collection (707) and Indian varieties (316) were evaluated for resistance against gall midge at Warangal. 91 accessions were found resistant.

In order to identify sources of resistance to BPH, 2,360 rice germplasm materials from different geographical regions were evaluated at AICRIP, Hyderabad (Sai Krishna, 1977). The material included 890 Assam Rice Collections, 110 from Mizoram, 175 varieties/cultivars from IRRI, 344 varieties from IRRI gene bank, 427 from AICRIP variety collection, 95 from Jeypore Botanical Survery (JBS) varieties, 275 varieties from Pattambi (Kerala) and 44 from Coimbatore (Tamil Nadu). The relative contribution of resistant varieties by various sources indicated that varieties from Pattambi (Kerala) contributed highest (21.46%) followed by ARC varieties (8.4%). It was significant to note that a large number of BPH resistant varieties from these sources also showed resistance to various other insects and diseases.

To identify sources of resistance to BPH, 297 cultivars from ARC, 52 from Pattambi and 23 high yielding elite resistance cultures developed at Directorate of Rice Research (DRR) were screened as well as mechanism of resistance studied. Highly resistant genotypes included Chennellu (Pattambi), ARC 6650 (indigenous), Rathu Heenati and Horanamawee (Sri Lanka) and 4 dwarf derivatives IET Nos. 6615, 6858, 6238 and 6209 which were utilized in further studies (Shobha Rani, 1989).

Initially, collection of rice germplasm was not a mandate of the All India Rice Improvement Project. Therefore, active participation in collection and conservation was limited while AICRIP was strong in utilizing the multilocation facility as well as artificial screening under glasshouse conditions for screening against major pests and diseases available at the headquarters in Hyderabad and elsewhere. Further, as part of the DRR mandate to serve as a major centre for exchange of research material and information, much of the germplasm either as elite lines or as local landraces when nominated by the cooperators was evaluated as reliance on narrow germplasm base is to be avoided not only to minimize the risk of epiphytotics but also to widen the genetic base in various rice breeding programmes. Such efforts led to the identification of donors over the last four decades with resistance to major insect pests and diseases, several abiotic stresses, for new plant type, hybrid rice and basmati quality materials (Anonymous, 1980-1985; Anonymous, 1988-1996; Anonymous, 1996-2001; Progress Reports (AICRIP), 1966-1985; Progress Reports (DRR), 1986-2007; Research Highlights, 1983-1994; Annual Reports, 1995-2006; Kalode, 1980; Bentur *et al.,* 2003) (Tables 3.6, 3.7, 3.8 and 3.9).

Table 3.6. Donors identified resistant/tolerant to insect pests in rice under AICRIP 1965-2007.

Brown planthopper
AC 5-1-142-462, ADR 52, ARC Nos: 5378, 5723, 5780, 5784, 5951, 5964, 5981, 5982, 5984, 6025, 6092, 6172, 6610, 6650, 7080, 7272, 7528, 10197, 10550, 10654, 10659, 10660, 10666, 10847, 10849, 11220, 11321, 11353R, 11704, 12620, 14529, 14766, 14771, 15634, 15680, Aganni, Aicharamoti, Anaikomban, Andrewsali, Babawee, BAM 3, BPT 6090, BPT 6866, BPT 6868, BPT 7350, Chemban, CO 25, Ganjeikalli, Heenati, Honderwala, Horanamawee, HR 19, Jeeragasala 3, K 1421, K 2351, K 2370, Kabari, Kudvalian, Leb Muey Nahng, Majila, Manoharsali, Marvel, MGL 1, MO 1, Mudgo, Mudukiriyal, N 22, Nagrasal, Nanhigora, Nonabokra, Orumundakan, Pandy, Podiwi A8, Ptb 18, Ptb 2, Ptb 21, Ptb 33, Rathu Heenati, S 2204, S 644, Safed Danwar, Sinna Sivappu, Sudu Honderwala, Suduru Samba, Swarnalatha, T 1415, T 1426, T 1432, T 1471, Velluthacheera, White ponni
Whitebacked planthopper
Acc 15859, Acc 162498, Acc 27770, ARC Nos: 5752, 5984, 6248, 6288, 6566, 6619, 6650, 10239, 10654, 10912, 11316, 11321, 11367, 15570-A, Aganni, Anai Komban, Babawee, Basmati 6129, Chennellu, Chuvanna Kumbalam, Eswaramangalam, Hondarawala, HR 22, Intan Gawri, IRGC 13605, IRGC 16249, Kula Peruvela, Luanhe, MO 1, Mudgo, Mudu Kiriyal, Nonabokra, Pandi, Ptb 21, Ptb 33, Ramjiwain, Ramsali, Rathu Heenati, Sinna Sivappu, Sonth, WC 1240, Vellai Ilankalayan, Velluthacheera
Green leaf hopper
ARC 6650, Culture 8756, Culture 8770, Lalbasmati, Ptb 19, Ptb 33(A), T 1425, T 1471, TKM 6, Valsarachampara, Velluthacheera, Ptb 2, ADT 14, ARC 6606, ADR 52, CO 9, SLO 4, ASD 7, ASD 8, Pankhari 203
Gall midge
AC 1224, AC 1224-2, AC 169, AC 3283, AC 355, AC 424, AC 630, AC 8010, ARC Nos: 1049A, 5378, 5723, 5810, 5833, 5834, 5848, 5911, 5912, 5918, 5939, 5951, 5959, 5984, 5988, 6001, 6079, 6103, 6157, 6158, 6221-1, 6605, 6606, 6607, 6618, 6619, 6632, 6650, 6855, 7255, 7293, 7328, 10040, 10227, 10272, 10331, 10360, 10377, 10460, 10557, 10627, 10654, 10817, 10854, 11704, 13516, 13564, 14787, 15680, 15831, ADR 52, Aganni, Balam, Banglei, BG 12-1, BG 275-1, BG 276-6, BG 304-1, BG 334-2, BG 402-7, Bhumansan, Calrose 76, Chakhaopoiretol, Chempan, Chenellu, CR 57-29, CR 57-49, CR 94-13, Eswarakora, Gurmatia, JBS 446, Jhipiti, Karuthavelletha, Kodiyan, Kulipidicham, Leaung 152, Moirangpheu, Moranga Kayan, MR 1523, Nagrasal, NHTA 8, Orumundakan, Pandi, Phodum, Podiwi A 8, Ptb 10, Ptb 18, Ptb 21, RD 4, RD 9, Ritiang, RP 9-4, RPW 6-12, RPW 6-13, RPW 6-15, RPW 6-17, RPW 6-18, RPW 12, RPW 13, RPW 15, S 2204, Sakti, Siam 29, T 10, T 1477, Tellagarika sannavari, Vellachenipan, Velluthacheera, W 11952, W 1253, W 1257, W 1263, W 12708, W 13362, W 13400, W 13418
Stem borer
ACC 11057, ACC 39473, ADT 2, ARC Nos: 5500, 5757, 5984, 6044, 6158, 6579, 6650, 10443, 11281, 11744, Aganni, Akiyudaka, ASD 10, Banih Kuning, BG 951, CB 1, Chemban, CO 18, CO 25, CR 52-3, Lalmati A & B, Leaung 152, Manoharsali, Podiwi A8, Ptb 15, SLO 17, Sornavazhai, Suinphul, TKM 6, Velluthacheera, W 1263

Leaf folder
Ac 1224-1, AC 710, AC 8895, ARC 10550, ARC 10660, ARC 11128, ARC 11281, ARC 16852, ARC Nos: 5982, 6626, 6650, 7064, Balam, Choorapandy, CO 29, Darukasali, Gorsa, JR-14, Lalmati A, Lalmati B, Pandy, Parukulam, Ptb 12, Ptb 19, T 1425, T 1432, T 2005, W 1263, Valsarachampara, Chemban, Vellai Ilankalayan, Velluthacheera
Hispa
ARC Nos: 5764, 5778, 6632, 6650, Chemban, Chittaraikar, Choorapundy, Darukasail, GS 531, Orumundakan, T 10, T 1426, T 2005
Whorl maggot
Aganni, ARC Nos: 5500 A, 7080, 13516, 14378, 14549, Chakhaopoiretol , Chemban, GS 531, Pandy, T 1477, Velluthacheera
Thrips
AC 124-1, Aganni, ARC Nos: 5378, 5778, 6157, 7080, 15831, Dahanala 2220, Dudmona, Heenati 662, Kalubalwee, Kaluwee 715, Mudu Kiriyal 783, T 2005, Vellai Ilankalayan
Gundhi bug
ARC Nos: 5778, 6107, 6157, 6619, 7255, 13516, Chemban, GS 531, Orumundakan, SLO 16

Table 3.7. Donors identified resistant/tolerant to major diseases in rice under AICRIP 1965-2007.

Blast
Africano Desconocido, ARC Nos: 5173, 6564,10372, 10572, 10599, 11353, 11354, 15460, ASD 4, ASD 5, Bansi, Barkat, BJ 1, BR 5, BR 6, BR 7, BR 8, BR 9, Carreon, China, Chinoor, CO 25, CO 4, Dawn, Dissipative, Dular, Farel, Hashi Kalmi, IAC 25, IAC 5544, IRAT 3, JC 99, Jeeragasala 2, Jhum paddy 2, Jhum paddy 3, Kataundam, Khudwalian, K 332, K 333, K 60, Katakthara, Kothamolagolukulu, Kubantes, Lacrosse-Zenith-Nira, Luchai, Miliyang 46, Milyang 30, Moroberkan, MTU 14, MTU 3, MTU 5, N 12, Namsagui 19, N 32, Nira, Oryzicallanos 4, Pai-kan-tao, Palman 246, Palman 46, Ponni, Raminad Str.3, Sanghuanzan 2, Sigadis, Simhapuri 49, Sriranga 48, Swarnamukhi, T 3, T 9, Tadukan, Ta-poo-cho-z, Taride 1, Tellavadlu, Tetep, Thapa chinni, TKM 1, Tox. 1785, VL 206, VL 8, Wase-aikoku, Zaithon, Zenith, Zongam
Bacterial leaf blight
AC 19-1, Andrewsali, ARC Nos: 5984, 7133, 7140, 10945, BJ 1, Brazil 407, Cauror, Cemposalak, Chinsurah Boro II, DV 85, Early Prolific, EH-IA-Chiu, Hashikalami, Horthoria-sarsari, Java 14, Kalimkeri, Kalimonch 64,KK 2, Kuntalan, Leuang 152, Malagkit Sungsong, Namsagui 19, NC 10844, NC 10980, NC 1184, Nigeria 5, Nira, Ponni, Rajakajar, Sayaphal, Sigadis, Sung song, Suphala, Tadukan, TKM 6, Wase Aikoku, Zenith
Rice tungro virus
AC 4236, ADT 21, Ambemohar 102, Ambemohar 159, ARC Nos: 4236, 5570, 5767, 5905, 5918, 5951, 5981, 5984, 6172, 6650, 6660, 9570, 10343, 10599, 10654, 11354, 11554, 13754, 13804, 14056, 14320, 14529, 14766 B, Balimau, Churanakati, K 118, Kaira Nawa, Kalimoonch, Kamod

253, Karanphool, Kataribhog, Latisail, MIC 5-1, Pankhari 203, Petali, Ptb 8, Ptb18, Soltiya, Sukulporo, Tadukan, Tiokirivel, TKM 6

Brown Spot

Akiyudaka, BAM 10, Cauvery, CH 13, CH 45, Geonchiew, GS 531, Jeeragasala 3, Karikagga, Katuandan, Mettasanna, Morangkayan, Nam Sugai 19, RP 9-4, Shimokita, Shrauni, Siam 29, Suweon 339, T 141, Taichung sen, Tetep, TKM 6

Sheath Blight

Ajan, AKP-1, Ambemohar 159, ARC Nos: 5912, 5984, 10315, 10365, 10378, 10531, 10532, 10535, 10555, 10573, 10600, 10606, 10616, 10632, 10635, 10654, 10662, 11214, 11220, 11262, 11274, 11275, 11283, 11284, 11292, 11329, 11347, 11357, 11359, 11361, 11375, 11381, 11744, 11829, 11865, 11920, 15368, 15762, 18119, 18275, 18545, ASD 1, ASD 2, B 321, Bahagia, BAM 10, BCP 3, BG 11-11, Bhuhjan, CO 22, Galleiganthi, Gopalbhog, H4, Jilo, JBS 40, JC 118, JC 126, JC 18, JC 220, JC 41, JC 5, JC 56, JC 66, JC 75, JHS 40, Jhumpaddy, JV 42Kavya, Koyalari, Mehran 69, MTU 17, MTU 6, Nermar, Nira, Orumundakan, OS 4, Phourel, R 68-1, Remadja, S 22, Saduwee, Saibham, Satya, Sigadis, STBN/9837, STTN, T 141, Ta-poo-chooz, Thakurbhog, Toga, Zinnia 31

Sheath Rot

ARC Nos: 6650,10372, Dasal, NES 63, Pedigera, Raimuniya

Table 3.8. Donors identified resistant/tolerant to abiotic stresses, new plant type, hybrid rice and among wild accessions under AICRIP 1965-2007.

Waterlogged & Deep water

ARC 6650, Badhon, Beldor, Bennjing, Benong III, Bhetnasia, Bogan, Chenob, D 25-4, Dhania, DWR 4107, FR 13 A, FR 43 B, Gajepsali, Ghew bora, GS 302Habiganj, Hastipagra, Jalmagna, Jhingasali, Kajallata, Kalambanka, Kalopakhi, Lothabar, Nagari Bao, Najami, NC 477, NC 487, NC 490, NC 491, NC 492, NC 493, NC 678, Rajepsali, Sagarsamba, Salamat, Siam 29, Sindhur mashuri, Tilakkachari

Submergence

Bans Cherali, Beldan, Chew bora, Dasal, Dolangi, Gandhi bora, Leaung 152, Leb Nahng nang, Nagueri boa, Ranachudi, Satri

Drought

AC 2119, ARC Nos: 10312, 10372, 10659, Assaulia, Badam Form, Bendi, Berii, Black Gora, BR 23, C 22, Chettaraikar, Chitarakar, Culture 2325-2-2, Danger ARC 11775, DJ 5 DR 92, Dular, Fine Gora, Gajgaur, Gavaisaru, Gour Lalred 140, IAC 125, IAC 25, JBS 446, JBS 575, JBS 841, Kagalikai, Kalakeri, Kalapidichin, Karanga Gora, Kathiwar, Kolombo 540, Lallo 14Lalnakanda 41, Lohandi, Ludkhan, M.Gora, Mettasanna, Mettasannavari, MTU 17, N. Gora, Nagina 22, Nagpur 22, Nandi, Numbyol, Oodasannalu, Rajbhog, Rushkadam, Sathi 34-36, Stagree 45, Tapachini, Ta-poo-cho-z, Tellavadlu, Tulajapur 1, Velluthacheera, White Gora, Yunlen

Saline/Alkaline

Bhawlot, Bhundi local, Champasali, Cherivirupppu, Daeng, Damodar, Dasal, Dikwee, Dindi 1, Divuee

Kalagora, Gavirsaru 1, Getu, Kadamakudu, Kadari 1, KalagoraKamamali, Kannamali, Khao Doeng, Khaw, Korget white, Muskon 7, Nonabokra, Nonasal, Orumundakan, Patnai 23, Patni local, Pokkali, Ratria, Sahadiya, Singhwar 3, Singhwar 4, Sitah, SR 26 B, Valiya kunju, Vyttila kuruku
Low temperature
Bamani, CH 1039, CH 988, Dhungansali, Hiykumochi, Jelia, Junjen 4, Kathwar, Koshihikari, Larbyol, Numbyol, Palman Kathwar, Thimpu, Yamadanishiki
Iron
ARC 10372, Brown Gora, Dular, Sathi, Tuljapur 1
Phosphorous
C 14-8, Mashla
Zinc
Bali, Banut, Beakgangas, Begawan, BG 3-4-5, Darialsali, Dima, Djawasendang, Kautitiserai, Latisail, Patnai 6, Sahdeya, Singhawar, Talay, Zhona 349
New Plant Type
ARC Nos: 14007, 14210, 14260, 14288, 14324, 14403, 14419, 14425, 14449, 14454, BS 1-109, Cheoleweon 44, Dumpolon, GP 1224, GP 1294, Nagpur 27
Hybrid Rice
Cytosterile stocks: DRW 21018 (*O. rufipogon*) **Restorers**: ARC 11353, MS 577A, Sanghuanzan 2, RPW 20001 (*O.sativa),* DRGR 30090 (*O. glaberimma)*
Wild Rices
Rice Tungro Virus*:** DRW 22003, DRW 22007, DRW 22017 (*O. rufipogon);* DRW-21019 (*O. nivara)* ***Leaf folder: IC 87968 (*O. nivara*), IC 87986, IC 87993-OS 13, IC 87993-OS 77 (*O. spontanea*), OS 23-2, WR 25, WR 57, WR 70 ***Rice Ragged Stunt Virus*:** WR-ON-7, WR-OR-6II ***Sheath blight*:** BM 47, BM 106,TRN 11-4, TRN 14-3, TRN 37, TRN-11-3, WC 2957, WC 3011, WC 3113, WC 532, WR 1-1, WR 1-2, WR 1-3, WR 1-4, WR 1-6, WR 24, WR 25, WR 57, WR 70, WR 71, WR 72

Table 3.9. Donors identified among basmati quality rices resistant/tolerant to biotic stresses in rice under AICRIP 1965-2007.

Basmati
Quality: Basmati 127A, Basmati 370, Basmati 5853, Basmati 5854, Basmati 5888, Basmati 6311, Basmati 334, Basmati 112, Basmati Kota, Basmati Surkh 161, Begami 2-8, Bong Cay, Gopalbhog, Jeeriga Samba, Kalasukhdas, Kariga Javile, Karpoorkranthi, Krishnabhog, Lakhanbhog, Laxmibhog, Pak Basmati, Tharunbhog

Blast: Basmati 112, Basmati 122, Basmati 123, Basmati 140, Basmati 242, Basmati 334, Basmati 376, Basmati 5854, Basmati 5888, Basmati 6311, Basmati Kota, Basmati Surkh 161, Basmati T-3, Basmati Kamon, Dehradun Basmati, DM 24, Kamod 118, Karnal local, Mussactrene

Bacterial Leaf Blight: Badshabhog, Basmati 242, Basmati 410, Basmati 93, Basmati 5888, Basmati Kamon, Basmati Sufaid 100, Govindabhog, Ranga Joha 2

High Density Grain: Basmati 5853, Basmati 5888, Basmati Bahar, Basmati Sufaid 100, Cemposelak, Jaomali, Kataribhog, Kawn Kikwai, Lakshmi Bilash, Luachan Vithaiphong, Siang Nue

White Backed Plant Hopper: Basmati 6129

Role of AICRIP in Varietal Testing and Release

The present Directorate of Rice Research (DRR), frist established as All India Coordinated Rice Improvement Project (AICRIP), was instrumental in successfully integrating two streams of research systems available under the central and the state governments and has extensively contributed to the development of appropriate rice varieties, production and protection technologies suitable for different seasons, situations and ecosystems and in bringing in the much needed gains to rice production. The unique concept of multi-location and multi-disciplinary testing with the underlying philosophy of 'cooperative effort' in problem identification and action played a catalytic role in adoption and spread of first generation of HYVs and appropriate package of practices for varied ecologies across the country. Elevation of AICRIP to the Directorate status in 1985 with added additional mandate to conduct strategic and applied research in the major thrust areas of irrigated rice was the beginning of second phase wherein the facilities were strengthened, new research networks initiated and transfer of technology and breeder seed production activities undertaken. This impressive growth culminated into the largest network of over 100 cooperating centres with mainly 200 rice scientists under AICRIP and smaller networks exclusively devoted to the development and use of hybrid rice technology, rice biotechnology, improvement of quality rice for export, evaluation of rice germplasm for biotic stresses and others. The mission of DRR as outlined in vision 2025 is to develop technologies to enhance rice productivity, resource and input use efficiency and profitability of rice cultivation without adversely affecting the environment. The mandate of DRR is as under

- To coordinate multi-location testing at national level to identify appropriate varietal and management technologies for all the rice ecosystems.
- To conduct strategic and applied research in the major thrust areas of irrigated rice aimed at enhancement of production, productivity and profitability and at preserving environmental quality.

- To initiate and coordinate research networks relating to problems of national and regional importance.
- To serve as major centres for exchange of research material and information.
- To accelerate the pace of technology transfer through frontline demonstrations, training programmes and ICTs.
- To develop linkages with national, international and private organizations for collaborative research programme.
- To provide consultancy services and undertake contractual research.

Varietal testing procedure under AICRIP

The Coordinated Variety Improvement and Testing Programme covers 46 cooperating centres in addition to around 90 voluntary centres in different agro-climatic regions of the country (Table 3.10). This programme helps in exchange and evaluation of the breeding material and germplasm quickly among cooperators. AICRIP testing of promising breeding material (varieties, hybrids, etc.) helps in identifying the most stable, high-yielding or superior genotypes suited for different agro-climatic conditions. In the process, varieties are identified and released, not only for the zones/regions where they have been bred, but also for other regions as well. This helps in complementing the efforts of relatively upcoming centres where the breeding programme is not strong enough to cater to the needs of the specific region. Also, common testing helps in avoiding any possible bias towards the promotion of a breeder's own varieties. In the process, only the best materials get identified for release.

Table 3.10. Funded and voluntary test locations for Breeding, Entomology and Pathology trials under AICRIP.

State	Districts	Breeding	Entomology	Pathology
Andhra Pradesh	Maruteru	*	*	*
	Rajendranagar	*	*	*
	Warangal	*	*	
	Bapatla	*		
	Jagtial	*	*	
	Machilipatnam	*		
	Nellore	*		*

	Pulla	*		
	Ragolu	*		
Andamans	*Port Blair*	*		*
Assam	Karimganj	*		
	Titabar	*	*	*
	Gerua	*		*
	North Lakhimpur	*		
Bihar	Patna	*	*	*
	Pusa	*	*	*
	Sabour	*		
Chhattisgarh	Jagadalpur	*	*	*
	Raipur	*	*	*
Delhi	*New Delhi*	*	*	*
Goa	*Goa*	*		
Gujarat	Nawagam	*	*	*
	Navasari	*		*
	Derol	*		
	Vyra	*		
	Danti	*		
Haryana	Kaul	*	*	*
	Karnal	*		
Himachal Pradesh	Palampur	*	*	*
	Katrain	*		
	Malan	*		*
	Surrendernagar	*		
Jammu & Kashmir	Khudwani	*	*	*
	Chatha	*	*	*
	Larnoo	*		
	Rajouri	*		
Jharkhand	Ranchi	*		
	Hazaribagh	*		*

	Ranchi			*
Karnataka	Mandya	*	*	*
	Mangalore	*	*	
	Mugad	*		
	Ponnampet	*		*
	Siriguppa		*	*
	Bangalore	*		
	Brahmavar	*		
	Gangavati	*		
	Mudigere	*		
	Mugad			*
	Kathalagere	*		
	Sirsi	*		
Kerala	Moncompu	*	*	*
	Pattambi	*	*	*
	Vyttila	*		
Madhya Pradesh	*Rewa*		*	
	Jabalpur	*		
	Waraseoni	*		
Manipur	Imphal	*		
	Wangbal	*	*	*
	Lamphalpat	*		
Maharashtra	Karjat	*	*	*
	Sakoli	*	*	
	Tuljapur	*		
	Panvel	*		
	Sindewahi	*		
	Radhanagari	*		
	Palghar	*		
	Shirgaon	*		
	Pondaghat	*		

Meghalaya	Upper shillong	*	*	
	Barapani	*		*
Mizoram	Kolasib	*		
Orissa	Chiplima	*	*	*
	Jeypore	*		
	Bhubaneswar	*	*	
	Cuttack	*	*	*
	Motto	*		
Pondicherry	Kurumbapet	*	*	
	Karaikal	*	*	*
Punjab	Kapurthala	*		
	Ludhiana	*	*	*
	Gurdaspur	*		
	Rauni	*		
Rajasthan	Kota	*		
	Kota		*	*
	Banswara	*	*	
Tamil Nadu	Aduthurai	*	*	*
	Coimbatore	*	*	*
	Ambasamudram	*		
	Paramakudi	*		
	Madurai	*		
	Trichy	*		
	Annamalainagar	*		*
	Tirur	*		*
	Gudalur	*		*
Tripura	*Arundhautinagar*	*		*
	Lembucherra	*		
Uttarakhand	Pantnagar	*	*	*
	Almora	*	*	*
	Majhera	*		

	Dehradun	*		
	Haridwar	*		
Uttar Pradesh	Faizabad	*	*	*
	Ghagharaghat	*	*	*
	Kanpur	*		
	Nagina	*		
	Varanasi	*		*
	Modipuram	*		
	Lakhoti	*		
	Allahabad	*		
	Lucknow	*		
	Hastinapur	*		
West Bengal	Bankura	*		*
	Chinsurah	*	*	*
	Bankura		*	
	Canning	*		
	Kalimpong	*		
	Pundibari	*		

Italics: Voluntary centres for the respective disciplines

Normally, it takes a minimum of three to four years to identify a promising variety. Cooperating centres nominate their promising material directly for the Initial Varietal Trial (IVT). In the earlier years, the promising materials were nominated to National Screening Nursery (NSN) for screening against pests/diseases. From NSN the material was flowing to the IVTs. IVTs are conducted for one year and material that show promise on the basis of either significant yield advantage over check varieties as per the norms or on the basis of better disease/pest resistance, drought tolerance, quality attributes, etc., are promoted to the next stage called Advance Variety Trial (AVT 1). Best entries are tested in AVT for two years (AVT 1 and AVT II). Simultaneously, agronomical and quality data are generated in respect of promising varieties and the reaction to disease and insect pests are recorded under hot spot conditions as well as under glass house screening at selected centres where competent expertise is available through all three years of testing under NSN 1 comprising of AVT entries and NSN 2 with IVT entries. Special nurseries are organized for soil stress, semi-deep water/deep water and for aromatic short grains to identify the promising breeding lines/donors nominated by the breeders (Fig. 3.1). The donors identified through this process along with the donors from general germplasm that was screened under AICRIP are listed in Tables 3.6, 3.7, 3.8, 3.9.

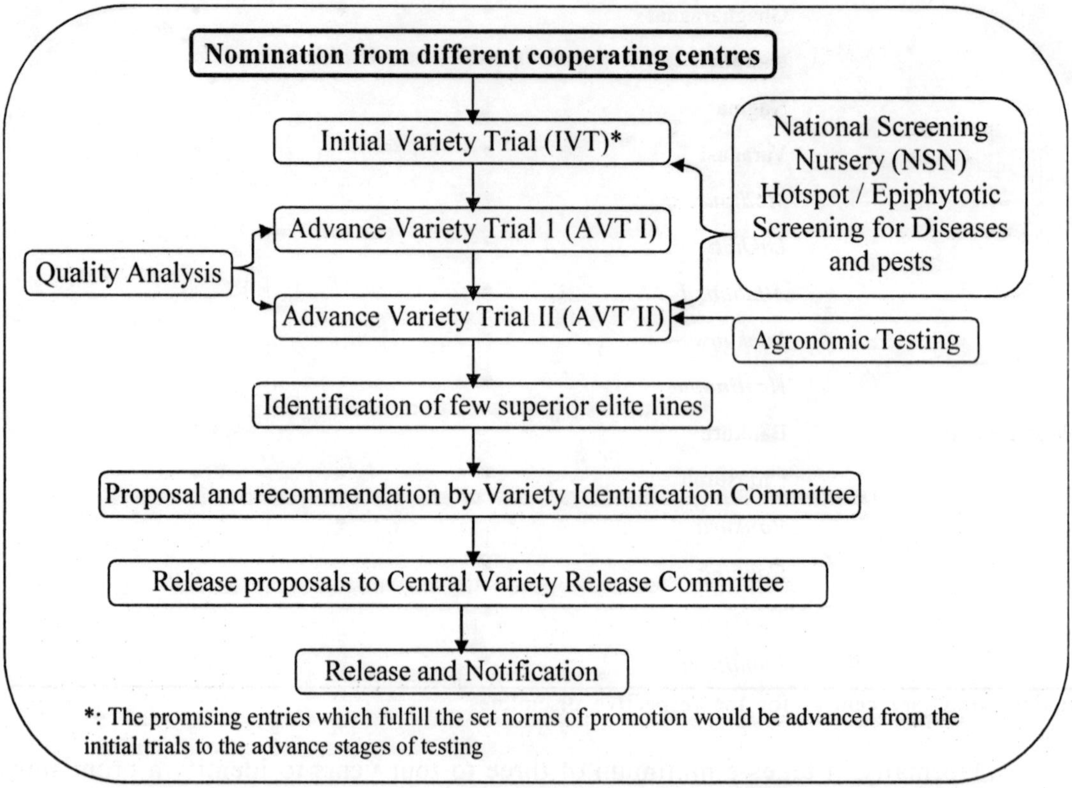

Fig. 3.1. Procedure for varietal testing and release under AICRIP.

Evaluation of breeding lines/germplasm and their utilization in varietal development under AICRIP

A total of 21,229 breeding lines which are included in the AICRIP since 1965-2007 were screened through the mechanism of NSN. In this extensive testing during the last four decades the screening for insect pests has gone up from 2 to 13 insect pests and diseases from 4 to 7 from 1970 to 2007 respectively and the test locations from 7 to 48 and 18 to 54 during the same period which is a great strength to the AICRIP (Table 3.11). Screening for major pests and diseases has been throughout while it was scanty for some of the pests like whorl maggot, thrips and gundhi bug. Maximum number of breeding lines were promising in their reaction to GM (1374), followed by BPH (1002), blast (955), WBPH (797), stem borer (794), BLB (629), RTV (406), sheath blight (396) and brown spot (421) (Fig. 3.2 and 3.3; Plate 3.1 and 3.2). Testing of this vast number of genotypes in the multilocation testing programme over the last four decades resulted in the identification of 812 elite breeding lines as eligible for release through central and state release committees for submitting the proposals by the breeders (Progress Reports (AICRIP), 1966-1985; Progress Reports (DRR), 1986-2007). A maximum number of 372 breeding lines including hybrids were found promising for irrigated ecology, 136 for upland, 89 for lowland, 67 for hills, 56 saline/alkaline situations, 48 for semi deep water, 24 for export quality basmati and short grain aromatic category, 17 for deep water ecosystem and 6 for *boro* (Fig. 3.4). Of these 209 elite materials were released and named as varieties, while the actual varietal releases are 852, the remaining 593 were state releases.

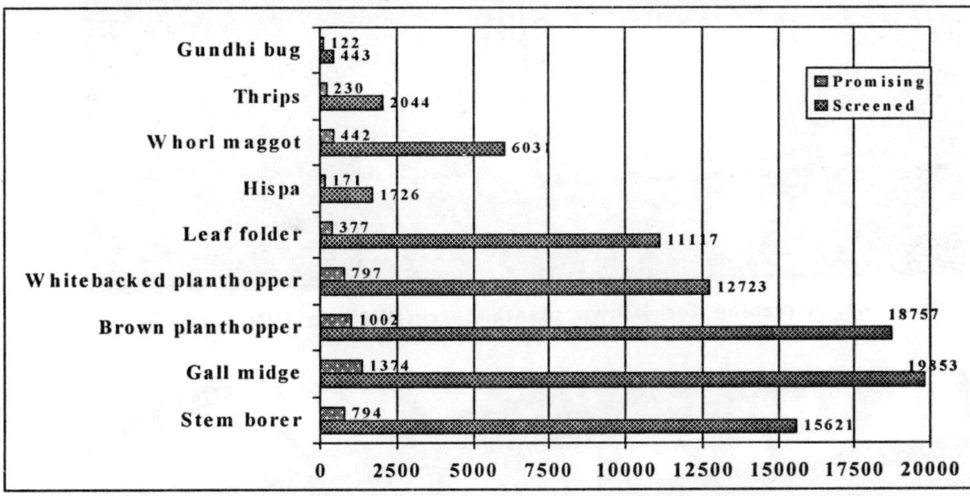

Fig 3.2. Number of entries in variety trials screened and found promising for insect pests in AICRIP Programme, 1965-2007

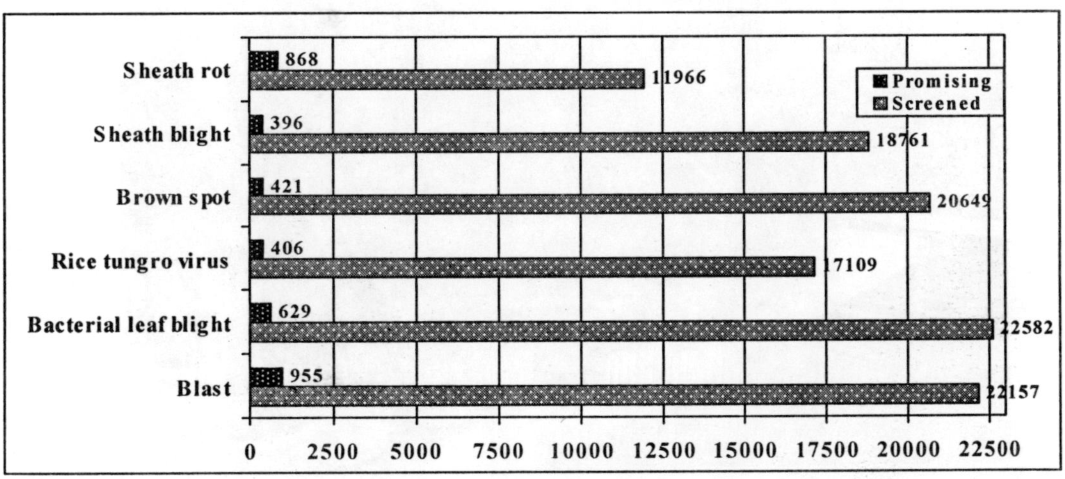

Fig 3.3. Number of entries in variety trials screened and found promising for diseases in AICRIP Programme, 1965-2007

Plate 3.1. Screening for Brown planthopper (BPH) in DRR glass house

Plate 3.2. Screening for Sheath blight in DRR glass house

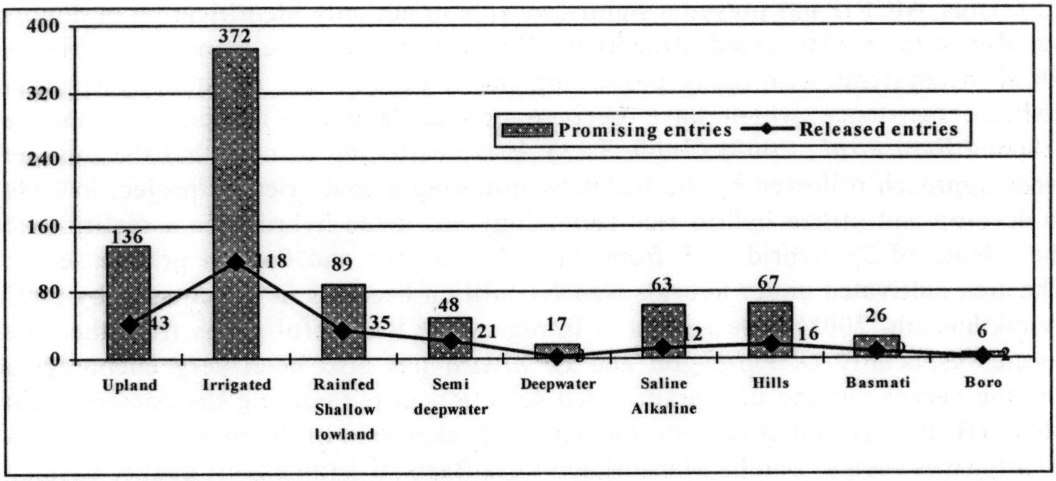

Fig. 3.4. Number of promising entries and released varieties in variety trials under AICRIP, 1965-2007

A study on the utilization of donors identified through the multilocation testing at the national level and through the breeding programmes of SAUs has indicated that a maximum of 155 donors including exotic ones were in the pedigree of 251 varieties released so far (Table 3.11). This list is taking into consideration the direct involvement of donors in the cross combination but, if the pedigree analysis is done, the number of donors involved in the development of the varieties released in India till 2007 and also the factual number of varieties would be much higher. The important donors which made significant contribution in the development of varieties include ARC 6650, Manoharsali, Ptb 33, Siam 29, W 1263, Ptb 10, TKM 6, Fine gora, N 22, M 63-83, Patni 23, Leb Muey Nahng, Jhingasail, Tadukan, Shenei, SR 26B and Basmati 370 (Table 3.12). ARC 6650 was involved in development of 7 varieties for three ecosystems in three states; W 1263 in 6 varieties for two ecosystems in six states; Ptb 33 in 6 varieties for two ecosystems in five states; TKM 6 in 13 varieties in four ecosystems in 5 states; N 22 in 9 varieties for two ecosystems in 6 states; Ptb 10 in 6 varieties all irrigated in the state of Kerala; Patnai 23 in 5 varieties for two ecosystems in two states and Basmati 370 in 10 varieties in six states. This process of intense multilocation evaluation of elite lines and germplasm materials and the utilization of identified donors in the rice improvement programmes throughout the country was instrumental in the development and release of 852 rice varieties including 33 hybrids till 2008 in the last forty two years. Of these varieties 517 for irrigated areas (456 including scented, saline, *boro*, irrigated hills), 115 for rainfed uplands, 158 for rainfed low lands, 37 for semi-deep, 14 for deep water situation, 37 for irrigated high altitudes, 7 for upland hills, 31 for saline and alkaline areas, 11 for *boro* situation, 49 aromatic slender grain varieties and 1 for aerobic condition (Fig.

3.5). Thus, AICRIP has played a significant role in not only identification of donors but also in their widespread utilization which resulted in development of varieties for all ecosystems with many possessing insect pest/disease resistance/tolerance/ multiple resistance which have become popular and are in active commerce (Shobha Rani *et al.*, 2008) (Table 3.13). It is heartening to note that the mission mode approach followed by the ICAR by initiating a goal oriented project in 1989 to develop and utilize hybrid rice technology has made hybrid rice a reality with the release of 33 hybrids, 25 from the public sector and 8 from private sector. The area cultivated under hybrids was 1.1 million hectares in the country by 2007 (Virakthamath, 2008). The results in introgressing the useful genes from the wild species especially *O. rufipogon* and *O. nivara* has also been very encouraging and the successful use of marker aided selection in pyramiding the bacterial leaf blight (BLB) resistant genes into the genetic background of the most widely grown quality rices such as Samba Mahsuri and Pusa Basmati 1 (an export quality basmati rice) have been some of the significant achievements in the recent years.

Table 3.11. Details of screening for insect pests and diseases in National Screening Nursery under AICRIP.

Year	Entomology		Pathology	
	Insect pests	Locations	Diseases	Locations
2007	13	48	7	54
2000	12	42	6	43
1990	11	39	6	35
1980	7	20	6	32
1970	2	7	4	18

Table 3.12. Donors utilized in the development of rice varieties in AICRIP.

Andrewsali	ARC 6650	Basmati 370	Bhasmanik
ARC 12422	Azucena	Beldar	BhuraRata 4-10
ARC 12751	Bamani	Belle Patna	BK 79
ARC 14529	Bangaru Teegalu	Benong 3	Black Gora
ARC 5984	Baran Basmati	Bhadshabhog	Bondu
Boyan	CH 1039	Cherali	China 971

BR 4-10	CH 988	Chettivirippu	Chiriviruppu
Brown Gora	Chambal	Chianing Sen Yu 47	Chucheng
C 14-8	Chandan	China 1007	Cica 4
C 22	Chenab	China 4	CN644
CNM 539	Dagi	Dee-Geo-Woo-Gen	Formosa
CR 1014	Damodar	Deval	FR 13A
CR 57-MR 1523	Danger	Dolangi	Gandhi Bora
Cuttack Basmati	Dasal	Edavaka	Gauchari
Dadani Kalam	Dasari	Fine Gora	GEB 24
Getu	HBJ DW 8	IRAT 8	Jelia
Ghew bora	Holamaldiga	Jaladhi 2	Jhingasail
Ginmsari	HPU 741	Jaldidhan 8	Jikkoku
Gora mutant	Intan	Jalmagna	Jun Jen-4
Halwasal 17	IRAT 138	Jeeragasamba	Kagalikai
Kajallata	Karjat 184	Khonorulu	Leb Muey Nahng
Kalakeri	Karnal Local	Lallo 14	Lohandi
Kalasukhdas	Katarni	Lalnakanda	M 63-83
Kalkheri	Kathawar	Latisail	Manoharsali
Karivennel	Khao Nahng Muey 11	Leaung 152	Millek Kunning
MR 1523	Nam sagui19	Oodasannalu	Pankhari 203
N 22	Nona Bokra	Orumundakan	Parewa
Nagheriboa	Nonasal mutant	OS 4	Patnai 23
Nagpur 22	*O. glaberrima*	Pak. Basmati	Patong 32
Najani	*O. rufipogon*	Palghar 84-3	Peta
Peta 3	PTB 10	Remadja	Seraup Kechil
Peta 5	PTB 2	Sadri	Shenei
Phul Patas	PTB 21	Safri	Shensi
Pokkali	PTB 33	Sathi	Shir Yeki
Prasadbhog	Rathu Heenati	Sathi 34-36	Siam 29
Sigadis	T 3	TKM 6	Waikoku
Sindhur Mukhi	Tadukan	Tongkai Rotan	Warangal 487

SR 26 B	Ta-poo-cho-Ze	Urang Urangan	WGL 12708
Srirendah Nehra	Tetep	W 1263	
Suweon 334	Thimpu	W 12708	

Table 3.13. Details of selected donors used in rice breeding programmes and their contribution towards varietal improvement in India.

Donor	CVRC/ State	Variety Name	IET No	DFF	Eco system	P/D
ARC 6650	AS	Chandrama	10419	170	Boro	R-Bl, MR-BLB, RTV, ShBl, SB, BPH, WBPH, GM
	KE	Makom	9268	80	IRE	MR-Bl, BLB, SB, BPH, GM
	AP	Deepti	8100	120	RSL	R-BPH
	AP	Nandi	8101	120	RSL	-
	AP	Pratibha	8106	120	RSL	R-BPH
	AP	Surya	15721	130	RSL	MR-BPH, WBPH
	AP	Vajram	8105	120	RSL	R-BPH, MR-Bl
Basmati 370	CVRC	Jamuna	1892	97	IRME	R-Bl
	KA	Kusuma	1906	90	SCR	-
	CVRC	Kasturi	8580	100	SCR	MR-Bl
	CVRC	Haryana Basmati 1	10367	110	SCR	R-Bl, WBPH, MR-SB
	JK	Ranbir Basmati	11348	95	SCR	R-WBPH, GM, MR-BPH
	MH	SKL 7	10651	100	SCR	-
	PU	Punjab Basmati 1	7313	95	SCR	-
	RA	BK 79	7028	100	SCR	MR-BLB, GLH
	RA	Mahi Sugandha	12601	100	SCR	MR-Bl
	UT	Pant Sugandh Dhan 15	14132	114	SCR	MR-Bl
	JH	Birsadhan 105	12050	60	RUP	R-Bl, R-BLB, MR-BS
N 22	UP	Ashwini	-	75	IRE	-

	AP	Prasanna	7564	75	RUP	R-BS, MR-LF
	BI	Kanchan	4815	70	RUP	R-Bl, BLB, GM
	CVRC	Narendra 97	9210	70	RUP	MR-BLB, ShBl
	CVRC	Kameshdhan	19253	77	RUP	R-WBPH, GM, MR-SB, LF
	OR	Ghanteshwari	9802	70	RUP	R-Bl, GLH, GM, MR-BLB, ShBl, SB
	UP	Narendra dhan 80	7626	75	RUP	R-Bl
	UT	VL Dhan 97	9210	85	RUP	-
	WB	Kiron	3226	75	RUP	-
Patnai 23	AS	Lakhimi	6666	120	RSL	-
	CVRC	Purnendu	10029	130	SDW	MR-BS, SB, GM, LF
	WB	Ambika	11901	130	SDW	-
	WB	Bagheerathi	11272	120	SDW	-
	WB	Mahananda	11910	135	SDW	R-Bl, MR-SB, WBPH, LF
PTB 10	KE	Annapurna 28	-	70	IRE	-
	KE	Aswathi	-	80	IRE	R-Bl, MR-GM
	KE	Bharathi	16707	90	IRE	-
	KE	Jyothi	2700	85	IRE	R-Bl, GM, MR-BPH
	KE	Rohini	_	70	IRE	-
	KE	Ahalya	15623	89	IRME	MR-Bl
	PU	PR 108	7318	100	IRME	-
PTB 33	KE	Aruna	9287	80	IRE	MR-ShBl, SB, BPH, GM
	KE	Remya	9266	84	IRE	R-GM, MR-Bl, ShBl, BPH
	PY	Bharathi dasan	9291	85	IRE	R-BPH, WBPH, MR-BLB, RTV, SB
	TN	ADT 37	10391	75	IRE	R-Bl, GLH, BPH
	WB	Jamini	12133	80	RUP	R-Bl, MR-RTV, ShBl, GM, LF
	KE	Kanakom	9380	100	IRME	R-BPH, MR-Bl,

						RTV, ShBl, SB, GM
TKM 6	CVRC	Cauvery	355	84	IRE	MR-SB
	CVRC	Ratna	1411	84	IRE	MR-Bl, SB
	KA	Madhu	2665	90	IRE	R-Bl, BLB, MR-SB
	KA	Suma	1762	85	IRE	-
	UP	Saket 4	1410	85	IRE	MR-BLB, GLH, SB
	CVRC	IR 20	1926	105	IRM	MR-BLB, RTV, SB
	GU	CR 138-928	4078	100	IRM	MR-BLB, SB, LF
	CVRC	Sasyasree	2815	95	IRME	R-Bl, MR-BLB
	CVRC	Vikas	3116	84	IRSA	MR-BS, SB
	UP	Narendra 2	4555	85	IRSA	MR-Bl, BLB, RTV
	OR	Ramakrishna	4155	100	RSL	R-BLB
	OR	Parijat	2685	70	RUP	R-BLB, GLH, MR Bl
	UP	Jal Lahari	13534	140	SDW	R-Bl
W 1263	KA	GMR 17	2900	80	IRE	R-GM
	AP	Kakatiya	2946	95	IRME	R-Bl, GM, MR-BS
	CH	Asha	6012	98	IRME	-
	TN	MDU 3		95	IRME	R-Bl, BPH, WBPH, GM, MR-LF
	KA	Karna	7209	105	IRM	R-Bl, GM, MR-SB
	OR	Rambha	7975	120	SDW	R-Bl, GLH, MR-BLB, BPH

SCR: Scented; HRIR: Irrigated hills; IRE: Irrigated early; IRME: Irrigated mid early; IRM: Irrigated medium; IRSA: Saline / Alkaline; RSL: Rainfed shallow lowlands; RUP: Rainfed uplands; SDW: Semi deepwater; DFF: Days to 50% flowering

In addition to the mandate in organizing multilocation testing of breeding lines emanating from cooperating centres within the country and also from International Rice Research Institute, IRRI, Philippines, DRR has also organized several externally funded network programmes related to germplasm evaluation and conservation in the last two decades. Some of the externally funded projects which were successfully implemented by the DRR where germplasm was evaluated with limited collection work undertaken are listed below (Table 3.14):

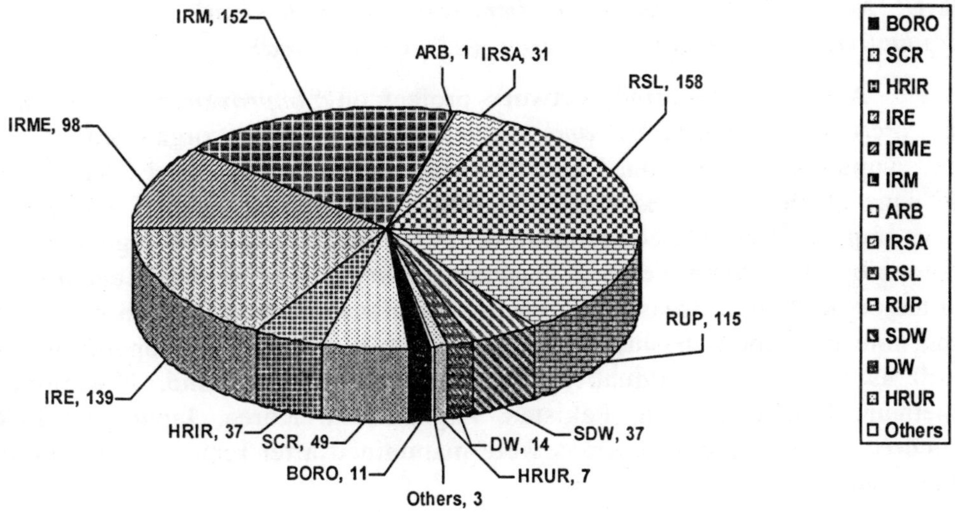

Total: 852 Irrigated: 517 Rainfed: 332 Others: 03

SCR: Scented; HRIR: Irrigated hills; IRE: Irrigated early; IRME: Irrigated mid early; IRM: Irrigated medium; ARB: Aerobics; IRSA: Saline / Alkaline; RSL: Rainfed shallow lowlands; RUP: Rainfed uplands; SDW: Semi deepwater; DW: Deepwater; HRUR: Rainfed upland hills;

Fig 3.5. Rice Varieties released for various Ecosystems

ICAR Network Programmes organized/implemented by DRR:

Sl No.	Project	Year	Centres	Budget (Rs. in lakhs)
1	Improvement of basmati rices for increased productivity and export purposes	1990-1993	6	Rs. 34.5
2	Evaluation of rice germplasm for biotic stresses	1993-1998	15	Rs. 108.2
3	Genetic enhancement of quality rices for higher productivity and export	1994-1999	14	Rs. 134.12
4	Evaluation of rice cultivars for rainfed rice production system (NATP-RRPS 10)	1999-2003	9	Rs. 131.07
5	Sustainable management of Plant biodiversity on mission mode (NATP-MM)	1999-2003	1	Rs. 18.77
6	Genetic enhancement of aromatic short grain rices for higher productivity and export	2002-2005	5	Rs. 46.06
7	Multi-location germplasm evaluation programme (NBPGR-EFC)	2004-continue	15	Rs. 172.05

ICAR Network programme on improvement of basmati rices for increased productivity and export purposes (AP Cess Fund)

An ICAR Research Network project on "*Improvement of Basmati Rices for increased productivity and export purposes*" was organized at 6 centres *viz*., Kapurthala, Ludhiana, Kaul, Pantnagar, Cuttack, IARI and DRR during 1990-93. One of the objectives was collection, evaluation of basmati and other quality rices acquired from the cooperating centres, study of exploitable genetic variability for various key components of basmati quality and to screen them for resistance to major pests and diseases. The achievements include collection of one thousand and nine indigenous basmati genotypes contributed by the cooperating centres as well as exotic scented/quality rices mainly from Thailand, Philippines, USA, Vietnam, Indonesia, Iran, Pakistan, Afghanistan, Korea, Japan and IRRI. Nine hundred eighty three accessions were maintained after removing the mixtures and duplicates from the collection.

Three hundred sixty accessions were evaluated for agro-morphological characters including yield components. Five hundred and thirty eight accessions were screened for the extent of variability for important quality parameters. Two hundred and thirty six were rated as strongly scented, 250 mildly scented and 52 non-scented. Karnal Local, Basmati 150, Basmati Kota, Basmati 5853 and Karnal Local semi-dwarf mutant 28 possess high kernel length after cooking (KLAC) (above 14 mm). Two hundred six accessions were evaluated for high density grains at DRR among which Dawag, Basmati 127A, Sams Faroz, Basmati Nahan 281, Sathi 34-36, Basmati 5853, Jao Mali, Xiang Jeng 3 and Kwan Kikwai recorded high density index of more than 50%. Four hundred accessions were screened for leaf blast, neck blast, bacterial leaf blight, whitebacked planthopper, stem borer, brown planthopper, gall midge biotype 1 & 2 and sheath blight at different centres. Promising donors possessing pest/disease resistance based on screening at various basmati growing locations and at DRR are listed in Table 3.14.

Table 3.14. Promising donors among indigenous and exotic scented/quality rices resistant/ moderately resistant to biotic stresses, (1990-1993).

Bacterial leaf blight	
DRR	Basmati 217, Basmati Norat 439, Bhoga Manik Madhuri, Domsiah, Gansal, Ranga Joha 2, RD 15
Kapurthala	Basmati bahar, Binirhen, Govindbhog, RD 15, Sadri

Ludhiana	Basmati 372, Hei Xiang-Jeng 3, Karjat 35-3, Karnal Local, NIRRI 138-5, NIRRI 750, Short Straw Star Bonnet, Sitabo-Tabo
Pantnagar	Guinata
Leaf blast	
DRR	Ambemohar159, Barah, Basmati 140, Basmati 405, Domsiah, Kasturi, Khao Dawk Mali, Khao Khodri, Leung Hawn, Sinagabe
Neck blast	
CRRI	Abor Joha, Badha Bhog, Basamti, Basmati 242, Basmati Kamon, Chufan Joha, Khom Joha 1
Kaul	Basmati, Basmati 5853, Basmati 5854, Basmati 6129, Basmati Kamon, Domsiah, Guinata, Sinagabe
Sheath blight	
IARI	Jaijai 77, Kataribhog, Khao Lo, Khao Luang, Mussatreen, Popot, Ram Sail, Sadri, Sathi Basmati
Whitebacked planthopper	
Ludhiana	Basmati 123, Bindli, Karnal Local, Pak Basmati, Ranbir Basmati
DRR	Basmati 140, Basmati 6129, Basmati bahar, Multani
Stem borer	
Kaul	Basmati 370, Basmati 410, Rodjolele 2, Rodjolele 4, Rodjolele 5
Pantnagar	Basmati 410, Basmati Aman, Basmati Dehradun, Basmati Jasmuna, Guinata, Raskadam, Sitabo-tabo
GM biotype1	
DRR	Badshabhog, Hansraj, Prasad bhog, Rajbhog, RD 15, S 1732, Sams Faroz, Sitabo-tabo
GM biotype 2	
CRRI	Basmati 6311, Basmati Surkh 161, Hasan Serai, Manipuri joha, Nepali Joha, Prasad bhog, S 1732, Sitabo Tabo
Brown planthopper	
DRR	Bong Cay, Karjat 13-21
CRRI	Basmati Dehradun, Black Puttu, Gopalbhog, Lalchaddu, S 1733

ICAR Network programme on genetic enhancement of quality rices for higher productivity and export (AP Cess Fund)

Under AP Cess Fund, an ICAR network programme on "*Genetic enhancement of quality rices for higher productivity and export*" was organized at 11 research and 3 testing centres with the Directorate as strategic and coordinating centre to address the important researchable issues in both basmati and non-basmati programmes from 1994-1999. The cooperating centres for (1) basmati programme were: Kapurthala, Kaul, IARI, New Delhi, Ludhiana, Pantnagar and R S Pura, and (2) non-basmati programme were: Adhuthurai, Raipur, Nellore, Faizabad with Banswara, Varanasi and Siruguppa as testing centres. The major objectives included (i) collection, evaluation and maintenance of scented/quality germplam, (ii) development of high yielding export quality basmati and non-basmati rices; (iii) stabilizing their yield by insulating them against target pests, and (iv) development of suitable agronomic practices for harvesting higher yields retaining the quality traits (Shobha Rani and Prasad, 2000).

Eight hundred and seventy six scented quality rice germplasm accessions were characterized for 19 morphological, 15 yield and agronomic characters and 12 milling, grain and cooking quality features, along with screening for pests/ diseases. 198 accessions possess typically intermediate amylose content. Some of the genotypes possessing intermediate amylose, with superfine grains coupled with scent include Basmati 397, Basmati 410, Basmati 370, Basmati 242, Basmati Jamuna, Azucena and Basmati 372.

Some of the accessions combining intermediate amylose content with resistance/tolerance to pests/diseases are listed in Table 3.15. The accessions which showed built-in resistance to leaf blast, bacterial leaf blight and whitebacked planthopper in reconfirmation test are presented in Table 3.16.

Table 3.15. Scented/quality germplasm accessions with intermediate amylose content with multiple resistance to target pests/diseases, *Kharif,* 1997-1998.

Multiple resistant		
Xiang Geng Dao: MR to Bl, R to WBPH	Domsiah : MR to LBL, NBL, BLB	Basmati 6129 : R to NBL, MR to BLB and WBPH
Serakap: MR to Bl, WBPH	Basmati Aman : MR to SB, BLB	Basmati T3 : R to LBL, NBL
Basmati: MR to blast, BLB	Basmati Gola : MR to BL,WBPH	Sonth : R to WBPH, MR to BLB
Karnal Local: MR to blast, WBPH and Stem borer	Kataribhog, Gopalbhog, Dehradun Local, Basmati 397: MR to WBPH, BLB	Govindbhog : MR to BLB, R to blast

Table 3.16. Scented/quality germplasm accessions resistant/moderate resistant to some diseases and pests in re-confirmation tests at DRR, Kharif 1994-1996.

Leaf Blast

ACC 27716	Badshabhog	Basmati	Basmati 122
Basmati 124-10	Basmati 134	Basmati 140	Basmati 150
Basmati 242	Basmati 370 (Punjab)+	Basmati 385+	Basmati 397
Basmati Aman	Basmati C621	Basmati gola +	Basmati Kamon
Basmati Kamon*	Basmati Mehtrah	Basmati U Zarka +	Govindbhog*
Basmati*	Bong Cay +	Domsaiah @	Khudi joha
Kamod 118	Karnal Local*	Khao Dwak Mali	Tulsibhog
Mussatrene	Sitabo-tabo	Sugandha	

+: Also resistant/Mod.Resistant to WBPH during kharif 1994 at DRR; * : Also resistant/Mod. Resistant to leaf blast at Kaul during kharif 1994; @ : Also resistant to leaf blast in 1996

Bacterial leaf blight

ARC 6011*	Basmati C 622	Gobindbhog	N 702
Basmati 334	Bong Cay	HKR 236	Pusa 615
Basmati 372	Chufan Joha	IR 841-85-1-1-2	Samudrawat
Basmati 376 Hai Duang	Domsiah	Khao Luang	Tam Xoan
Basmati 6129*	Du Thom-Thai Binh nai phong	Kon Joha	Chhatri

* Also resistant to BLB in kharif 1994 at DRR

Whitebacked planthopper

Chomsa Bora++	Basmati Aman	Dehradun Local*	Karnal Local Semi dwarf mutant-28	Padi Emphat Bulan 2*
Basmati 123	Basmati Lamo++	Gopalbhog++	Karnal Local*	Ramsail**
Basmati 370*	Basmati+	HBC 5+**	Kataribhog**	Seeta Sail**
Basmati 397**	BD 319**	HBC 98	Khao Dawk Mali+	Sonth**++
Basmati 405++**	Bong Cay	HKR 125	Lua Nhe	Sugadas
Basmati 6129++	Ambemohar 159	Jai Jai++	Multani	Sukhwal++**
Basmati 43 A	KLSDM 28	HKR 120,	NP 114++	Basmati Tall

*: Also resistant/mod. Resistant to WBPH at Ludhiana during 1994/kharif 1996; +: Also resistant to leaf blast at DRR during kharif 1995; ++: Also resistant/mod. Resistant to WBPH during kharif 1994 at DRR; **: Also resistant/mod. Resistant to BLB at DRR & Ludhiana in kharif 1995

The entire characterized data on 481 scented/quality germplasm indigenous to India and those procured from exotic sources were digitalized in the form of a catalogue (unpublished) giving the details of passport data on accessions, characterization and evaluation data for agro-morphological, quality and pest/ disease, frequency distribution for various qualitative and quantitative traits, statistical analysis of variability parameters with respect to yield and quality characters etc., The promising accessions for yield and agronomic characters are given Table 3.17 and the frequency distribution for quality traits in Fig. 3.6. The valuable germplasm was also deposited at NBPGR, New Delhi with the indigenous accession numbers assigned IC Nos: 323554 to 323981 and the exotic accession numbers EC Nos: 490209 to 490348.

Table 3.17. Promising genotypes in scented/quality germplasm for yield and agronomic characters, 1994-1999.

IC No	ACCN	Value	IC No	ACCN	Value
Plant Height (cm) (tall accessions)			**Panicle length (cm)**		
IC 323586	Basmati Bahar	174	IC 323592	Badshahbhog	35
IC 323638	BR 9	174	IC 323609	Basmati 1	35
IC 323585	Basmati Mehtrah	173	IC 323576	ARC 14372	34
IC 323660	Basmati 93 B	170	IC 323611	Basmati 43 A	34
IC 323872	Randhunipagal	170	IC 323709	Hansraj	34
Plant Height (cm) (dwarf accessions)			**Total Grain Number**		
IC 323710	Pusa 615-14-9-8	80	IC 323760	Kusuma	405
EC 490242	IR 841-85-1-1-2	78	EC 490297	IR 62873-405-1-1	391
EC 490339	YAR 96V13	77	IC 323763	Vijaya Mahsuri	383
EC 490321	Koshihikari	76	IC 323732	Kon Joha 2	347
EC 490320	Hinohikari	73	IC 323723	Katarni	344
Tiller Number			**Sterility Percentage**		
IC 323849	Badshahbhog	23	IC 323671	Basmai 376	2.2
IC 323707	HBC 85	22	IC 323566	Abor Joha	2.1
EC 490300	IR 62873-417-4-3	22	EC 490211	Ayepyaung	1.8
IC 323580	ARC 14760	21	IC 323636	Basmati 370 B	1.8
IC 323699	Gopal Bhog	21	EC 490236	Hung mi hsiang ma tsan	1.4

Flowering Duration (days) (late accessions)			Single Plant Grain yield (gm)		
EC 490249	Khao Jao Hawm	140	EC 490224	C4-63G	9.99
EC 490219	Bong Cay	137	IC 323567	ARC 7229	9.99
IC 323850	Pusa Basmati 1	131	IC 323671	Basmai 376	9.99
IC 323758	Kusuma	127	IC 323762	Akkuliu	8.33
IC 323622	Basmati 124-10	126	IC 323565	AKP 1	7.99
Flowering duration (days) (early accessions)				**Test weight (gm)**	
IC 323897	RPSC 148	92	EC 490270	Naipon No.1	34.7
IC 323833	RP 2144-19-6-2	90	IC 323828	Pusa 743	31.7
EC 490322	Hiyokumochi	90	IC 323885	RP 3120-15-12-9-4	31.7
EC 490329	Hei Bao	88	IC 323687	Basmati T 370	31.2
EC 490320	Hinohikari	87	EC 490337	SPRLR 841-84-9-5-2-1-3	30.8

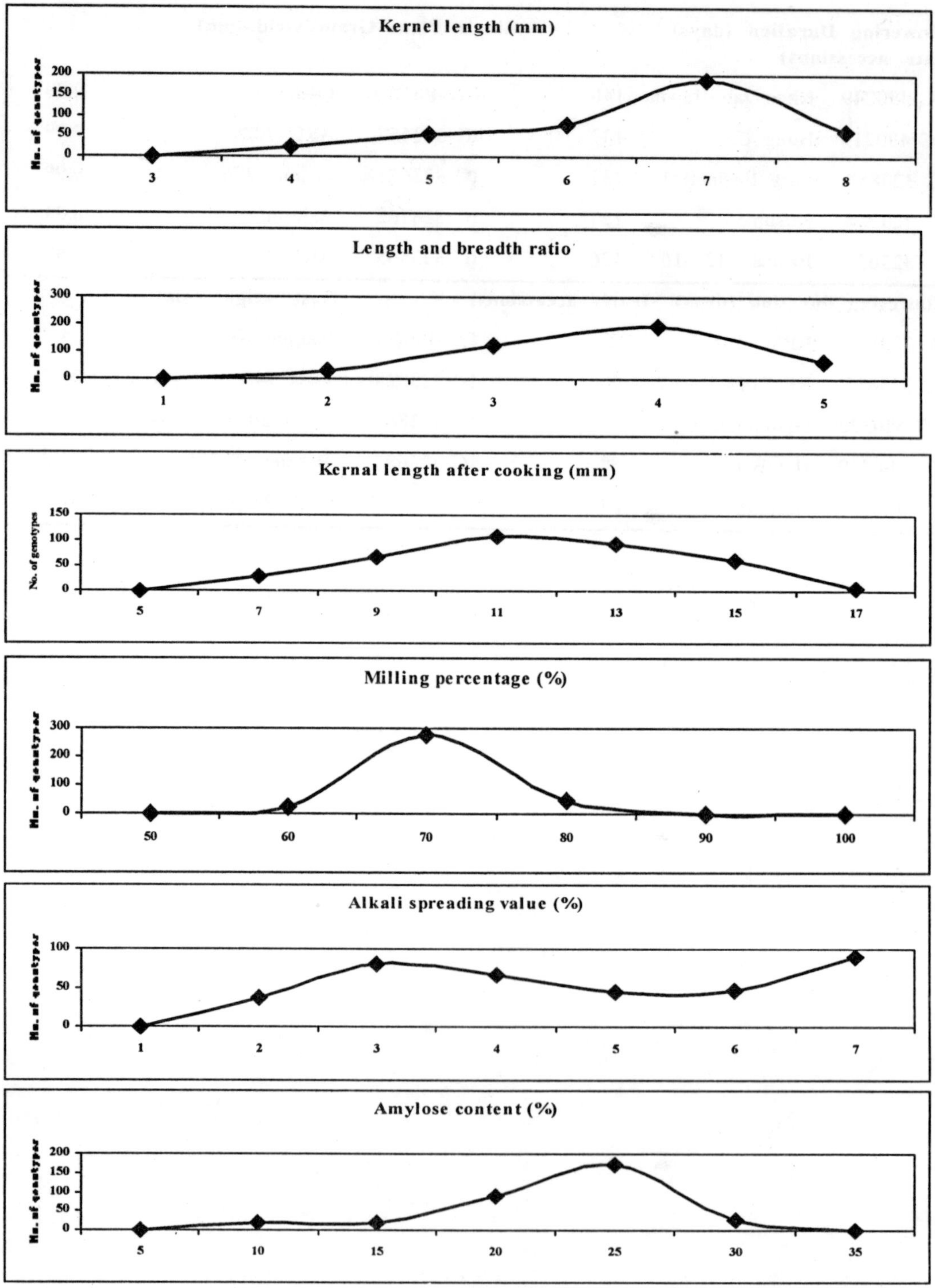

Fig 3.6. Frequency distribution of scented / quality germplasm for quality parameters, 1994-1999.

Evaluation of rice germplasm for biotic stresses

An ICAR network programme on "*Evaluation of rice germplasm fo, biotic stresses*" was implemented during 1993-98 with a budget of Rs. 108.2 lakhs (DRR 1993-1998). The objectives of the programme were evaluation of rice germplasm for major pests/diseases, conservation, documentation and exchange of potential gene sources for utilization in the breeding programmes.

The network coordinated by the Directorate of Rice Research comprised of two lead centres *viz.,* NBPGR, New Delhi and IGKVV, Raipur and 12 hot spot centres viz., Almora, Aduthurai, Kapurthala, Sambalpur, Raipur, Jagdalpur, Pondicherry, Maruteru, Ragolu, Mandya, Ponnampet and Moncompu.

The germplasm comprising 13,620 accessions obtained from Raipur (10620) and CRRI (3000) were evaluated for 3 major diseases (blast, bacterial leaf blight, sheath blight) and 3 insect pests (brown planthopper, gall midge and stem borer). All the accessions were evaluated for 21 morphological descriptors.

A large number of accessions were found to be resistant to gall midge biotype 1 (395) and gall midge biotype 4 (787), BPH (282), blast (218), stem borer white earheads (SBWE) (74), BLB (42), sheath blight (23) and a few multiple resistant donors were also identified (Table 3.18).

Table 3.18. Promising multiple resistant accessions under evaluation of rice germplasm for biotic stresses, 1993-1998.

Pest / disease	INRC Numbers
GM1, GM4, LBL	4275 (Jalpa), 4516 (Atma Shital), 9161 (Batasi), 9199
GM1, SBWE, BLB	25
GM4, LBL	4275 (Jalpa), 4516 (Atma Shital), 9006 (Ptb 32), 9161 (Batasi), 9199, 9241 (P 3), 10776 (Kondi)
BLB, SHBL	4580 (Laliajon), 5674 (Budhabudhi)
BPH, GM4, SBWE	13251 (Jawaphool)
BPH, GM4, LBL	9006 (Ptb 32), 916

Seven thousand three hundred accessions were deposited with National gene bank, New Delhi for long term storage. The programme has been successful in documenting the data and bringing out a catalogue for ready reference on all accessions for their pest/disease reaction as well as on morphological characters scored. In addition, the potential identified donors were multiplied and made available to breeders engaged in resistance breeding programme for utilisation (DRR, 1993-1998).

Genetic enhancement of aromatic short grain rices for higher productivity and export

India has a wealth of specialty rices, the prominent among them are the aromatic rices including both basmati and a wealth of short grain indigenous types which have become part of our heritage. Almost every state of India has its own collection of non-basmati short and medium grain aromatic rices and many of these are cultivated in other than designated basmati growing areas of northwest India mainly in Uttar Pradesh, Bihar, Orissa, West Bengal, Assam, Chhattisgarh, Madhya Pradesh, Maharashtra and also in other states. The cultivation of these rices is limited to small areas in specific locations. These are generally consumed locally and are widely used for making *kheer* (sweet rice) on religious and festive occasions. Often it is very difficult to even retain the produce for seed purpose. Since long time, farmers are using their own saved seeds which has resulted in seed deterioration in many cases due to non-replacement of seeds through strong maintenance and breeding efforts. Farmers still seem to grow these cultivars mainly for personal consumption despite being aware of their poor productivity and limited markets. There are scattered conservation efforts and seed production or improvement programmes for these rices. As a result in the last several years many valuable aromatic rice varieties have either disappeared or are in the process of disappearing which requires to be arrested. Therefore, giving special emphasis for improvement of aromatic short grain types, an ICAR network programme on "*Genetic enhancement of aromatic short grain rices for higher productivity and export*" was implemented during 2002-2005 with budget of Rs. 46.06 lakhs. DRR, Hyderabad coordinated this network at 5 funded centres (IGAU, Raipur; OUAT, Bhubaneswar; RAU, Pusa; NDUAT, Faizabad; IICT, Hyderabad (for aroma estimation and quantification) and 5 voluntary centres (Karaikal, Radhanagari, Siruguppa, Moncompu, Gerua) with the main objectives being (a) characterization of indigenous aromatic rices for agromorphological, quality and pest/disease resistance, (b) estimation of aromatic compounds and (c) documentation and conservation of the germplasm (Shobha Rani, 2007).

Characterization of indigenous aromatic rices for agromorphological, quality and pest/disease resistance

A collection of 252 aromatic short grain landraces were collected from the cooperating breeders of four centres Raipur, Bhubaneswar, Faizabad and Pusa. 110 genotypes were evaluated during 2002, 61 in 2003 and 81 in 2004. Agro morphological descriptors were recorded for all three years at DRR and Raipur; two years at Faizabad and one year at Bhubaneswar. The characters observed varied from 13 to 22 at different centres. Superior aromatic short grain entries for grain yield, quality parameters and with resistance to moderate resistance to leaf blast, bacterial leaf blight and whitebacked planthopper were identified (Plate 3).

Promising genotypes for yield: Acharmati, Dhobaluchi, Malaysia, Sonachoor, RAU 3037, Thukrasuna were promising at 2-3 locations. On the overall basis, Acharmati, Shukla phool, RAU 3037, Sonachoor, Ghansal, Malaysia, Kheerasai and NDR IRRI 3131 were superior to popular short grain rices such as Dubraj and Kalanamk in yield (Table 3.19).

Table 3.19. Promising aromatic short grain rices with high yield potential at various locations, 2002-2005.

Location	Aromatic short grain genotypes
DRR	Acharmati, Ambemohar 157, Baranagomati, Chhabiswa, Dhobaluchi, Dhuriawa, Gatia, Ghansal, IGSR 3-1-1, Kalanamak 2, Lectimachi A, Maguraphulla, Malaysia, RAU 3037, RAU 3080, Sonachoor, Tulsiphuila
Raipur	Acharmati, Chhabiswa Dhobulachi, Gangabarud, Ganjeikalli, Gatia, Heerakani, Kalajauvan, Kalakanhu, Kheerasai, NDR IRRI 3131, NDR IRRI 67, NDR IRRI 69, NDR IRRI 75, Parijatak, Pimpudibasa, RAU 3027, RAU 3037, Shuklaphool, Sonachoor, Thukrasuna
Bhubaneswar	Deulabhog, Dudaga, Dudh Nag, Ganjo, Gatia, Jawaphool, Khosakani, Malaysia, Neelabati, ORS 199-5, Rajin 7, RAU 3014, RAU 3037, RAU 3087, Soanachoor, Thukrasuna, Tulsi Bhog
Pusa	Basnadhan, Beribhog, Ramdhuri, RAU 3037
Faizabad	Badshapasand, Kheerasai, Kalanamak 3

Promising genotypes for pest/disease resistance: Parbatjira is resistant to blast and bacterial leaf blight; while Muhulakuchi, Acharmati, Deulabhog, Ganjeikalli, Barikunja, RAU 3014 and Barangamali are resistant/moderately resistant to blast and brown plant hopper. As many as 82 genotypes are resistant/moderately resistant to leaf blast as listed in Table 3.20.

Table 3.20. Promising aromatic short grain rices with resistance/moderate resistance to pests/ diseases, 2002-2005.

Resistant to Blast:					
Acharmati	Bairgaj	Dhaniya-B2	Jala Jaiphulla	Kankjeer-A	Maguraphulla
Ambemohar 159	Bansphool-B	Dhoiabankoi	Jiraphool	Kesar	Modak-B

Amrit bhog	Beribhog	Dindli	Kali Muchh	Lokti muchhi	Muhulakuchi
Atmashital	Deulabhog	Dubraj		Magura	Parbatjira
RAU 3014	RAU 3041	RAU 3061	RAU 724-48-33	Sitabhog	
RAU 3027	RAU 3044	RAU 3080	Seetakeshari	Sonachoor	
RAU 3030	RAU 3051	RAU 3084	Shankar chini-B	Tilakchandan	
RAU 3033	RAU 3054	RAU 741-1	Shyamjeera	Tulsi Prasad	
***Moderately resistant to blast*:**					
Baranagomati	Bhulasapuri	Ganjeikalli	Juhibengal-B	Karpurakalli	RAU3079
Barangamali	Dubraj	Gatia	Kalajira	Mirzapur	RAU 3081
Basnadhan	Dubrajsena	Gopal bhog	Kalanamak	Nagri Dubraj	RAU 3087
Bhatagundi	Durgabhog	Jhilipanjarai	(Bordpur)	Nalidhan	RAU 738

***Moderately resistant to bacterial leaf blight*:** Barangamali, Parbatjira

***Resistant to brown planthopper*:** Kalikati, Muhulakuchi, Parijatak

***Moderately resistant to brown planthopper*:** Acharmati, Basmati, Barikunja, Deulabhog, Ganjeikalli, RAU 3014, RAU 3037

*evaluated under controlled conditions at DRR

Promising genotypes for quality characters: Many genotypes with medium slender, short bold or short slender grain type with desirable starch properties and quality features especially alkali spreading value, amylose content and gel consistency have been identified after intensive testing for the entire array of quality parameters (Table 3.21). Use of these genotypes would greatly enhance the quality features in the development of high yielding aromatic short grain rices.

Table 3.21. Promising aromatic short grain rices with excellent cooking quality traits, 2002-2005.

Cultivar	ASV	AC (%)	GC (mm)	Aroma	Cultivar	ASV	AC (%)	GC (mm)	Aroma
				Medium slender grain					
Bans patri	5.0	22.34	57	MS	Malaysia	5.0	25.02	42	SS
Bans Phool A	5.0	22.41	63	MS	NDR IRRI 3131	3.5	23.04	63	SS
Champaran Basmati 4	5.0	21.60	63	SS	NDR IRRI 67	3.0	22.71	63	SS
Dhanprasad	3.0	22.05	53	SS	Raja Bhog	5.0	24.42	49	SS

Durga Bhog	5.0	20.80	45	MS	Ramdhuri	4.0	24.48	76	MS
Ganga Baru	5.0	24.36	68	SS	Dubraj	5.0	23.40	27	SS
Juhi Bengal-B	5.0	21.18	59	MS	Kadam Phool	4.0	23.76	37	SS
Kalanamak 3	4.0	22.95	60	MS	Shukla Phool	5.0	22.68	74	SS
Kali Muchh	5.0	22.98	57	SS	Sonachoor	5.0	22.97	28	MS
Kanak Jeer	5.0	20.43	55	SS	Tulsi Bhog	5.0	24.48	53	SS
Kataribhog	3.5	22.59	61	MS	Tulsi Prasad	5.0	22.09	57	SS
Kubrimohar	4.0	23.43	58	SS					
				Short bold grain					
Acharmati (B)	4.0	21.69	48	SS	Lectimachi-B	5.0	22.32	67	SS
ASGPC-19	5.0	23.04	59	SS	Lokti muchhi	5.0	22.79	42	MS
Barang	4.0	22.09	50	SS	Maguraphulla (B)	4.0	22.86	60	SS
Bishnu bhog	5.0	21.58	61	MS	Modak-B	5.0	23.60	56	SS
Chatianaki	5.0	23.29	55	MS	Nagri Dubraj	5.5	23.77	58	SS
Chini Kapoor	5.0	22.68	56	SS	Nanu	5.0	24.36	49	SS
Dindli	5.0	20.52	78	MS	Neelabati	5.0	23.52	47	MS
Dudgi	5.0	24.66	53	MS	Shankar chini-B	5.5	23.52	48	SS
Kalajira	5.0	22.93	45	MS	Sitabhog	5.5	23.35	46	SS
Kalanamak (Birdpur)	5.0	22.42	61	SS	Thakurabhog	5.0	21.47	33	MS
Kanak Jeera	4.0	22.05	50	SS	Til Kasturi	5.0	22.34	52	MS
Kubri Mohar	5.0	22.26	67	MS	Tilakchandan	5.0	21.92	90	MS
				Short slender grain					
Chhatri Bhog	5.0	23.31	69	SS	RAU 3031	5.0	22.68	49	MS
Lalmati	5.0	22.23	56	MS	RAU 3054	5.5	24.02	42	SS

ASV: Alkali spreading value, AC: Amylose content, GC: Gel consistency

The entire data on yield and yield components, morphological characters, quality and pest/disease reaction of the entries is being documented and the seed would be deposited for conservation with NBPGR, New Delhi.

Sustainable management of plant biodiversity on the mission mode under NATP

With objectives being collection, evaluation, conservation and documentation, the research project on *Sustainable management of plant biodiversity on the mission mode under national agricultural technology project* (NATP) was initiated in 1999 for a period of five years upto 2003 with National Bureau of Plant Genetic Resources (NBPGR) as lead institution for implementing the programme. The Directorate of Rice Research (DRR) is one of the cooperating centres under the Xth zone (South East Coastal) of the mission and the programme on rice involved a budget outlay of Rs. 18.77 lakhs.

The objectives of the programme are survey and collection of the rice germplasm from unexplored and under explored areas, evaluation of germplasm for agro-morphological descriptors and for various biotic stresses, deposition of seed of collected germplasm for conservation in national gene bank and providing documented passport and evaluation data to NBPGR.

A collection of 663 landraces/native cultivars from Bastar (272) and Raigarh (61) in Chhattisgarh and Adilabad (165), Nizamabad (38), Khammam (70) and Medak (57) in Andhra Pradesh was made from seven collection expeditions during 1999-2004. While collecting the passport information it was brought to light that several valuable landraces like Paccha Ganneru, Gadda Rekkalu, Dhakad Govda, Kappa Sara, Potti Kasulu, Erra Ganneru, Chinta Puvvu, Nizam Govda and Nallari are believed to have disappeared.

630 accessions were characterized for 21 morphological and 9 agronomic characters and screened for major biotic stresses like leaf blast, bacterial leaf blight, rice tungro virus, brown spot and sheath blight, brown planthopper, gall midge and stem borer. Some of the promising accessions with high yield potential coupled with biotic stress tolerance are given in Table 3.22 (Plate 3.4) (Shobha Rani *et al.*, 2004).

Table 3.22. Promising accessions with high yield potential with biotic stress tolerance under NATP on Plant Biodiversity, 1999-2003.

IC 346857 (Assam chudi)	High productive tillers, High yield, Resistant to leaf blast
IC 346953 (Dharaur)	High yield, Resistant to brown planthopper, Moderately resistant to planthoppers
IC 347619 (Bade safri)	Long panicle with low spikelet sterility; Resistant to BPH, Moderately resistant to WBPH, PH
IC 347620 (Luchai)	Low spikelet sterility, High yield, Resistant to WBPH, LBL, Moderately resistant to PH
IC 347604 (Luchai)	High yield, Resistant to LBL
IC 34632 (Chandan bhans)	Good panicle exsertion, Resistant to WBPH, Moderately resistant to PH
IC 347635 (Luchai)	Very low spikelet sterility, Resistant to LBL
IC 347646 (Jona phool)	Low spikelet sterility, Long panicle, Resistant to LBL, SHR
IC 310256 (Gaurani dhan)	Semi tall, Mid duration, Resistant to PH, GM, LF, Moderately resistant to SB

Plate 3.3. Paddy and rice samples of promising short grain aromatic rices under AP cess fund programme

Plate 3.4. Promising accessions identified in Bastar II under NATP on Plant Biodiversity

All the 663 rice germplasm accessions have been deposited with NBPGR, New Delhi and IC Number has been assigned as shown in Table 3.23. The entire digitalized passport and evaluation data was submitted to NBPGR, New Delhi and efforts are being made at DRR to bring out a catalogue with easy retrieval mechanism with coloured plates.

Table 3.23. Details of submission of seed and allotment of IC Number by NBPGR under NATP on Plant Biodiversity.

Time of submission	Expedition	No. of accessions	IC Numbers
February, 2000	Bastar I	123	264135-257
April, 2001	Adilabad	165	310251-415
January, 2002	Bastar II	149	346808-956
	Raigarh	61	347596-656
December, 2002	Nizamabad	38	375765-802
February, 2004	Khammam	70	426219-288
February, 2004	Medak	57	426162-218

Thus, the efforts made under this mission mode project has led to the collection of rice germplasm from unexplored tribal areas. Evaluation resulted in identification of unique landraces possessing high variation for key agro-morphological traits and also showing resistance/tolerance to major pests/diseases. The information would be catalogued and the identified germplasm could be good donor source for geneticists and breeders in pre-breeding and varietal improvement activities.

Evaluation of rice cultivars for rainfed production system (NATP RRPS-10)

In order to identify genotypes suitable for rainfed uplands, lowlands and medium lands in the states of Chhattisgarh, Jharkhand and Orissa, promising cultures pooled from CRRI and DRR were evaluated under "On-farm adaptive research (OFAR)" in participatory mode with the farmers under rainfed uplands, lowlands and medium lands in Chhattisgarh, Jharkhand and Orissa. In addition 102 accessions of the germplasm collected from these centres were pooled and evaluated at Raipur, Jagdalpur and the promising entries on the basis of grain yield, silver shoot per cent and blast scale were : Laxmibhog-2, Sonachur-1, Jhingiparas-2, Dhaniyaphool, Bhurkund, Kusumphul, Laxmibhog-1, Kheersai, Badshabhog and Jawaphool (Table 3.24). 126 scented and non-scented landraces from Raigarh region were collected and evaluated at Regional Agricultural

Research Station, Boirdadar. The top yielders with important morphological traits were Dayadhan (19.9 g), Annapurna (18.85 g), Sariya (20.8 g) and Bhursidhan (18.41 g). In other words the collected germplasm possess greater variability for maturity, morphology, seed size and yield (Subba Rao and Shobha Rani, 2003).

Table 3.24. Yield and yield components of promising germplasm accessions evaluated under NATP RRPS 10, 1999-2003.

Accession No.	Entry name	Plant ht.(cm)	Days to 50% flowering	Panicles per plant	Panicle length (cm)	Yield (kg ha^{-1})	SS (%)	Blast (0-9)
AR 1062	Laxmibhog-2	153	111	14.0	28.2	4297	10.14	1
AR 1040	Sonachur-1	154	113	8.2	25.8	4200	12.40	1
AR 1060	Jhingiparas-2	161	102	11.4	27.6	4169	2.30	1
AR 1015	Dhaniyaphool	161	109	8.4	25.2	4125	4.80	1
AR 1105	Bhurkund	146	107	10.6	24.2	4107	3.58	1
AR 1046	Kusumphul	165	115	9.4	28.6	4096	1.08	1
AR 1052	Laxmibhog-1	153	101	10.4	26.0	4096	2.40	1
AR 1002	Kheersai	149	105	8.6	25.0	4083	11.23	1
AR 1067	Badshabhog	158	109	10.0	27.2	4050	4.54	1
AR 1090	Jawaphool	153	113	9.4	21.0	4003	5.38	1

Agro-morphological characterization of rice germplasm under NBPGR-EFC (2004-2007)

One of the mandates of NBPGR is to promote safe conservation of plant genetic resources on a long-term basis using ex *situ* methods and *in situ* on-farm conservation of landrace diversity and in this regard it is a repository of about 70,000 rice germplasm accessions and much of it is yet to be evaluated and characterized for utilisation by the rice breeders. In order to augment this process NBPGR initiated a project entitled "Multi-location evaluation of rice germplasm" in four crops and rice is one among them. Since the multi-location testing facility is in-built in the structure, the task of coordinating this activity was given to DRR under which 1000 accessions are multiplied every year and another 1000 accessions are evaluated since the commencement of the project in December 2003 (Plate 5&6). The Project has the following objectives:

(i) to study agro-morphological characters of improved and acquired germplasm and to identify the promising lines for different traits, (ii) to evaluate for major biotic stresses at hot spot locations and to identify new resistant sources,

Plate 3.5. Agro-morphological evaluation of rice germplasm under NBPGR-EFC Programme at DRR, Hyderabad

Plate 3.6. Agro-morphological evaluation of rice germplasm under NBPGR-EFC Programme at Titabar, Assam

and (iii) to utilize the promising accessions in various breeding programmes.

Agro-morphological characterization was undertaken at 7 centres and the accessions were screened at 13 hot spot locations for plant diseases and 10 hot spot locations for insect pests. A total of 3502 germplasm accessions have been evaluated for 12 quantitative characters and 22 qualitatative traits, since the commencement of the project in 2004 (Annual Report, 2004-2006). Good variability was observed among the accessions for the agro-morphological characters (Fig. 3.7). The range of variability encountered for various agronomic and yield components and promising accessions for important yield contributing traits are presented in Tables 3.25 and 3.26 (Annual Report, 2004-2006).

Table 3.25. Range of agronomic and yield components of germplasm accessions under NBPGR-EFC, 2004-2007.

Range Character	Days to 50% flower	Leaf length (cm)	Leaf width (cm)	Plant height (cm)	No. of tillers/ plant	Effective tillers/ Plant	Panicle length (cm)	No. of grains/ panicle	No. of filled grains/ panicle	100 grain wt. (g)	Grain yield/ pt. (g)
Minimum	66	13.2	0.4	62.1	3.0	2.0	10.0	56	11	0.93	2.61
Maximum	126	88	2.9	160.4	27.0	25.0	29.5	250	165	3.70	29.46
Range	60	74.8	2.5	98.3	24	23	19.5	194	154	2.77	26.85

Table 3.26. Promising accessions with high yield and yield components under NBPGR-EFC, 2004-2007.

Character	Promising accessions
Very Early to Early duration	IC Nos 114016, 114061, 114064, 114035, 114036, 114439, 115166, 115315, 114107, 115112, 115128, 115132, 115579, 116051,115520, 115449, 115455, 115514, 115519, 115532, 115429, 115370, 115426.
Medium to late duration	IC Nos 114239, 114237, 114732, 114740, 114889, 114169, 114262, 114429, 114703, 114432, 114815, 114874, 252121, 311857, 330790, 332896, 334179, 334222, 337151, 347597, 350289, 342876, 252121.
Effective tillers	IC Nos 114327 (25.0), 114338 (25.0), 114421 (25.0), 114426 (25.0), 114840 (23.8), 114852 (23.2), 114851 (23.0), 114834 (22.6), 114835 (22.2), 114334 (21.0), 114427 (21.0), 114966 (21.0), 320825 (15), 353014 (15).
Panicle Length (cm)	IC Nos. 114495 (29.5), 114058 (29.32), 114026 (29.14), 115921 (29.00), 115917 (29.00), 311850 (28.9), 114515 (28.80), 114937

	(28.60), 114118 (28.28), 115574 (28), 114402-1 (27), 115558 (26), 115544 (26), 115536 (26)
Grain Yield (gm)	IC Nos. 319435 (29.46), 346932 (28.90), 252040 (27.50), 331770 (27.50), 115203 (26.00), 115795 (26.00), 115079 (25.00), 115560 (25.00), 336913 (25.00), 346808 (25.00), 114468 (24.50), 115007 (24.50), 115279 (24.50).

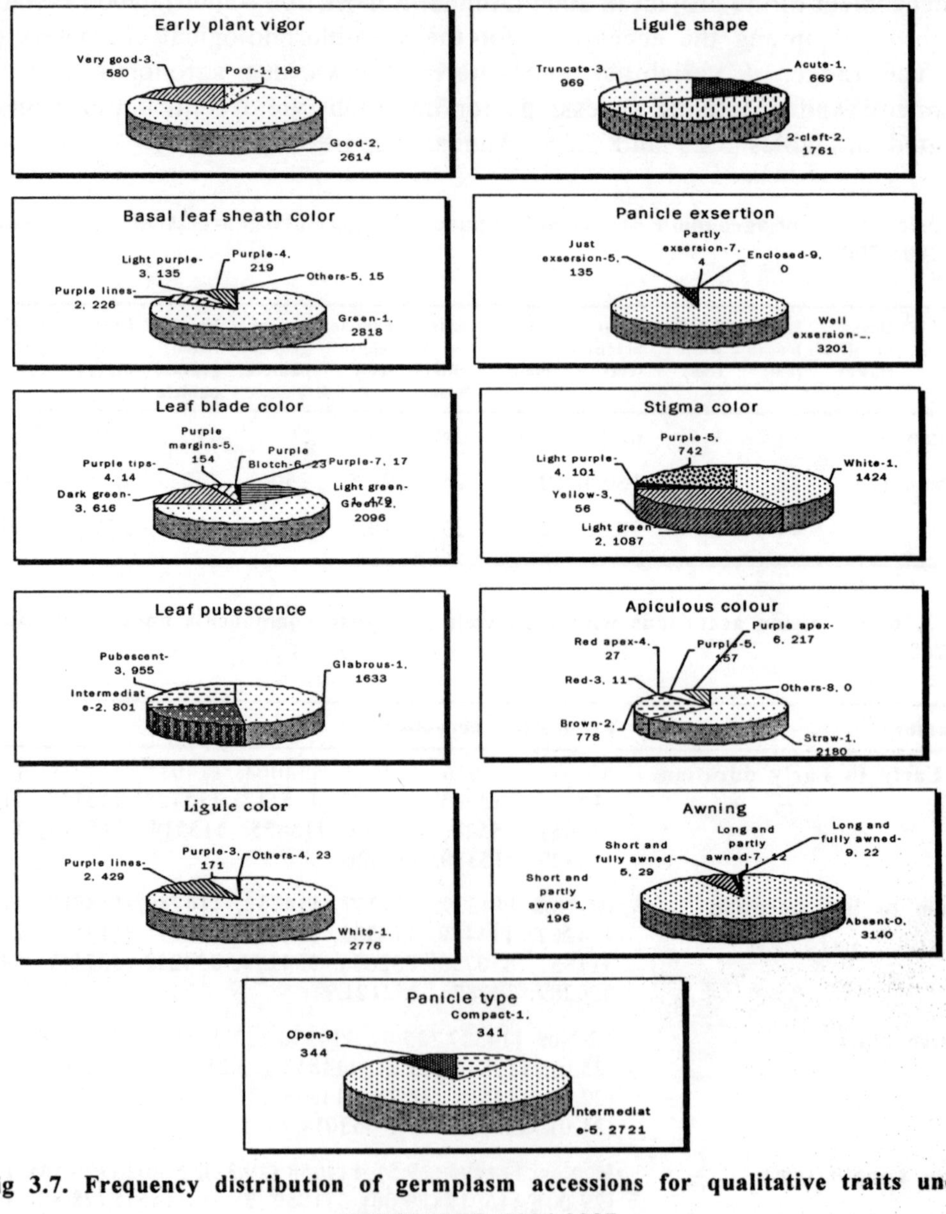

Fig 3.7. Frequency distribution of germplasm accessions for qualitative traits under NBPGR-EFC, 2004-2007

Apart from agro-morphological characterization, the accessions were screened for various plant diseases like leaf blast, brown spot, bacterial leaf blight, sheath blight, bacterial leaf streak and rice tungro disease at DRR, Coimbatore, CRRI, Hazaribagh, Ludhiana, Malan, Maruteru, Nellore, Pantnagar, Ponnampet, Raipur, Tirur, Titabar; and for various insect pests like stem borer, plant hoppers, leaf folder and gall midge at DRR, Maruteru, Mandya, Karjat, Ludhiana, Sambalpur, Jagtial, Ragolu, Moncompu and Raipur. Promising accessions identified against various plant diseases and insect pests along with those accessions which have multiple resistance are presented in Table 3.27.

Table 3.27. Promising accessions for pest/disease resistance under NBPGR- EFC, 2004-2007.

Diseases /Insect Pests	IC Nos		
Leaf blast	IC Nos: 15032, 113991, 114725, 115474, 114313, 114335, 114289, 114294, 351689, 337557, 252073, 352861, 114995, 332885, 352896, 114838, 330421		
Bacterial leaf blight	IC Nos : 114340, 252142, 114335, 114546, 114653, 252043, 115734, 115738, 356428, 252128, 252083, 252087, 311124, 311176, 332914, 332987		
Sheath blight	IC Nos: 115565, 252079, 330201, 336375, 252026, 252229, 115970, 311851, 333026, 115986, 337151, 347669, 346932		
Brown spot	IC Nos: 114315, 114780, 252028, 114970, 114972, 115099, 115330, 115310, 115311, 115349, 311128, 114055, 114087, 115851, 311176, 346866,		
Rice tungro	IC Nos: 114231, 114880, 114180, 114306, 114787, 336081, 334377		
Insect Pests			
Planthoppers	IC Nos: 114455-1, 114458, 114786, 114835, 252132, 114520, 115117, 115431, 116065, 252194, 252229, 319435, 330695, 330796		
Stem borer	IC Nos: 114040, 114733, 115065, 115632, 115633, 115921, 252180, 252182, 320859, 114267, 332938, 332972, 334161, 349679, 353013		
Gall midge	114694, 114695, 114799, 115007, 115076, 115210, 115220, 115232, 115234, 115241, 115272, 115293, 115530		
Leaf folder	114189, 114614, 114978, 115164, 115737, 115873, 331654, 332014, 332015, 332774, 332875, 350254		
Promising germplasm accessions with multiple resistance			
IC 114087	LBL, BS	IC 115550	BPH, GM, SB
IC 114273	LBL, BLS	IC 115565	ShR, BLS
IC 114335	LBL, BLB	IC 311176	BLB, BS

IC 114507	LBL, BLB	IC 316276	LBL,RTV
IC 114653	BLB, RTV	IC 321504	BPH, WBPH, PH, SB,
IC 114787	BLB, RTV LF,	IC 331779	BPH, WBPH, GM, SB,
IC 114936	BPH, GM, SB	IC 333010	LBL, RTV
IC 115026	BPH, GM, SB	IC 334132	ShR, RTV
IC 115211	BPH, GM, SB	IC 337578	BPH, SB, LF
IC 115503	GM, SB, LF,	IC 352896	LBL,BLS

Genetic Diversity Using Molecular Markers

In a group of 30 aromatic varieties, Sujatha *et al.,* (2006) reported 2 major clusters, the short grain rices and long grain rices clustering separately. Genetic divergence study in 29 aromatic rices facilitated clustering of genotypes into seven, five and four clusters by using observations from physicochemical, ISSR and SSR markers respectively (Richa Upadhay, 2004). Phylogenetic relationship among wild Oryza species based on 159 pooled Inter Simple Sequence Repeat (ISSR) and Simple Sequence Repeat (SSR) markers grouped the species into 2 clusters; one consisting of the A genome species and the other consisting of B, C, BC, CD and HJ genome species. *O. australiensis* (E), *O. branchyantha* (F) and *Porteresia* (HK) remained separate (Sujatha *et al.,* 2006).

Genetic diversity among 86 Indian landraces including wild species and varieties of rice clearly delineated wild accessions, *japonica* varieties, improved *indica* varieties, hybrids and landraces. Wild accessions of *O. nivara* and *O. rufipogon* were together distinct from the landraces than from varieties while the landraces were distinct from the cultivated varieties.

Of 25 accessions of *O. rufipogon*, 5 accessions each of *O. nivara* and *O. sativa* studied, *O. rufipogon* accessions formed 2 major groups one that clustered along with *O. sativa* and *O. nivara* and the other remained distinct. Two accessions of *O. rufipogon* from Bangladesh did not cluster with any other accessions. Two accessions from Thailand clustered loosely with the two major groups. *O. sativa* cultivars demonstrated a high degree of relatedness while among *O. rufipogon*, it varied widely. Similarity coefficients between *O. nivara* and *O. sativa* were relatively higher than between *O. rufipogon* and *O. sativa*. Of 24 accessions of *O. nivara* representing 11 states, two accessions each from Uttar Pradesh and Bihar are genetically distinct from those representing other states (Siddiq, 2003).

Epilogue

The plant genetic resources occupy a unique place within the overall purview of biological diversity and are of paramount importance in a crop improvement programme. Indian sub-continent is very rich in genetic diversity as it is one of the 12 mega centres of biodiversity, wherein more than 160 major crop species and 320 species of their wild forms and close relatives have originated (Rana, 1993). This constitutes an invaluable reservoir of genes that are needed by plant breeders for improving crop varieties and the exploitation of such identified genetic resources led to the development of high yielding varieties which saved the world from the scourge of hunger and contributed to the green revolution. Earlier it was thought that the biological diversity of our planet was inexhaustible. But a different scenario suggests that it is finite, yet a renewable resource; when wisely managed it can support the world's burgeoning population into the foreseeable future. But these resources are increasingly threatened on account of the continuing degradation of the habitat and also due to rapid replacement of locally adapted indigenous cultivars by modern high yielding varieties in all crops including rice. India is considered as a secondary centre for origin of rice with about 30,000-50,000 landraces. Hence, scientific management of the valuable rice genetic resources is of prime importance and conservation efforts gained remarkable momentum in 1960s and 70s. Establishment of Central Rice Research Institute (CRRI), helped in the collection activity as well in pooling of the national collections. Rice varieties from different experimental stations in India were collected and varieties from other countries were introduced. National Bureau of Plant Genetic Resources initially named as National Bureau of Plant Introduction was set up in 1976 with the mandate of introduction, collection, conservation, evaluation and safe exchange of plant genetic resources at national level with Exploration Base Centre at Cuttack within CRRI Campus; Regional Stations at Shillong and Thrissur and crop based National Active Germplasm Sites at CRRI and IGAU, Raipur which augmented a great deal in collection and conservation of rice genetic resources (Dhillon *et al.,* 2004). Most of the rice research stations under SAUs too collected and maintained their own local collections of rice germplasm with most of them depositing the same with NBPGR. With the awareness that the genetic resources need to be wisely managed, exhaustive collections were made, conserved, partly evaluated and utilized in the last 30 years. This activity was revisited under the National Agricultural Technology Programme (NATP) in the recent years to collect rice germplasm from the unexplored and under-explored areas. As the pioneers of PGR movement say, most of what is collectable has been collected and highest priority should now be given to the evaluation and utilization of the collections. Towards meeting such as an objective, the All India Coordinated Rice Improvement Project (AICRIP) of Directorate of

Rice Research, Hyderabad has accomplished it's mission by making laudable efforts.

The multi-disciplinary and the multi location testing process of AICRIP has become a model to the agricultural systems across the world, especially at IRRI, Philippines where it was replicated as International Rice Testing Programme (IRTP) now changed as International Network for Genetic Evaluation of Rice (INGER). This unique system has been the biggest strength of DRR in evaluating the breeding lines and genetic stocks. Initially starting with Assam Rice Collection, many other collections made elsewhere were screened under AICRIP to identify donors for biotic and abiotic stresses. A vast number of breeding lines that were developed by the cooperating breeders utilizing many of the donors identified were evaluated in the breeding and variety trials on multi-location basis and simultaneously in National Screening Nursery where screening is done at hot spot locations as well as under artificial glass house conditions. This well laid out procedure for testing has resulted in the release of 852 rice varieties as central and state releases for all the rice growing ecosystems with biotic and abiotic tolerance which is a testimony to the role played by AICRIP in varietal testing, identification and release of varieties. The analysis of obvious parentage of the released varieties revealed involvement of 155 donors for all ecosystems – irrigated, rainfed uplands, lowlands, semi-deep water, aromatic, hills and saline situations. Many of the varieties thus released have become extensively popular with the farmers and are in breeder seed production chain indicating their involvement in active commercial cultivation.

DRR made outstanding contribution in the evaluation of rice germplasm as it was one of the objectives in seven externally funded projects. A collection of scented/quality germplasm and aromatic short grain rices have been characterized for agro morphological, yield, pest/disease and quality parameters, under three AP Cess fund projects. A huge collection of 13,620 accessions were evaluated for pest and disease resistance and yield contributing traits and catalogued while another set of 3502 accessions were evaluated for 12 quantitative and 22 qualitative descriptors on multi-location basis both coordinated by DRR, the former as an AP Cess Fund and the latter as a part of NBPGR-EFC programme. These concerted efforts resulted in unraveling enormous genetic variability for various traits which will be utilized in the breeding programmes for widening the genetic base, in order to break the yield barrier as plateauing of yields is a matter of great concern. The NATP on Plant Biodiversity was a success story with all objectives of the programes accomplished. With the looming threat of depleting natural resources and the destabilising factors of climate change imposing severe limitations, increasing the production and productivity of rice has become once more a major challenge. Search for new rice genetic resources and innovative tools to harness the identified genes has become vary crucial against this backdrop. It is important that the evaluation of germplasm in identification of new genes not only for the set descriptors relating to morphology, yield, pest/disease and quality, but the screening should also encompass many of the abiotic stresses especially heat, submergence and cold and physiological parameters such as high photosynthetic efficiency, total dry matter accumulation, enhanced input responsiveness, high harvest index, low spikelet sterility, etc. Identifying donors for developing wide choice of parental lines to increase the frequency of heterotic hybrids and a host of such important attributes is equally important. A mission mode approach need to be put in place so that the identified gene sources are exchanged with the rice breeders for their utilization in pre-breeding activities under the approved Material Transfer Agreements (MTA). This would be abiding to the IPR requirements but at the same time ensuring the free flow of germplasm for research purpose. Efforts towards development of core collections which covers the entire genetic variability of the actual collection; digitalized database with appropriate security systems and catalogues prepared and made available to all by placing them in public domain; registration of elite breeding lines and genetic stocks which are unique and have potential attributes of academic, scientific or commercial value; classifying the accessions on the basis of DNA polymorphism to recognize the genetically diverse materials and also in development of molecular IDs for their documentation and protection are the other key issues which will be effectively addressed by DRR in the years to come.

References:

Annual Reports. 1995-2006. Directorate of Rice Research, Hyderabad.

Annual Reports. 2004-2006. Directorate of Rice Research, Hyderabad.

Anonymous. 1973. *India's Rice Revolution, The Role of AICRIP, a Beginning*. All India Coordinated Rice Improvement Project, Hyderabad. 76 pp.

Anonymous. 1980-1988. *Report of the Quinquennial Review Team*. Directorate of Rice Research, Hyderabad. 24 pp.

Anonymous. 1981-1985. *Quinquennial Report for 1981-85*. Directorate of Rice Research, Hyderabad. 359 pp.

Anonymous. 1988-1996. *Rice Research and Production in India. Present Status and a Future Perspective*. Quinquennial Review Committee, Directorate of Rice Research, Hyderabad. 216 pp.

Anonymous. 1996-2001. *Quinquennial Review Committee Report*. Directorate of Rice Research, Hyderabad, 97 pp.

Anonymous. 1990-1993, *ICAR Network programme on Improvement of Basmati Rices for Increased Productivity and Export Purposes.* AP Cess Fund Project, Final report 1990-1993. Directorate of Rice Research, Hyderabad. 61pp.

Bentur JS, Pasalu IC, Sarma NP, Prasada Rao U, Mishra B. 2003. *Gall Midge Resistance in Rice: Current Status in India and Future Strategies.* Directorate of Rice Research, Hyderabad. 20 pp.

Dhillon BS, Tyagi RK, Arjun Lal and Saxena S. 2004. *Plant Genetic Resources Management.* Narosa Publishing House, New Delhi. 433 pp.

Directorate of Rice Research. 1993-1998. *ICAR Research Network Programme on Evaluation of Rice Germplasm for Biotic Stresses. Final Report 1993-1998.* Directorate of Rice Research, Hyderabad. 341 pp.

Hakim KL. 1986. *Genetic Resources Activities in Rice.* Report submitted in partial fulfilment of the requirements. In: *Genetic Resources Conservation and Management Training Course, January 27, 1986.* IRRI, Los Banos, Philippines.

Kalode MB. 1980. The rice gall midge varietal resistance and chemical control in rice. Pages 173-194 in *Improvement in China and Other Asian Countries.* International Rice Research Institute, Philippines.

Progress Reports. 1966-1985. All India Coordinated Rice Improvement Project, (AICRIP), Hyderabad.

Progress Reports. 1986-2007. Directorate of Rice Research (DRR), Hyderabad.

Rana RS. 1993. India takes a lead in conservation of plant genetic resources. *Indian Farming* 43(7):37-45.

Research Highlights. 1983-1994. Directorate of Rice Research (DRR), Hyderabad.

Richa Upadhay. 2004. *Genetic Diversity of Aromatic Rices Based on Quality Traits and Molecular Markers.* M Sc Thesis, Allahabad Deemed University, Allahabad.

Sai Krishna T. 1977. Studies on genetic mechanism of resistance to rice brown planthopper, *Nilaparvata lugens* (Stal). Ph D Thesis, Osmania University, Hyderabad. 211 pp.

Seetharaman R, Sharma SD and Shastry SVS. 1972. Germplasn conservation and use in India. Pages 187-200 in *Rice Breeding.* International Rice Research Institute, Los Banos, Philippines.

Shastry SVS, Sharma SD, John VT and Krishnaiah K. 1971. New sources of resistance to pests and diseases in the Assam rice collections. *International Rice Commission Newsletter* 20(3):1-16.

Sheshu DV, Prasad Rao U, Srinivasan TE and Shastry SVS. 1984. New source of dwarfism in the Assam rice collections. *Indian J Genet* 34(3):390-394.

Shobha Rani N. 1989. Genetic basis of breeding for brown planthopper *Nilaparvata lugens* Stal resistance in rice *Oryza sativa* L. PhD Thesis, Osmania University, Hyderabad. 244 pp.

Shobha Rani N and Prasad GSV. 2000. *ICAR Network Programme on Genetic Enhancement of Quality Rices for Higher Productivity and Export.* AP Cess Fund Project) Final Report 1994-1999. Directorate of Rice Research, Hyderabad. 147 pp.

Shobha Rani N, Subba Rao LV, Mishra B, Prasad GSV, Pasalu IC, Muralidharan K, Rama Prasad AS and Prasada Rao U. 2004. *Sustainable Management of Plant Biodiversity*. Final Report on National Agricultural Technology Project on Mission Mode Rice. Directorate of Rice Research, Hyderabad. 60 pp.

Shobha Rani N. 2007. *ICAR Network Programme on Genetic Enhancement of Aromatic Short Grain Rices for Higher Productivity and Export.* Final Report 2002-2005, AP Cess Fund Project, Directorate of Rice Research, Hyderabad. 113 pp (unpublished).

Shobha Rani N, Prasad GSV, Subba Rao LV, Sudharshan I, Pandey Manish K, Babu VR, Ram T, Padmavathi G, Muralidharan K, Pasalu IC and Viraktamath BC. 2008. *High Yielding Rice Varieties of India.* Directorate of Rice Research, Hyderabad, India. 188 pp.

Siddiq EA. 2003. *Enhancement of Yield through Innovative Genetic Approaches.* Rice National Professor Project, Final Report (1997-2002). Indian Council of Agricultural Research, New Delhi.

Subba Rao LV and Shobha Rani N. 2003. *Evaluation of Rice Cultivars for Rainfed Production Systems.* Final Report on National Agricultural Technology Project on Rainfed Rice Production Systems-10. Directorate of Rice Research, Hyderabad. 54 pp (unpublished).

Sujatha K, Upadhay R, Kaladhar K, Shobha Rani N and Sarala N. 2006. Genetic relationship among aromatic short grain and basmati rice based on ISSR and SSR markers. *Rice Genetic Newsletter* Vol. 21.

Sujatha K, Rai VP, Kaladhar K, Ram T and Sarala N. 2006. Selected ISSR and SSR markers reveal phylogenetic relationship among wild Oryza species. *Rice Genetic Newsletter* Vol. 21.

Viraktamath BC 2008. Current status and future prospects of hybrid rice technology in India. Pages 49-65 in *Training Programme Manual on Hybrid Rice Production Technology.* 18-25th November, 2008, Directorate of Rice Research, Hyderabad.

Genetic Resources of Rice in India : 135-150
Editor : S. D. Sharma
Today & Tomorrow's Printers and Publishers, New Delhi - 110 002 (India)

Chapter 4

Role of Indian Agricultural Research Institute in Collection, Acquisition, Evaluation, Enhancement, Utilization and Conservation of Rice Germplasm

V.P. Singh and A.K. Singh

The Indian Agricultural Research Institute (IARI) is the premier national Institute for agricultural research, education and extension. It was first established in the year 1905 in Darbhanga District (Bihar) and was named as Imperial Agricultural Research Institute. In 1947, its name was changed to Indian Agricultural Research Institute. Two persons played an important role in the establishment of this pioneer institute. The first one was Lord Curzon, the then Viceroy of India, who realized the importance of creating a separate research institute solely devoted to development of agriculture in this country and foresaw its possible impact. The other one was Mr. Henry Phipps, an American philanthropist, who made a munificent contribution for construction and development of this Institute. The location of the Institute got its own name (Pusa) there and carries this name even today in the names of varieties of different crops bred by the Institute.

Due to a devastating earthquake at Pusa in January 1934, the Institute was shifted to its current location in New Delhi. His Highness Marquis of Linlithgow, the then Viceroy of India, inaugurated the Institute's library building on November 7, 1936, which has become the landmark of the Institute. The total area of the institute is 1250 acres including laboratory, research farm and the residential area for staff and the students. In 1958, the IARI was declared a deemed university for conducting M. Sc. and Ph. D. courses. At present, the institute offers M. Sc. and Ph. D degrees in 24 disciplines of agriculture.

The mandate of the Institute is to conduct basic and strategic research leading to crop improvement and sustained agriculture productivity in harmony with environment, to provide national leadership in agricultural research, extension and technology assessment and transfer by developing new concepts and approaches and serving as a national referral point for quality and standards.

The research mandate of the institute cuts across the five schools, namely, Crop Improvement, Crop Protection, Basic Sciences, Resource Management and Social Sciences and hosts many multidisciplinary centres like Nuclear Research Laboratory, Water Technology Centre, a National Research Centre on Plant Biotechnology and headquarters of several All India Coordinated Research Projects with a strength of 500 scientists. The Institute has also established as many as 12 regional stations/centers located throughout the country to carry out the research and developmental activities specific to the regions.

IARI has served the cause of science and society with distinction through first-rate research, generation of appropriate technologies and development of human resources. In fact, the Green Revolution was born in the fields of IARI and our graduates constitute the core of the quality human resource in India's agricultural research and education system. The Institute has all along been adjusting and improving its policies, plans and programmes to effectively respond to the needs and opportunities of the nation. During the fifties, the advancement of scientific disciplines constituted the core programme and provided the base for its fast expansion in the 1960's and 1970's in all its three interactive areas, namely, research, education and extension. Besides, basic research, applied and commodity research gained great importance resulting in the development of several popular high yielding varieties of almost all major crops and their associated management technologies, which brought about an unprecedented increase in the national food and agricultural production.

Rice research at IARI was initiated as a post-graduate school programme by Dr. M.S. Swaminathan in 1958 and expanded to basic and strategic research in the areas of genetics, cytogenetics and breeding. Genetic variability is the basis of plant breeding. This variability can be created by collection of the local landraces and varieties being grown by farmers, acquiring already collected material from the national and international plant breeding institutions, through hybridization followed by selection and mutagenesis. In this chapter, an effort has been made to compile the role played by IARI in collection, acquisition, evaluation, enhancement, utilization and conservation of rice germplasm since the beginning of rice research programme.

Collection

In the late sixties of the 20th century, Dr. Motiramani who was posted in Assam brought to the notice of Dr. M.S. Swaminathan that *jhoom* cultivation[1] of rice practised in Assam will result in loosing the rice germplasm. In this context, Dr Swaminathan developed a project for collection of traditional varieties of rice from northeast India under a PL-480 grant by USDA. Exploration and collection of local varieties of rice from this region was done under the leadership of Dr. S.D. Sharma then posted at the IARI Regional Research Station, Hyderabad. The other scientists who assisted him in this programme were late Dr. R.K. Singh, Dr. J. Mohan Rao Velanki and Dr K.L. Hakim. Dr Sharma gave preference to states of Arunachal Pradesh, Nagaland, Manipur and Meghalaya as these states were unexplored till then with regard to rice germplasm. Mizoram could not be covered due to political disturbances at that time. Within the state of Assam, only four districts were selected: one hill district (Mikir and North Cachar Hills), one from upper (North Lakhimpur), one from central (Guwahati) and one from lower (Goalpara) Assam. The programme was for a period of five years from 1967-1972. A total of six thousand six hundred and thirty local rice varieties were collected.

Late Shri Phool Singh, one of our colleagues, collected a set of traditional basmati rice varieties from the villages (Raigarh, Mohbawala, etc.) near Dehradoon. All the collections represented either Basmati-370 or Type-3. Late Pandit Jai Bhagwan Sharma posted at IARI Regional Station, Karnal collected eight Basmati rice samples from farmers' fields of villages near Karnal in Haryana. Four were identical with Basmati-370 and the rest were similar looking but had longer grains than Basmati-370. They were also late by 15 days in total growth duration. One of them was named as Karnal Local. Later, this (Karnal Local) became a national check for quality in the national basmati trials replacing Basmati-370.

Dr. M.S. Swaminathan (then Director, IARI, New Delhi) brought Mahsuri, Jarabali, Manipuri and Randhunipagal. During the monitoring tours, Dangar, Patharia, Hinatisal, Gharbharan, Kapoorchini, Usha Barabanki and Anjani were collected. During November 2007, on an invitation from Sir Dorabji Tata Trust to evaluate SRI (system of rice intensification) method of cultivation, the senior author had an opportunity to interact with the farmers of the districts of Purulia (West Bengal), Nalanda and Gaya (Bihar). Farmers shared the seed of Dobeeja (having two kernels in one spikelet) and Lakhansal a local rice variety having very sturdy stem.

---

[1] In *jhoom* cultivation, the farmers clear the bush, cultivate rice in that location for a few seasons, then shift the location by clearing the bush in another location. This is also known as slash and burn method of cultivation. It is prevalent in tribal areas of many states in India.

Acquisition

It was again Dr. M.S. Swaminathan who wrote to Dr. C.R. Adair in USA for sending a set of aromatic rices. In response IARI received five aromatic lines of Iran origin. Based on our indents, a large number of accessions of rice germplasm representing aromatic rices, released varieties, breeding lines, wild rice species, *japonicas*, *javanicas*, bacterial leaf blight differentials, isogenic and pyramided lines were received from International Rice Research Institute (IRRI), Manila. Late Dr. R.P. Puri visited IRRI during 1969-70 in connection with the Genetic Evaluation and Utilization programme of that Institute and brought a large number of breeding lines with resistance to bacterial leaf blight. Due to limited resources, a few promising lines only were maintained and used in the hybridization programme. IARI always believed in free flow of germplasm. The material received from and given to various research stations is given in Tables 4.1, 4.2 and 4.3.

National and international evaluation trials of rice are a rich source of important rice germplasm. Among the entries of Basmati trials, IET 18415, an awnless, sturdy stem mutant of Pusa Basmati-1 with different aroma has been included in the list of working germplasm. From International Fine Grain Aromatic Observational Nursery, Heibao (China) and Shahpasand (Iran) were retained for completely dark purple caryopsis of the former and for very long, robust, profuse, colorless awns and a different alkali-spreading pattern of the latter.

Table 4.1. Number of aromatic rice germplasm received from IRRI.

Country of Origin	1976	1986	1990
Afghanistan	-	01	-
Bangladesh	02	09	09
China	-	11	01
India	-	23	47
Indonesia	24	12	27
Iran	-	02	17
Japan	-	01	-
Korea	-	03	02
Laos	-	03	02
Malaysia	-	-	17
Myanmar	-	03	05
Nepal	-	01	15
Pakistan	58	74	05

Philippines	-	13	-
Sri Lanka	-	-	03
Thailand	-	02	01
U.S.A.	-	01	-
Vietnam	-	03	01
Total	**84**	**162**	**152**

Table 4.2. Rice germplasm received and shared with national programmes.

Scientist / Institution	Year	No.	Remarks
Dr. J.P.S. Malik, S.R.O. (Rice Breeding) Dept. of Genetics & Plant Breeding G.B. Pant Univ. of Agri & Tech Pantnagar 263 145 (India)	1992	10	Received five cms lines along with their B lines.
Dr. D.V.S. Panwar, Sr. Rice Breeder Rice Research Station (HAU), Kaul Dist Kaithal, Haryana (India)	1990	300	Received Haryana Basmati Collection (HBC) series and rest from IRRI aromatic world rice collection
Dr. M.P. Pandey Rice Breeder Dept. of Plant Breeding G.B. Pant Univ of Agri & Tech Pantnagar 263 145	1990	16	Received induced semi-dwarf Tilak Chandan mutants and other aromatic traditional and improved lines.
The Project Director Directorate of Rice Research Rajendranagar, Hyderabad	1994	150	Shared selected aromatic rice germplasm.
Dr. R. G. Saini, Head Department of Genetics & Plant Breeding PAU, Ludhiana	1994	150	Shared selected aromatic rice germplasm.
Dr. B. Mishra, Head Division of Genetics & Plant Breeding, CSSRI, Karnal	1992	60	Shared short and fine grain aromatic germplasm.

Table 4.3. Rice lines shared with IRRI.

Designation	Parentage	Plant height (cm)	Growth duration (Days)	Milled rice kernel length (mm)	Aroma score*
Pusa 150-21-1	IR22 / Pusa 140	95 - 100	125 - 130	6.95	3
Pusa 150-9-3-1	IR22 / Pusa 140	95 - 100	135 - 140	7.80	3
Pusa 150-9-4-1	IR22 / Pusa 140	95 - 100	135 - 140	7.80	3
Pusa 167-120-3-2	Type 3 / Ratna	95 - 100	125 - 130	6.90	3
Pusa 367-4-2-1	Pusa 167-120-3-2/ Type 3	95 - 100	140 - 145	7.50	3
Pusa 367-4-2-2	Pusa 167-120-3-2/ Type 3	95 - 100	140 - 145	7.60	3
Pusa 388-4-4	Sona / Jarabali Dwarf Mutant	95 - 100	145 - 150	6.80	2
Pusa Basmati-1 (IET 10364)	Pusa 150 / Karnal Local	95 - 100	140 - 145	7.60	1
Pusa Basmati 1121 (IET 18004)	Pusa 614-1-2 / Pusa 614-2-4-3	100 - 110	145 - 150	8.30	3
IARI 7447	Pure line from Assam Rice Collection	175 - 180	150 - 155	5.40	3

* Aroma score: 1 = Mild, 2 = Medium, 3 = Strong

Evaluation

A representative set of Assam rice collections was screened for amylose content, gelatinization temperature and nutritive quality in terms of crude protein content and distribution pattern of protein bodies in the endosperm. Among the lines screened IARI 5754, IARI 5933, IARI 5908 and IARI 6191 have better distribution of protein bodies in the entire endosperm[2] (Kaul *et al.*, 1969).

The collection represented a rich variability for morphological traits. Thirty spontaneous dwarfs were identified from the entire collection. These dwarfs were studied for (a) classification based on the quantitative morphological traits and

[2] The accession numbers of Assam Rice Collections were assigned the prefix 'ARC' by the All India Coordinated Rice Improvement Project (AICRIP) but the same ARC accession numbers were assigned the prefix 'IARI' by the Indian Agricultural Research Institute (IARI). Though they bear different prefixes, the accession numbers were the same. — Editor

physiological response to growth promoters, (b) stability, and (c) mode of inheritance and allelic relationship for plant height. Classification of these spontaneous dwarfs on the basis of multivariate analysis on the one hand and relative response in terms of increase in height to giberellic acid (GA) treatment on the other grouped them into 8-10 clusters. Composition of the clusters observed in two analyses was entirely different. Presence of dwarf with high (IARI 5842, IARI 5825, IARI 5293) and low (IARI 5924, IARI 5753B, IARI 6612, IARI 5901-1) response to GA indicated that there could be different genetic pathways governing the dwarf stature in rice (Singh *et al.,* 1977; Singh *et al.,* 1978; Singh *et al.,* 1983).

The spontaneous dwarfs were crossed to a common tall line. The segregation pattern for height indicated three major groups: (a) single recessive gene allelic to Dee-geo-woo-gen (DGWG) with a large number of modifiers predominantly with negative effect, (b) single recessive gene probably non-allelic to DGWG in association with equal number of modifiers of positive and negative effects, and (c) single recessive gene non-allelic to DGWG in association with equal strength of positive and negative modifiers (Singh *et al.,* 1979). Combining ability analysis of these dwarfs showed that IARI 5842, IARI 5901-2, IARI 5923, IARI 7312B and IARI 10560 to be better general combiners compared to DGWG and other induced dwarfs. Based on this information IARI 5901-2 was crossed with IR 8 and resulted in the development and release of Pusa 44-33 (IET 5722) in 1994, which is still the most popular high yielding non-basmati variety with long slender grain and highly suitable for combine harvest. Pusa 44 has been extensively used in hybridization programme. Stability experiment indicated that IARI 5972 and IARI 6636 had wider adaptability with high mean grain yield. IARI 5824, IARI 5823, IARI 10536, IARI 10537 and IARI 10688 were found promising for developing parental lines for rice hybrids. Considerable variability for stigma size and glume opening period was also observed in the lines (Singh *et al.,* 1979a; Parmar *et al.,* 1979; Parmar *et al.,* 1979a; Parmar *et al.,* 1979b; Parmar *et al.,* 1980). Mahsuri, Jarabali, and Manipuri were irradiated and a number of mutants for height and other morphological characters were induced (Siddiq *et al.,* 1982).

The main mandate of rice breeding programme of IARI has been the development of high yielding and high quality Basmati rice with resistance to major diseases of rice in the Basmati region. All the collections obtained from IRRI and collected from different rice research centres were analyzed for their amylose content, alkali spreading value, gel consistency, linear cooked kernel elongation ratio, aroma, head rice recovery and milled rice kernel dimensions (Sood and Siddiq, 1980). Traditional Basmati rice varieties are characterized by long slender grain (milled rice length 6.6 mm and above) intermediate amylose, intermediate GT, medium gel consistency, pleasant aroma, longer shelf-life after cooking, high volume expansion and easy digestibility. Besides, they remain non-sticky and tender after cooking and have good mouth-feel and appeal. In brief aroma, taste and high volume made up by lengthwise elongation during cooking with minimum lateral swelling are major factors Basmati consumers look for. Rice must give a fine look before and after cooking (Singh, 2000).

All the traditional Basmati rice varieties are highly thermo-photo-sensitive. Due to very weak stem, they are highly prone to lodging in fertile soil even without fertilizers. Inheritance studies of cooking quality reveal that all these characteristics are governed by many genes and their modifiers. Once a cross is made with a non-Basmati good agronomic base, all the quality characters get distorted. Linear cooked kernel length elongation, kind and intensity of aroma, taste and grain dimensions suffer maximum distortion. In hybridization programme, Basmati-370, Type-3 and Karnal Local (which was released as Taraori Basmati by C.C.S. Haryana Agricultural University, Hissar in 1996) were extensively used. When vegetative parts of the hybrids of the crosses between Basmati and non-aromatic high-yielding varieties were analyzed for aroma, all were found to be non-aromatic. This suggested that aroma in rice is a recessive character. Single-grain aroma analysis of F_1 produce, show a segregation of 3:1 (3 non-aromatic to 1 aromatic). However, the aromatic grains vary in kind and intensity of aroma. Though 2-acetyl-1-pyrroline is the major compound, more than 100 chemical compounds are responsible for aroma in rice. Traditional rice varieties also differ in kind and intensity of aroma. Unlike other characters, aroma is more influenced by environmental factors. A temperature range of 25-32°C with moderate humidity, clear bright sunny days, soil fertility status and quality of water help more and better aroma development. Any biotic or non-biotic stress between flower initiation and complete grain development stage affects the grain yield, cooking and eating quality in various ways.

Enhancement

Genetic enhancement for key quality and agronomic characters, disease resistance and its maintenance is the basis of future rice breeding programme. Grain appearance is the major factor consumers look for in basmati rices. Among the popular traditional basmati rice varieties, Basmati-370 which was released in 1933 remained popular till mid-seventies with farmers, traders and rice breeders. In the domestic as well as international market, there is a preference for longer grain. This became evident when Karnal Local that has longer grains than Basmati-370 caught the attention of basmati rice exporters. Keeping in view the importance of longer grain, Pusa Basmati-1 was developed in 1989. The success of Pusa Basmati-1 due to its slightly longer grain than Karnal Local and better linear cooked kernel elongation further confirmed the need for longer grains in basmati rices.

Encouraged by the impact of Pusa Basmati-1, efforts were made to develop longer grains. Figure 4.1 depicts how, through step by step improvement, grain length has been enhanced. The major breakthrough was achieved with the development of Pusa 1121 which has a milled rice kernel length of 8.3 mm and cooked kernel elongation ratio of 2.5 (Singh *et al.*, 2002). Recently, it has been given the status of Basmati. IRBB-55 and IRBB-60 were crossed with elite enhanced Basmati lines. Pusa-1460, Pusa-1526 and Pusa-1527 series carrying two to four genes for bacterial blight resistance and having typical Basmati quality characteristics have been developed (Table 4.4). Grain length has been increased from 7.5 mm in Pusa Basmati-1 to 10.5 mm in Pusa-1302 and Pusa-1561 series (Figure 4.1). Very early maturing lines like Jaldi Dhan-6, Jaldi Dhan-8 and Jaldi Dhan-14 have been developed.

Table 4.4. Bacterial Leaf Blight resistant promising lines with better grain, cooking and eating quality.

Designation	Parentage	Gene Status			
		Xa4	*xa5*	*xa13*	*Xa21*
Pusa 1460-04-67-3	Pusa Basmati-1/IRBB 55	-	-	++	++
Pusa 1460-04-67-7	Pusa Basmati-1/IRBB 55	-	-	++	++
Pusa 1526-04-25-12	Pusa 1121-92-8-3-3/IRBB 60	++	++	++	++
Pusa 1526-04-25-69	Pusa 1121-92-8-3-3/IRBB 60	++	++	++	++
Pusa 1527-04-56-13	Pusa 2512/IRBB 60	++	++	++	++
Pusa 1527-04-56-40	Pusa 2512/IRBB 60	++	++	++	++

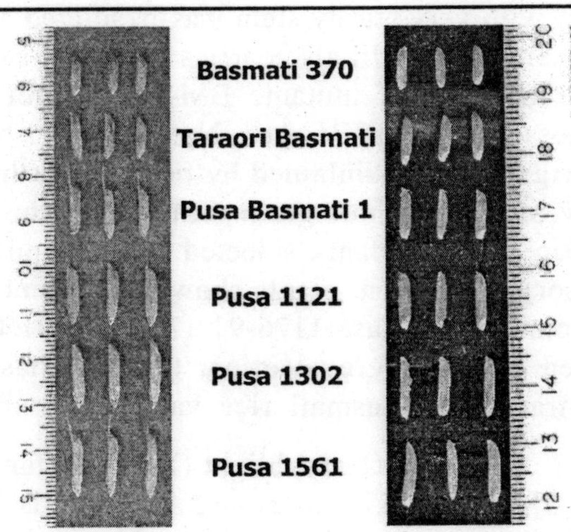

Fig. 4.1. Rough and brown rice length of enhanced aromatic rice.

Utilization

Based on their specific features, Pusa 2-21, Pusa 5, Pusa 33, Pusa 44, Pusa 140, Pusa 150, Pusa 167, Pusa 169, Pusa 245, Pusa 598, Pusa 705, Pusa 743, Pusa 1121, Pusa 1302, Improved Sabarmati, Pusa Basmati-1, Pusa Rice Hybrid-10, PNR 162, PNR 381, PNR 519, PNR 555-2 have been used in rice improvement programme (Table 4.5). PRH-10 has been used to develop doubled haploid to capitalize on conserving hybrid vigor along with quality. Pusa Basmati-1, the first semi-dwarf high yielding Basmati variety, has been very widely used for developing quality rice through hybridization, mutagenesis and secondary selections. It has been used in 86 cross combinations in the national programme. At IRRI, it has been used in 193 cross combinations in the name of IET 10364 and Pusa 615-140-10-1. Pusa 167-120-3-2 was identified as a perfect maintainer by the team led by Dr. S.S. Virmani in Hybrid Rice Project at IRRI. Through eight back crosses with IR 48483A, a new CMS line (IR 58025A) was developed. This CMS line has very high stability for sterility and wide adaptability. This is at present the female parent in majority of hybrids in India and abroad. Its B line has superfine grain with rare chalky spots on the caryopsis, pleasant aroma and good taste. The only demerit is the sticky nature of the raw rice. To overcome this problem, the commercial produce of the hybrids with IR 58025A is being used for parboiling (Table 4.6). Some farmers of Haryana and western Uttar Pradesh have realized a yield of 10 tonnes per hectare in a total duration of 110 days and sold at prices ranging from Rs 8,500-15,000 a tonne of paddy in recent years. Apart from that, a large number of traditional and improved rice lines/varieties have been used in the breeding programme (Table 4.5).

To create sturdy stem lines in the background of traditional aromatic rice, a local variety, Bindli, with very sturdy stem was irradiated and semi-dwarf sturdy stem mutants were developed. Bindli also carries genes for rice tungro virus (RTV) resistance. One of its semi-dwarf mutants BM-68 was found resistant to RTV. This mutant was crossed with IRRI Acc No 16136 (Assamia Basmati) – a collection of Nepal origin being maintained by IRRI. Both the parents are strongly aromatic, have sturdy stem and short grain. Since both the parents in this cross were strongly aromatic, all the plants selected in F_2 population were aromatic. However, during sensory evaluation, plants showed different kind and intensity of aroma. One highly aromatic line Pusa 1176-91-1-3-2-1-1 (IET 17582) was selected from this cross. Based on sensory evaluation, this line has been assessed to be more aromatic than traditional Basmati rice varieties.

With a view to develop bacterial blight (BB) resistance lines with Basmati quality, crosses with BJ-1, Malagkit Sungsong, Ajay, IRBB-55 and IRBB-60 were carried out. F_2 population was inoculated with a highly virulent BB strain from

Rice Research Station, Kaul, Haryana. Promising lines based on their resistance reaction were selected. They were further evaluated in every generation for quality characteristics and resistance to BB until these lines became uniform for plant height, flowering duration, grain dimensions and quality characters.

Three CMS lines Er-Chiu-Nan 1A, Nan-You 6A and Zhen-Shan 97A received from IRRI were evaluated for their stability at New Delhi. These lines were very early, highly susceptible to diseases and had very poor grain and cooking quality. Utilising them in the hybridization programme, new CMS lines Pusa 33A, Pusa 3A, Pusa 4A, Pusa 5A and Pusa 6A have been developed. Among them, Pusa 4A, Pusa 5A and Pusa 6A were found to be stable. Utilising these lines, a number of hybrid combinations were developed. Pusa Rice Hybrid-10, a combination of Pusa 6A/PRR 78 was released for commercial cultivation in 2001 (Zaman *et al.,* 1995; Abraham *et al.,* 1995). In combining ability experiments, Gharabharan was identified to carry wide compatibility genes. It was further used in the hybridization programme and in developing promising lines for high fertility and moisture stress.

Table 4.5. Donors used for different characters.

Parameters	Varieties
Short Stature	T(N) 1, IR 8, IR661, IR844, Jarabali mutant, Bindli mutant 68
High tiller number	Ratna, Mahsuri mutant, IR 50, Pusa 1401, Pusa 1173
Stiff stem	Peta, Improved Sabarmati, H 104, IR 26, NP 49, Pusa 44, Pusa 317, Pusa 318, Pusa 360-7-2, Pusa 456, Seong Pelan Dok, MR 219, MR 220, Pusa 1460
Leaf orientation	Pusa 44, Pusa 1460, Pusa 1401
Dark green leaf color	Sakotira-55 (*japonica*), IARI 5901-2, 81B-25
Stem angle	Pusa 205, Pusa 426
Longer grain length	Pusa 1121, Pusa 1302, Pusa 1484, 81B-25
Attractive grain shape	Belle Patna, NP 130, Sona, Karnal Local, Dumsiah, Pusa 147, SL0-16, Pusa 167-120-3-2, CP 231, Pusa145, Pusa 371, Pusa 506, Pusa 388-4-2
High grain density	Type 9, IR 5929, IR 36, Pusa 510
High grain number	IR 127, Mahsuri, NPT 11, Pusa 2511
Protein content	Chang-Chang, Santo, Omrit-39, IR 480-5-9, Improved Sabarmati,

Alkali spreading pattern	Shahpasand, Pusa 1484, Pusa 1176
Non-chalky grains	Pusa 1401, Type 3, Basmati 370, Type 9, Shahpasand.
Aroma	Basmati 370, Improved Sabarmati, Bindli, Karnal Local, Dumsiah, NP 49, N.P. 114, Randhunipagal, Pusa 317, Pusa 318, Khalsa-7, Pusa 1176, IRRI Acc No 16136
Linear cooked kernel elongation	Pusa1301, Karnal Local, Pusa 367-4-2-1, Pusa 751-87-7-1, Pusa 614-1-2, Pusa 614-2-4-3, Pusa Basmati-1, Pusa 1121, Pusa 1401
Bacterial leaf blight	IR 68-469, BJ-1, Neera, Malagkit-Sung-Song, IR 2071-625, IR 2061-464, Pusa 37, IR 20, IET 4141, Ajay, IR 72, IR 64, IRBB-55, IRBB-60
Rice blast	Tadukan, Tetep, IR 64
Stem borer	TKM-6, IR 20
Brown planthopper	IR 64, Ptb-33
Plant type	IARI 10560, IARI 5901-2,Pusa 167-120-3-2,Pusa 150-21-1, Pusa 44, Pusa 1460
Grain length	AC 5636, 81-B-25, Pusa 140, Pusa 150-9-4, Pusa 751-87-7-1, Pusa1121, Pusa 1302, Pusa 1484
Panicle type	BP 1-76
Very early duration	Jaldi Dhan-6, Sattari, AD 7711 and Teen-pakhia
Early duration high yield	Pusa 2-21, Pusa 33, IR 28, IR 22, AD 7711, IR 30, IR 50 and IR 64
Direct seeding and moisture stress	Gharabharan and Nagina 22
Early seedling vigour	Pusa 2511 and Sathi (maturing in 60 days and panicles remain enclosed in leaf sheath)
Allelic relationship for plant height	Manipuri mutant, Mahsuri mutant, Kalimonch mutant, BM-68 and KLM-2
Bright husk colour or lustre on husk	Pusa 1176, IARI Acc No 16136, Bindli
High head rice recovery	IRRI Acc. No. 27797, Type-3
Taste	Pusa Basmati 1121, Pusa Basmati 1, Karnal Local
Brown rice colour	Basmati 370, Karnal Local, Type 3, Shahpasand, Gharabharan
Glabrous husk	Dawn, Della, IARI 6579
High biomass	Bhawani, Type-9, H 104, Bindli, Sitabhog, Pusa 743, Hawm Setti, AR 102, IRRI Acc. no. 16136
High spikelet fertility	Tainan 3, Taipei 309, Siong Pelan Dok

Intermediate amylose and gelatinization temperature (GT)	IR 64, Pusa 1484, IR 72, Basmati 370, Type 3, Pusa 1302
Intermediate amylose and low GT	Pusa Basmati-1, Pusa 1121, Pusa 1401, Pusa 1301
Very short grain & high grain number	Sonasal, Na-ma-tha-lay, Jarabali

Table 4.6. IARI rice varieties used as a parent in breeding programme.

Varieties	Cross Combinations	Mutants Induced
Pusa 2-21	05	-
Pusa 33	21	03
Improved Sabarmati	02	-
Pusa 44	18	-
Pusa Basmati-1	86+193[@&]	26
Pusa Basmati-1	02*	-
Pusa 167	1+1[@&]	-
Pusa 1121	04	-
Pusa 140	01	-
Pusa 150	04	-
Pusa 245	01	-
Pusa 743	03	-
Pusa 705	01	-
Pusa 169	02	-
Pusa 1302	02	-
Pusa 598	02	-
Pusa Rice Hybrid-10	01**	-
Pusa 5	01	-
PNR-162	01	-
PNR-519	01	-
PNR-381	03	-

* Secondary selections made

** Double-haploid developed

[@&] Used at IRRI

Conservation

As part of the genetic stocks, three CMS and seven enhanced lines have been registered with NBPGR (Table 4.7). Pusa 33, Pusa 2-21, Improved Sabarmati, Pusa 1173, Pusa 1174, Pusa 1176 and a set of induced mutants of Karnal Local and Bindli have been deposited with NBPGR. Semi-dwarf mutants of Bindli have sturdy stem and as strong aroma as their parent. Similarly some of Karnal Local mutants are twenty-day early in flowering (compared to their parent) with as good aroma and cooking quality as their parent (Singh *et al.*, 1989). Apart from these, around two hundred enhanced lines are being maintained as working germplasm.

Future Scope and Plan

Mankind will continue to depend on rice as a staple food. It can be successfully grown in a wide range of environments. Nutritionally, it is superior to other cereals. Rice bran has unique nutritional properties. Rice varieties differ in bran colour and their chemical profile. With modern techniques of biological sciences, now it is possible to improve the nutritional value of rice. Transfer of genes responsible for the production of beta-carotene, which is a precursor of Vitamin A in the popular rice varieties, has reached an advanced stage. Similarly, iron-rich rice varieties are being developed. Though protein content of rice is highly determined by the environmental conditions, there is scope of seriously working on developing high-protein rice varieties with uniform distribution in the endosperm. This will minimize the protein loss during the process of milling.

Biotic and abiotic stresses are all the more important factors in rice breeding. In the *kharif* season of 2008, rice blast and brown planthopper created history by damaging rice crop in farmers' fields up to 100%. Work is already in progress for incorporating gene(s) for resistance to rice blast and bacterial leaf blight. Isogenic lines are being developed in the background of Pusa Basmati 1 and Pusa Basmati 1121. The work on pyramiding of genes for BB and blast in the popular and upcoming rice lines is a priority at IARI, retaining all the Basmati grain and cooking quality characteristics.

Impending shortage of manpower and water is another challenge. Allele mining for genes for moisture stress using Nagina 22 is methodically in progress. Direct seeding experiments for Basmati cultivation have been laid out in Punjab, Haryana and western Uttar Pradesh. Dry-direct seeding and under-puddle conditions both are under evaluation. The produce of such experiments will be evaluated for key characteristics such as milling, processing, grain characteristics, cooking quality and eating quality. A large number of enhanced germplasm for plant type, sturdy stem, active leaves and roots, different growth durations, fertilizer inputs, shallow water with different grain type and high brown and head rice

Table 4.7. Rice germplasm registered in National Bureau of Plant Genetic Resources, New-Delhi.

NBPGR IC No.	Name of Variety	INGR No.	Year	Parentage	Specific feature
IC296757	PNR 519 (IET 9944)	99035	1999	Tainan 3 Mutant/Basmati 370//PNR 417-3	Dwarf, high yielding, fine grain with wider adaptability to irrigated as well ass rainfed conditions
IC296763 (IC471849)	Pusa 3A & 3B	00004	2000	(IR58025-A x Pusa Basmati-1) x Pusa Basmati-1	CMS with basmati quality
IC296764 (IC471850)	Pusa 5A & 5B	00005	2000	IR58025A/Pusa 150-21-1-1-2	CMS in different genetic back ground
IC296800 (IC471851)	Pusa 4A & 4B	01021	2001	IR 58025A/IET 10649 (Selection)	CMS line basmati type in different genetic back ground
IC296825	Jaldi Dhan 8 (IET 11405)	01046	2001	Dular x Nagina 22 (PNR 550-1-2)	Extra early (80 days)
IC296890	Pusa 1121 (IET 18004)	02001	2002	Pusa 614-1-2/ Pusa 614-2-4-3	Highest cooked kernel elongation
IC522199	Pusa 1460 (IET 18990)	05002	2005	IRBB 55/Pusa Basmati-1//Pusa Basmati-1	Resistance to bacterial blight with basmati quality
IC548399	Pusa 1401 (IET 18005)	07022	2007	Pusa Basmati-1/ Pusa 1121-92-8-2-7-1	Exceptionally good cooking quality
IC548340	BM 68	07023	2007	An induced mutant of Bindli	Resistance to rice tungro virus
IC546941	VSR 8 *	06002	2006	IR 64/Pusa 1121-92-8-2-13	Resistance to rice blast

* In collaboration with Vivekananda Pravatiya Krishi Anushandhan Sansthan, Almora

recovery will be developed.

References:

Abraham MJ, Zaman FU, Natarajan US, Anju Mahendru, Mohammed F. 1998. Developing Pusa 5A, a stable *indica* CMS line with outcrossing potential. *IRRN* 23(2):15.

Kaul AK, Dhar RD and Swaminathan MS. 1969. Microscopic screening of rice grains for protein characteristics. *Curr Sci* 38(22):529-531.

Parmar KS, Siddiq EA, Swaminathan MS. 1979. Chemical induction of male sterility in rice. *Indian J Genet* 39(3):529-541.

Parmar KS, Siddiq EA, Swaminathan MS. 1979a. Variation in anther and stigma characters in rice. *Indian J Genet* 39(3):551-559.

Parmar KS, Siddiq EA, Swaminathan MS. 1979b. Variation in components of flowering behaviour of rice. *Indian J Genet* 39(3):542-550.

Parmar KS, Siddiq EA, Swaminathan MS. 1980. Variation in reproductive organs with reference to male incompatibility index. *Indian J Genet* 40(1):262-271.

Singh VP, Swaminathan MS, Siddiq EA. 1979. Mode of inheritance of dwarf stature and allelic relationships among various spontaneous and induced dwarf of cultivated rice (*Oryza sativa*). *Theor Appl Genet* 55:169-176.

Singh VP, Siddiq EA, Mehra RB, Swaminathan MS. 1979a. Divergence among dwarfs of cultivated rice. *Indian J Genet* 39(2):316-323.

Sood BC, Siddiq EA. 1980. Studies on component quality attributes of basmati rice, *Oryza sativa* L. *Z Pflanzenzuchtg* 84:294-301.

Siddiq EA, Sadananda AR, Singh VP, Zaman FU. 1982. Genetics of some of induced and spontaneous dwarfs of rice and their utilization in cross breeding. Pages 197-208 in *IAEA Technical Document – 307.*

Singh VP, Siddiq EA, Swaminathan MS. 1983. Response of spontaneous and induced dwarfs of rice to gibberellic acid. *Indian J Genet* 43(3):350-355.

Singh VP, Zaman FU, Sadananda AR. 1989. Induced variations in aromatic rice cultivars. *Intern Rice Res Newsletter* 14(3):3:14.

Singh VP Singh AK, Atwal SS, Joseph M, Mohapatra T. 2002. Pusa 1121: a rice line with exceptionally high cooked kernel elongation and Basmati quality. *IRRN* 27(1):25-26.

Singh VP. (2000). *Basmati Rice of India.* Pages 135-154 in Singh RK, *et al.* (eds) *Aromatic Rices.* Oxford and IBH Publishing Co, New Delhi.

Zaman FU, Abraham MJ, Sadananda AR, Faiz Mohammed, Yog Raj. 1995. Identifying restorers and maintainers for five cytoplasmic male sterile lines. *IRRN* 20:(3)6.

Genetic Resources of Rice in India : 151-184
Editor : S. D. Sharma
Today & Tomorrow's Printers and Publishers, New Delhi - 110 002 (India)

Chapter 5

Management of Rice Genetic Resources: Role of National Bureau of Plant Genetic Resources

Sanjeev Saxena, S.K. Sharma, Kalyani Srinivasan, D.K. Hore, J.C. Rana, D.R. Pani, N. Dikshit, and A.D. Sharma

The National Bureau of Plant Genetic Resources had its humble beginning as a section within the Division of Botany (later named as Division of Genetics) of the Indian Agricultural Research Institute, New Delhi. In 1961, it was elevated to a separate division named as Division of Plant Introduction under the leadership of Dr. Harbhajan Singh. This Division launched several expeditions for germplasm collection in collaboration of scientists from various countries. This also institutionalised the mechanism for assembly of global germplasm in India through a single-window system. At the same time, the different institutions and universities spread over the country maintained the genetic resources in crops as per their mandate. However, during these years, PGR management was mainly confined to collection of landraces in important crop species from different parts of the country, their classification, developing varieties through pure line selection and studying inheritance of important traits for identification of appropriate breeding strategies for gene introgression (Dhillon *et al.*, 2006).

The National Bureau of Plant Genetic Resources (NBPGR) was established with a mandate to cater to the PGR in all agricultural and horticultural crops for the country in a network mode. The network comprises 10 Regional Stations under its direct administrative control and active collaboration with other ICAR institutes, Project Directorates, National Research Centres (NRCs), All India Co-ordinated Research Project (AICRP), State Agricultural Universities (SAUs) and other stakeholders, such as NGOs, professional societies and International Agricultural Research Centres. The regional stations established in different phytogeographical regions of the country are mandated to undertake the activities of PGR collection, characterisation and preliminary evaluation of diverse germplasm of various agricultural and horticultural crops for the region. Thus, NBPGR Regional Stations at Shimla and Srinagar cater to the temperate and cold arid region; Bhowali for the sub-temperate region; Jodhpur for the arid region; Akola and Hyderabad for the semi-arid region; Thrissur for the humid tropical region; Cuttack and Ranchi for the semi-humid, subtropical region; and Shillong for the semi-sub-tropical region. Further, as the crop improvement programmes in the country gained momentum, the access to germplasm for improvement of potential genetic material from important centres of diversity scattered around the world became a continuous requirement. This trans-boundary exchange of material, however, was beset with the danger of introduction of pest and pathogens of respective crops along with the plant material, and therefore plant quarantine in India was initiated. The NBPGR also was designated the nodal organization in the country and vested with the authority to issue Import Permit and Phytosanitary Certificate and undertake quarantine processing of all seed materials and plant propagules of germplasm including transgenic planting material exchanged for research purposes. The NBPGR headquarters at New Delhi and NBPGR regional station, Hyderabad, provides these services. The NBPGR regional station at Hyderabad was specially established for catering to the quarantine needs of International Crop Research Institute for the Semi-Arid Tropics (ICRISAT) for introduction and distribution of required germplasm of ICRISAT mandate crops.

The genetic diversity of rice is spread across the country and NBPGR in association with its regional stations particularly located in Bhowali, Cuttack, Ranchi, Shillong, Shimla and Thrissur in collaboration with other institutes undertaking rice crop improvement work have made significant efforts in collection and conservation of rice genetic resources from their regions.

Survey and Collection of Rice Diversity

Survey and collection of germplasm is the first activity required to be undertaken by any PGR management institute. Rice being an important food security crop, various rice research stations were established in diverse agro-ecological zones of the country since 1911 up to 1975 i.e. before the establishment of the NBPGR. These stations had already surveyed, collected, evaluated and utilised rice germplasm of the country to some extent. With the establishment of NBPGR these activities got a further boost.

NBPGR: Early collections

After its establishment, in India, NBPGR has played an important role in collection of rice germplasm across the country in various crop specific or multi-crop collaborative explorations. In collaboration with CRRI, Cuttack, State Agricultural Universities and its regional centres, the Bureau collected 8,353 cultivated and 87 wild rice germplasm lines from different parts of the country during 1976-84. Furthermore, 6,214 germplasm lines of cultivated and about 800 lines of four wild species were collected from Orissa, Bihar, Uttar Pradesh, Haryana, Rajasthan, Andhra Pradesh Kerala, West Bengal, and Northeast India. A special drive for upland paddy varieties under cultivation in Andhra Pradesh, Karnataka, Maharashtra, Madhya Pradesh, Uttar Pradesh, Orissa and West Bengal resulted in collection of 1,938 cultivars. In addition, joint explorations by NBPGR in collaboration with State Agriculture Universities during 1978-80 and Central Rice Research Institute, Cuttack during 1985 resulted in collection of about 7000 and 447 accessions (from Sikkim, South Bihar, and parts of Orissa) respectively. Vivekanand Parvatiya Krishi Anusandhan Sansthan (VPKAS), Almora explored the germplasm of hilly region of Uttar Pradesh and 1,247 cultivars were collected. The ICAR made a massive collection of 6,439 accessions with the collaboration of State Agricultural Departments and Universities during 1978-80. In Punjab, Rice Research Station, Kapurthala, collected about 1,178 landraces. Most of these collections were of *indica* types with some japonica types. Haryana Agricultural University, Kaul Station has also collected a total of 960 accessions comprising 600 basmati and 360 non-basmati type of germplasm.

In addition, international collaborative explorations for wild rice were also undertaken. The CRRI-IRAT-ORSTOM explorations resulted in the collection of 79 accessions of *O. nivara* and *O. officinalis* from western India during 1984. In 1986 French teams from Institut de Recherche Agronomiques Tropicales (IRAT) and Office de Recherche Scientifique et Technique en Outre Mer (ORSTOM) in collaboration with the ICAR made sizeable collection of *O. nivara*, *O. rufipogon* and *O. officinalis* from Goa, Karnataka, Maharashtra and Gujarat. Additional 75 samples comprising *O. granulata*, *O. rufipogon*, *O. nivara*, *O. officinalis*, *O. spontanea* were collected from coastal, mid-land and mountainous areas of Kerala during 1987. Further, joint explorations by NBPGR-IRRI in Orissa, West Bengal, Bihar, Madhya Pradesh and Uttar Pradesh resulted in the acquisition of 725 accessions of different wild relatives of cultivated rice (Rana and Sharma, 1991; Saxena and Gautam, 2000).

Through these collaborative efforts, over 21,000 accessions of both cultivated and wild species of rice have been collected. The important areas surveyed included Northeastern hills, Assam, Tripura, Sikkim, Maharashtra, Andhra Pradesh, Jammu and Kashmir, Punjab, Himachal Pradesh, Haryana, Uttar Pradesh, Rajasthan, Gujarat, Madhya Pradesh, West Bengal, Bihar, Jharkand, Orissa, Kerala and Tamil Nadu (Malik *et al.*, 2001).

Collections under NATP

The NBPGR has conducted a total of 2,308 explorations in different phyto-geographical regions of the country, of which about 1500 explorations were conducted in the last 10 years (NBPGR, 2007). This was achieved with the enhanced support from the National Agricultural Technology Project (NATP), a World Bank-aided project, implemented by the ICAR and the Department of Agriculture and Co-operation (DAC). Under this project, the sub-project on Plant Biodiversity was implemented by NBPGR during 1999-2004. While NBPGR was the lead institution, 131 co-operators from diverse stakeholder groups were involved in activities mainly focussing on collection, conservation and characterisation of genetic diversity in all agricultural and horticultural crops. The NBPGR in collaboration with other organisations have collected a large number of rice genetic resources during this period. The variability captured during different exploration missions represents the varietal groups that are adapted to various agro-ecologies.

In the project, exploration missions from the regions harbouring the more concentrated diversity of the crop were systematically planned and executed in collaboration with other crop-based institutes located in the region. Emphasis was laid on the collection of landraces along with their wild relatives. Broadly, two kinds of exploration were planned: based on priority crop(s) (crop specific) or priority region (region specific). In addition, in case of severe threats due to natural calamities (cyclones, droughts, floods etc.) and other forms of human interference (building up of dams, infrastructure development etc.), mission-oriented explorations were also undertaken to collect the germplasm diversity of the specific regions in a time-bound manner. The team for carrying out an exploration mission generally included crop curator(s) along with the local guide. The team was equipped with the important accessories (such as exploration kits, camera, global positioning system, vasculum, etc.) required during the tour and a proper means of transportation. In order to effectively handle the large number of accessions collected during the period, a germplasm handling unit (GHU) comprising a multi-disciplinary team of scientists to assess the quality, quantity and health of the seed samples was established at the NBPGR headquarters. The seed material collected was handled mostly in the way shown in the flow diagram (Fig. 5.1). Each collected accession was assigned a specific collector number at the time of collection. In addition, information was recorded during the explorations on minimum passport data (regarding the habitat, nature of the plant, its growth behaviour, socio-economic values, ethno-botanical information etc and other unique features recorded at the site). Generally the sample size during the collection was 10-15% of the population of a particular type of germplasm/landrace growing at one site. Each sample collected was divided into three subsets while the collaborator and the NBPGR regional station located in the explored zone retained one sub-set each for further characterization and evaluation, one sub-set was sent to the GHU at NBPGR Headquarters.

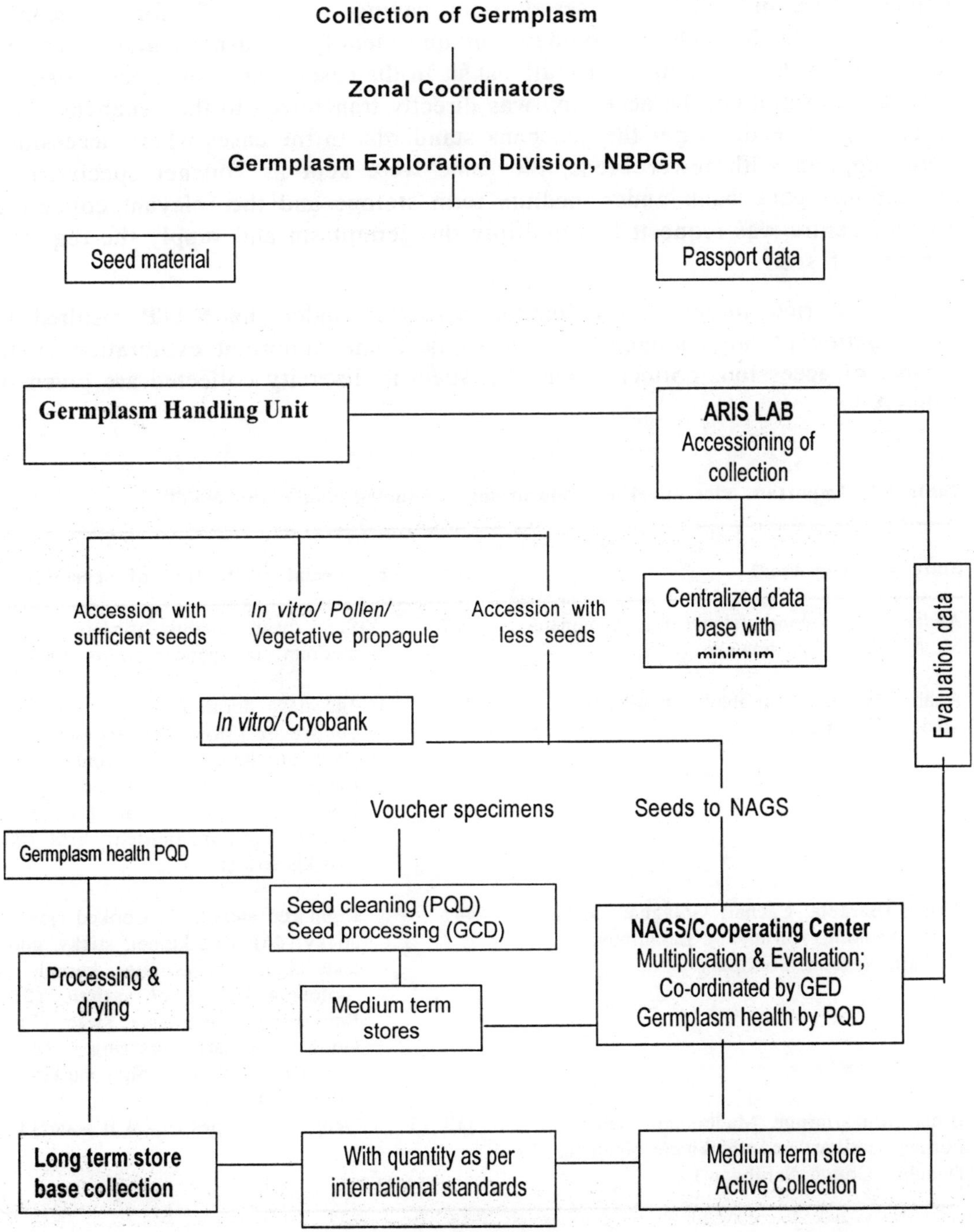

This list and the passport data accompanying the accession were provided to the Agricultural Research Information Services (ARIS) of the NBPGR for documentation in a database and assignment of a national identity number. These numbers were provided in a serial order having a prefix IC, for indigenous collections. The IC number provides a unique identity for a particular accession that should never be changed or reallocated. In the cases where sufficient quantity of seed was supplied, the accession was directly transferred to the genebank after processing of seeds as per the genebank standards. In the cases where accessions were supplied with fewer seeds, the same were kept as voucher specimen in the national gene bank under medium term storage and the relevant collecting institute/centre was requested to multiply the germplasm and supply the required quantity of seed.

The rice survey and collection activities under the NATP resulted in accumulation of large number of accessions. Some important explorations with number of accessions collected and trait specific diversity collected are given in Table 5.1.

Table 5.1. Important rice specific explorations conducted under the NATP.

State (District/Area)	Total Acc	Special traits (no. of accessions)
Andaman & Nicobar Islands (N. Andaman, M. Andaman, S. Andaman)	167	Semi dwarf, medium duration, medium size slender grains (64)
Andhra Pradesh (Adilabad, Khammam, Medak, Nizamabad)	330	145 days duration, scented, very good taste (38);Local landraces that grow on their own, without irrigation, fertilizers and pesticides, heavy shattering of grain (70); 6 months duration, rainfed, yield 700-840 kg/acre (57)
Assam (Barpeta, Cachar, Golaghat, Jorhat, Karbi Anglong, Karimganj, Lakhimpur, Nalbari, Nagaon, Sonitpur)	441	Light red stem (51); Cooked rice sticky (46); Red kernel, sticky, good taste (11); *Ahu* varieties, kernel colour red (15); *Sali* varieties (21); Red kernal, good taste, sweet (7); Good taste, early maturing (19); Scented, good taste, high market value (22)
Bihar (Muzaffarpur, Madhubani, Saharsa, Purnea, Katihar) and Jharkhand (Dumka, Deoghar, Godda, Sahibganj)	113	Deep water types (3 - 4 ft water) (3)

Chhattisgarh (Bastar, Dantewada, Jagdalpur, Raigarh)	395	Less susceptible to BPH, blast (149); 5 months duration (61).
Kerala (Alappuzha, Ernakulam, Kannur, Kasargod, Kozhikode, Palakkad, Malappuram, Thrissur, Thiruvananthapuram)	598	Photo-insensitive, medium duration, saline resistant (13); Photo-insensitive (27)
Madhya Pradesh (Narsinghpur Jabalpur, Damoh, Sagar, Mandla, Dindori, Seoni, Panderwani), Chhattisgarh (Kawardha, R'gaon, Balaghat, Bastar)	398	Local landraces
Manipur	44	Local landraces and improved cultivars
Tamil Nadu (Kanyakumari)	7	Local landraces
Uttar Pradesh (Allahabad, Amethi, Azamgarh, Bahraich, Balia, Basti, Bhadohi, Chandauli, Deoria, Faizabad, Ghazipur, Gonda, Gorakhpur, Jaunpur, Maharajganj, Mau, Mirzapur, Naugarh, Pratapgarh, Sant Kabir Nagar, Sant Ravidas Nagar, Siddharthnagar, Sonbhadra, Sultanpur, Varanasi) and adjoining Madhya Pradesh (Rewa, Satna)	851	Grain black with blackish medium awns (91); Grains yellowish short bold (84); Grains medium short (105); Tall, late, grains long bold, golden colour (21); Tall, grain short fine, straw colour (95); Grain golden brown, grow in shallow water (70); Grain straw colour, short fine grain scented rice (41); Wild species (201)
West Bengal (Bardhaman, Chinsurah, Cooch Behar, Hooghly, Jalpaiguri, Maldah, Medinipore, North and South Dinajpur, Nadia, 24 Parganas, Purulia,) and adjoining Orissa (Mayurbhanj)	312	Semi-deep (41-70 cm water) (24); wild species (113), local landraces (35)

Source: GHU, NBPGR, New Delhi

The wild rice accessions comprising *O. rufipogon* and *O. nivara* were collected from Uttar Pradesh (Ambedkar Nagar, Balrampur, Barabanki, Faizabad, Rai Bareilly, Sant Kabir Nagar, Shravasti, Sultanpur districts), Bihar (Sonepur district) and West Bengal (Birbhum, Burdwan, Hoogly, Malda, Midnapore, Murshidabad, Purulia, North 24 Parganas, South 24 Parganas, North Dinajpur, South Dinajpur districts). Around 450 accessions were collected from Orissa (Angul, Balangir, Boudh, Cuttack, Dhenkanal, Gajapati, Ganjam, Jagatsinghapur, Kalahandi, Khordha, Koraput, Mayurbhanj, Nayagarh, Nuapada, Puri, Rayagada, Sambalpur, Sundargarh districts).

Pattern of Rice Diversity

The survey and collection missions for rice conducted by NBPGR and its collaborators recorded a rich genetic diversity among rice genotypes and augmented trait specific genotypes found in various regions (Table 5.2).

Table 5.2. Trait-specific genotypes found in various States/Regions of India.

Plant Type/Trait	States/Regions
Short stature	North Cachar, Manipur, Arunachal Pradesh
Biotic stress	
Stemborer	Meghalaya, Assam, Arunachal Pradesh, Tamil Nadu
Gall midge	North-Eastern States (Garo, Mikir, Khasi and Jaintia hills)
Tungro virus	Assam, Meghalaya, Tamil Nadu, Kerala
Abiotic stress	
Submergence	West Bengal, Eastern Uttar Pradesh, North Bihar, Orissa
Drought	Tamil Nadu, Assam (Garo hills), Jharkhand (Chhotanagpur), West Bengal (Purulia), Uttar Pradesh (Nagina), Orissa (Koraput, Kalahandi), Gujarat
Salinity	West Bengal (24 Pargana area), Coastal regions (East and West)
Low temperature	Uttarakhand, Kashmir and Himachal Pradesh (higher elevations)
Deep water (floating)	West Bengal (Midnapore, Birbhoom, Hoogly, Howrah), Eastern Uttar Pradesh, Northwestern Bihar, parts of Assam
Quality	
Low amylose (*Japonica*)	Arunachal Pradesh (Kamang),
Waxy endosperm	Assam (Garo hills) and other North eastern hill states
Basmati	Uttarakhand, Jammu & Kashmir, Haryana, Punjab
Scented	Uttarakhand, Orissa, Assam, Eastern Uttar Pradesh, Bihar, Madhya Pradesh
Soft cooking	Assam
Speciality	Kerala, Tamil Nadu, Andhra Pradesh, Bihar, Madhya Pradesh, Chhattisgarh
Medicinal Properties	Kerala, Chhattisgargh

Source: Siddiq *et al.*, 2005

Diversity of collection from Himalayan region

Northeastern Himalayan region of India is a high rainfall and humidity area with varied topography and altitude. The region has been found to be the richest reservoir for genetic variability of rice. About 0.5 million-farmer families are practicing shifting cultivation (locally known as 'jhoom') over 16% of land area at any given time. Rice is the principal food crop of the region and is extensively cultivated (72% of the total cultivated area) in upland, lowland and deep-water conditions. It is estimated during the germplasm survey and collection made so far by the NBPGR from major rice ecologies that about 2,000 local landraces are available and this comprise about 60% of all rice grown on a small scale by the marginal farmers. The rest of the cultivation is mostly of HYV (Hore, 2005). Rice diversity pattern of the northeastern states of India based on the exploration and collection of rice germplasm by NBPGR Regional Station, Shillong have been summarized in Table 5.3.

Table 5.3. Patterns of rice diversity of the northeastern states of India and collections made by NBPGR Regional Station, Shillong.

State	Important varietal diversity (areas of distribution)	Wild relatives (areas of distribution)	Accessions collected (no)	Areas still to be explored (area of distribution)
Arunachal Pradesh	Seventeen diverse landraces, some resemble Japonica type for culm, tillering, panicle and grain characteristics		931	South Siang, Dibang valley, Changlang, Tirap (bordering Myanmar, Nagaland)
Assam	Landraces of eco-genetic groups *Bao, Ahu, Sali, Boro, Joha, Lahi*; Landraces named on husk color *Tulsibao* (blackish); *Bogibao* (brownish–white) and *Kekuabao* (reddish); Winter *Sali* varieties *Malshira, Monohar, Kartiksali, Bornidhan, Paroma, Bhola, Nepali Bhog* (Dhubri and Goalpara districts)	*Oryza granulata* (Kamrup, Hathgaon, Goalpara, Sonaimukh, Bonaibuchi, Diphu); *O. officinalis* (Lakhimpur); *O. rufipogon* (Kamrup Neematighat); *O. meyeriana* (Lumding)	1580	Parts of Barpeta, Nalbari, Dhubri, Darrang, Sonitpur, Lakhimpur, Dhemaji, Karimganj, Nagaon, Morigaon, Cachar, Hailakandi districts
Manipur	*Chahao* black kernelled aromatic (higher elevation of Ukhrul district)	*O. rufipogon* (Loktak)	93	

Meghalaya	*Manipuri* and *Ngoba* (Sticky varieties), *Mirikrak*, *Chanmouri*, *Kbathugmaw*, *Batlong*, *Ryllobed*, *Kbaswarit*, *Abor red*, *Khonorullu*, *Ngoba*, *Dullo*, *Thermeru red*, *Lyngsi*, *Kuki*, *Tongla*, *Nong-lwai*, *Maiku Tsuk* (Red-kernelled varieties, used in preparation of *Jadoh* during ceremonies and soft rice varieties used in flaked or raw form as breakfast in Khasi-Mikir Hills border areas).	*O. officinalis* (Nongpoh, Burnihat); *O. rufipogon* Griff. (Jaintiapur); *O. nivara* (Burnihat)	296	Pockets in West Khasi hills bordering Bangladesh
Mizoram	Nearly 52 predominant local landraces Tai (lowland or wetland, matures early and aromatic) in nature. At least 10 rice landraces are sticky rices. *Hmawrhang* (drought resistant) *Buhtawi* and *Kasaba* (cold tolerant).		362	Areas bordering Myanmar in the east and Bangladesh in the west
Nagaland	At least 65 landraces with variability in grain size, shape, awn character, glume and kernel colour	*O. officinalis* (Dimapur)	460	Remote areas of Wokha, Tuensang, Mon districts
Sikkim	*Buidhan*, *Lama*, *Thapachini* (Directly sown); *Attey*, *Mansara*, *Jhapaka*, *Dut-kalami*, *Basmati*, *Krishna-Bhog*, *Talasi*, *Poudyal* (Transplanted); *Brimphul*. *Attey*, *Mansara*, *Marshi* (early maturity, up to 1500 msl); *Phudange*, *Takmaru*, *Darmali* (above 1500 msl, cold-tolerant); Krishna Bhog, Brimphul,	*O. officinalis* (Terai region)	90	

	local Basmati (scented); *Phorel* (glutinous); *Takmara*, and *Lama* (red-kernel); *Phulpatta* (very small grain similar to *japonica*)		
Tripura	At least 65 types of wetland and 32 *jhoom* types and 5 types of deep-water and 5 types of aromatic rice. *Boro* rice cultivation in large scale is a recent trend	244	South Tripura bordering Bangladesh

Source: Hore, 2005

The specific adaptability to the ecological conditions and consumer preference has led to evolution of unique eco-genetic groups in the northeastern region. These are (i) autumn rice (*Ahu*); (ii) *kharif* or winter rice (*Sali*); (iii) spring or summer Rice (*Boro*); (iv) shallow water (1–2 m) rice (*Asra*); (v) deepwater (2–5 m) rice (*Bao*) and (vi) hill Rice. The hill rice is again classified into two categories according to altitude, i.e. mid-hill (800–1200 m) and high hill (above 1200 m) (Borthakur, 1992). The *Boro* is grown mainly as irrigated rice. The short duration *Aus* varieties are photo-insensitive and can be grown in lowland as well as in upland areas. The *Bao* varieties are generally very long duration types (170 to 215 days) with very bold grains and high shattering nature. Winter rice is again classified as *Lahi* (fine grained); *Joha* (Scented); soft rice (used as parched rice, parched paddy, flaked rice); *Boro* rice (breakfast type) and cooked rice (staple food). Within each group, there are a number of landraces that have been nurtured and cultivated by the farmers. The diversity in wild rices includes *Oryza granulata* Nees et Arn. ex Watt and *O. meyeriana* (Zoll et Mor. ex Steud.) Baill. belonging to Meyeriana complex; *O. officinalis* Wall. ex Watt of the Officinalis complex; *O. rufipogon* Griff. of the Sativa complex and *O. nivara* Sharma et Shastry of the Sativa complex. In addition, *Zizania latifolia*, a wild relative of rice is available in the Loktak area of Manipur state (Hore, 2005).

In the Western Himalayan region, rice is generally cultivated below 1000 msl; above this altitude (up to 2300 msl) the cultivation is in terraces where source of irrigation is melting snow and average soil and water temperature remain below 15^{0}C. About 55% of the total area is lowland and valleys which invariably have assured irrigation and receive plentiful rainfall; the rest about 45% of the area is devoted to upland/rainfed rice where production is highly dependent on the amount of rainfall received during cropping season. In lowlands along the river basins and valleys having assured irrigation, predominantly, the HYVs are grown by raising the nursery in May-June and transplanting the seedlings at 3 to 5 leaf stage in puddled fields. In uplands, predominantly, broadcasting method is used for sowing the landraces, after the pre- monsoon showers, in the month of May-June. After about one month, as the seedlings attain height of 8-10 cm, and when heavy rains are received, the field is ploughed; the practice is locally called *Halod*, *Hodd*, *Pandair*. It requires higher seed rate (almost two times) than that of transplanting. Terraces require well-repaired bunds so that it can hold most of the rainwater during crop season. To avoid the problem of *reesa* (previous year's seed germinate, escape weeding, mature early and spike remain unfilled/partially sterile) farmers rotate varieties with purple colour strains. Further, as the cultivation areas changes from lower to higher elevations, the crop duration increases from 120 days to 150 days.

Consequent to the cultivation of rice in such high-risk environments and the wide range of micro-climatic niches in the hills, the region has nurtured and conserved many landraces that yield under stress situations. In order to assemble rice germplasm from these regions, NBPGR regional stations, Shimla, Bhowali and Srinagar have collected more than 2,500 germplasm accessions in the last 20 years. Important areas in the northwestern Himalayan region that harbour rich diversity for various traits are given in Table 5.4.

This region is not much known for occurrence of wild species and only *O. rufipogon* has been observed sporadically in the foothill areas but there are about 200 named landraces that contribute nearly 50-60% of the total rice production of the region. These landraces have unique features that make them useful raw material for utilisation in crop improvement programmes particularly quality traits and difficult agro-climatic/stress situations. Landraces such as *Jaulia*, *Thapachini*, *Jamali*, *Lalsati*, *Chinbhuri*, *Anjan*, *Jhini*, *Safed-Brez*, *Shahie*, *Siga*, *Madumalthi*, *Katyoor* are predominantly grown over a large area but by few farmers only while landraces such as *Boran*, *Khazia*, *Gajaiya*, *Ruwat*, *Syont*, *Ghesu*, *Jhumariya*, *Safed syal*, *Mushaqbudji*, *Qadir beigh*, *Qadir Ganaie*, *Til-Zag*, *Jatto*, *Barpasso*, *Sela*) were grown in limited area but by many farmers. Some landraces preferred for particular trait(s) are given in Table 5.5. In addition, the western Himalayan region harbours one of the best quality Basmati rice, famous for its grain size and shape and aroma, the world over (Singh *et al.*, 2000)

Landraces of rice had evolved through generations and farmers have nurtured them for particular trait(s) that suit local conditions and also reflect socio-cultural preferences. During the explorations, farmers' preference was noted for a number of landraces because of a specific trait(s) (Table 5.6).

Table 5.4. Areas in the northwestern Himalayan region rich in trait specific rice diversity.

Area	Traits
Himachal Pradesh	
Dhauladhar (mountain range), Palamapur, Rait, Shahpur & areas bordering Chamba and Karsong	Fine grains basmati types
Changer belt and Kandi area	Coarse grain, drought tolerant
Chidgaon, Rohru, Jagatsukh, Nauradhar, Jogindernagar	Red grain, cold tolerant
Jammu and Kashmir	
Quazigund, Halluen Kund, Kulgam, Gawaran, Hapatnaar, Khansahib, Krimshore, Ratthan, Khang, Shopian, Tral, Keller, Gabarpora, Kunzer, Kralweth, Tangmarg, Hatchmarg, Hafrada, Meleyal, Voile, Kangan, Gund	Aromatic, cold tolerant
Uttarakhand	
Ram Ganga valley, Garur valley, Charkata, Tapovan, Maletha, Kotdwara	Aromatic and fine grains
Johar, Danpur, Darma valley, Kali, Tamali-Sore, Gangolihat, Thalisain, and Alaknanda	Coarse amd red grain, cold tolerant

Table 5.5. Some landraces preferred for cultivation for particular traits.

Trait	Preferred landrace
Aroma and taste	*Hansraj*, *Mushaqbudgi*, *Basmati*, *Larbeoul*, *Qadir begh*, *Shahie*, *Barpasso*, *Katyoor*
Grain length	*Begumi*, *Ramjwain*, *Thapachini*
Grain colour	Brown: *Drangbara*, *Prenie* Yellowish white: *Ramjawain*, *Mushaqbudgi* White: *Larebeoul*, *Sigae safed*; Red: *Kard*, *Chiratau*, *Jatoo*, *Ambirkhan*
Stickiness on cooking,	Non-sticky: *Jhini*, *Ramjwain*, *Baber*, *Basmat;* Very sticky: *Ital*, *Kamed*, *Numbeoul*
Good swelling upon cooking	*Fulpatash*, *Baber*, *Kathwar*
Better milling recovery	*Razuli*, *Meerzag*, *Prenie*, *Baber*, *Seal*, *Begumi*
Medicinal properties	*Katheri*, *Kafalaya*, *Kalon safed*, *Kalon lal*, *Gulzag*, *Lalsati*, *Madguri*, *Bhabri*
Religious purposes	*Sela*, *Sawa*, *Barpasso*, *Jamlkhijia*, *Tilkakchandan*, *Lalchandan*

Table 5.6. Number of landraces cultivated based on farmers' preferences for particular traits.

Preferred trait	Landraces (no.)			
	Himachal Pradesh	Jammu & Kashmir	Uttarakhand	Total
Scented/ aromatic	16	25	7	48
Good taste and cooking quality	26	17	18	60
Red pericarp	14	11	5	30
Upland type	13	4	15	27
Cold tolerant	11	6	10	27
Bold/round/coarse grain	11	27	14	52
Long/slender/fine grain	19	7	13	39
Glutinous	9	8	1	18
Medicinal	4	5	7	16
Early maturing	3	16	1	20
Easy to milling	4	9	5	18
High yield	9	19	1	29

Source: Rana *et al.* 2008

Diversity of collection from Eastern and Central region

The NBPGR Regional Station at Cuttack has been actively involved in the management with the responsibility for exploration and collection of indigenous agri-horticultural crops of Orissa and adjoining areas along with the characterisation and conservation of plant genetic resources of northern and eastern plain regions of India.

The region is rich in genetic diversity of rice and forms a part of the region where the cultivated rice possibly has originated. The agroclimatic condition of the region is also diverse. Rice is grown in rainfed low lands, irrigated lands, rainfed uplands and flood prone areas. The region is ethnically very diverse; much of it is hilly area with its coastal region bordering Bay of Bengal. Rice collection missions were initiated by NBPGR Regional Station, Cuttack since its establishment in 1986 and a total of 3,185 accessions were collected from Orissa, West Bengal and Madhya Pradesh, including 328 accessions from special exploration mission viz., super cyclone affected area of Orissa (231 accessions), drought affected area of Orissa (52 accessions) and Sunderban area (45 accessions) (Table 5.7).

Table 5.7. Exploration and collection of cultivated rice form eastern region.

Area	No. of accs.	Year
Puri, Keonjhar (Orissa)	226	1986
Koraput, Kalahandi, Ganjam, Mayurbhanj, Phulbani, Bolangir, Sambalpur (Orissa)	300	1987
Santhal Pargana (Jharkhand)	109	1988
Mayurbhanj, Parts of West Bengal, & Bihar (Eastern Ghats)	581	1989
Ganjam	5	1992
Orissa (17 districts)	337	1996
Bastar (Chhattisgarh)	609	1997
Sundargarh, Sambalpur and coastal districts (affected by super cyclone)	528	1999
Puri, Cuttack and Nayagarh	375	2000
Similipal (northern Orissa)	18	2001
Southern Orissa	52	2002
Sundarbans	45	2003 & 04
TOTAL	**3185**	

The primitive upland cultivars of the Jeypore tract are short-duration and photoperiod-insensitive. Similar types of cultivars often cultivated as rainfed upland crop, especially in unbunded fields in the southeastern region include *Tikradhan* (Chhatisgarh region), *Bhathadhan* (western Orissa) and *Goradhan* (Jharkhand region). These landraces are collectively referred to as Southeast Indian hill rices or 'seihr' by Sharma *et al.* (1997). Some important traditional landraces like *Sathka*, *Sathia*, *Kuliha*, *Kusuma*, *Mundisanra* etc. for early maturing type, *Vanduasali*, *Thengabaula*, *Patolimadhoi* for very late, *Nandimachhakanta*, *Hundamakara*, *Kedargouri*, *Jagannath ballav*, *Champa*, *Mainsitinka*, *Bhutia*, *Kalasura*, for deep water, *Khajara*, *Putia*, *Ravena*, *Champeisali*, *Meghabadal* for submergence tolerance, *Sitabhog*, *Basuabhog*, *Nadiaphula*, *Krushnabhoga*, *Chinikamini*, *Samaleibhog*, *Maharagi*, *Lectimachy*, *Laisari*, *Haladigundi*, *Karpurakaati* for scented, *Sunakathi*, *Kakudimanji*, *Jhilli*, *Sarumalli*, *Patharmukta*, *Betakanta* for fine grain type, *Nalidhusura*, *Manjara*, *Somanath*, *Kalmabanki*, *Nalifarm*, *Mayurkantha*, *Rajamalli and Nalibanki* for popping and puffing quality were collected through different exploration missions in the region. Among the scented rices *Kalajeera*, *Badsabhog*, *Neelabati*, *Krushnabhog*, *Govindabhog*, *Padmakeshari*, *Tulasiphoola* are predominant in coastal belts while a few additional scented varieties like *Pimpudibasa*, *Karpurakeli*, *Kalikati*, *Laxmibilas* and *Makarakanda* are common in the plateau regions of the state. Orissa is also rich in genetic diversity of wild rices. The species like *O. nivara*, *O. rufipogon*, *O. officinalis*, *O. granulata*, *O. sativa spontanea*, *Porteresia coarctata*, are grown well in different eco-systems of the state. It is assumed that Asian cultivated species *O. sativa* has originated from *O. nivara*.

Diversity in medicinal rices

Ancient Indian records speak of the existence of rice varieties of curative value for various ailments as detailed in the Ayurvedic Treatise (Indian *Materia Medica*) of the 15th and 16th centuries AD. Varieties like '*Njavara*' and '*Gathuran*', for instance, were used in the treatment of arthritis. Knowledge regarding the nutritional or medicinal value of some of the landraces is prevalent among the folklore. Earliest records list cultivars like *Krishnavrihi* as nourishing rice, and several such as *Kalama*, *Pundarika*, *Panduka*, *Shakunahrit*, *Sugandhaka*, *Kardamaka*, *Kanchanaka*, *Mahasali*, *Mahishamastaka* and *Lodhrapushpaka* as of medicinal value (Chunekar and Pandey, 1986). Rice collection surveys done in Chhattisgarh state led to the identification of many varieties of medicinal value. They include *Aalcha* (against pimples), *Karhani* (against paralysis), *Maharaji* and *Liacha* (against weakness in pregnant ladies), etc. Cultivars like *Bhejri* fed to cows are believed to help in quick release of placenta during calving (Das and Oudhia, 2001). Similarly, Ayurvedic centres in Kerala carry out a therapeutic measure '*Navarakizhi*' (so named because of the fact that the boluses of cooked Navara rice are used in the process) is usually done for all skeletal and muscular diseases, paralysis, sciatica, etc. Even today, certain varieties with medicinal value are used in Karnataka, Madhya Pradesh, Kerala, Tamil Nadu, Uttar Pradesh, the Western Ghats, and Himachal Pradesh to treat skin diseases, blood pressure, fever, paralysis, rheumatism, and leucorrhoea, as well as a health tonic and for lactation (Ahuja *et al.*, 2008). Western Himalayan region also grows different landraces of red rice (red pericarp) known for their aroma, cold tolerance and medicinal properties such as curing high blood pressure, fever, leucorrhoea, etc. For instance, landrace *Katheri* is given to women as sweet boiled rice just after delivery for a few days with the belief that it strengthens the body and increases lactation, *Kafalya* (curing leucorrhea), *Sawa* (for worship), *Jamali* (makes excellent porridge) and *Khayija* (for popping). Ancient Ayurvedic literature (Indian *Materia Medica*) from the fifteenth and sixteenth centuries AD also described different rices, particularly scented varieties with medicinal and curative properties (Nene, 2005).

Augmenting genetic resources form exotic sources

Although India is rich in rice genetic diversity, plant breeders need diverse germplasm carrying different desirable genes to combine together their genes and develop better cultivars. Hence, acquisition of diverse and elite germplasm from exotic sources is critical in this regard. The NBPGR is the nodal institution for introducing germplasm at its own initiative as well as on the basis of specific requests received from the scientists of various research institutes in the country including private sector and non-governmental organisations. This single window system for germplasm exchange facilitates the import and export and quarantine of small samples, including those of transgenics, meant for research. The introductions include whole array of germplasm, namely improved cultivars, elite strains, genetic stocks, breeding materials etc. The introductions also include the IARCs international breeding trials and nurseries. Rice breeding materials introduced from IRRI during 1960s including Taichung Native (1) (EC 305391), IR-8 (EC 288054), IR-36 (EC 297702) laid the foundations of Green Revolution. Further, the rice varieties, such as *Mahsuri* and *Adt 27* were developed as a result of introduction of *japonica* rice for inter sub-specific (*indica* x *japonica*) hybridization to recombine high fertilizer responsiveness and non-lodging habit of the *japonica* rice with the grain quality and adaptation of *indica* rice. In addition, trait specific introductions have been made including cold tolerant (EC 121003), blast resistant (EC 121807-15, EC 307145-61), bacterial blight resistant (EC 121816-23), leaf folder resistant, salinity tolerant (EC 121828-53, EC 122935-40, EC 132900-10), sheath blight and stem borer resistant (EC 199465-66), drought tolerant and bacterial leaf blight resistant (EC 201878-975), soil stresses tolerant germplasm (EC 268456-931) and lines having wide compatibility genes (EC 469595) (Singh *et al.*, 2006).

Earlier germplasm was considered as the common heritage of humankind. There existed practically no regulatory instrument affecting access to the germplasm except plant quarantine. Consequently, there was a free access to PGR found in centres of origin/diversity or other regions. The germplasm was collected without restrictions by scientists/explorers and utilised as the 'raw materials' in plant breeding programmes throughout the world. However, with the advent and wide application of modern plant breeding techniques during 1960s, most countries, witnessed a formidable participation of private sector in plant breeding programmes. This development led to international debates that were polarised towards the increasing demand for PGR and the need to control or maintain their free flow. While on one hand, concern of intellectual property rights (IPR) for the plant breeders was raised, on the other, issues like the inequitable and contradictory nature of free access to PGR were raised. This has lead to a highly complex situation for the management of genetic resources particularly from issues emanating from Convention on Biological Diversity (CBD) and the Agreement on Trade-Related Aspects of Intellectual Property Rights (TRIPS) administered by the World Trade Organisation. The CBD has emerged out of the global concern for conserving biodiversity and was signed by 150 countries at the United Nations Conference on Environment and Development in 1992 in Rio de Janeiro (the 'Earth Summit'), and entered into force on 29 December 1993. It recognised the sovereign rights of nations over their biological diversity and has three main objectives; (i) conservation of biological diversity, (ii) the sustainable use of its components, and (iii) fair and equitable sharing of the benefits arising from the use of genetic resources, including by appropriate access to genetic resources and by appropriate transfer of relevant technologies. There are currently 175 Parties to the Convention, 174 countries and the European Union; notable exceptions include the USA. This obligates adherence to Material Transfer Agreements (MTA), Material Acquisition Agreement (MAA) for genetic resource accessed from institutions or genebanks and prior informed consent of farmers or communities if material is obtained from them. The WTO came into being in June 1994 and the TRIPS provisions that became operational from 1st January 2005, for the first time established enforceable universal minimum standards of protection for all the major intellectual property rights of individual or corporate bodies for new inventions, technologies and products developed by them. More importantly, it extended the application of the IPR regimes to living things. The necessity of access and flow of genetic resources have mandated respective nations to enact legislation and develop suitable regulatory mechanisms to ensure the enactment of the commitments made in the various negotiations particularly related to variety protection and providing quality seed to the farmers (Saxena and Singh 2006a).

Characterization and Evaluation

Though the role of germplasm in crop improvement is well recognised, lack of sufficient information on the performance of germplasm collections has led to its limited use. Thus, proper characterization and evaluation of germplasm and dissemination of the information to crop breeders and others is a priority area in any germplasm management programme. Characterization of germplasm is based on a set of easily identifiable but stable morpho-physiological traits that enable establishment of the identity of an accession while evaluation provides complete information on traits of economic importance. In rice, field evaluation is undertaken to record data on various morphological, physiological and agronomic characteristics including some special features, such as stress tolerance, insect-pest, disease resistance, etc. It is followed by studies in laboratory on the quality and other biochemical traits. Molecular techniques are also being used to analyze genetic diversity.

Characterization, largely in qualitative terms of collections made by different institutions/scientists has been done and catalogues have been published. The descriptors given emphasis include traits like stature, culm strength, sensitivity to day length, tillering habit, panicle size and shape, grain size and shape, grain and kernel colour, maturity, etc. However, there is less information on resistance/ tolerance to biotic and abiotic stresses, yield determining morpho-physiological features, grain type and cooking and nutritive quality which are of interest to breeder.

The NBPGR Regional Station, Cuttack characterised a set of 7,897 accessions of rice germplasm for 28 morpho-agronomic traits as per IRRI-descriptors (Table 5.8). While 1,520 accessions screened for BLB and RTV resulted with moderate resistant to RTV (211 accessions) and BLB (35 accessions), screening of 474 accessions against gall midge resulted in identification 83 accessions that were resistant/moderately resistant. The list of germplasm showing multiple resistances are given in Table 5.9 (Dikshit *et al.*, 2004).

Table 5.8. Rice germplasm lines showing resistance or moderate resistance against insect pests and diseases.

Stress	Year	Accessions (no)	Resistant lines	Resistant/moderately resistant
Gall midge (studied under field condition)	1988 1989 1992	474 456 1802	83 14 08	*Boita, Nagara, Kharkoula, Makarakanda, Balamanji, Makaraakam, Minja, Langudi, Lache, Mahipal, Parijat desi, Bhataaburai, Benibhog-I, Tigaamanjia, Chandrakant, Baramalata, Ratnapakhia, Nalibainshi, Nanimundi, Kalajira,*
Bacterial leaf blight (inoculated artificially with the inoculum of *Xanthomonas campestris* pv. *Oryzae*)	1989 1990 1992	1036 1732	05 01 02	*Kalajira, Gopalbhog, Kulania, Dhusuri, Dudhasar*
Sheath blight (artificially inoculated with sheath blight fungus in wet season)	1990	1732	04	*Bubeimeta, Badametra, Rajmuhachudi*

Rice tungro virus (RTV) (studied during wet season)	1989	353	08	*Ratnapakhia, Sunakhila,* IC-86023, *Baniamalata, Kanthakamala, Ratna chudi, Bhutia, Basmata, Fhsbkuji, Jsnhiphool, Jagar*
	1990	324	01	
	1991	211	29	
	1992		02	
Blast (studied during wet season)	1989	481	35*	*Dudhasara, Kalpa, Pakhia, Arnakhumba, Gourangasali,* D00067, *Dhalikauiti, Malata, Kalakalam, Sanra, Matiasuluri, Kakuri, Dhobtusku, Nunbaha, Khuntia, Chinamal, Mahumachhi, Bahalmalli, Durgabhog, Gopalbhog, Jhalka late, Machhakanta, Tumbadhan, Nadikanda, Badhandhan, Haldikalama, Nankikalama, Khirasar, Kumrah, Laldhan,* SR26B, *Kundaphaphuli*
	1990	228	46*	
Brown Plant Hopper (BPH)	1989	252	14	*Sarubhujni Gourangsali, Dudhasara, Pakhia*
	1990	530	16	
	1991	305	12	

*Resistant to *Pyricularia oryzae*

In addition to evaluation using morphological characters, molecular markers are being increasingly used to reveal differences among accessions at the DNA level as they provide a more direct, reliable, and efficient tool for genetic analysis (Ni *et al.*, 2002). In rice, development of isozyme and, later, DNA marker technology such as restriction fragment length polymorphism (RFLP), random amplified polymorphic DNA (RAPD), amplified fragment length polymorphism (AFLP) and microsatellite or simple sequence repeat (SSR), have been used as an efficient tool not only to facilitate identification of accessions but also to determine the genetic structure and pattern of diversity of accessions and to optimize the collections for developing and restricting core collection sets (Siddiq, 2000).

Table 5.9. Rice germplasm lines showing multiple resistance to pests and diseases.

Lines	GM	BPH	BLB	Sh. Blight	Blast	RTV
Sarubhujni	MS	R	S	S	MR	S
Karirinau	S	S	S	S	MR	R
Kalajira	MR	S	MR	S	S	S

Ratnapakhia	MR	S	S	S	S	R
Boita	MR	R	S	S	S	S
Nagara	MR	S	S	S	MR	S
Dhabkhuji	MS	MR	S	S	S	R
Tegannamanjia	MR	S	S	S	MR	S
Lache	MR	R	S	S	S	S
Basmata	S	R	S	S	S	R
Sonakathe	S	MR	MS	MS	S	S
Careful	R	R	-	-	-	-

GM - Gall midge, BPH - Brown Plant Hopper, BLB - Bacterial Leaf Blight, Sh. Blight - Sheath blight, RTV- Rice Tungro Virus

The International Rice Genome Sequencing Project that was established in 1997. By December 2004, rice became the first cereal crop to be completely genetically decoded. The stupendous task was carried out by a consortium of publicly funded laboratories, comprising members from Japan, the United States of America, China, Taiwan, Korea, India, Thailand, France, Brazil and the United Kingdom. The high-quality map-based sequence of the entire rice genome is now available in the public domain. With the completion of sequencing, the next challenge to the scientific community would be to determine the function of about 37,000 predicted genes in rice and to utilize this information in identifying agronomically important genes not only in cultivated rice but also in wild relatives of rice and in other cereal crops. (Sasaki *et al.*, 2008)

Germplasm Registration

The crop improvement programmes operating in the country have paid rich dividends and the improved cultivars have played pivotal role in agricultural progress. Experimental cultivars and hybrids are evaluated in state or zonal/national coordinated trials and those showing better performance than check cultivars are released at state or zone/national levels by the State Seed Sub-Committees or Central Sub-Committees on Crop Standards, Notification & Release of Varieties for Agricultural and Horticultural Crops. Such varieties are notified through a Government gazette (the performance of the cultivars released by State Seed Sub-Committee is also examined by the Central Sub-Committee before notification). However, there are many materials, which give good performance but do not show the required superiority to get released. In addition, there are germplasm, developed and/or identified by scientists with one or more outstanding traits, which may serve as genetic stocks in crop improvement programmes or are of academic importance. Such value-added materials generally remained under-utilised or were even lost. In order to recognize the accomplishments and efforts of scientists who identified or developed these, and to encourage the sharing of such germplasm with other workers in crop improvement programmes, ICAR developed a system of germplasm registration with NBPGR as the nodal institute. Provisions to conserve and use research generated genetic stocks and improved/preferred germplasm, which broadly include new alternate sources of resistance to biotic and abiotic stresses, unique quality, male sterility, induced mutants, cytogenetic stocks, etc has been made use of by rice scientists. So far, 40 such germplasm accessions have been registered and conserved in the National Gene Bank at NBPGR (Table 5.10).

Table 5.10. Research generated and unique rice germplasm registered and conserved.

S. No.	Designation	Pedigree	Traits
1	INGR 98004PBNR87-8, Sugandha	Pusa Basmati1/Basmati-370	Dwarf, non-lodging, iron chlorosis, blast tolerant aromatic rice
2	INGR 98009NBPGR-TGMS1	Single plant selection from IC99422	Thermo-sensitive genetic male sterility
3	INGR 98019Taraori Basmati	Pureline selection from HBC-19	Aromatic, thermo-insenstitive, photoperiod sensitive, extra long kernelled, stem rot resistant
4	INGR 98032GAURAV, Basmati A54	Induced mutant of Basmati-370	Fine grain, early maturing, dwarf
5	INGR 99013IC137532 Munda Dhan	Landrace (Munda Dhan) collected from Orissa	Tungro virus disease resistant
6	INGR 99014IC137542, Malsundri	Landrace (Malsundri) collected from Orissa	Tungro virus disease resistant
7	INGR 99035PNR519 (IET9944)	Induced mutant from Basmati-370	High yielding, fine grained & dwarf stature
8	INGR 00004PUSA3A	IR58025A/Pusa Basmati-1	Cytoplasmic male sterile
9	INGR 00005PUSA5A	Pusa150-21-1-1-2 as recurrent parent	Cytoplasmic male sterile

10	INGR 00007KHO-CHI	Collected from northern Andaman (Andaman & Nicobar Islands)	Less tiller number, low spikelet
11	INGR 00008Red Burma	Collected from northern Andaman (Andaman & Nicobar Islands)	Bold grain with striped husk, black violet in colour, dehusked grains violet in colour with high nutritive value
12	INGR 00009 Black jeera	Collected from northern Andaman (Andaman & Nicobar Islands)	Short, slender, scented grains, black stripes on husk
13	INGR 02014 Boga Chokuwa	Landrace Chokuwa-1 (Assam)	High 1000 grain weight (40 g), long bold grain
14	INGR 02015 Boka Chokuwa	Landrace Chokuwa-2 (Assam)	Red kernel with low amylose
15	INGR 02016 Saru Chokuwa	Landrace Chokuwa-7 (Assam)	Long bold grain, low to intermediate alkali spreading value and intermediate amylose content
16	INGR 02017 Kalamdani Chokuwa	Landrace Chokuwa-9 (Assam)	Long slender grain Chakuwa rice
17	INGR 02018Misimi Chokuwa	Landrace Chokuwa-15 (Assam)	Red kernel with low amylose content
18	INGR 02019Nepali Chokuwa	Landrace Chokuwa-16 (Assam)	High 1000 grain weight (40 g), small bold grain
19	INGR 02001Pusa 1121	Pusa 1121-92-8-1-3-3	Highest cooked kernel elongation
20	INGR 01019 Black Burma	Indigenous germplasm of middle and South Andamans (Andaman and Nicobar Islands)	Salt and aluminium toxicity tolerant
21	INGR 01020Nona	Indigenous cultivar of North, middle and South Andamans (Andaman & Nicobar Islands)	Tolerant to salinity and iron toxicity
22	INGR01021Pusa 4A	IR58025A/IET10649 – selection	Basmati quality male sterile line
23	INGR01022UPRI95-140 TGMS	Spontaneous mutant of UPRI95-140, single plant selection	Thermo sensitive genic male sterility
24	INGR01023UPRI195-167 TGMS	UPRI95-140/UPRI95-117 selection from in vitro unpollinated F_1 ovary culture	Thermo sensitive genic male sterility
25	INGR01046Jaldi Dhan 8	Induced mutant of Dular/ Nagina-22 (PNR550-1-2)	Extra early (80 days)

26	INGR 03029 DRR–2A & 2B	V20A/RNR 18953/RNR 18953	CMS in different genetic back ground
27	INGR 03030 DRR-3A & 2B	V20A/IET 9792/ IET 9792*8	CMS in different genetic back ground
28	INGR 03048 DMS-4A (RPMS 1-4), DMS-4B	*Oryza rufipogon* (VN1) x PMS 2B// PMS 2B	CMS from *Oryza rufipogon*
29	INGR 03049 DMS-3A (RPMS4), DMS-3B	*Oryza nivara* (DRW 21018) x IR 66 // IR 66	CMS from *Oryza nivara*
30	INGR 04001 PD 27 (Khoda) CRRI Acc No. 36470	Landrace from Orissa	Landrace tolerant to complete submergence
31	INGR 04002 RP-3135-97-1-11-5	RP 3135-97-1-11-5, PR 109/ Pakistan Basmati (IET 15833)	Semi-dwarf basmati
32	INGR 04003 Samridhi	IR 22 x W 1263	New Gene *Gm1* for resistace to rice gall midge disease
33	INGR 04004 Surekha	IR 8 x Siam-29	New Gene *Gm2* for resistance to rice gall midge disease
34	INGR 04005 RP2068-18-3-5	Swarnadhan x Velluthacheera	New Gene *gm3* for resistance to rice gall midge disease
35	INGR 04006 Abhaya	CR 157-392 x OR 57-21	New Gene *Gm4* for resistance to rice gall midge disease
36	INGR 04007 ARC 5984	ARC 5984	New Gene *Gm5* for resistance to rice gall midge disease
37	INGR 04008 RP2333-156-8	Ratna x ARC 10659	New Gene *Gm7* for resistance to rice gall midge disease
38	INGR 04009 Jhitpiti	Germplasm J 461	New Gene *Gm8* for resistance to rice gall midge disease
39	INGR 04010 Line 9	Jaya x Dubraj	New Gene *Gm9* for resistance to rice gall midge disease
40	INGR 04064 Clustered Spikelet Mutant	Selection from natural population	Clustered spikelets mutant

Conservation

In view of still large uncovered genetic potential available in the wild species and primitive landraces, and the likelihood in the near future of their disappearance, their conservation have become urgent. Two major approaches in vogue for conservation of rice germplasm diversity are *in-situ* and *ex-situ*. These are not mutually exclusive but complementary to each other.

In situ *conservation*

In situ conservation involves forest area reserves, national parks and on-farm conservation by farmers. *In situ* conservation of wild relatives of rice is being carried out in several forest reserve areas such as Bhoothathankettu Forest Reserve, Kerala (*O. rufipogon*), Parambikulam Game Reserve, Kerala (*O. officinalis, O. granulata*), Karulai Range Teak Plantation/Forest Reserve, Kerala (*O. nivara, O. granulata, O. malampuzhaensis*), Agasthya Malai Forest Reserve, Tamil Nadu (*O. granulata*), Mudumalai Wildlife Sanctuary, Nilgiris, Tamil Nadu (*O. granulata*) and Simlipal National Wild Life Sanctuary, Orissa (*O. nivara, O. granulata, O. officinalis*).

On-farm conservation of PGR is a complementary strategy for *in situ* conservation of valuable plant varieties (developed and protected by farmers for centuries in certain areas). It involves continuous cultivation and management by farmers of a diverse set of crop populations in the agro-systems where the crop has evolved or in secondary centres of diversity (Bellon *et al.*, 1997). The strategy allows for crop populations to continue their evolution in response to natural and human selection. Initiatives have been made for *in situ* conservation of rice jointly by the NBPGR, IRRI, and IGKV, in the tribal region comprising Raipur and Bastar districts of Chattisgarh during 1997-1999. Bastar is a potential region where tribal farmers maintain enormous diversity. Such germplasm-rich regions where diverse landraces are still under cultivation by natives and in most cases adopting traditional cultivation practices include, Jeypore-Koraput region in Orissa, several pockets in the northeastern Indian hills, deep water rice areas in Sunderbans of West Bengal and Assam and several saline areas along the coast such as Bhitarkanika tidal forest of Orissa. The potential areas for *in-situ* on-farm conservation include Phulbani and Koraput districts of Orissa and South Bihar Plateau (for upland types), northeast and Bastar district of Chattisgharh (upland/lowland types), Midnapore district of West Bengal and Balasore, Jajpur, Bhadrak, Jagatsinghpur, Puri and Cuttack districts of Orissa (for lowland types), eastern Haryana, western Uttar Pradesh including Dehradun Valley and foothills of Himalayas (for quality scented rices), higher elevations (>1800 msl) of western Himalaya (for cold tolerant, red grained types), eastern Uttar Pradesh, Apatani plateau and Khampti Valley of Arunachal Pradesh (for unique landraces) and some lower parts of West Bengal (for *Boro* rice). It is imperative to have a systematic monitoring activity to assess the drift in cultivation of particular landraces over time and make concerted efforts to conserve them.

Traditional practices of managing and developing the vast wealth of genetic resources by farming families within rice based cropping systems should form an important part of the strategy for *in situ* on-farm conservation and sustainable use of rice genetic resources.

Farmers and communities play a pivotal role in on-farm conservation largely because the food security and self-sufficiency particularly in the marginal areas depends largely on the availability of genetic diversity. The communities, usually, have multiple interests or concerns and are confronted with numerous challenges of sustenance. To overcome these, farmers in diversity rich areas often plant several crop varieties in one season and adaptive complex of crop genetic diversity enables them to adopt crops suited to their ecological niches and cultural food production systems and practices. The wider environmental adaptability of agricultural diverse crops and varieties also provides the farmer with a risk aversion measure. The farmers transform or abandon particular varieties to suit their needs as the conservation of these resources are not the only concern for them (Saxena and Singh, 2006b). The on-farm conservation of genetic resources can only be achieved if maintaining crop populations and varieties is made advantageous to farmers. Various economic and cultural incentives for farmers to encourage them to continue growing the crop genetic resources are to be developed (Smale and Bellon, 1999; Gupta, 1995).

Ex situ *conservation*

The concern for conservation led to considerable worldwide efforts for *ex situ* conservation in gene banks. The *ex situ* gene bank plays an extremely important role in conservation of diversity for posterity. In some of the extreme cases, the *ex situ* gene banks have also helped in rehabilitating the *in situ* diversity. *Ex situ* conservation, that is particularly important for crop gene pools, began on a wide scale about three decades ago and is now practised, to some extent, in almost all countries.

Ex situ seed storage is the most convenient and widely used method for conservation of rice genetic resources. Based on the storage behaviour the seeds of rice are classified as 'orthodox seeds' since their moisture content can be reduced and the dried seeds can be stored at low temperature without affecting the viability. Thus these are conveniently stored in the seed gene banks. In essence the operation of a gene bank for the conservation of PGR is simple, that germplasm is received, recorded and maintained in storage under appropriate conditions. The sub-samples are withdrawn and dispatched as and when required. However, genebank management encompasses a multitude of activities that require proper planning, execution and a long term commitment to provide constant scientific and technical support for operation and maintenance of the facilities (Saxena *et al.*, 2003). The time scale of concern for conservation depends upon the mandate of the genetic resource centre and based on the requirement of the programmes the seed storage could either be for short, medium or long term. The seeds are hygroscopic in nature and the ambient relative humidity (RH) and temperature have a direct bearing on seed moisture content. The duration of storage decides the conditions to be maintained and the requirements become more exacting as the duration increases. The two most important factors that influence the seed longevity in the conserved accessions are the seed moisture and the storage temperature.

The short-term storage facilities are required by the various crop improvement centres generally for storage of the working collections for a time period up to 5 years. In the arid and cool agro-climatic zones, seeds of rice maintain a high germination capacity for as long as 1 to 2 years without much efforts and facilities. However, under warm and humid environments the stored seeds are liable to lose their viability very rapidly. Therefore, under these conditions storage facilities having the temperature of $20 \pm 2^{\circ}C$ and RH of around 35 per cent are required.

The value of collected and conserved germplasm can be realised only after proper characterisation and evaluation, complemented by biosystematic studies of the wild species. The responsibility of evaluation, supply for utilisation and maintenance of the germplasm for medium term are entrusted with the "active sites" and the collections are called as active collections. The active collections are stored in the medium term storage modules, where the conditions should be adequate for maintaining the viability of the seeds for 5-25 years without substantial change in genetic integrity and seed viability. To achieve this, seeds are stored in medium-term storage modules where, a temperature of $5^{\circ}C$ and 35-40 per cent RH are maintained. The seeds, after proper drying are stored in various types of containers such as cloth bags, metal cans or glass jars and kept in the storage racks of the modules. The active collections include national rice collections of about 12,000 accessions maintained at CRRI, Cuttack; 22,000 at IGKVV, Raipur; and 10,000 at DRR, Hyderabad. In addition the NBPGR Regional Stations also maintain the active collections of rice germplasm collected by them.

Base collections refer to those collections that are being conserved on long-term basis for posterity. These are distinct accessions that are closest to the original samples and are not to be disturbed except for regeneration of active collections. In India the responsibility of maintaining the *ex situ* collections of rice genetic resources on long-term basis is entrusted to the National Gene Bank at NBPGR. This National Gene Bank facility is the largest in Asia and ranks 4th in size of total germplasm holding in the world, which operates in the network mode to have an effective linkage with others engaged in rice improvement programmes and to make conserved PGR easily accessible to plant breeders. The conservation activities were initiated in 1980s with storage and frequent regeneration of germplasm and strengthened with the provisions of cold storage module facilities at NBPGR headquarters under the INDO-UK (Overseas Development Agency) project. The module had a capacity to hold 30000 accessions in two compartments maintained at -10°C and +4°C that catered to the needs of long- and medium-term storage. The storage capacity was enhanced to 2,50,000 accessions during 1986-1990, with the installation of four more long-term seed storage modules. This marked the initial period of acquisition of germplasm accumulated in the various rice research institutes and these were deposited in the National Gene Bank (Fig. 5.2). In 1996 under the joint Indo-United States Agency for International Development (USAID) project, one long-term and 12 medium-term seed storage modules were added, thus, creating facility for conservation of about a million accessions. At the same time 11 NAGS and 7 NBPGR Regional Stations, were also provided with medium-term seed storage modules and, thereby, establishing a network.

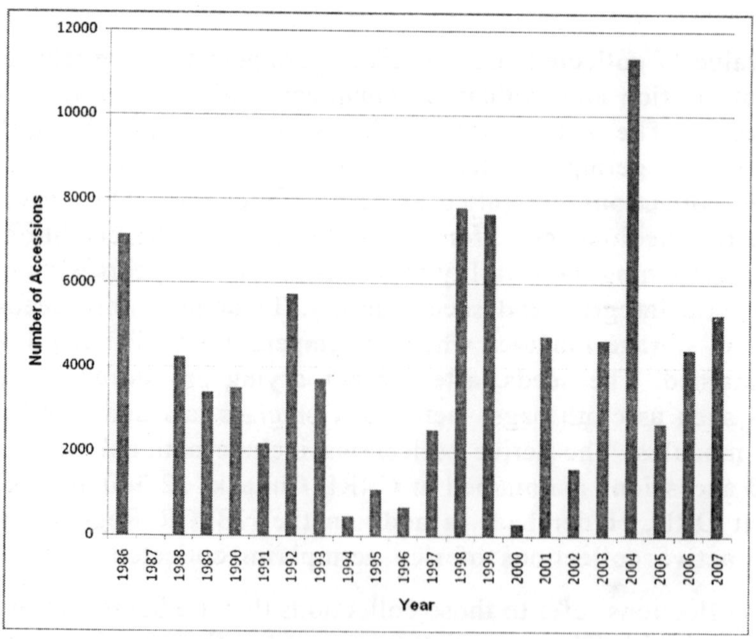

Fig. 5.2. Conservation of rice germplasm in the National Gene Bank over the years.

The conditions for maintaining base collections require that the seeds could be stored for a period of more than 50 years. To minimise the alteration in genetic structure and loss of viability in germplasm accessions during storage, the National Genebank preferably follows the genebank international standards as recommended by Biodiversity International (erstwhile International Plant Genetic Resources Institute) (IPGRI, 1994). The base collections are, therefore, being stored in modules maintained at -18^{0}C. For base collections to be put under long-term conservation, the preferable size of an accession in rice is 2,000 seeds. Maintenance of high viability of accessions during storage in long-term module is the basic requirement. Thus, all accessions before storage are tested for viability and seed moisture content. The most common and reliable method of testing seed viability is the germination test for which a minimum of two replicates using 100 seeds per replication is recommended for testing. The seeds to be tested are placed under optimal germination conditions, and if required treatment(s) to break dormancy are also given. The germination count is taken after a fixed period (generally before the seedlings exhaust the nutrients present in the seeds) and the normal seedlings as well as fresh ungerminated, hard seeds are counted to calculate the germination percentage. If the results of the replicates fall within the tolerance range, the average of the replicates is recorded as the percentage germination of that accession. The preferred standard for acceptance of germplasm for long-term storage is a minimum of 85 per cent germination. The preferred standard is 3-7 per cent moisture content for long-term storage. The desired seed moisture is achieved by equilibrating the seeds in special seed dryers maintaining temperature of 15°C and a 15 per cent RH. The dried seeds are hermetically sealed in tri-layered aluminium pouches and stored at -20°C. Apart from providing samples for replenishing the active collections, the base collections are not routinely disturbed except for the monitoring of viability of stored seeds. The preferred standard for monitoring is after 10 years of storage. Eventually, the conserved accessions are regenerated for replenishment when their viability drops below 85 per cent.

So far 85,254 accessions of rice are being maintained at the National Gene Bank at NBPGR as base collection. Of these collections 1,6224 accessions have been contributed by NBPGR, Regional Stations (Table 5.11).

Table 5.11. Important donor institutes which have contributed to Base collection conserved in NBPGR.

Donor Institute	Accessions (No.)
Central Rice Research Institute, Cuttack	30097
Indira Gandhi Krishi Vishwa Vidhyalaya, Raipur	22618

Directorate of Rice Research, Rajendra Nagar, Hyderabad	8828
National Bureau of Plant Genetic Resources Regional Station, Cuttack	7691
National Bureau of Plant Genetic Resources Regional Station, Thrissur	3495
National Bureau of Plant Genetic Resources Regional Station, Shillong	2499
Punjab Agricultural University, Ludhiana	946
National Bureau of Plant Genetic Resources Regional Station, Bhowali	846
National Bureau of Plant Genetic Resources Regional Station, Ranchi	732
Participatory Rural Development Foundation, Gorakhpur	722
Banaras Hindu University, Varanasi	680
Kerala Agriculture University, Pattambi	549
National Bureau of Plant Genetic Resources Regional Station, Hyderabad	522
Rice Research Station, Tamil Nadu, Aduthurai	490
ICAR Research Complex for NEH Region, Barapani	413
Sher-e-Kashmir University of Agricultural Sciences &Technology, Srinagar	350
Assam Agriculture University, Jorhat	340
Vivekanand Parvatiya Krishi Anusandhan Sansthan, Almora	300
College of Agriculture, JNKVV, Jabalpur	291
G.B. Pant University of Agriculture and Technology, Pantnagar	280
National Bureau of Plant Genetic Resources, New Delhi	275
Foundation for Development Integration, Guwahati	273
MS Swaminathan Research Foundation, Chennai	153
Narendra Dev University of Agriculture and Technology, Faizabad	144
National Bureau of Plant Genetic Resources Regional Station, Shimla	132
Rajendra Agricultural University, Pusa	109
T.M. Bhagalpur University, Bhagalpur	83
Orissa University of Agriculture and Technology, Bhuwaneshwar	77
Defence Agricultural Research Laboratory, Sonitpur	76
Gujarat Agricultural University, Vijapur, Gujarat	71
SASRD, Medziphema, Nagaland	64
Bidhan Chandra Krishi Viswavidyalaya, Kalyani	63
Regional Agricultural Research Station, Kerala Agricultural University	63

ICAR Research Complex, Tripura	60
Acharya N.G. Ranga Agricultural University , Rajendranagar, Hyderabad	56
Tamil Nadu Agricultural University, Coimbatore	52
Others	814
Total	**85,254**

Future Trends and Conclusions

The scientific and technical efforts in the field of PGR over the past three decades have focused on their conservation, use and management. Besides technical issues, the present day management of PGR encompasses, the international developments related to the intellectual property rights and trade, as they impinge upon the utilization of genetic resources, and the financial and cultural benefits accrued out of their use, particularly their access to the rural and native farming communities. In addition to technical developments the challenges for the future in the area of PGR management now include focused political, economic and ethical issues and in finding acceptable pragmatic norms both at global and national levels for the overall benefit to our society. The emerging threats to biodiversity and the regulated access to germplasm, requires emphasis on crop/ trait specific germplasm acquisition. Further, new tools, such as geographical information system and remote sensing, need to be deployed so as to supplement the existing ground data available with the genetic resource centres to explore the PGR particularly in difficult and inaccessible areas. The availability of improved techniques for gene transfer across species or genera or even kingdom and the threat of extinction of the wild populations suggest that the demand for wild germplasm will increase markedly and emphasis should be to collect and conserve wild and related species of crop plants that possess many useful genes for resistance/tolerance to biotic and abiotic stresses. A success of the genebank shall depend upon the diversity represented by the accessions it conserves rather than the number of accessions. It is essential for ensuring the availability of genotypes for promoting their utilization for present and future needs. It is of prime importance therefore, to estimate the diversity in the collections, take decision as to when the collection is complete, to identify the duplications if any and to monitor the genetic stability of the stored accessions. However, the utility of a genetic resource centre can only be realised if the germplasm is clearly defined and documented. The information regarding the germplasm identity including the passport, data and on the specific traits carried by these accessions requires extensive evaluation and characterisation. These tasks require substantial resources and inputs. A decentralised, coordinated systematic evaluation of the genetic wealth is of utmost importance. The advent of biotechnology as a potent tool in crop improvement has provided an opportunity that the genes an accession contains, rather than just the phenotypes exhibited, could be characterized. Using the molecular techniques it is now possible to flag accessions with major genes they carry, such as disease or pest resistance genes, male sterility factors, or growth habit genes. This would allow the breeders to utilise the genetic resources for pyramiding the disease resistance genes to obtain a more broadly based resistance (Rana, 2004).

The utilization of germplasm shall be considerably increased if the potential users of germplasm collections are able to find the information they are looking for. At present, only a few collections have a complete and user-friendly documentation. A network approach for sharing of information, involving all the stakeholders, is essential. Therefore, it is imperative to have harmonisation of documentation systems involving the genetic resource centres in the country and also at the community level so as to enable effective exchange of data and other information. The recent developments in information technology offer interesting opportunities for collation, retrieval and dissemination of information on PGR.

A long-term political agenda and commitment will have to be developed to assure long-term public-sector investment in *in situ* and *ex situ* conservations. This would require awareness generation of the people at various levels (policy makers, scientists, administrators, farmers etc.) about the value of PGR wealth. The emergence of new Intellectual Property Rights regimes, in relation to PGR call for working out modalities for benefit sharing between the private and public sectors to ensure continuity of germplasm exchange and synergy between the two sectors. Further, models will have to be developed for sharing of benefits with farmers and communities as well.

References:

Ahuja Uma, Ahuja SC, Thakrar Rashmi, Singh RK. 2008. Rice: A Nutraceutical. *Asian Agri-History* 12:93-108.

Bellon MR, Pham LJ, Jackson MT. 1997. Genetic conservation- a role for rice farmers. Pages 263-289 in Martel N, Ford-Lloyd BV and Hawkes JG (eds) *Plant Conservation: The In situ Approach.* Chapman & Hall, London.

Choudhury, BP, Subudhi HN, Biswal AK. 1999. *Bhitarkanika: The Wonderland of Orissa.* Pages 19-29 in Behura BK (ed)

Chunekar K, Pandey G (eds.) 1986. *Bhavaprakcsanighantu of Bhavamisra* (in Hindi). 7th Edn. Chaukhambha Bharati Academi, Varanasi, India, 984 pp.

Das DK, Oudhia P. 2001. Rice as medicinal plant in Chhattisgarh (India): A survey. *Agricultural Science Digest* 21:204–206.

Dikshit N, Malik SS, Tomar JB. 2004. *Evaluation Studies on Genetic Resources of Rice.* National

Bureau of Plant Genetic Resources, New Delhi, 159 pp

Gupta AK. 1995. Compensating local communities for conserving biodiversity: How much, who will, how and when. Pages 180-189 in Lakshman D, Guruswamy and McNeely JA (eds) *Converging Strategies*. Duke University Press, Durham and London. [Also published as Rewarding local communities for conserving biodiversity: The case of honey bee in protection of global biodiversity. Pages 32-39 in Hawkes JG, Maxted N and Ford-Lloyd BV (eds) *The Ex Situ Conservation of Plant Genetic Resources*. Kluwer Acad Pub, Dordrecht]

Hore DK. 2005. Rice diversity collection, conservation and management in northeastern India. *Genetic Resources and Crop Evolution* 52:1129-1140.

IPGRI. 1994. *Genebank Standards*. Food and Agriculture Organisation for United Nations. International Plant Genetic Resources Institute, Rome, Italy.

Khush GS. 2000. Taxonomy and origin of rice. Pages 5-13 in Singh RK, Singh US and Khush GS (eds) *Aromatic Rices*. Oxford and IBH Publishing, New Delhi.

Malik SS, Srivastava Umesh, Tomar JB, Bhandari DC, Pandey Anjula, Hore DK, Dikshit N. 2001. Plant exploration and germplasm collection. Pages 31-68 in Dhillon BS, Varaprasad KS, Srinivasan K, Singh Mahendra, Archak Sunil, Srivastava Umesh and Sharma GD (eds) *National Bureau of Plant Genetic Resources: A Compendium of Achievements*, NBPGR, New Delhi,

Misra, S, Choudhury A, Chattopadhyay S, Ghosh A. 1987. Sterols and terpenoids of the seagrass *Porteresia coarctata* grown in two different mangrove habitats of Sunderbans. *J Mar Biol Assoc India* 29(1-2):365-366.

NBPGR. 2007. Perspective Plan Vision 2025. In: Khetarpal RK, Anuradha Agrawal, Kavita Gupta and Rajan (eds). National Bureau of Plant Genetic Resources Pusa Campus, New Delhi, India. 66 pp.

Nene Y L. 2005. Rice Research in South Asia through Ages. *Asian Agri-History,* **9**:85-106.

Ni Junjian, Colowit Peter M, Mackill David J. 2002. Evaluation of genetic diversity in rice subspecies using microsatellite markers. *Crop Sci* 42:601-607.

Rana JC, Negi KS, Wani SA, Pradheep K, Saxena Sanjeev, Kak Anjali. 2008. Genetic resources of rice in the Western Himalayan Region of India. *Genetic Resources and Crop Evolution.* (submitted)

Rana RS. 2004. Emerging trends in managing and using rice genetic resources. *Asian Biotechnology and Development Review* 7(1):49-66.

Sasaki Takuji, Jianzhong Wu, Hiroshi Mizuno, Baltazar A. Antonio, Takashi Matsumoto. 2008. The rice genome sequence as an indispensable tool for crop improvement. Pages 3-12 in Hiro-Yuki Hirano, Yoshio Sano, Atsushi Hirai and Takuji Sasaki (eds) *Rice Biology in the Genomics Era*. Springer Berlin Heidelberg.

Saxena S, Gautam PL 2000. Rice germplasm diversity and conservation. *Rice India* 10:1-10.

Saxena S, Singh Anurudh K. 2006a. Legislations for providing protection to plant varieties and quality seed in megadiverse countries. Pages 64-78 in Verma DD, Arora S and Rai RK (eds) *Perspectives on Biodiversity-Vision for Megadiverse Countries*. Ministry of Environment and Forests (GOI), New Delhi.

Saxena S, Singh AK. 2006b. Revisit to definitions and need for inventorization or registration of landrace, folk, farmers' and traditional varieties. *Curr Sci* 9:451.

Saxena S, Chandak V, Ghosh SB, Sinha R, Jain N, Gupta AK. 2003. Cost of conservation of agrobiodiversity in India. Pages 137-173 in Virchow D (ed) *Efficient Conservation of Crop Genetic Diversity: Theoretical Approaches and Empirical Studies*. Springer-Verlag, Berlin, Heidelberg, New York.

Sharma BD, Rana JC. 2005. *Plant Genetic Resources of Western Himalaya – Status and Prospects*. Bishen Singh Mahendra Pal Singh. Dehrdun, India.

Sharma DK Sahu GR, Shrivastava MN 1987. Present status of rice germplasm in Madhya Pradesh (India). *Oryza* 24:127-131.

Sharma SD. 1982. Collection and evaluation of rice from NE India. *Plant Genet Resour Newsl* 50:62–69.

Sharma SD, Vellanki JMR, Hakim KL, Singh RK 1971. Primitive and current cultivars of rice in Assam - a rich source of valuable genes. *Curr Sci* 40:126–128.

Siddiq EA. 2000. Rice. Pages 1-85 in Chopra VL (ed.) *Breeding Field Crops*. Oxford and IBH Publishing Co, New Delhi.

Singh AK, Srinivasan K, Jain SK, Saxena S. 2001. Germplasm Conservation: Seed Genebank. In: Pages 183-208 in Dhillon BS, Varaprasad KS, Srinivasan K, Singh M, Archak S, Srivastava U and Sharma GD (eds) *National Bureau of Plant Genetic Resources: A Compendium of Achievement*. NBPGR, New Delhi.

Singh RK, Gautam PL, Saxena S, Singh S. 2000. Scented rice germplasm: Conservation, evaluation and utilisation. Pages 107-133 in Singh RK, G Khush and US Singh (eds) *Aromatic Rices*. Oxford and IBH, New Delhi.

Singh RV, Deep Chand, Tyagi Vandana, Gupta Sandhya. 2006. Exchange of Crop Germplasm: Achievements and Challenges. Pages 63-82 in Singh AK, Kalyani Srinivasan, Saxena S and Dhillon BS (eds) *Hundred Years of Plant Genetic Resources Management in India*. NBPGR, New Delhi.

Smale M, Bellon MR. 1999. A conceptual framework for valuing on-farm genetic resources. Pages 387- 408 Wood D and Lenné JM (eds) *Agrobiodiversity: Characterization, Utilization and Management*. CABI Publishing, London.

Whyte RO. 1972. The Gramineae. Wild and cultivated plants of monsoonal and equatorial Asia. 1. Southeast Asia. *Asian Perspective* 15:127-151.

World Conservation Monitoring Center (WCMC). 1992. Global biodiversity: Status of the earth's living resources. Pages 1-3 in *National History Museum*, London in association with IUCN, UNEP, WWF, WRI. Chapman and Hall, London.

Genetic Resources of Rice in India : 185-227
Editor : S. D. Sharma
Today & Tomorrow's Printers and Publishers, New Delhi - 110 002 (India)

Chapter 6

Collection, Evaluation and Characterization of Rice Germplasm for North East India

A. Pattanayak, B. K. Sarma, J. K. Singh, M. Borthakur, T. P. Devi and P. Khate

The wide variation in topography coupled with favourable agro-climatic conditions of the NE region has given rise to enormous crop genetic diversity in the region (Prasad *et al.*, 1993). The variability observed by different rice workers in the collections from North East India is quite enormous.

Collection and Maintenance

The first systematic collection of rice germplasm from NE India was executed under a PL 480 Project on 'Collection of Rice Germplasm' during 1968-73 conducted jointly by ICAR with the financial support from USAID. The rice collection in NE states of India which are supposed to be nearer to the center of origin of rice, resulted in a collection of 6,630 local cultivars and landraces designated as Assam Rice Collections (ARC) from all the districts of Arunachal Pradesh, Nagaland, Manipur, Tripura, Meghalaya and from Lakhimpur, Kamrup and Goalpara districts of Assam (Sharma, 1982). This region deserved priority for several reasons: (a) collections from NE region of India were poorly represented in the national collection, (b) topographical variations on the region has caused climatic variations, (c) different types of rice cultivars have been introduced in the region due to ethnic immigrations, (d) isolation of the region has caused restriction of varietal exchange, and (f) the region falls in the fringe of the center of origin of rice (Sharma, 1987). The collection was exhaustive from Arunachal Pradesh, Meghalaya, Manipur and Nagaland. The present Assam and Tripura were covered only partly whereas Mizoram and Sikkim were left out. The collections made under this Indian Programme were a valuable addition to International Rice Research Institute (IRRI) genebank. Since 1984 sporadic programmes of rice germplasm collection from NE India by NBPGR and ICAR Research Complex for NEH Regon (ICAR-RC-NEHR) resulted in 593 and 587 accessions (Sarma *et al.,* 1987). The collaborative efforts were accelerated with the establishment of NBPGR and, till 1995, a collection 3,477 accessions of rice germplasm were made from Assam, Tripura and Sikkim. While collecting germplasm, major emphasis was laid on tapping such areas where native diversity occurred and genetic erosion was evident.

Asthana and Majumdar (1981) collected hill rice germplasm from NE hills. Rice collections from Sikkim, Arunachal Pradesh, Tripura and Mizoram were also reported by various workers (Arora, 1982; Kocchar, 1987; Sasikumar and Sardana 1987). In addition, Shillong centre of NBPGR did regular explorations to various sites of NE India during 1978-1995 and collected 1698 accessions. The Assam Agricultural University (AAU) also maintains more than 3000 accessions of indigenous rice germplasm in its rice research stations at Karimganj (glutinous rice), Titabor (*Sali* rice), Diphu (hill rice) and North Lakhimpur (*bao* or deep water rice) (Das, 1987). About 25% of the rice germplasm collected in India is from NE region. ICAR-RC-NEHR, collected 2,300 rice germplasm under plant biodiversity programme of National Agricultural Technology Project (NATP-PB) during 1999-2005. In 2007 and 2008, explorers from ICAR Research Complex made specific trips to Assam and Arunachal Pradesh and collected about 150 germplasm of various categories including deep water and *Khamti* rice. The rice germplasm maintained at different locations and districts explored so far are shown in Table 6.1 and Table 6.2.

Table 6.1. Rice germplasm maintained at rice research stations of NE India.

State/Institute	Location	Number of accessions maintained
NBPGR	Umiam, Meghalaya	2,600
AAU, RRRS	Titabor, Karimganj, Lakhimpur	3,299
Manipur	Wangbal	240
NEH Complex	Barapani, Tripura	1,875

Table 6.2. Districts covered under various exploration programmes.

State	Districts covered
Arunachal Pradesh	West Kameng, East Kameng, Changlang, Tirap and Lohit districts.
Assam	Majuli Island, Barpeta, Bongaigaon, Jorhat, Dhubri, Lakhimpur, Dhemaji and Tinsukia.
Manipur	Imphal East, Thoubal, Manipur and Bishnupur.
Meghalaya	East Khasi Hills, Ri-Bhoi and Jaintia Hills
Mizoram	Aizawl, Kolasib, Champhai, Serchip and Lunglei
Nagaland	Kohima, Dimapur, Okhwa and Paren
Sikkim	East, West, North and South Sikkim districts
Tripura	North and South Tripura districts
West Bengal	Darjeeling

Krishnamurthy and Sharma (1970) collected rice germplasm from Manipur and recorded wide diversity in the collection. Sasikumar and Sardana (1987) collected 134 landraces of submergence tolerant and aromatic rice from Tripura. Kocchar (1987) collected 129 upland local cultivars of Arunachal Pradesh from shifting cultivation system. Sharma and Hore (1989) collected 252 rice germplasm from Assam (112), Meghalaya (48), Tripura (28) and Nagaland (57). However, after the PL-480 project, major collection effort was made under the NATP-PB programme. Sarma *et al.* (2000) collected 572 landraces of rice from different parts of NE states. In addition, 20 diverse rice germplasm from Arunachal Pradesh (Singh *et al.*, 2002), 53 from Upper Assam and Lohit and Dibang valley of Arunachal Pradesh (Rathi *et al.*, 2002) which included scented rice, soft rice, sticky rice, etc. with long, bold, fine, grains having yellow, red, brown and black colours, 75 from Changlang and Tirap districts of Arunachal Pradesh (Sarma *et al.*, 2004), 68 from Mamit district of Mizoram (Pal *et al.*, 2004) and 53 by Singh *et al.* (2002) were collected under NATP-PBD. Besides ICAR Complex for NEH Region, collections were also made by NBPGR Regional Station, Assam Agricultural University, Nagaland University and Central Agricultural University. Srivastava and Gautam (1999) suggested systematic collection based on priority areas and, accordingly, listed priority areas for rice collection (Table 6.3).

Preliminary Studies

Studies have been made earlier on the rice collections made in the NE region under PL 480, which were assigned ARC accession numbers. Some collections made earlier from Assam and Manipur were included in the catalogue

Table 6.3. Priority areas for rice germplasm collection of NE India.

Trait	Areas to be explored
Short stature types	North Cachar district of Assam, Manipur and Arunachal Pradesh
Glabrous husk forms	Along Myanmar border in Arunachal Pradesh
Waxy endosperm types	Mizoram and Garo Hills (Meghalaya)
Japonica types	Kameng and Subansiri in Arunachal Pradesh
Drought tolerant types	Garo Hills, (Meghalaya)
Gall midge and green leaf hopper resistant types	Karbi Anglong district (Assam), Khasi, Jaintia and Garo Hills (Megalaya)
Tungro virus resistant forms Blast resistant types	Kamrup and Goalpara district of Assam, Meghalaya
Bacterial leaf blight resistant forms and wild rice	Assam

of National Germplasm Collection of Rice (Padmanabhan *et al.,* 1971). Sikkim rice was described by Kihara and Nakao (1960) and rice from Naga Hills was described by Kulkarni (1962). ARC material was studied at DRR, Hyderabad and CRRI, Cuttack. Collections made by ICAR Complex for NEH Region and its centres, collections made by NBPGR, New Delhi and a good number of collections maintained by AAU were characterized for various agro-botanic characters and stress tolerance. The collections made by ICAR Complex during 1999 to 2003 under NATP-PBD were evaluated at Umiam, Meghalaya for various morphological and agronomical characters. Malik and Sharma (2005) listed the indigenous high quality non-aromatic and aromatic types collected from Assam. These types are: Manoharsali, Aijani (Mahsuri), Rangaduria, Lalahu, Kekuabao (non-aromatic), Gunibora (glutinous), Badshabhog, Prasadbhog, Malbhog, Ketekijoha (aromatic).

The detailed study of ARC shows that they possess many desirable characters. One hundred ninety land races were studied for genetic divergence and these were classified into 9 clusters (Vairavan *et al.,* 1973). The study showed that some clusters resembled *japonica* plant and grain type whereas others occupied intermediate position between *japonica* and *indica.* Plant height followed by leaf area was the most important characters for primary differentiation and 100-grain weight and amylose content were important for secondary differentiation. According to them, the first two characters were subjected to natural selection while the other two were subjected to artificial selection. They also could not find any relationship between geographical distance and genetic diversity. Later, Srivastava (1978) classified 52 varieties from NE India into three groups and attributed to gene-ecological differentiation for a constellation of characters identifying each group, viz., i) thick culm types with few tillers, moderate to high 100-grain weight and medium to long coarse grains; ii) thin culmed types with more profuse tillering, short to long fine grains and good grain quality and iii) types with beaded arrangements of grains on the panicle branches, low to moderate 100-grain weight and fairly profuse tillering. Not only isolated characters but whole constellations of *japonica* characters such as thin culm, shy tillering, short panicle etc. in ARC was reported (Seetharaman, 1974).

Genetic Studies

The variability observed by different rice workers in the NE material is quite enormous. The wide variation in topography coupled with favourable agro-climatic conditions of the NE region has given rise to enormous crop genetic diversity in the region (Prasad *et al.*, 1983). Sharma *et al.* (1971) reported maximum variation in grain yield in cultivars of Assam rice. High variabilities were also recorded in panicle weight, plant height, grain/panicle and 100 seed weight. Characters like days to flowering, panicle height and days to maturity showed low variability. They reported high heritability for the characters like days to flowering, grain yield, 100 seed weight, panicle weight and grain/panicle. Moderate heritability was observed for panicle length, days to maturity and plant height. They also found high GA associated with high heritability for days to flowering. Ghosh *et al.* (1981) reported high phenotypic and genotypic variability for grain breadth and length/breadth ratio and low variation in grain length, milling and hulling percent in the indigenous rice varieties of Meghalaya. Debchoudhury and Das (1997) recorded moderately high genotypic and phenotypic coefficient of variation in characters like grain yield, grains/panicle and plant height while it was low for the parameters like days to flowering, days to maturity and panicle length in deep water rice. High heritability was observed for days to maturity, days to 50% flowering and grains/panicle, while high genetic advance was observed in grain yield and grains/panicle. Borbora and Hazarika (1998) reported that panicle characters like sterile spikelets, fertile spikelets, grain yield/panicle, 1000-grain weight, secondary branches/panicle and panicle number/plant showed high to moderate genotypic coefficient of variation (GCV). They recorded high heritability for days to 50% flowering, plant height, panicle length, number of primary and secondary branches, number of chaffy and filled grains/panicle, 1000-grain weight and grain yield/panicle. Pandey and Gupta (1991) reported significant genetic variability amongst the genotypes for yield contributing traits with high GCV in yield /plant in iron toxic soils at high altitudes of Meghalaya. Talukdar and Talukdar (1998) studied the genetics of photoperiod insensitivity in indigenous scented rice of Assam and suggested involvement of two dominant duplicate genes for control of photoperiod insensitivity.

Forty-three upland and 90 lowland rice cultivars were evaluated at Umiam (Meghalaya) during 2003 to estimate their genetic variance, heritability, GA, GA as % of mean, correlation coefficients, path coefficients and genetic divergence (Singh, 2007; Barthakur, 2007). In upland germplasm, high PCV and GCV were recorded for grains/panicle, leaf area, panicle weight, grain yield/plant and plant height indicating variability in these characters was genetic and interaction of environment was low. Days to maturity, days to 50% flowering, spikelet fertility (%) and panicle length exhibited low magnitude of phenotypic and genotypic variation (Table 6.4).

Table 6.4. Genetic variance in 43 upland rice germplasm.

Character	SE	PV	GV	PCV	GCV	H	GA	GA as % of mean
LA	2.2	258.53	244.06	33.08	32.14	94.4	31.27	64.33
DF	1.21	126.04	121.62	11.26	11.06	96.5	22.32	22.38
DM	1.34	126.04	120.67	8.22	8.04	95.7	22.14	16.21
PH	4.73	844.37	777.38	27.23	26.13	92.1	55.11	51.64
PL	1.27	16.01	11.16	16.8	14.03	69.7	5.74	24.12
BP	0.65	6.6	5.32	23.76	21.32	80.6	4.26	39.42
PW	0.14	0.6	0.55	29.47	28.06	90.7	1.45	55.04
EBT	0.25	1.62	1.43	19.87	18.65	88.2	2.31	36.08
SW	0.04	0.26	0.26	20.29	20.15	98.6	1.04	41.22
GP	4.98	1232.83	1158.48	36.91	35.78	94	67.97	71.44
SF	0.9	77.29	74.84	11.92	11.73	96.8	17.54	23.77
GLBR	0.1	0.48	0.45	21.32	20.65	93.9	1.34	41.22
GY	0.68	19.21	17.84	29.33	28.26	92.9	8.38	56.1

LA: leaf area (cm^2); DF: days to flowering; DM: days to maturity; PH: plant height (cm); PL: panicle length (cm); BP: no. of primary branches/panicle; PW: panicle weight (g); EBT: no. of effective tillers/plant; SW: 100 seed weight (g); GP: grains/panicle; SF: spikelet fertility (%); GLBR: grain length breadth ratio; GY: grain yield/plant; SE: standard error; PV: phenotypic variance; GV: genotypic variance; PCV: phenotypic coefficient of variation; GCV: genotypic coefficient of variation; H: heritability in broad sense; GA: genetic advance

In lowland rice germplasm, it was recorded that grain yield/plant exhibited maximum PCV followed by panicle weight and ear bearing tiller. Lowest PCV was observed for days to flowering followed by leaves/tiller. Maximum GCV was estimated for panicle weight followed by yield/plant; the lowest GCV was for panicle length (Table 6.5).

Table 6.5. Genetic variance in 90 lowland rice germplasm.

Character	GV	PV	GCV	PCV	Heritability	GA (%)	GA as % of mean
DF	84.63	88.17	8.8	8.98	0.96	18.57	17.77
PH	252.35	293.74	15.51	16.74	0.86	30.33	29.62
LT	0.19	0.22	9.29	9.98	0.87	0.83	17.87
EBT	0.28	1.17	9.49	19.52	0.24	0.53	9.51
PW	0.22	0.46	21.76	31.42	0.48	0.67	31.03
PL	2.43	7.29	6.94	12.02	0.33	1.86	8.26
BP	0.75	1.83	8.77	13.71	0.41	1.14	11.56
FLL	16.53	28.94	13.8	18.26	0.57	6.33	21.48
FLB	0.01	0.04	7.87	14.2	0.31	0.12	8.99
LL	24.1	40.6	13.59	17.65	0.59	7.79	21.58
LB	0.01	0.03	9.97	13.84	0.52	0.17	14.79
GY	2.99	13.49	18.77	39.83	0.22	1.68	18.21
SW	0.03	0.09	7.96	12.97	0.38	0.23	10.06
GR	0.15	0.16	13.52	13.93	0.94	0.78	27.02

EBT: ear bearing tiller; FLL: flag leaf length; FLB: flag leaf breadth; LL: leaf length; LB: leaf breadth; GR: grain ratio.

Heritability estimates were recorded high for all the characters in upland germplasm except for panicle length, which ranged from 69.70% to 98.60%. The heritability estimates were found to be more than 90% in traits like panicle weight, plant height, grain yield/plant, grain length breadth ratio, grain/panicle, leaf area, days to maturity, days to 50% flowering and spikelet fertility percent. In the lowland germplasm, maximum heritability percent (in broad sense) was observed for days to flowering, closely followed by grain ratio, leaves/tiller and plant height whereas the lowest heritability was recorded in grain yield/plant.

Genetic Divergence

De *et al.* (1988) studied the genetic divergence in 77 rice strains and grouped them into 13 clusters. They reported the association of ARC strains with non-ARC in the same group had great significance. The 67 ARC strains showed high level of genetic diversity, indicating the importance of northeastern region of India as rich source of rice germplasm diversity. The ARC strains also generated noteworthy variability among them with regard to intra and inter cluster distance. Two characters viz., 100-seed weight and number of grain/panicle was the highest contributor to D^2 values. They concluded that genetic similarity between strains might be due to the application of directional selection pressure for realizing high yield. They suggested that the genetic drift and selection in different environments might have caused greater genetic diversity than the geographical distance. Sardana *et al.* (1997) studied genetic divergence in rice germplasm of Tripura and grouped the 82 genotypes into 18 clusters. They also reported that the number of grains/panicle, leaf area, grain weight/panicle, effective tillers/plant and grain yield/plant were the major components contributing to the genetic diversity. Ahmed and Borah (1999) studied genetic advance in 85 glutinous rice germplasm of Assam and grouped these genotypes into 12 clusters on the basis of 13 agro-morphological characters. The 85 genotypes grouped into different clusters irrespective of their place of origin exhibited genetic drift and they were of the view that human selection in different environments caused greater genetic diversity than geographical distance. Consideration of relative contribution of different characters revealed that tiller number and grain yield contributed the highest to total divergence. These two traits along with panicle/hill, grain fertility and grain/panicle accounted for about 50% of the total diversity. Contribution of grain length, grain breadth and days to 50% flowering to total diversity was low. Pandey and Gupta (1991) while subjecting 50 rice genotypes of Meghalaya to D^2 analysis reported six clusters where days to flowering, plant height, branches/panicle and 100-seed weight were major contributors to total divergence. Singh (2007) grouped 43 upland rice germplasm of NE Hills (core collection) into 9 clusters (Table 6.6). The study revealed absence of parallelism between geographical diversity and genetic diversity.

Table 6.6. Cluster composition of 43 upland rice germplasm.

Cluster	Number of entries	Cluster composition	Area/location
I	8	Kawnlong, Sunidhan, Mizo local., Trivuban, Rungtai, Buhpui, Sanyak, Taikawnganpui	Mizoram, AP
II	10	Nilhari, Bangorice, Rungfai, Mangbu, Col. 2, Majhera 3, Keieng, Pangra, Idaw 4, Koetodhan	AP, Mizoram, Manipur, Sikkim
III	9	Sasaing, Take, Bhalum 1, Mangthan, Samidhan, Nunmam, Maophoh, Tinmahina, Ngokwai	AP, Sikkim, Meghalaya, Mizoram
IV	1	Zichum	Mizoram
V	1	Chakachang cha	Nagaland
VI	6	Fazaitai, Tialte, Bali, Akatan, Tai 1, Sakhi	Mizoram, AP
VII	1	Buhbawn	Mizoram
VIII	1	Paukhidhan	Sikkim
IX	6	Tondonba, Palidhan, Kalamunia, Galadhan, Lahudhan, Col. 3	Manipur, Sikkim, Meghalaya, Tripura

Borthakur (2007) subjected ninety lowland genotypes (indigenous and breeding lines) to D^2 analysis and derived 9 clusters (Table 6.7). The pattern of distribution of genotypes within various clusters was independent of geographical distribution. The components, which contributed maximum to the total divergence, were days to flowering followed by grain ratio and leaf/tiller.

Table 6.7. Clustering pattern of 90 lowland rice germplasm on the basis of genetic divergence.

Cluster	No. of genotypes	Genotypes
I	47	CN 881-261, CR 796-12, DKP 14, IR 45970-110-3-2, PNA 1121, F4-150-1, P 6975F-MR-1-1-1, KS 1, KS 2, LT 5, LT9, LP 8, LP 9, TRC 8776, TRC 64-104, TRC 87-252, V 88-1011, Zli, RCPL 1-3pp, RCPL 1-4p, RCPL 3-5, RCPL 3-6, RCPL 3-2, RCPL 3-4, RCPL 3-8, RCPL 4-12, RCPL 4-18, Bagonbishi, Govind, Lanten, Marrow, Malaria, Bapnah, Manipur, Prasad, Punsi Mutant, Akatan, Kanchang 4, Kemeyha, Pant Dhan 4, Ngoba (Nagaland), Ngoba, DR 92, RCPL 1-87-8

II	1	LP 7
III	8	Miakrilha, Manipur, Manipur (Eithali), RCPL 1-2, Lakang, Kemehya Sopa, Mandri 1, Mandri 2
IV	22	VRDKP, CU 5, IET 7614, LT 3, LP 5, LP 6, RCM 5, RCM 6, SP 1, B 15-1, TRC 10, TRC 87-251, UPR 89-1190, RCPL 1-1pp, RCPL 1-8, RCPL 4-2, Boro 1, Sarassa, Ghavri, Rasi, Grouping Tsuk, CV 4
V	4	Tierei, Kaulosing, Krishna, La Pyngad (NC)
VI	5	HPU 626, LT 4, Dhali Boro, UPR 88-1-18, RCPL 1-6
VII	1	UPR-103-80-1-2
VIII	1	VL 1179
IX	1	RCPL 7-5

Glaszmann *et al.* (1987) studied variations of 289 cultivars from northeast India at 14 isozyme loci. The study inferred that the morphological diversity seen in the germplasm of NE India is due to very high genic diversity and the co-existence of most varietal types. However, the diversity seen in the region was not different from the diversity seen for Asia as a whole. Sarma and Bahar (2005) studied genetic variations in *bora* rice of Assam using RAPD markers. Their study revealed existence of considerable variations among the *bora* rice genotypes although several duplicates were identified. Bhuyan *et al.*, (2007) studied genetic variations among 24 lowland rice using RAPD and ISSR markers. Their study suggested that the lowland rice of Assam can be broadly classified as scented and non-scented. The non-scented group can be further sub-classified as coarse/ bold grain and medium-slender grain. They concluded that their study actually supported farmers' classification of *sali* (coarse grain), *lahi* (medium fine grain) and *joha* (scented fine grain). Classification of scented (*joha*) rice based on seed storage protein was also done at Gauhati University. Fourteen *joha* cultivars of Assam were evaluated for their nutritive value and, on the basis of seed protein profile, molecular characterization was done as well as their phylogenetic relationship was established. (J. Datta Roy, personal communication).

Characterization

Sharma (1987) described the characteristics of rice cultivars collected under ARC. The rice cultivars from Tawang area of Arunachal Pradesh exhibited *japonica* characters like globose grain, thin culm, narrow and dark green leaves, drooping flag leaf and thermo sensitivity. Many of these types were phenol positive and their kernels stain light brown with potassium iodide. The *japonica* characters were less prominent in the collections of east areas. The pubescent husk types were common in western Arunachal Pradesh whereas glabrous types were common in the eastern parts. About 40% of the cultivars collected from Nagaland were intermediate between *japonica* and *indica* types. Nagaland was considered as the centre of distribution of glutinous and glabrous varieties because of highest percentage of occurrence of such varieties. The hill varieties of Manipur and varieties from Tripura were also intermediate between *japonica* and *indica* with glabrous husk and glutinous kernel. However, the collections from Tripura lack glabrous types and contain a few glutinous types but numbers of fine and scented types were more. The varieties of the three districts of Meghalaya showed many distinctive features. Collections from Garo Hills showed *japonica* characters and many of them were fine types. The glabrous husk character was found in some Khasi and Jaintia varieties. Blast resistant types and bacterial blight resistant types were identified from the collections of Meghalaya. The green leaf hopper resistant types were more in Garo Hills collection while stem borer and gall midge resistant types were distributed randomly in the state. The collections from Garo Hills showed highest percentage of varieties resistant to drought despite a good rainfall in the area. A large number of collections from the two hill districts of Assam were found to be dwarf like Taichung (Native)-1.

Studies with ARC

Seetharaman *et al.* (1974) studied 46 short statured cultures from NE India for nine characters and divided them into two broad groups based on tillering, panicle size, flag leaf angle, senescence and grain density. They indicated that shy tillering habit, slow leaf senescence and wide leaf angle are desirable characters for direct seeding in hill areas while high tillering and narrow leaf angle are desirable for low altitudes. The two groups formed by Seetharaman *et al.* (1974) with 46 short statured cultivars from NE India were: (a) Varieties with moderately high to high tillering habit, relatively long panicles with varying grain number and different grain size, narrow leaf angle and early senescence, and (b) varieties with *japonica* type grain, shy tillering, short panicle, dense grain, flag leaf thicker, horizontal or descending and late senescence. Some landraces from northeastern region with *glaberrima* like characters viz., truncated ligule, absence of secondary branches and red kernel were also reported (Seetharaman and Ghorai, 1976). These studies strengthen the observation made by Kihara (1968) that racial differentiation in *O sativa* took place in this region. Some of the collections had high yield potential, particularly at low phosphorus soil conditions (Rao and Srinivasan, 1978). The frequency of fine grain types of Arunachal Pradesh landraces except of Kameng and Subansiri districts ranged from 9-32% (Sharma, 1982). Vairavan and Siddiq (1973) studied the relative sexual affinity in the cross of TKM 6 (*indica*) and 14 different types of rice from NE India. They found a wide range of sterility in the hybrids and were of the opinion that the collections differed from typical *indica* types from plains in varying degrees. This variation may be related to the region being located on the periphery of center of origin of *Oryza sativa*. However, they suggested that the degree of sterility is in no way related to sub-specific differentiation and the intravarietal hybrid sterility might be largely as a result of disharmonious gene combinations rather than detectable structural differences.

Srivastava and Nanda (1977) found higher protein content in some of the *japonica* grain types and semi dwarf types. Hakim and Sharma (1974) recorded a good number of drought tolerant types particularly in the collections from high rainfall area. Shastry *et al.* (1971) in their studies on resistance to disease and insect pests listed several ARC entries as donors. Some of the landraces possess high yield potential particularly under low phosphorus and low temperature (Rao and Srinivasan, 1978). Apart from this, some of the entries showed higher protein content in some of *japonica* grain types and semi-dwarf types of *indica*. Hakim and Sharma (1974) analyzed the geographic distribution of the various morphological and physiological traits in the germplasm and found that most of them were either localized in distribution or showed mosaic distribution (Table 6.8).

Table 6.8. Distribution of accessions according to some morphological characters in ARC.

State	Percentage of landraces showing the traits					
	Japonica plant type	Semi-dwarf type	Glabrous husk	Waxy endosperm	Fine grained	Scented kernel
Arunachal	24	1	11	8	16	2
Nagaland	3	16	22	18	10	6
Manipur	1	0	5	12	16	2
Tripura	5	0	0	4	11	3
Meghalaya	0	1	2	11	17	5
Assam (Hills)	2	5	1	11	12	4

Shastri *et al.* (1971) identified several ARC cultivars as donors for resistance to disease and insect pests (Table 6.9). Seshu *et al.* (1974) while studying new alternate source of dwarfism in ARC materials identified 3 promising dwarfs, viz., ARC 5981A, ARC 12805 and ARC 5929. At Pusa, one cultivar ARC 11218-2 exhibited resistance to 5 pests, 3 landraces (ARC 5932-2, 10598 and 11205) showed resistance to 4 insect pests while 11 other ARC varieties showed resistance to 3 insect pests (Datta, 1987).

Table 6.9. ARC cultivars identified as donors for resistance to disease and pest.

Character	Entries
Resistance to blast	ARC 5805 (Horu Sali), ARC 6225 (Khamti Sali), ARC 7105 (Sill Sali)
Resistance to bacterial blight	ARC 6114 (Kishore Bardhan), ARC 1524 (Khamti Sali)
Resistance to TRV	ARC 5846 (Badal Sali), ARC 5988 (Betguti Sali), ARC 6561 (Kola Sali), ARC 6568 (Haladhar Sali)
Resistance to gall midge	ARC 5810 (Laudubi Sali), ARC 5947A (Ekara Sali), ARC 6163 (Bogi Sali), ARC 6232 (Kula Sali), ARC 6234 (Ranga Sali), ARC 6238 (Jiton Sali), ARC 6557 (Haladhar Sali), ARC 6631 (Taraboli), ARC 7213 (Boni Sali), ARC 7308 (Hati Sali), ARC 1210 (Hatimul Sali), ARC 1307 (Badshabhog)
Resistance to stem borer	ARC 6033 (Ban joha), ARC 6045 (Chakua dhan), ARC 7080
Resistance to GLH	ARC 7128 (Bangal Sali), ARC 1054 (Kanchi Sali), ARC 1307 (Badshabhog)

State-wise collections

Assam

The earlier collections of rice made by the Rice Research Station, Assam included *kharif* or *sali* rice, *ahu* or summer rice and some other photosensitive *khaif* rice such as *bora* (pastry rice), *chakoa* (soft rice), *joha* (scented and fine grained rice) and all the collections were described in detail (Mitra *et al.*, 1942). The cultivars were classed according to maturity group and size of the grain. Some of the cultivars were of specific interest. The earliest variety, which was photo-insensitive and suited for direct sowing under rainfed condition was Dumai. It took less than 80 days to mature, and hence suitable for late planting and for direct sowing as a pre-flood crop in summer (Borthakur, 1993). The cultivar, Badshahbhog was a fine-grained aromatic variety. Das *et al.* (1981) in a study of 136 tall *indica* rice cultivars from *sali* germplasm collection of AAU grouped the cultivars into three categories on the basis of length/breadth ratio (l/b) of grain size.

Das *et al.* (1981) also reported substantial variability in *ahu* (autumn) rice of Assam. *Ahu* rice cultivation starts with the help of pre-monsoon showers in March-April and the crop is harvested in June-July. They studied 686 traditional *ahu* accessions for days to flowering, plant height, panicle number/ hill, grain weight/panicle, grain weight and grain yield/ hill and classified the varieties into 96 groups and of these 21 were identified as promising in yield. Altogether, 131 traditional *ahu* accessions were found as high yielders (4 t/ha or more). Among the three locally recognized classes of *ahu*, viz., *dumai* (DM), *murali* and *chengri* (MC) and *aus* (AS), DM was clearly distinguished from AS as having less duration, height and grain weight/panicle. MC was intermediate in these characters. The following *ahu* cultivars were identified for their special characteristics.

Manipuri Dumai: resistant to drought, stem borer, root-knot nematode, early maturing

Koimurali: resistant to drought, early maturing

Maibi: high drought tolerance and recovery

Garem: resistant to root-knot nematode, white kernelled

Nilaji: early maturing and selected for growing in flood affected areas

In Assam, glutinous rice is used in breakfast, for preparation of dishes (*pitha*) and a local sweet rice beer. In certain areas of Upper Brahmaputra valley, it is used as a regular breakfast diet. Studies on glutinous rice of Assam revealed certain diagnostic characteristics, which are more or less common in all glutinous rice varieties of the world (Das, 1987). The glutinous rice kernels are opaque and cooked rice is sticky and glossy and retains its shape with very little volume expansion. Glutinous rice varieties of Assam, locally known as *bora* or *birain*, are tall and photoperiod sensitive and are grown along with *sali* rice (winter rice) as transplanted crops. Therefore, this group of rice is considered as a special category of *sali* rice only. The study of 85 glutinous rice varieties of Assam (Das, 1987) showed substantial variability in quantitative characters (Table 6.10). The presence of anthocyanin pigmentation in different plant parts is an important feature of glutinous rice of Assam. Only two varieties were found to be non-pigmented; the remaining 83 showed pigmentation in one or more plant parts. About 50% of the varieties showed simultaneous pigmentation in inner glume, apiculus and inside of leaf sheath. One accession B 119 showed pigmentation in as many as 11 plant parts.

While there is a distinct tendency for simultaneous pigmentation in different plant parts, there appears to be no association between pigmented plant parts and any quantitative characters. Genetical studies have revealed linked association of glutinous endosperm with anthocyanin pigmentation in any plant part. Close linkages of endosperm types have, however, been observed with pigmented outer glume and pigmented internodes with cross-over percentages varying from 6.55 to 13.0 and 10.44 to 13.26 respectively. The studies have also showed that inheritance of endosperm type is virtually independent of any quantitative character thereby indicating the possibility of combining glutinous character with the desired level of any yield or fitness component in breeding for varietal improvement (Das, 1987).

Table 6.10. Variability in 19 characters in 85 glutinous cultivars of NE India.

Character	Mean	Range	GCV (%)
Grain yield (kg/ha)	1741	517-3133	1.24
Days to flower	115	107-129	5
Days to maturity	143	133-156	4.22
Plant height (cm)	133	119-156	6
Tillers/hill	6.65	3.80-12.00	31
Panicles/hill	5.73	2.80-11.80	27
Grain weight/panicle (g)	1.27	0.55-2.84	36
Panicle length (cm)	22.07	19.40-28.30	8
Grains/panicle	70	33-106	22
Sterility percentage	25.93	6.70-56.85	37.82
250 grain weight (g)	6.19	3.60-8.20	10
Spikelet density	4.15	2.21-8.98	22.73
Grain length (cm)	0.89	0.78-1.10	19
Grain breadth (cm)	0.31	0.25-0.37	10
L/B ratio	2.89	2.07-3.79	18.62
Flag leaf length (cm)	37.03	26.80-50.60	13
Flag leaf breadth (cm)	1.58	1.10-2.30	13
Penultimate leaf length (cm)	57.61	45.00-47.60	9
Penultimate leaf breadth (cm)	1.31	0.90-2.00	17

Northeastern region of India, especially Assam, had immense wealth of aromatic rices (Table 6.11). But a lot has already been lost as an aftermath of the residual effect of green revolution. This is despite the fact that some of the non-Basmati scented rices are much superior to Basmati types with respect to traits like aroma, kernel elongation after cooking, fluffiness, taste, etc. Nevertheless, due to the special importance of these varieties in the tradition of the region, a large numbers of them are still in existence.

Table 6.11. Indigenous aromatic rice landraces of NE India.

State	Landrace
Assam	Banali Joha, Bhaboli Joha, Bhugui, Boga Joha, Boga Manikimadhuri, Boga Tulasi, Bogi Joha, Bokul Joha, Barjoha, Borsal., Cheniguti, Kunkuni Joha, Goalparia Joha 1 and 2, Gobindabhog, Joha Bora, Kaljeera, Kamini Joha, Kataribhog, Khorika Joha, Koli Joha, Kon Joha 1 and 2, Krishna Joha, Manikimadhuri Joha, Ramphal Joha, Ronga Joha, Keteki Joha, Badsabhog, Chenisakkar, Tulsibhog (all small grain)
Manipur	Chahao Amubi (black), Chahao Angangbi (pink/red) – both medium grain
Tripura	Khasa, Kalikhasa, Binni, Binni Guria, Binni Korma, Binni Mebang, Binni Bongai, Biral Binni 1, Biral Binni 2, Maimisahai

Studies made at IARI, CRRI, ICAR Complex and AAU Research stations on the indigenous rice of NE India have helped in identifying a good number of cultivars possessing traits for tolerance/resistance to physical, physiological and biological stresses (Table 6.12).

Table 6.12. Some promising indigenous rice varieties. (Part 1)

Variety	Remarks	Reference
ARC 10372	Drought resistance at vegetative stage, 90-100% drought recovery at 5 bar SMT	IARI, 1985
Maibi	A variety from Karbi Anglong with excellent drought recovery	Annual Report, RARS, Diphu
Amiong	----------	Annual Report, ICAR Complex
Addy	Good yield	-do-
Changpalnam	Good yield	-do-
Pyare	Good yield	-do-
Nemo	Good yield	-do-
Maidangho	Good yield	-do-
Amo	Tolerance to Fe-toxicity	-do-

Pangnkla	Tolerance to Fe-toxicity	-do-
Leitalbuhban	Tolerance to Fe-toxicity	-do-
Maupa	Tolerance to Fe-toxicity	-do-
Anker	Tolerance to Fe-toxicity	-do-
Kalingbuh	Tolerance to P-deficiency	-do-
Pangnkla	Tolerance to P-deficiency	-do-
Pawnbuh	Tolerance to P-deficiency	-do-
Ngoba	Tolerance to P-deficiency	-do-
Pangnkla	Cold tolerance at flowering stage	-do-
Khonorullo	Cold tolerance at flowering stage	-do-
Kbathangme	Cold tolerance at flowering stage	-do-
Basantibahar (As 24-1)	Cold tolerance at flowering stage	Das *et al.,* 1981
Badshabhog	Fine grain	AR, RARRS, Karimganj
Kalijira	White kernelled ahu variety	-do-
Sahebsali	Joha variety of sali class	-do-
Konjoha	Joha variety of sali class	AR, RARRS, Titabor
Manipuri Dumai	Joha variety of Sali class	AR, RARRS, Karimganj
Koimurali	Joha variety of Sali class	-do-
Chanmouri	80 days duration under direct seeding	AR, ICAR Complex
Rangagellong	90 days duration under direct seeding	-do-
Kangpui	Early duration	-do-
Bethi	Early duration	-do-
Ngoba	Early duration	-do-
Khonorullo	Early duration	-do-
Ryllored	Early duration	-do-

Source: Das (1987)

Table 6.12. Some promising indigenous rice varieties. (Part II)

Variety	Remarks	Reference
Ryllored	Early duration	-do-
Abor red B	Non-lodging tall	-do-
Nikhunkatai	Non-lodging tall	-do-
Kanchalmang	Non-lodging tall	-do-
Khuctio	Short stature	-do-

Maophoe	Short stature	-do-
Leitalbuhban	Short stature	-do-
D 208-2	Short stature	-do-
AS 789	Short stature	Das *et al.*, 1981
ARC 5756	Short stature, Dumai group of ahu rice	-do-
Lal ahu	Short stature, Aus group of *Ahu* rice	Sidhu *et al.*, 1978
Kolongi bao	Res. to bacterial leaf blight (BLB)	-do-
Moina sali	Res. to bacterial leaf blight (BLB)	-do-
Kartic kolma	Res. to bacterial leaf blight (BLB)	AR, RARRS, Titabar
Dholamula	Res. to BLB: *Sali* variety of Assam	-do-
Tarabali	Res. to BLB: *Sali* variety of Assam	-do-
Japora Sali	Res. to BLB: *Sali* variety of Assam	-do-
Dholamula	Res. to sheath blight, *Sali* var. of Assam	-do-
Supkhera	Res. to sheath blight, *Sali* var. of Assam	-do-
Basikalam	Res. to sheath blight, *Sali* var. of Assam	-do-
Maimonsingia bao	Res. to sheath blight, *Sali* var. of Assam	-do-
Salpona	Res. to sheath blight, *Bao* var. of Assam	-do-
Kalamdani	Res. to sheath blight, *Bao* var. of Assam	-do-
Tarabali	Res. to blast, *Sali* variety of Assam	-do-
Mugisali	Res. to blast, *Sali* variety of Assam	-do-
Madri	Res. to blast, *Sali* variety of Assam	-do-
Mirikirk	Res. to blast, *Sali* variety of Assam	-do-
Pawnbuh	Res. to blast, *Sali* variety of Assam	AR, ICAR Complex
ARC 5905	Res. to blast, Hill rice	-do-
ARC 11554	Res. to blast, Hill rice	AICRIP, 1980
ARC 13560	Res. to rice tungro virus	-do-
ARC 13820	Res. to rice tungro virus	-do-
Kataribhog	Res. to rice tungro virus & also GLH	-do-
ARC 7140	Res. to rice tungro virus & also GLH	-do-
ARC 6064	Res. to rice tungro virus & also GLH	-do-

Source: Das (1987)

Table 6.12. Some promising indigenous rice varieties. (Part III)

Variety	Remarks	Reference
ARC 10963	Res. to rice tungro virus & also GLH	IRRI Ann. Rep, 1985
ARC 12596	Res. to rice tungro virus & also GLH	-do-
ARC 12778	Res. to rice tungro virus & also GLH,	-do-
ARC 13349	Resistance to white beaked plant hopper with monogenic dominant resistant gene Wbph	Saini *et al.*, 1982
ARC 10239	-do-	
ARC 6650	Resistance to white beaked plant hopper with monogenic recessive resistant gene Wbph	Angeles *et al.*, 1981 Hernandez & Khush, 1981
ARC 6564	Resistance to white beaked plant hopper with monogenic dominant resistant gene Wbph	- do -
ARC 14342A	-do-	
Manoharsali	Resistence. BPH, *Sali* rice of Assam	-do-
Balam	-do-	AR, RARRS, Titabor
Garem	Res. to root knot nematode, *Ahu* rice	-do-
Dumai	-do-	Nematology Dept, AAU
Negheribao	Res. to flood & water stagnation, *Bao* rice of Assam	RARS, North Lakhimpur
Kekoabao	-do-	-do-
Salibadal	Resistance to flood & water stagnation, *Asra* rice of Assam	RARS, Karimganj
Dholabadal	-do-	-do-

Source: Das (1987)

Rice is life to the people of NE India. Obviously, farmers of the region are very careful about selection of optimum rice growing situation and, out of necessity, have to grow their rice crop in every available situation wherever the crop has a chance to survive and produce grains. Rice is grown in widely varied land situations such as hill terrace, drought affected upland, flood prone lowland and deep-water situations. Likewise, rice-growing seasons are *ahu* (March/April-June/July), *sali* (June/July – November/December) and *boro* (November –May).

These together have created vast diversity in rice growing situations. Das (1987) reported some possible sources for specific traits of Assam rice (Table 6.13).

Table 6.13. Sources of specific traits of Assam rice.

Traits	Possible source
Drought tolerance and recovery at vegetative stage	*Ahu* germplasm particularly upland dry broadcast types and ahu rice from rain shadow areas of Karbi Anglong district
Tolerance to flood and water stagnation	*Asra* and *Bao* rice
Earliness	*Dumai* group of ahu rice and hill rice
Adaptation to low light intensity	*Sali, Asra* and *Bao*
Cold tolerance	*Boro* rice and hill rice of Meghalaya
Fine and aromatic quality rice	*Joha* group of *Sal*i rice
Resistance to pests and diseases	All agricultural classes of rice across the region

Meghalaya

Asthana and Majumdar (1981) evaluated 465 landraces collected from NE region and 56 characters for 287 cultures were catalogued. Variability present in these collections was studied in 56 characters and promising lines were selected for desirable traits (Table 6.14). Different ecotypes were identified based on distinguishing characters with the help of metroglyph. Based on morphological characters, the cultures were divided into 33 major groups and 245 sub-groups. On the basis of pigmentation, the landraces were further divided into 9 groups and 49 sub-groups. The studies, based on morphological, genetic and some chemical characters, indicated that large number of germplasm was intermediate type between *japonica* and *indica.* Ngoba was identified as a new alternate source of dwarfing gene (Ghosh *et al.,* 1981).

Table 6.14. Variability in rice germplasm from NE India (Asthana and Majumdar, 1981).

Character	Range
Plant height (cm)	56-220
Flag leaf length (cm)	16.8-56.0
Flag leaf width (cm)	1.1-2.8

Leaf area/plant (cm^2)	180-6433
Days to flower	94-163
Number of ear bearing tiller	2.2-20.2
Panicle length (cm)	14.3-31.1
Number of branches/panicle	Mar-16
Sound grains/panicle (%)	Mar-99
Number of grains/panicle	31-329
Grain density/cm	1.7-15.7
1000 grain weight (g)	11.6-34.0
Grain yield/plant (g)	0.4-24.0
Leaf blast (0 – 9 scale)	03-Sep
Neck blast (0 – 9 scale)	03-Sep

Verma (1985) screened over 500 local cultivars of NE India for their resistance to blast (*Pyricularia oryzae* Cav.). The cultivars resistant to leaf blast were Amdang, Maibung, Amte-Boro, Vira-Bhon, Mau Phou, Boro, Namyi and Ampakhi whereas the cultivars resistant to neck blast were Dullo, Ryllored 5, Nirkhumkate, Khongman, Ngoba, Kongphou, Kbajogma and Pawnbuh. Pandey and Gupta (1995) evaluated 62 rice cultivars at high altitude (< 1300 meter) in Meghalaya and recorded that local cultivars possessed high tolerance to low temperature at reproductive stage. It was found that almost all the characters showed high amount of variability (Table 6.15) and identified cultivars with desirable characters for cold tolerance (Table 6.16).

Table 6.15. Variability in agronomic characters of rice germplasm evaluated at high altitude of Meghalaya (1987-1990).

Character	Mean	Range	SE	CV(%)
Days to 50% flowering	110	99-179	1.03	2.21
Days to maturity	165.2	135-185	1.77	1.07
Plant height (cm)	93.8	55.7-108.0	11.24	11.98
Number of total tillers	4.5	3.2-7.4	1.04	23.22
Number of ear bearing tillers	4.0	28-5.5	0.6	1.21
Panicle length (cm)	19.7	11.9-23.2	0.85	4.31
Panicle weight (g)	2.3	0.5-3.2	0.35	15.58

Number of spikelet/panicle	83.4	31.9-128.6	11.37	22.83
Chaffy grains/panicle	7.8	1.9-44.4	4	51.34
1000 seed weight (g)	24.9	19.1-28.8	1.41	5.68
Grain yield/plant (g)	8.0	2.8-18.0	1.48	18.52

Table 6.16. Cultivars with desirable characters for cold tolerance at high altitude of Meghalaya.

Character	Score	Cultivar
Days to germination	11	Khonorullo, Abor B, Kbathangmaw, Nonglwai
Tillering ability	5	Ryllo Red 4, Ganam, Eiko, K 332, K 333
Panicle exertion	1	Khonorullo, Abor B, Kbathangmaw, Namyi
Leaf colour at seedling stage	5	Eiko, Ngoba, Zenith, Meghalaya 1
Leaf colour at tillering stage	5	Eiko, Ngoba, Yam Nami, Namyi
Cold tolerance at reproductive stage	5	Meghalaya 1, Khonorullo, Abor B, Kbathangmaw
Resistance to sheath rot	5	Lynsi, Ryllo Red 6, Tangla, Namyi

NB: Days to germination is in number of days; others are scores in 1 to 5 scale where 1 is the least and 5 the most desirable score

Arunachal Pradesh

Wide diversity of rice landraces for many characters in collections made from Arunachal Pradesh was reported (Kocchar, 1997; Sharma, 2002). These collections were intermediate between *indica* and *japonica* types. Some landraces and primitive cultures prevalent in various districts of Arunachal Pradesh are presented in Table 6.16. Local landraces suited to high altitude conditions (more than 1500 m above msl) are predominantly grown in Tawang and West Kameng.

Majority of the germplasm from Lower Subansiri, Upper Subansiri, West Siang, East Siang and Dibang valley were collected from mid-altitude conditions (450-1500 m msl). The germplasm prevalent in Apatani Plateau located at 1500 m provide a unique set of intermediate types, perhaps introgressed forms, between *indica* and *japonica* types, which possess a high degree of location specific adaptability. These cultures have evolved under one of the most systematic indigenous farming systems in the NE region of India. The Apatani cultivars were predominantly late and required additional 90 days to mature when tested at low elevation (Pasighat) and mid-elevation (Basar). The predominant cultivars of

Apatani rice-based farming system are: Emo, Pyat and Payapin. The Apatani system presents one of the most elaborate examples for logical classification of rice germplasm (Sharma, 2002). The varieties were so named that represented their maturity duration, grain size etc (Table 6.17). The on-farm conservation of these valuable local rice germplasm landraces is highly desirable for preserving the ethical, scientific and cultural values of Apatani tribe of Arunachal Pradesh.

Table 6.17. Local landraces prevalent in different districts of Arunachal Pradesh. (after Kocchar and Chandel, 1996).

District	Name of rice cultivars
Tawang	Olem
West Kameng	Chuksing Pa, Nabeye, Ne Sa, Rabalangbu, Raw, Sung, Sungha, Thongbar, Thumbara, Thumpi
East Kameng	Lonrow, Pakke Amm, Pany, Paru, Rallow
Lower Subansiri	Anka, Betguti, Khilly Amo, Nilony, Thirti, Yamuk
Apatani plateau	Emo, Emo Elang, Emo Empu, Emo Enkhe, Emo Are, Emo Hat, J Pyat, Nipya Pyani, Mipya Pyat, Mishing, Owe Impo, Pyaping, Pyankha, Pyaping Pyat, Pyaping Pyani, Pyaping Pyare, Pyat kogya, Pyani, Pyat Pyapu, Rade Amo
(Lower Subansiri)	Bali, Bee, Buing, Daji, Dazing , Gapu (Yapu), Gyopung, Hiber Hirpu, Ke, Che, Ke Ling, Lite. Pana, Pinini (penin), Pomin, Yapu Pumik, Zingmick (Zickmik)
Upper Subansiri	Amka, Amkel, Jorang, Amlong, Ampang, Amtun, Boling, Bura, Jajum, Jiko, Amo, Jorang, Joya, Kasiang, Kolom Amo, Laye Amo, Letung, Longkang, , Nipung, Mipung, Moping, Motum, Bora, Oiky, Pashing, Pehi, Pelu, Royila Amo, Tasha, Tajah, Takel, Tuinu, Tumtum, Yaling, Yamuk
East Siang	Amte Kimin, Anker, Bali, Damgam, Jorang, Karpu, Khamti Yali, Khilly Amo, Kimin, Lahi, Lelli, Lemuk* Lemuk Mukte, Lipu, Lite, Mede Deli, Mede Depu, Mipun*, Nami*, Neuly, Ningtal., Plu Amo, Pupi, Soiki, Tabor, Irri, Tharti Ahu, Yabor, Yamul,
West Siang	Yayo*
Dibang Valley	Ahu Mali, Amdang, Amte Boro, Amyong, Asiamkel*, Beera, Birabonga*, Dajum, Dard, Keopu, Taying
Lohit	Ahu, Aphu, Babang, Bor, Bora, Chomcha, Chomju, Chomkhu, Chompat, Kamplung, Mang, Khamti Joha, Kachal Mang, Kayum Mang, Ke Bord, Ke Huse, Ke SA, Ke Shi, Khaunou, Khwamakhen, Khwaji, Loilang, Mahjang, Maibum, Maidang, Mesong, Miyo, Nyang, Patom, Patong, Poyong, Setiseds, Singkhan, Tahang, Tulu

Tirap	Bakbak, Betguti, Gamlam, Gangtah, Honai, Joha, Kanyak, Khopsan, Lahi, Longdiak, Maichong, Mate, Patam, Pelu, Ramlo, Sampu, Sasang, Sasian, Simoi, Tahsa, Taling, Tuinu, Wangkan, Wangkham, Wangsam, Yasing, Yammuk

The *Khamti* rice of low altitude of Lohit and Tirap districts of Arunachal Pradesh are generally famous for their softness, glutinous character and aroma. Out of 21 landraces of rice grown by *Khamti* tribe of Lohit region, 16 were found long slender, one long broad, one medium slender and two short slender. The kernel colour varied from white, milky white to reddish white. 1000-seed weight varied from 17.2 to 28.4 g and the kernel test weight ranged from 14.4 to 24.9 g. Bali is the most important landrace in Siang district, which serves as a typical example of adaptation through phenotypic plasticity and tolerance to multiple diseases and pests. Some of the desired traits of Khampti rices are shown in Table 6.18.

Table 6.18. Some *Khamti* rice cultivars with desired traits.

Character	Cultivar
Long slender grain	Joha, Bondhu Bora, Kala Hali, Rongalahi, Bogalahi, Khamptilahi, Kalajaha
Long broad grain	Lakhimilahi
Long bold grain	Bahadur
Medium slender grain	Borgha
Short slender grain	Khawnoo, Khawan
High yielding	Khawjee

Mizoram

Amongst different tribes of Mizoram, the Lakhers are one of the important tribes. They migrated from the Chin Hills in Myanmar in the 11th century, passed through the Chindwin valley and the Chin Hills and finally came to the present Mizoram. The Lakher agriculture is mainly *jhum* but the main crop is rice (Sarma and Sharma, 2000). There are many rice varieties – both white and red kernelled. The most common varieties grown by the Lakhers are shown in Table 6.19.

Table 6.19. Rice varieties grown by the Lakhers.

Variety	Characteristics
Livawneng	Small, fine, long, white grain; grown on bamboo ground
Batanong	Round grain, white or red in colour
Botinong	Long grain, white or red in colour
Bupinong	Roundish grain, white or red in colour
Chario	Long grain, usually red in colour
Zaibenong	Dark ear, long grain, white, grows well on rocky soil
Saibinong	Red ear, long white grain
Sapazupa	Both ear and grain red, sticky
Tlaikokenleu	Round, large white grain, ear spiny
Matukeuleu	Small and long white grain
Sasaipa	Red ear, roundish white grain
Sacharapa	Very small, long, white or red grain, early maturing
Iakasa	Sweet, medium sizes white grain

Sikkim

Arora (1982) reported that local variability in rice collected from Sikkim was mainly of the *indica* types – tall, medium-coarse-grained, maturing in about 120 days. Much variation was observed in morpho-agronomic characters in panicle type, awn/awnless character, glume colour (black, buff, red) glabrous/hairy nature and in kernel colour (red/white). The collected cultivars were characterized into various categories (Table 6.20).

Table 6.20. Nature and extent of diversity in Sikkim rice collection.

Character	Cultivars
Direct sown types	Buidhan, Lama, Chinichampa
Transplanted types	Atte, Mansara, Jhapaka, Dut-kalami, Basmati, Krishnabhog, Tulasi, Poudyal., Brimphul
Early types suited to low/medium elevation (up to 1,500 m)	Atte, Mansara, Marshi
Early types for higher elevation (above 1,800 m) and cold tolerant	Phudange, Takmaru, Darmali

Fine grain types	Champai, Chinichampa, Champasiri, Phorel
Medium grain types	Brimphul, Tulasi, Satoyangsiri, Mashi
Scented types	Krishnabhog, Brimphul, local Basmoti
Glabrous types and akin to *japonica* form	Phulpatta
Red kernel types	Takmara, lama
High yielders (about 4 t/ha)	Dutkati, Basmati, Krishnabhog, Mansara

Tripura

Sardana (1997) classified some local rice germplasm of Tripura having dwarfness, early maturity, scent and resistance to leaf blast infection and estimated their agronomic qualities (Table 6.21, 6.22 and 6.23). A few cultivars were also identified as resistant to iron toxicity such as Singrijoli, Galongkufodu, Latisail, Matrimangal., Maloti, Binni Marang and Kalahatia. These cultivars yielded 30-65% more.

Table 6.21. Some local dwarf rice germplasm of Tripura and their agronomic traits.

Cultivar	Plant height (cm)	Days to maturity	Production potential (g/plant)
Kangpui	73.1	130	9.6
Khurpi	78.1	128	8
Badiya	83.3	128	9.9
Badiyabisi	85.3	131	5.7
AR 11	71.6	93	9
Jungli Boro	80.1	133	7.9
Dhupami	83.9	123	7.4
Dhupgallong	89.1	124	19.3
Gallongkufro	89	122	15.5
Horpi	100	114	8.3
Sarendaima	99.3	108	7.5
Kartick Sail	90.5	122	10.6
Erimala	94.9	123	7.1
Maipla	95.3	103	11.4

Table 6.22. Early maturing local cultivars of Tripura.

Cultivar	Days to maturity	Plant height (cm)	Production potential (g/plant)
Kalahatiya	104	169.8	7.4
Singrijoli	104	111.1	7.7
Maipla	103	95.3	11.4
AR 11	93	71.6	9
Joli	109	114.3	6.4
Garo	106	111.5	14.6
Bilrong	103	117	10.9
Sanmuri	103	120.3	11.1
Sonamukhi	103	116.9	9.7

Table 6.23. Local scented rice of Tripura.

Cultivar	Plant height (cm)	Days to maturity	Production potential (g/plant)	Length of grain (mm)
Khasa	125.7	135	12.2	4.1
Kalikhasa	143.3	134	12.2	3.85
Binni	145.6	136	5.7	6.62
Binni Guria	141.2	144	5.3	6.2
Binni Korma	135.9	137	7.7	6.45
Binni Mebang	123.7	137	4.7	6.62
Binni Bongai	136.5	137	6.8	7.05
Biral Binni 1	105.7	106	10.2	6.97
Biral Binni 2	140.7	141	6.7	6.07
Maimisahai	148.2	148	12.4	6.08

Nagaland

Exclusive study of the rice germplasm of Nagaland has been taken up very recently and is underway at present. The rice germplasm of Nagaland can be broadly classified into 3 major groups: glutinous or sticky rice (mainly *japonica* type), brown rice (*japonica* and *javanica* types) and aromatic rice (*joha* group). In general, the local rice genotypes of Nagaland are tall, photosensitive, medium-long duration and poor-medium yielders. Majority of the *jhum* varieties are intermediate between *indica* and *japonica*. Wide variations in grain size and colour are also seen in the germplasm. World's tallest paddy popularly known as Miracle Rice is also reported from Nagaland. Department of Agriculture, Govt. of Nagaland through its concerted efforts collected and compiled a list of 867 germplasm of Nagaland. The collections were designated district-wise with their local names and village from where collected. Detailed characters of about 100 important collections and several specialty ones were also listed (Anonymous, 2007).

Manipur

Although the state is rich in the genetic diversity of rice, very few studies, exclusively on the germplasm of Manipur, have been made. Probably the only published systematic study is by Singh and Sharma (1997) who collected and characterized 300 local land races of rice (*Oryza sativa* L.) of Manipur including two wild rice locally known as *Wainuchara* (*Oryza rufipogon*) and *Murshi* (*Oryza sativa* f. *spontanea*). Other studies on the rice germplasm of Manipur were conducted on the morphological characters (under NATP) and biochemical characters which are reported in various sections of this chapter.

Studies with the Materials Collected under Plant Biodiversity Management Programme

Detailed study of the rice germplasm collection made during the first 5 years of this century was done at ICAR Research Complex and NBPGR at Umiam, in Meghalaya. Observations on 16 qualitative (morphological) characters in core collections of 210 upland (Table 6.23) and 160 lowland (Table 24) rice germplasm were recorded. These lines were selected after preliminary studies on the entire collection and elimination of duplicates. The core collection was designed in such a way that more than 95% of the variability is retained. Characterization was done using a minimal list of descriptors designed and adopted by NBPGR, New Delhi.

Out of 210 germplasm, 58 accessions showed very good early plant vigour (EPV) and 43 accessions had poor EPV. Basal leaf sheath colour was green in 149 collections. Fifty-six germplasm showed intermediate, 117 pubescence and 37 glabrous leaf pubescence. Coleoptile colour was green in 198 accessions and purple in 12 accessions. Leaf blade colour ranged from purple to green, maximum being green. Panicles were well exerted in 79 accessions, moderately exerted in 58 and just exerted in 46 accessions. Stigma colour was white in 111 accessions followed by purple in 41, light green in 27 and light purple in 22 accessions. The apiculus was straw coloured in 77, white in 59, purple in 26, brown in 19, purple apex in 15 and red and red apex in 7 accessions each. Panicle types were varying degrees open to compact types. Most of the accessions had awned spikelets. Although white seed coat colour dominated amongst the germplasm, 37 collections showed brown, 39 light brown, 16 sparkle brown, 13 red, 8 variable purple and 5 purple colour seed coat. The hull colour was observed to be straw in 77, golden in 58, brown furrow on straw in 30, golden brown in 15, brown in 12 and purple in 3 accessions. Threshability was found to be intermediate in 99, easy in 77 and difficult in 47 cellections. Only 10 accessions showed high susceptibility to brown spot and 13 accessions showed high susceptibility to blast although none of the genotypes were highly resistant to blast (Table 6.23).

The observations recorded for seven qualitative (morphological) characters in 160 lowland germplasm of NE region in 2002 indicated that a few of the characters showed almost similar trends for most of the genotypes. Plant vigour was good for majority of genotypes. Coleoptile colour was green in all the germplasm. Basal leaf sheath colour was green for most of the entries while a few showed purple lines and purple colour leaf sheath. Leaf colour varied from different shades of green to purple margins and purple blotches. Leaf texture was mostly glabrous and intermediate. Stigma colour was mostly yellow with a variation of white and light green in a few. With regard to variation in presence and absence of awn, some entries had very long awn while awn was totally absent in others. Awn length ranged from 0.1 to 3.6 cm. Varied apiculous colour of straw, brown, purple, red and black was observed in the collection (Table 6.24, 6.25).

Table 6.24. Range of variability in 18 characters of 210 upland rice germplasm of NE India (2001-2003).

Character	2001(150)	2002 (210)	2003 (43)
Leaf length (cm)	25.1 – 59.2	23.9 – 61.1	24.3-60.3
Leaf width (cm)	0.8 – 2.2	1.1 – 2.2	1.0-1.9
Leaf area (cm²)	27.5-79.3	26.8-81.0	26.9 – 82.8
Days to 50% flowering	80 - 119	74.4 – 120.4	77.3 – 120.0
Days to 80% maturity	116 – 159	108 – 159	112 – 156
Plant height (cm)	70.0 – 151.5	38.3 –148.8	44.7 – 148.8
Panicle length (cm)	16.4 – 31.9	14.7 –32.1	16.3 – 29.7
Primary branches/plant	5.5 – 15.5	6.3 – 16.9	6.6 – 15.6
Panicle weight (g)	1.2 – 5.6	0.8 – 5.3	1.3 – 4..9

Effective tillers/plant	3.4 – 8.3	4.2 – 9.5	4.1 – 9.2
100 seed weight (g)	1.6 – 4.1	1.2 – 3.6	1.4 – 3.7
Grain yield/plant (g)	4.7 – 30.3	4.0 – 27.0	7.3 – 24.4
Filled grain/panicle	40.5 – 162.8	34.8 – 215.3	39.1-187.2
Chaffy grain/panicle	8.8 – 81.2	9.4 – 60.2	8.9-76.3
Grains/panicle	60.3 – 233.1	53.0 – 263.6	47.8 – 184.3
Spikelet fertility %	51.9 – 88.2	42.7 – 90.7	55.8 – 85.4
Seed length (cm)	6.7 – 10.7	5.5 – 10.8	6.4-10.9
Seed breadth (cm)	1.8 – 3.8	2.0 – 3.6	1.9-3.7
Grain length-breadth ratio	2.0 – 4.4	1.7 – 4.8	2.0 – 4.6

Figures in parenthesis indicate number of accessions evaluated

Table 6.25. Range of variability in 14 characters of 160 lowland rice germplasm of NE India (2001-2003).

Character	2001 (150)*		2002 (160)*		2003 (90)*	
	Mean	Range	Mean	Range	Mean	Range
DF	105.6	71.0-144.0	112.6	82.0-130.0	105	89-126
PH (cm)	106.1	65.3-145.7	113.8	75.0-157.8	102	72.3-144.1
LT	4.9	3.2-7.2	4.5	3.4-6.0	4.7	3.7-5.4
EBT	7	3.5-17.0	8.9	3.8-16.8	5.6	4.3-7.5
PW (g)	1.9	0.3-5.7	2.2	0.2-4.4	2.2	1.1-3.3
PL (cm)	21.8	13.2-27.7	21.9	16.3-27.4	22.5	16.3-29.5
BP	11.2	6.4-21.2	9.7	1.0-12.8	9.9	7.0-11.9
FLL (cm)	32.3	20.4-54.3	34	22.7-52.3	29.5	22.1-47.0
FLB (cm)	1.4	0.7-2.3	1.6	0.8-2.2	1.3	1.0-1.6
LL (cm)	41.4	24.5-90.0	44.7	31.4-71.0	36.1	26.8-56.2
LB (cm)	1.1	0.6-1.7	1.2	0.6-1.7	1.2	0.9-1.4
GY (g)	10.4	1.4-24.0	9	1.1-25.2	9.2	4.1-14.7
SW (g)	2.4	1.0-3.1	2.3	0.5-3.1	2.3	1.7-2.7
GR	2.7	1.4-4.3	2.7	1.8-4.0	2.9	1.9-3.8

* Figures in parenthesis indicate number of accessions evaluated

Some promising genotypes/cultivars of upland and lowland rice were identified on the basis of three years evaluation of the core collection. The landraces/cultivars with desirable characteristics are presented in Table 6.26 and 6.27.

Table 6.26. Promising upland landraces with desirable attributes.

Character	Genotype
High yielder	Satya, Pangla, Fazaitai, Charimpok, Zerusalem and Bubawn
Longer leaf	Faziatai, Keotadhan, Manipuri,
Wider leaf	Chakhchang cham, Bubawn, Nomdumli,
Early flowering and maturing	Pangla, Pagra, Mycassak
Dwarf plant type	Tompuing, Tondoniba, Bhousak Khagola and Lamdhan
Longer panicle length	Kayamong, Charimpok, Thaikawnganpui and Khatimachiya
More primary branches/panicle	Khatimachiya, Akatan, Biralbinni and Kayamong
More effective tiller/plant	Dhan 9, Nampokdhan, Dhan 10, Palidhan and Lamdhan
Bold seed	Maphhaw, Ngarempaoo, Kawnlong and Dhan 11
More filled grains/panicle	Yaling, Jouchou, Chhalayang and Taisa local
High spikelet fertility	Kpalum, Betkutechameli, Idaw, Kawnlong and Colonel 4
High grain ratio	Nunman, Chakhchang cham, Idaw 5

Table 6.27. Promising lowland landraces with desirable attributes.

Trait	Genotype	Range
Early flowering (days)	Basmati 370, Sarengma, RCM 5	71-89
Short plant height (cm)	Manipur, Seetabhog, Kemehya sopa	65.3-75.0
Higher number of leaves/tiller	Lalatai, Suagmoni, RCPL 1-1pp	5-7
More effective tillers/plant	Dhali boro, RCPL 4-12, Ghavri	7-17
Heavy panicle weight (g)	RCPL 1-6, RCPL 1-29, LaPyngad	3.29-5.70
Long panicle (cm)	RCPL 3-6,RCPL 7-5, Manipuri (Eithali), TB 15-1	27.4-29.5
More branches/panicle	RCPL 4-12, Lanten, RCPL 3-6	12-21

Longer flag leaf (cm)	RCPL 3-6, Bondhu Bora, Mandri	47.0-54.3
Broader flag leaf (cm)	TRC 87-252, Dhali boro, Sangmang Dsuck, La Pyngad	1.6-2.6
Long leaved (cm)	PNA 1121, Khaiwai, RCPL 7-5	56.2-90.0
Broad leaved (cm)	TRC 87-252, Sangmang Dsuck, Minil Doka, Kemehya sopa	1.4-1.7
Bold seeded (g)	Manipuri (Eithali), Miakriha, RCPL 1-87, RCPL 1-6	2.7-3.1
Slender grained (grain ratio)	BRDKP, Beroing, RCPL 1-12, VL 1179	3.8-4.3
High yielding (g)	Marrow, Manipur yellow, RCPL 3-6	14.7-18.8

Nutritional Quality of Rice

Rice is generally not very rich in protein like other cereals but its nutritional quality is claimed to be markedly superior as judged by the amino acid composition, growth promoting properties and biological values of its proteins. Dutta and Barua (1981) studied nine varieties of rice grown in Assam to estimate the effect of processing on the essential amino acids and vitamin contents and reported uneven distribution of amino acids in different fractions of proteins. When the values of eight amino acids of polished rice either raw or parboiled are compared with those of dehusked rice, it is observed that the polishing increased the percentage of the amino acids in rice grain. The two methods of rice processing yielded significantly different amounts of vitamins irrespective of varieties. However, when the values of vitamin contents of polished rice are compared with that of simple dehusked rice, it is observed that on polishing there is a considerable loss of vitamins. Vitamins are present in higher concentration in the grain and outer bran layer in comparison to the starch endosperm (Table 6.28).

Biochemical characterizations of the germplasm were done by several workers at ICAR Research Complex, Manipur University and Gauhati University. At ICAR Research Complex, Devi and Pattanayak (2008) analyzed group of medicinal rice (Table 6.29) genotypes endosperm constituents. Most of the genotypes were rich in protein and ash (Table 6.30). Devi *et al.* (2008) analyzed a set of germplasm from the NEH states. Few genotypes showed seed protein in excess of 12%. Pajat (2008) analyzed biochemical characteristics of 29 indigenous rice germplasm of Meghalaya. Grain physical characteristics, cooking quality and nutritional quality were studied. Most of the genotypes were bold grain type and head rice recovery was low although milling percentage was more than 65%. Most of the varieties showed high volume expansion and water uptake. As

Table 6.28. Aminoacid and vitamins in some Assam rice.

Variety	Processing	Protein (%)	Histidine (g/16gN)	Lysine (g/16gN)	Methionine (g/16gN)	Tryptophan (g/16gN)	Valine (g/16gN)	Thiamine (ig/g)	Niacin (mg/100g)	Pantothenic acid (ig/g)
Monoharsali	RPR	8.55	2.6	4.52	2.24	1.28	6.4	0.86	1.63	5.08
	PPR	8.66	2.5	4.38	2.11	1.17	6.29	1.61	3.01	6.35
	DR	9.35	2.47	4.12	2.04	0.92	5.99	3.14	4.16	8.98
Jaya	RPR	7.55	2.95	9.34	2.24	1.12	6.58	0.89	1.79	6.34
	PPR	7.66	2.9	8.92	2.11	1.04	6.29	1.68	3.06	6.06
	DR	8.35	2.8	4.12	2.25	0.87	6.23	3.26	4.33	11.39
China 63	RPR	10.5	2.2	4.37	2.02	1.07	6.12	1.16	2.84	7.31
	PPR	10.71	2.12	4.44	1.9	0.96	5.9	2.31	5.16	9.35
	DR	11.46	2.04	4.05	1.76	0.77	5.47	4.65	6.7	13.64
Prasadbhog	RPR	8.3	2.38	4.75	3.17	1.35	7.31	1.02	2.26	5.34
	PPR	8.42	2.34	4.77	3.16	1.19	7.1	1.97	4.33	7.35
	DR	9.11	2.22	4.22	3.15	0.93	6.92	3.84	5.75	10.15
Badsabhog	RPR	9.14	2.04	3.81	2.48	1.02	5.98	1.43	2.1	4.57
	PPR	9.27	1.88	3.61	2.4	0.95	5.82	2.53	3.95	5.44
	DR	10.01	1.87	4.3	2.28	0.71	5.42	4.47	5.37	7.83
Sorusokuwa	RPR	8.07	1.89	4.69	2	0.94	6.44	0.98	2.18	7.83
	PPR	8.13	1.89	4.57	1.89	0.85	6.2	2.4	4.44	10.82
	DR	8.85	1.72	4.2	1.79	0.65	5.89	4	5.85	14.06
Ampakhisokuwa	RPR	8.26	2.43	4.63	2.1	1.36	6.32	0.98	2.68	8.27
	PPR	8.36	2.4	4.51	1.9	1.21	6.17	2.53	5.22	11.64
	DR	9	2.27	4.15	1.94	1.1	5.81	4.16	6.61	15.9
Janganibora	RPR	8.77	2.22	4.61	2.93	1.08	6.52	1.21	2.53	7.17
	PPR	8.92	2.12	4.53	2.89	0.99	6.21	3.12	5.15	10.31
	DR	9.49	1.92	4.1	2.88	0.75	6	5.07	6.2	13.56
Gheebora	RPR	8.38	2.22	4.33	2.74	1.17	6.34	1.16	3.22	8.86
	PPR	8.49	2.12	4.2	2.65	1.1	6.14	3.1	5.87	12.62
	DR	9.11	2.05	3.81	2.55	0.85	5.71	5.22	7.4	16.8

RPR: raw polished rice; PPR: parboiled polished rice; DR: dehusked rice

Table 6.29. Ethno pharmacological uses of rice in Manipur and Nagaland.

Sl.No.	Disease	Predominant variety	Preparation of rice used	The powder is given orally
1.	Dysentery	Sticky rice varieties e.g. Moirangphou, Phouren phoujao, Chahou, Chahou poireiton	Rice kernel is roasted on pan and powdered with . pine resin	The powder is given orally.
2.	Muscular sprain	Sticky rice e.g., Chahou, Chahou poireiton, Moirangphou, Phouren phoujao.	Sticky rice is cooked. Wood charcoal is powdered and smashed with the cooked rice.	The paste resulting from this mixture is applied on the area covered with banana leaf and bound to fix and left overnight.
3.	Dog bite	Any type of rice	Raw polished rice is chewed	Chewed rice is applied on the bitten area. Then one coin is placed on it for sometime.
4.	Reduced eyesight	Any type of rice	Rice husk is rubbed with ripe banana; the resulting paste is trained through muslin cloth.	The filtrate is soaked in cotton and applied on the eyelids.
5.	Mouth ulcers	Chahou, Chahou poireiton, Moirangphou	Chewing	For curing mouth ulcers raw milled rice is chewed.
6.	Dental problems (Hypersensitiveness of teeth after biting sour things)	Chahou sticky varieties	Chewing	Chewing
7.	Easy labor and healthy baby	Chahou poireiton	Cooked rice	Included in diet of the pregnant mother.

8.	Scanty lactation	Chahou	Cooked rice	Cooked rice is given with sugar. Rice water is boiled with some herbs and applied to hair and washed after few minutes
9.	Cuts and wounds, intestinal ulcer, nutritional drink	Lumen	Pounded, cake, raw starch or drink	Pounded raw or dry cake form applied to cuts, wounds and external injuries. For suspected intestinal ulcer, rice is consumed in starch form. The powder is also used as nutritional drink.
10.	Gastritis, stomach disorders, body ache, fever, loose motions	Rifulu	Soft rice	Soft rice in small dosages taken orally.
11.	Substitute for tea or additive to tea	Lopkhatso, Setsarima, Jumumusuma, Kuthunyilha, Lamthongma, Chomi matse, Imsutsuk, Sungo tsuk, Calha	Brew or leaves added to brew	Grains / powder used for preparation of drink like tea. Leaves added to tea for additional flavor.

Source: Devi and Pattanayak (2008), Anonymous (2007). (Cultivars in 1–8 are from Manipur and 9–11 are from Nagaland)

expected, ash fat and crude fibre contents were low. Amylose content ranged from 4.04 to 28.13%. Three genotypes *viz.* Mipun, Ngoba and Jaya were reported to contain high iron where as zinc content was reported to be high in three genotypes *viz.* Pangkas, Shasarang and Bapada-704 (Table 6.30).

Table 6.30. Biochemical parameters of some medicinal rice varieties of Manipur.

Sl. No.	Name of the variety	Total crude protein (%)	Total - carbo hydrate (%)	Amylose (%)	Total fat(%)	Ash(%)	Crude fibre (%)
1.	Moirangphou	10.55	70.44	13.06	2.8	0.88	0.12
2.	Phouren phoujao	11.65	67.59	4.29	1.37	2.56	0.61
3.	Chahou amuba	11.65	67.59	6.7	2.6	2.56	0.61
4.	Chahou poireiton	12.15	72.43	8.27	4.8	1.87	0.71

Germplasm Conservation

Rice germplasm conservation activities in the region can broadly be classified into three specific periods: The pre-ARC period, ARC period and NATP period. In the pre-ARC period (including pre-independence period) germplasm conservation activities were basically concentrated at Upper Shillong Experimental Farm and Rice Research Station at Karimganj. After 1923, a rice station at Titabar was established to work on plain rice varieties while another research station at Habibganj (now in Bangladesh) was established to work on deep water rice and *boro* rice. Among these stations, Karminganj station maintained the largest collection of germplasm. With establishment of Assam Agricultural University in 1969, all rice research stations of Assam except the station at Upper Shillong were brought under it. In the ARC period the conservation efforts were spearheaded by AAU (Jorhat) and CRRI (Cuttack). As mentioned earlier, a set of the ARC was also conserved at IRRI, Philippines. In the NATP period, germplasm collection and conservation activities in the NE region were funded and coordinated exclusively by ICAR through National Bureau of Plant Genetic Resources. All collected germpplasm were characterized and processed for long-term storage. A number of short-term storage facilities were created in the region and several regeneration centres were identified to maintain the collections.

The only attempt for *in vitro* conservation of rice of NE India was made at ICAR Research Complex for NEH Region. Two wild rice species *Oryza rufipogon* and *O. officinalis* along with *O. sativa* (var. Lumpnah) were used. Both short-term and long-term conservation methods were standardized. In the short-term method, calli and clonally propagated shoots of all the three species were maintained under slow growth condition for more than two years without loss of fertility. Molecular and agrobotanic analysis of the regenerated plants indicated that variations induced were insignificant.

Cryo-preservtion was used for long-term conservation. Attempts were made to cryo-preserve germinating seedlings and calli of the above three species. Calli of *O. rufipogen* and *O. sativa* were preserved successfully. Molecular analysis of the regenerated plants showed that although callusing induced a large amount of variation, cryo-preservtion did not affect the genetic stability much. *O. officinalis* could not be regenerated from cryo-preserved calli. Similarly, germinating seedlings also did not tolerate the shock of cryo-preservtion.

Germplasm Utilization

Considering the germplasm wealth available in the region, its utilization in the region is not extensive. Rice breeding in the region is mainly undertaken at two organizations, Assam Agricultural University and ICAR Research Complex for NEH Region. Starting from 1920 a large number of local germplasm were evaluated at Karimganj and Titabor in Assam and about 50 cultivars were recommended for cultivation. Fourteen varieties were also developed through hybridization among the local cultivars (Sharma and Ahmed, 2004). However, during the later years only 6 varieties developed in Assam had local rice cultivar as one of their parents. The local parents used were Monohorsali and four *bora* genotypes. At ICAR Research Complex, three local genotypes were mainly used for contributing cold tolerance (Khanorullo and Nonglowai) and dwarf stature with cold tolerance (Ngoba). Khonorullo was used for developing two high yielding varieties with cold tolerance at reproductive phase. The genotype also possessed the unique combination of cold tolerance, shadow tolerance and prolonged seed dormancy. The genotype 'Ngoba' is one of the parents of the most popular variety 'Shasarang'. Ngoba possess a unique combination of cold tolerance, iron toxicity tolerance, high tillering and high spikelet fertility. The genotype is also dwarf which is a very uncommon character in the local genotypes. 'Bali', an upland genotype from Arunachal Pradesh has been used for developing a series of genotypes for *jhum* areas. Although, these genotypes are still under testing, they are expected to perform well in the most difficult ecology.

Biotechnological research carried out with the germplasm of rice from northeast India can be broadly classified into three classes – research for genetic improvement, research for conservation and research on assessment of diversity utilizing nucleic acid and biochemical characterization. Research on genetic improvement was done at ICAR Research Complex for NEH Region while research on assessment of diversity through nucleic acid and biochemical characterization were carried out at ICAR Research Complex and universities in Assam and Manipur.

Biotechnological research on rice germplasm of the region was initiated in 1982 at ICAR Research Complex with the establishment of a tissue culture laboratory. Work started with anther culture and protocols were standardized for microscopic identification of suitable microspores and anther culture of local cold tolerant germplasm (Gupta and Borthakur, 1987). Later, Khonorullo, a local cold tolerant rice and IR 70 were used to develop new varieties through anther culture. Two lines DH 7 (later designated as RCPL 1-30) and DH 21 were identified which recorded more than 7% yield increase over parents and best checks. However, the most significant achievement of rice anther culture came from development of two varieties for upland. Anther culture was done from selected F_2 plants of a cross between local germplasm and 29 selected doubled haploids which were evaluated throughout the hill region. Later two lines were identified and released as Bhalum 1 and Bhalum 2 in 2002. These lines are highly resistant to blast, early maturing and yielded nearly 70% more than the best local upland cultivar Bali. The work also demonstrated the relative advantage of F_2 derived anther culture for developing new varieties (Pattanayak *et al.,* 2000).

After the success of anther culture, works on protoplast culture and genetic transformation were initiated. Protoplast culture systems were standardized for cold tolerant lines RCPL 1-2C, 1-3C, etc. (Gupta *et al.,* 1993, Bhattacharjee and Gupta, 1995). Protoplast culture protocols for *indica* lines IR 36 and IR 65 were also standardized (Bhatacharjee *et al.,* 1998). The protocols were then standardized for transfer of wild abortive (WA) cytoplasmic male sterility to RCPL 1-2C, which was identified as a restorer for WA cytoplasm (Bhattacharjee *et al.,* 1999). The protoplast system was also utilized for genetic transformation and an early flowering gene (OSMADS 1) was transferred to RCPL 1-2C by electroporation (Bhattacharjee *et al.,* 1998). With the availability of *Agrobacterium* strains for rice transformation, work on using this method was initiated and a *Bt* gene *CryIAc* was introduced into several genotypes.

Conclusion

Germplasm collection efforts made to collect the rice biodiversity is extensive and it appears that a major part of the diversity is already collected.

However, collection for some specific types like *jhum*, small grain types, medicinal, submergence tolerant, iron toxicity tolerant, shadow tolerant, etc. need to be undertaken as separate programmes. Compared to the number of collections, utilization of germplasm is very less. The main reason for low utilization of this vast wealth is non-specific characterization of the germplasm. With the advent of new techniques it has now become possible to handle a large number of germplasm at molecular level. The rice germplasm of northeast India is a potential candidate for large scale exploitation through bio-prospecting. For this to happen, major emphasis should be given on characterization of the germplasm for specific characters at target environments. Considering the microclimatic diversity in the region, it is expected that the already collected germplasm may prove to be a goldmine for useful genes.

References:

Ahmed SA, Barua I, Sarkar CR, Thakur AC. 1996. Volatile component (2-acetyl-1-pyrroline) in scented rice. *Proc Sem Problems and Prospects of Agricultural Research and Development in North East India.* Assam Agricultural University, 27-28 Nov 1995, Jorhat, Assam.

Ahmed T, Borah P. 1999. Genetic diversity in glutinous rice germplasm of Assam. *Oryza* 36:74-75.

Anon. 2005. National Bureau of Plant Genetic Resource Newsletter. NBPGR, New Delhi

Arora RK. 1982 Collecting rice germplasm in Sikkim Himalayas. *IBPGR Newsletter*

Asthana AN, Majumdar ND. 1981. *Studies in Rice Germplasm of North Eastern Hill Region. Res Bull No 11.* ICAR Research Complex for NEH Region, Shillong.

Barthakur M. 2007. Genetic studies on some rainfed lowland rice (*Oryza sativa* L.) cultivars of northeast India. Ph. D. thesis, Gauhati University, Guwahati.

Bhattacharjee B, Gupta HS. 1995. Fertile plant regenerated from protoplasts of cold tolerant rice line RCPL1-2c. *J Plant Biochem Biotechnol* 4:61-65.

Bhattacharjee B, Pattanayak A, Gupta HS. 1998. Fertile plant regeneration from suspension culture-derived protoplasts of two *indica* rice lines and field evaluation of the seed progeny. *J Genet & Breeding* 52:135-141.

Bhuyan N, Borah BK, Sarma RN. 2007. Genetic diversity analysis in traditional lowland rice (*Oryza sativa* L.) of Assam using RAPD and ISSR markers. *Curr Sci* 93: 967-972.

Borbora TK, Hazarika GN. 1998. Study of genetic variability, heritability and genetic advance for panicle characters in rice. *Oryza* 35:19-21.

Borthakur DN, Majumdar ND, Varma Arun. 1983. Germplasm resources of field crops of northeast India. *Proc Intern Genetic Congress*, New Delhi.

Das GR. 1987. Collection, maintenance, evaluation and utilization of *ahu* rice germplasm of NE India. Proc Summer Inst on *Prospects of Utilizing Plant Genetic Resources of North East*

India. Assam Agricultural University, June 15-July 4, Jorhat.

Das GR. 1987. Prospects of utilizing rice germplasm of NE India for resistance to diseases and pests, quality attributes and stress situations. *Proc Summer Inst on Prospects of Utilizing Plant Genetic Resources of North East India.* Assam Agricultural University, June 15-July 4, Jorhat.

Das GR Ahmed T, Bhattacharya HC. 1981. Associations between flag leaf and grain dimension in rice. *J Res Assam Agric Univ* 2:96-97.

Das GR. 1987. Prospects of utilizing rice germplasm of NE India for resistance to diseases and pests, quality attributes and stress situations. *Proc Summer Inst on Prospects of Utilizing Plant Genetic Resources of North East India*. Assam Agricultural University, June 15-July 4, Jorhat.

De RN, Setharaman R, Singh MK, Banerjee SP. 1988. Genetic diversity in rice. *Indian J Genet* 48:189-194.

Devi PT, Durai A, Singh TA, Gupta S, Mitra J, Pattanayak A, Sarma BK, Das A. 2008. Preliminary studies on physical and nutritional qualities of some indigenous and important rice cultivars of north eastern hill region of India. *Journal of Food Quality* 31:686-700.

Devi PT, Pattanayak A. 2008. Medicinal rice of Manipur, their ethnopharmacological uses and preliminary biochemical studies. *Asian J Microbiol Biotechnol and Env Sci* 10:129-141.

Dutta D. 1987. Collection, maintenance, evaluation and utilization of *sali* rice germplasm of NE India. *Proc Summer Inst on Prospects of Utilizing Plant Genetic Resources of North East India.* Assam Agricultural University, June 15-July 4, Jorhat.

Dutta L, Barua JN. 1981. Effect of processing on nutritional quality of rice. Part II. *J Res Assam Agric Univ* 2:11-15.

Ghosh AK, Bhattacharya PK, Asthana AK .1981. Genetic variability in indigenous rice varieties of Meghalaya. *Indian J Agric Sci* 51:281-282.

Gupta HS, Borthakur DN. 1987. Improved rate of callus induction from rice anther culture following microscope staging of microscopic staging of microspores in iron alum-haematoxylin. *Theor Appl Genet* 74: 95-99.

Gupta HS, Sarma BK, Pandey DK, Prasad RN. 1995. Collection, characterization, evaluation and conservation of rice germplasm in northeastern hills of India. *Indian J Hill Farmg* 8:111-115.

Gupta JN, Pattanayak A, Gupta HS. 1993. Plant regeneration from protoplasts of cold tolerant rice varieties. *Rice Genet Newsletter* 10:113-116.

Hakim KL, Sharma SD. 1974. Localized distribution of certain characters of rice cultivars of northeast India. *Indian J Genet* 34:16-21.

Kihara H.1968. History of biological and other sciences in Japan in retrospect. *Proc XII Int Genet Congr* 3:49-50.

Kocchar S. 1987. Variability analysis in upland paddy germplasm collection from Arunachal Pradesh and their proposed utility. *Proc Int Symp on Crop Improvement*. Punjab Agricultural University, Ludhiana, Punjab.

Kocchar S. 1997. Variability analysis in upland paddy germplasm collection from Arunachal Pradesh and their proposed utility. *Proc Intern Symp on Crop Improvement.* Punjab Agricultural University, Ludhiana.

Krishnamurthy A, Sharma AC. 1970. Manipur rice in rice germplasm. *Oryza* 7:45-50.

Kulkarni VA. 1962. Studies on some rice varieties from Naga hills, Tuensang areas. *Sikkim and Nepal Rice Newsletter* 10:22-26.

Malik SS, Sharma SD. 2005. Conservation and utilization of genetic resources of rice. Pages 255-281 in Sharma SD and Nayak BC (eds) *Rice in Indian Perspectives.* Today and Tomorrow's Printers and Publishers. New Delhi.

Mitra SK, Ganguli PM. 1942. Studies on Barak valley rices and their classification. *Indian J Agric Sci* 2:571-606.

Padmanabhan SY *et al.* 1971. *National Germplasm Collection of Rice.* Central Rice Research Institute Cuttack. 235 pp.

Pajat J. 2008. *Biochemical Characteristics of Indigenous Rice Germplasm of Meghalaya.* M. Sc. Thesis, Allahabad Agricultural Institute, Allahabad, India.

Pal AK, Singh JK, Pattanayak A, Hore DK. 2004. Survey and collections of multi-crop germplasm from Mamit district of Mizoram. *Indian J Hill Farmg* 17:145-149.

Pandey DK, Gupta HS. 1991. Evaluation of rice cultivars at high altitude in Meghalaya. *Indian J Hill Farmg* 8:196-199.

Pandey DK, Gupta HS. 1995. Evaluation of rice cultivars at high altitude in Meghalaya. *Indian J Hill Farmg* 8:196-199.

Pattanayak A, Gupta HS. 2000. Evaluation of anther culture derived lines under upland conditions for the North Eastern Hills of India. *IRRN* 25:9-10.

Prasad RN, Sarma BK, Pandey DK. 1993. Collection, characterization, evaluation and conservation of rice genetic resources – a review and progress of work in north eastern hill region of India. *Proc National Group Discussion on the Issues of Rice Germplasm.* IGKVV, 4-5 October, Raipur.

Rao UP, Srinivasan TE. 1978. Evaluation of Assam dwarfs – suitability under low P and N conditions. *Madras Agri J* 65:626–627.

Rathi RS, Barua U, Das G, Hore DK .2002. Survey and collection of multi-crop diversity from parts of Upper Assam and Lohit and Dibang valleys of Arunachal Pradesh. *Indian J Hill Farmg* 15:56-60.

Sardana S. 1997. Submergence tolerance and elongation ability in newly developed rice for Tripura. *Indian J Hill Farmg* 10:111-112.

Sardana S, Borthakur DN, Lakhanpal TN. 1997. Genetic divergence in rice germplasm of Tripura. *Oryza* 34:201-208.

Sarma BK, Sharma UC. 2000. Agriculture of the Lakhers of Mizoram. *Asian Agri-History* 4:21-30.

Sarma BK, Abraham MJ, Borthakur DN.1987. Collection, evaluation, maintenance and utilization of high altitude rice germplasm of north east India. *Proc Summer Inst on Prospects of Utilizing*

Plant Genetic Resources of North East India. Assam Agricultural University, June 15-July 4, Jorhat.

Sarma BK, Singh JK, Pattanayak A. 2004. Collecting crop diversity from Changlang and Tirap districts of Arunachal Pradesh. *Indian J Hill Farmg* 17:150-152.

Sarma BK, Singh JK, Verma DK, Annadurai A, Pattanayak A, Laxminarayan K. 2000. Collection of multi-crop diversity from the eastern Himalayan region. *J Assam Sci Soc* 41:326-335.

Sarma RN, Bahar B. 2005. Genetic variations in *bora* rice (glutinous rice) of Assam as reaveled by RAPDs. *PGR Newslett* 144:34-38.

Sasikumar B, Sardana S. 1987. Genetic resource of rice in Tripura. *IBPGR Plant Genetic Newsletter* 71:31-33.

Seetharaman R, Srivastava DP, Ghorai DP. 1974. Preliminary studies in rice cultivars from NE India. *Indian J Genet* 34:143-149.

Seshu DV, Rao UP, Srinivasan TE, Shastri SVS. 1974. New source of dwarfness in the Assam Rice Collections. *Indian J Genet* 34: 390-394.

Setharaman R, Ghorai DP. 1976. Occurrence of types with characters of *Oryza glaberrima* in Assam Rice Collection. *Curr Sci* 45:67.

Sharma BD, Hore DK. 1989. Diversity of rice germplasm of north and northeast India. *Indian J Hill Farmg* 2:25-30.

Sharma SD. 1982. Collection and evaluation of rice germplasm from northeast India. *IBPGR Plant Genetic Resource Newslett* 50:62-69.

Sharma SD, Vellanki JMR, Hakim KL, Singh BK. 1971. Primitive and current cultivars of rice in Assam – a rich source of valuable gene. *Current Sci* 40:126-128.

Shastry SVS, Sharma SD, John VT, Krishnaiah K. 1971. New sources of resistance to pests and diseases in Assam Rice Collection. *IRC Newsletter* 22:1-16

Singh BD. 2000. *Plant Breeding*. Kalyani Pub, New Delhi.

Singh JK. 2007. Studies on High altitude rice germplasm of northeastern hill region of India. Ph D thesis, Gauhati University, Guwahati.

Singh K, Barua U, Srivastava R, Hore DK. 2002. Collection of multi-crop germplasm from central part of Arunachal Pradesh. *Indian J Hill Farmg* 15:87-90.

Singh RK, Sharma R. 1997. *Indigenous Rice Cultivars of Manipur Hills*. Manipur Science Society, Imphal. 72 pp.

Srivastava DP. 1977. Gene-ecological variations from north east India and their importance in rice breeding programme. *Proc Indian Sci Congr Part III: Agricultural Science*. Indian Science Congress Association.

Srivastava DP. 1978. Diversity in the late duration cultivars from northeast India. *Oryza* 5:26-33.

Srivastava DP, Nanda BB. 1977. Variation in grain protein in some groups of rice varieties from the collection of northeast India. *Oryza* 14:45-46.

Talukdar A, Talukdar P. 1998. Inheritance of photoperiod insensitivity and genetic utilization in India. *Proc Symp on Rice Breeding*. IRRI, Los Banos, Philippines.

Vairavan S, Siddiq EA. 1973. Cytogenetic studies on hybrids between rice varieties from northeast Himalayas and TKM 6, a typical *indica* variety. *Indian J Genet* 33:355-361

Vairavan S, Siddiq EA, Arunachalam V, Swaminathan MS. 1973. A study of the nature of rice germplasm of Assam Rice Collection. *Theor Appl Genet* 43:213-221

Verma RN. 1985. Research highlights of plant pathology of ICAR Complex. *Proc Workshop in Agrl Res*, Aizawal, Mizoram.

Genetic Resources of Rice in India : 229-277
Editor : S. D. Sharma
Today & Tomorrow's Printers and Publishers, New Delhi - 110 002 (India)

Chapter 7

Collection, Evaluation and Utilization of Rice Genetic Resources for Salt Tolerance by Central Soil Salinity Research Institute in India

R.K. Gautam, R.K. Singh, B. Mishra, Vinod Verma and Sharik Ali

Historical Perspective

Salinity has been a threat to agriculture in some parts of the world for over 3000 years and the threat has grown more in recent times (Flowers, 2006). It is estimated that about 20% of all irrigated lands in the world are affected by excessive salts (Pitman and Lauchli, 2002). The occurrence of salts in soil and water poses a serious constraint to agricultural productivity especially in those areas and crops where irrigation is essentially required. Agricultural salinity affects about 1000 million hectares of land globally which limits agricultural production and productivity of these areas. Broadly speaking, salt affected soils include two types i.e. sodic and saline soils. Sodic or alkali soils are characterized by excess of sodium ions on exchange sites and high concentration of carbonate and bicarbonate anions. These are associated with high pH (8.5 up to 10.8), high sodium absorption ratio (SAR) and high exchangeable sodium percentage (ESP) which lead to poor soil structure. On the other hand, saline soils also possess excessive sodium but with dominant anions of chloride and sulphate resulting in higher electrical conductivity (> 4dS/m). The diagnostic features of salt affected soils are given in Table 7.1.

The Government of India constituted an Indo-USAID Committee in 1967 to set up a Water Management Research Program under ICAR leadership for the IV Plan period. This team recommended the establishment of a national centre for research on salinity and alkalinity in India. The Government of India accepted these recommendations in 1968 and established Central Soil Salinity Research

Institute (CSSRI) in 1969 at Karnal. Along with the establishment of CSSRI, the Central Saline Research Station, Canning (West Bengal) which was formerly a constituent of Central Rice Research Institute, Cuttack was transferred under the administrative control of CSSRI, Karnal in 1970 to address the research needs of coastal salt affected areas. In 1988, another research substation was established at Anand in Gujarat with the aim to tackle the salt related problems of heavy vertisols. Recently, this has been shifted to Bharuch in Gujarat. Subsequently in 2000, a third research substation was established at Lucknow for managing sodic soils of central and eastern Gangetic plains.

Table 7.1. Diagnostics characteristics of salt-affected soils.

Salt-affected type	Electrical conductivity *(EC in dS/m)	Exchangeable Sodium Percentage *(ESP) (%)	pH
Saline	> 4	< 15	< 8.8
Sodic	< 4	> 15	8.5-10.7
Saline-Sodic	> 4	> 15	Variable

* at 25°C

The rice breeding work at CSSRI started with the inception of the institute. In view of lack of any information on salt tolerance of crop plants, initial emphasis was given for the screening of crops for their suitability and limits of crop tolerance. Dr. M.N. Sarin, the first Head of the erstwhile Division of Genetics and Plant Physiology and Dr. B. Mishra initiated the research programme on rice and found rice as a suitable crop for the salt affected soils. Several thousand rice germplasm from various parts of the country and abroad including ARC (Assam Rice Collection) was collected and evaluated under salinity and sodicity but only a few could be identified as promising. Initially it was presumed that the landraces or varieties good for salinity would not perform well under sodicity. Dr. R.K. Bhattacharya, who was initially looking after the coastal salinity program of rice at Canning Town, joined rice program at the main institute, Karnal in 1974. He made several selections from the locally grown salinity tolerant rice varieties Damodar, Dasal and Getu of Sunderban area and renamed them as CSR 1, CSR 2 and CSR 3, respectively. These selections from local types, when tested at the main institute Karnal, were found highly tolerant to sodicity too and could grow in virgin sodic soils having pH_2 of 10.2-10.4 without any chemical amendment. Interestingly, this observation provided a fillip to the hypothesis that genotypes which perform better under salinity could also perform better under sodic soils. Significant to mention, Dr. Mangala Rai, the present Director General, ICAR joined the rice program at CSSRI, Canning Town station in late seventies and made significant contributions in the salinity program. Dr. T.S. Sinha and Dr. A.B. Mondal continued to work for the development of salinity tolerant rice varieties for coastal ecosystem at Canning Town until Dr. Sinha left the research station and rice programme in mid 1980s. At the same time, Dr. R.K. Singh joined the rice breeding project at the main institute, Karnal and gave a major fillip to research programme. After the superannuation of Dr. R.K. Bhattacharya in 1987, Dr. B. Mishra led the rice program with Dr. R.K. Singh as associate for a long time and made tremendous contributions. Later, Dr. V.R. Babu joined at Anand Station in Gujarat in 1991 and worked on rice for saline vertisols till 2001. Dr. R.K. Gautam joined at Lucknow station in 2000 to look after the Institute's rice program for sodic soils of Uttar Pradesh.

Successful development of CSR 10 turned out to be a major event which attracted many national and international collaboration in rice programme. Research collaborations with IRRI, Philippines (1987-2000 and 1996-2000), University of Sussex, UK (1993-2000) and World Bank funded project with UP Land Development Corporation, Lucknow (1994-2000) expanded the work and the best infrastructure facilities of the country were created for salt related research. Rice breeding program at Karnal has also the proud privilege that Dr. B.Mishra, Rice Project Leader was selected as Project Director of Directorate of Rice Research, Hyderabad in 2000 and then Project Director of Directorate of Wheat Research, Karnal to lead the country's most important cereal coordination programmes. After serving this national responsibility, he joined the prestigious position of Vice Chancellor, Sher-e-Kashmir University of Agricultural Sciences and Technology, Jammu (J&K). CSSRI achieved another feather to its cap when Dr. R.K. Singh was selected as International Research Fellow in 2004 at International Rice Research Institute, Philippines and then elevated to international recruited scientist to look after the global rice salinity program. To look after the rice program at main Institute, Dr. R.K. Gautam joined at Karnal in 2005.

Salt-affected Areas: Challenges and Impacts

India's human population at the time of independence stood at about 33 crores which has now crossed 1 billion mark and the food production has increased from about 53 m tones to about 200 m tones but the cultivated land has not concomitantly increased. To meet the challenge of more food for rapidly expanding population, food has to come from either the same or shrinking cultivated area which is an uphill task. One of the possible and achievable ways is to get the production from the problem soil areas which otherwise remain barren.

Furthermore, these are the areas where population densities and poverty levels are high and the prevailing stresses affect the livelihood of a large population. Any breakthrough in this production environment can help in achieving substantial poverty alleviation and larger impacts.

Salt-affected Areas and their Distribution in India

Current figures of salt affected area estimated through remote sense mapping studies indicated that in India, about 6.74 m ha land area is salt affected out of which 3.77 and 2.96 m ha are afflicted, respectively, by sodic and saline soils (Sharma *et al.*, 2004). The state-wise distribution of these soils is given in Table 7.2. As per the estimates, India incurs about 12% loss to the production system due to abiotic stresses (Singh *et al.*, 2003).

Table 7.2. Extent and distribution of salt-affected soils in India (Sharma *et al.*, 2004).

State	Saline	Sodic	Total
Andhra Pradesh	77598	196609	274207
A & N Islands	77000	0	77000
Bihar	47301	105852	153153
Gujarat	1680570	541430	2222000
Haryana	49157	183399	232556
Karnataka	1893	148136	150029
Kerala	20000	0	20000
Madhya Pradesh	0	139720	139720
Maharashtra	184089	422670	606759
Orissa	147138	0	147138
Punjab	0	151717	151717
Rajasthan	195571	179371	374942
Tamil Nadu	13231	354784	368015
Uttar Pradesh	21989	1346971	1368960
West Bengal	441272	0	441272
Jammu & Kashmir	-	17000	17000
Total	**2956809**	**3787659**	**6744468**

NB: Figures are areas in hectares

Importance of Rice in Salt-affected Areas

Approximately 100 million ha in South and Southeast Asia are covered by problem soils where rice (*Oryza sativa*) is the staple crop. The year 2004 was declared by UNO as International Year of Rice to mark the importance of this crop commodity in meeting major food requirements of the world population. It is naturally the most adapted crop for the coastal saline soils and also for sodic soils. Generally, in the coastal saline areas this is the only feasible crop. Rice crop due to many advantages is ideally suited to start with the reclamation of sodic soils. Since the nursery is grown in normal soils, rice planting in salt affected soils also permits early stage stress escape. Unlike other crops rice culture involves water stagnation which allows the dilution and percolation of excessive salts thus making hospitable soil environment.

Harnessing the potential of prudent management of such salt afflicted areas can play a significant role in increasing and sustaining the food security particularly in view of increasing population and shrinking cultivable land area. According to one estimate India can experience 40% productivity decline by the year 2080 due to climate change. The rise of sea level due to global warming is expected to increase water-logged and salt prone areas. The secondary salinization arising from canal networks in irrigated lands is also perceived to cause formation of new saline areas. Such situations are compelling the human efforts to explore, collect, evaluate and breed for more salt tolerant genetic resources of adaptable crops such as rice.

Why do we need salt tolerant cultivars?

The concept of prudent deployment and adoption of different management approaches, individually or in combinations depending upon the site problem and specificity, is required to overcome salinity constraints. Reclaiming problem soils by chemical amendments and drainage interventions, however, invariably involves higher costs which are generally beyond the economic access of poor and marginal farmers inhabiting such areas. Moreover, such pilot projects are often operative in larger areas and individual farmers with smaller land holding size may not be enthusiastic about adoption in a participatory manner. Therefore, another attractive approach could be genetic tailoring of the crop plants in the form of salt tolerant varieties to adapt to these agro-edaphic conditions. The approach is simple and economical to adopt and also prevents environmental degradation. There could be a third approach i.e. synergistic approach, based on harnessing the synergies between the environment modifying technologies and genetically enhanced plant types. This approach is considered to be more practical, economically viable, less pollution causing and efficient with tremendous potential.

The biological phenomenon of salt tolerance

Maas and Hoffman (1977) categorized different crops in 4 classes (tolerant, moderately tolerant, moderately sensitive and sensitive) based on the performance of one or few genotypes/varieties. However, their conclusions for individual crops rallied around few genotypes/rootstocks and obviously did not account for tremendous intra-specific variation for this trait in rice. This fact is substantiated by the study conducted by Gupta and Sharma (1990) for the response of crops to sodicity expressed in terms of ESP. Distinct genotypic differences were found in rice for threshold ESP (ESP_1) as well as ESP at which the yield is reduced to 50% (ESP_{50}). Salt tolerant rice variety CSR 1 was found with an ESP_1 of 20.63 while MI 48, a sensitive genotype, with an ESP_1 of 8.02. Corresponding values of ESP50 for these varieties were 81.6 and 30.6 respectively. This study clearly signifies the importance of genetic variability for salt tolerance. Rice is found to have the maximum range of genetic variability for alkalinity tolerance while barley has the maximum for salinity tolerance.

Morphological and physiological effects of salts on crops

Though material tolerant to salinity is also tolerant for sodicity or vice versa, yet sporadic variations from this rule have also been experienced. The major visible symptoms in plants suffering from salinity/sodicity stress at field level include leaf tip burning or whitening (salinity), leaf browning and necrosis (sodicity), stunted plant growth, reduced tillering, spikelet sterility, low harvest index, less florets per panicle, less 1000-grain weight, low grain yield, altered flowering duration, leaf rolling, white leaf blotches, poor root growth and patchy growth in field. Often the symptoms of salt stress also find overlap with those of other abiotic stresses; therefore, their diagnosis needs careful attention.

Regarding physiological and biochemical effects, it is mentioned that though numerous changes take place under altered stress environments, only some register marked differences and thus also contribute a lot to the salt tolerance mechanism. These changes have been understood to govern the solute and water balance and their distribution on whole plant and tissue basis. Based on the studies it was observed that most of the crop plants and varieties encounter high Na^+ transport to shoot, preferential accumulation of Na in older leaves, high Cl uptake, reduced K^+ uptake, lower fresh and dry weight of shoot and roots, low P and Zn uptake. Bio-chemically, change in esterase isozyme pattern, glyoxalase, increase of non-toxic organic compatible solutes like abscissic acid, prolein, ethylene, glycine betaine, sugars and polyamines etc are influenced.

Screening and evaluation methods

For effective and error proof screening the germplasm for knowing their true performance, a homogeneous plot or field representing salinity/sodicity is required. However, salt affected fields often suffer from field heterogeniety or spatial variability. This offsets the reproducible screening and response of genotypes in salt affected soils. Therefore, making a homogeneous soil environment representative of a particular stress situation is a key to successful screening and breeding process. A site or field intended for evaluation should be characterized for the type (salinity or sodicity) and severity of stress (pH or ESP for sodicity and EC for salinity) with proper soil sampling techniques and analysis before, during and after crop experimentation. This is required for knowing consistent performance of genotypes across stress levels, locations and years. Reliable and reproducible screening techniques constitute the core of any successful crop improvement programme. Though screening techniques vary with crop species, growth stage and type of stress imposed but ideally it should be rapid, reproducible, easy and less expensive. There are different methods being used to screen crop germplasm for their performance under salt affected conditions at CSSRI, Karnal. These include saline hydroponics/solution culture, sodic/saline trays, sodic/saline microplots or lysimeters and sodic/saline fields. Each method has both advantages and disadvantages in terms of precision and space. Therefore a combination of different methods depending upon the objective will help in making a right choice. Different screening methods are discussed below:

Field evaluation

This is the real site for validation of crop germplasm or varieties for salt tolerance for making practical impact. However, it suffers from inherent problem of uneven distribution of salts in natural field conditions resulting in spatial variability to adversely affect the screening efficiency and reliability. The original screening work started in the naturally occurring barren fields, which remained uncultivated for an unknown period. That is why not much success could be achieved in this endeavour in the past. Considering this problem, modifications in the layout plans, designs of experiments and replicated experiments along with other screening aids enhance the screening efficiency. For field experiments, it is desirable to have 2 to 5 long rows of 10 to 30 meters with space planting for each genotype to minimize the effect of spatial variability. For lesser number of genotypes (<30), randomized block design with at least 3 replications is ideal. For more number of genotypes, we should prefer the incomplete block design or augmented design in which a set of check varieties is repeated many times without repeating test entries. The initial soil sampling for stress monitoring within a field allows the "local control" which is a pre-requisite for any experiment. Accordingly within replication, stress variation should be minimum compared to between replications.

Screening in micro plots/lysimeters

To obviate the problems of soil heterogeneity and spatial variability, a state of art facility of mini-salt fields called as microplots or lysimeters was created at CSSRI, Karnal. It is possible to create and maintain desired levels of sodicity and salinity in these microplots simulating field conditions minus the soil heterogeneity. The plot size in these plots is kept small because of a good "local control" over micro-environment and to accommodate more number of genotypes. This constitutes the landmark facility for salinity research and crop improvement. Currently, the breeding programs at CSSRI, Karnal are actively using this facility in rice, wheat and mustard improvement for salt tolerance. However, large segregating populations or more lines cannot be screened due to space limitation, though the most promising material should be preferred under this facility. These facilities also form excellent platform for phenotyping the stabilized mapping populations for molecular tagging of genes controlling sodicity/salinity tolerance.

Screening in pots

For more controlled study of the individual plant response under a constant stress, round porcelain pots of 20 or 30 cm diameter, with a capacity of 8 or 16 kg soil with a provision to allow or plug off leaching from bottom, are used. Pot studies also allow for periodic uprooting of plants at a given growth stage for specific studies.

Screening in saline solution culture

Solution culture technique is being used in two ways, one for mainly screening up to seedling stage and second up to maturity. In the first category, 7 days old seedlings grown on non-stress modified Yoshida culture solution (Table 7.3) are transferred to the desired level of stress either in the bread box with perforated lid or perforated Styrofoam with mesh. This modified culture solution has KH_2PO_4 and K_2HPO_4 in place of NaH_2PO_4 as sodium salt within culture solution may increase the Na ion concentration (Prof. T.J. Flowers, personal communication). Initial salinisation of culture solution with 50 mM NaCl (equivalent to about 4.9 dSm^{-1}) for three days followed by increase in stress level upto 70 to 100 mM (approx. 6.8 – 9.5 dSm^{-1}) is good enough for screening of genotypes for salinity at seedling stage. After 10 to 15 days, treated seedlings are used for the morphological, physiological or biochemical analyses. Singh *et al.* (2002) have also devised a solution culture technique for screening for sodicity tolerance in rice.

Table 7.3. Rice culture solution (Modified Yoshida Solution).

Stock	Reagent	g l^{-1}
1.	NH_4NO_3	91.4
2.	K_2SO_4	71.4
3.	(a) KH_2PO_4	23.1
	(b) K_2HPO_4	4.3
4.	$CaCl_2.6H_2O$	175
5.	$MgSO_4.7H_2O$	324
6.	Minor Nutreients	
	(a) $MnCl_2.4H_2O$	1.5
	(b) $(NH_4)_6.Mo_7O_{24}.4H_2O$	0.074
	(c) H_3BO_3	0.93
	(d) $ZnSO_4.7H_2O$	0.035
	(e) $CuSO_4.5H_2O$	0.03
7.	or FeNaEDTA*	10.5
8.	$FeSO_4$ (made fresh)	2.5

* to be replaced regularly
NB : 1.25 ml each stock solution per litre culture solution
pH of 4.5 of the culture solution to be adjusted with HNO_3 regularly

Screening in trays

This method is particularly useful for rapid screening of large number of lines but is restricted only up to seedling stage. For this, salt affected soil is filled in shallow depth in the wooden or metal germination trays provided with a polythene sheet lining on the inner side. These are quite handy for screening lines with appropriate control on salinity, sodicity and moisture. The trays can also be conveniently shifted from one growth chamber to another.

Screening and selection criteria

The screening parameter or trait which shows clear difference between genotypes subjected to salt stress in a reproducible manner qualifies for a reliable screening criterion. In addition, the ideal screening parameter should also preferably influence the final economic yield. The more the intimate relationship between the morphological, physiological or biochemical trait with final yield, more indispensable would it prove for any successful breeding programme. Unfortunately, no single clear cut trait or criterion other than economic yield has emerged so far which is linked with yielding ability under salt stress. However, these screening criteria perceived to contribute to overall tolerance are described below:

Morphological parameters

Survival of the plant: It is mainly limited to the seedling studies; however, in some of the adult plant studies it has also been considered. Under moderate stress, plant survival is not a problem. But under high stress, it is a good selection criterion to select few robust tolerant segregants from large populations or lines from large collections.

Injury score: Individual plant or group of genotypes are scored usually on 1 to 5 or 1 to 9 scale where lower score indicates tolerant and higher score denotes sensitive genotypes.

Phenotypic expression: Excessive tip burning especially in younger leaves, spikelet sterility and stunted growth are considered for the overall phenotypic expression of the genotype under stress environment.

Root/shoot growth: Salt stress severely hampers the root/shoot dry as well as fresh weight. In rice, sodicity affects both root and shoot fresh and dry weights but genotypic differences were observed for the degree of reduction.

Grain yield: It is the major selection criterion usually followed for screening and evaluation of the collections and advanced breeding materials. Generally, in the absence of any simple and reliable selection criteria, grain yield performance of the genotypes under salt stress remains the practical measure of tolerance.

Stress susceptibility index (SSI): Stress susceptibility index is a better criterion of adjudging the tolerant genotypes than stress performance *per se* of genotype because it accounts for the differences in yield potential too. Through this method, better performance under stress can be traced back either to better yield potential or to true stress tolerance. Tolerance to salt stress can be worked out by stress susceptibility index (S) on the principle of yield minimization under stress compared to non-stress environment (Fischer and Maurer, 1978). The S values for individual genotype are calculated as:

$$S = \frac{1\text{-}Y_s/Y_p}{D}$$

Where S= Stress susceptibility index,

Ys = Mean grain yield of a genotype under stress

Yp = Mean grain yield of the same genotype under non stress

D = Stress intensity, which is calculated as under

$$D = \frac{1\text{- Means } Y_s \text{ of All genotypes}}{\text{Mean } Y_p \text{ of all genotypes}}$$

The higher value of 'S' means more stress susceptibility and vice versa. The low values of 'S' indicates those genotypes which perform well under stress with sufficient plasticity to respond to the potential environment.

Physiological parameters

Ionic parameters: Numerous physiological mechanisms are perceived to contribute to the overall ability of rice plant to cope up with excess salts (Yeo *et al.*, 1990; Flowers, 2004; Blumwald and Grover, 2006; Ismail *et al.*, 2007). Physiological mechanisms influencing tolerance include various attributes like Na exclusion, tissue tolerance, low Na/K ratio, efficient salt partitioning ability within plant to retain harmful salts in functionally less active organs like vacuoles and older plant parts (Singh *et al.*, 2004). Negative effects of salts are also offset by higher uptake of K, Zn and P. Since earlier results showed that no single mechanism could confer the absolute tolerance, pyramiding genes for diverse physiological mechanisms into one genetic background has been initiated and good breeding populations have been generated. A wide spectrum of germplasm in rice has been evaluated and categorized based on tissue tolerance, Na^+ exclusion, K^+ uptake, Na^+/ K^+ ratio and reproductive stage tolerance for using these as donors for strengthening breeding program as shown in Table 7.4 (Singh and Mishra, 2004). In the study involving 79 rice genotypes, Na^+/K^+ ratio varied widely. It ranged from as low as 0.81 (PR108 - a salt sensitive variety) to as high as 8.66 (Bas 370 - salt sensitive variety). No definite trend in Na^+/K^+ ratio for salt sensitive or salt tolerant genotypes could be observed (Singh, 1996). However, in another study, the lower Na/K ratio in general showed better tolerance.

Table 7.4. Sodium and potassium accumulation sorted by per day Na basis.

Variety	Na	K	Na/K	Days of Treatment	Per day accumulation Na	K
	Mmol/g dwt		ratio		mmol/g dwt	
Hathwan	7.20	3.03	2.38	4	1.799	0.757
Carps Clarke	4.51	1.23	3.66	3	1.504	0.411
Panvel 23	8.06	1.73	4.66	6	1.343	0.288
Swarn Dhan	6.65	2.73	2.44	5	1.331	0.546
Indrasan	3.88	1.52	2.55	3	1.292	0.507
Ravi	3.76	1.11	3.40	3	1.255	0.369
Udaya	3.61	1.28	2.81	3	1.203	0.428
Pusa 1298-1	7.01	2.27	3.09	6	1.169	0.378
IR 2031	8.97	2.55	3.52	8	1.121	0.319
P 1176-91-1-2-2	5.40	1.89	2.86	5	1.081	0.378
PR 106	5.38	1.82	2.95	5	1.076	0.364
CSR 891R-4	8.43	2.43	3.46	8	1.054	0.304
IR 61311-3B-8	8.42	2.46	3.42	8	1.052	0.308
Nam Sagui	4.74	1.54	3.09	5	0.949	0.307
PRH 107	6.13	1.97	3.11	7	0.876	0.282
CSR 21	5.20	2.65	1.96	6	0.867	0.442
Panvel 2	5.20	1.33	3.90	6	0.866	0.222
IR 58	4.32	1.82	2.38	5	0.864	0.363
CSR 8	5.74	1.53	3.74	7	0.819	0.219
Panvel 33	4.78	1.95	2.45	6	0.797	0.326
Pusa 1317-3	3.90	1.52	2.57	5	0.780	0.304
RP 144	3.70	2.13	1.74	5	0.740	0.426
Majhera 3	4.34	2.17	2.00	6	0.724	0.362
CSR 26	4.97	1.37	3.64	7	0.710	0.195
93109	6.61	2.57	2.57	10	0.661	0.257
ML 47-22	5.71	1.99	2.87	9	0.634	0.221
Pusa 150	6.90	2.52	2.74	11	0.627	0.229

Kalinga 2	3.73	1.31	2.85	6	0.622	0.219
BC-I	6.20	2.23	2.78	10	0.620	0.223
Pak 803	3.60	1.00	3.61	6	0.600	0.166
SR 26B	5.76	1.96	2.94	10	0.576	0.196
VKL 39	1.70	1.27	1.33	3	0.566	0.425
Sarbati	3.37	0.96	3.50	6	0.561	0.160
IR 4630	5.04	3.82	1.32	9	0.560	0.425
Barkat	2.78	1.21	2.29	5	0.557	0.243
CSR 11	3.79	1.84	2.06	7	0.541	0.263
IR 2153	4.20	1.07	3.92	8	0.525	0.134
T 23	3.66	1.42	2.59	7	0.523	0.202
Sukhvel	1.99	1.07	1.87	4	0.499	0.266
CSR 22	2.99	0.93	3.21	6	0.498	0.155
Prakash	1.95	1.39	1.40	4	0.488	0.349
Ratnagiri	1.87	0.88	2.12	4	0.466	0.220
GR 11	4.63	1.98	2.34	10	0.463	0.198
CSR 23	3.24	1.81	1.79	7	0.463	0.258
Salivahana	1.83	1.47	1.24	4	0.458	0.368
IR 42	1.76	0.96	1.84	4	0.441	0.240
Muskan	2.20	2.25	0.98	5	0.439	0.449
Hasan Sarai	2.19	1.77	1.23	5	0.438	0.355
Achhi	1.74	1.86	0.94	4	0.436	0.465
Bas sathi	2.17	1.02	2.13	5	0.434	0.204
Mangla	2.60	1.23	2.12	6	0.433	0.204
IR24	1.70	1.45	1.17	4	0.426	0.363
IR 36	3.36	1.69	1.99	8	0.420	0.211
Majhera 7	2.06	0.82	2.52	5	0.411	0.163
SLR 51214	2.05	1.33	1.54	5	0.410	0.267
MTUAR 2038	1.62	1.32	1.22	4	0.404	0.331
Prasad	3.06	1.25	2.45	8	0.383	0.156
Vandana	2.29	0.90	2.54	8	0.382	0.151

HBC 19	3.35	1.40	2.40	9	0.372	0.155
ADT 36	1.40	1.15	1.22	4	0.350	0.286
CSR 20	2.77	1.82	1.52	8	0.346	0.228
Dehraduni	2.63	1.40	1.88	9	0.292	0.156
Pokkali	4.26	1.39	3.08	15	0.284	0.092
Vikramarya	2.53	0.78	3.23	9	0.281	0.087
P 203	1.62	1.76	0.92	6	0.270	0.294
HB 101B	1.04	1.00	1.04	4	0.259	0.249
Narendra 80	1.26	0.92	1.36	5	0.251	0.185
Jaya	2.26	0.44	5.13	9	0.251	0.049
CSR 13	2.25	0.80	2.83	9	0.250	0.088
HKR 128	2.71	0.49	5.56	12	0.226	0.041
CSR 1	2.46	1.88	1.31	13	0.189	0.144
PAC 831	0.93	1.14	0.82	5	0.187	0.228
Bas 370	2.05	0.24	8.66	19	0.108	0.012
Ml 48	1.39	0.26	5.45	13	0.107	0.020
CSR 10	1.48	0.88	1.68	15	0.098	0.059
Bas 385	1.12	0.39	2.86	12	0.094	0.033
CSR 18	1.11	0.81	1.37	12	0.092	0.068
CSR 19	1.39	1.49	0.94	16	0.087	0.093
PR 108	0.73	0.89	0.81	9	0.081	0.099

Overall mean revealed higher magnitudes of Na/K ratio in alkali (0.45) and saline (0.46) than in non-stress soil (0.39). Based on the lower magnitudes, the best varieties identified were CSR 10, CSR 11, CSR 13, CSR 1, BR 4-10 and Pokkali for both alkali and saline soils. It clearly shows that the Na^+/K^+ ratio for the screening and identification of the genotypes could also be used but not as a sole criterion. That the lower Na/K ratio relates to higher salt tolerance has also been reported by Mishra *et al.* (1990).

Germplasm grouping for important attributes associated with salt tolerance: CSSRI, Karnal has grouped different rice genotypes based on the presence of particular physiological mechanism controlling salt tolerance (Table 7.5). Classifying these genotypes or donors for salinity tolerance mechanisms, inter-mating of the genotypes with high degree of expression of the contrasting salinity tolerance mechanism and identifying/screening of the recombinants for pooling of the attributes is being followed to enhance the further level of salt tolerance. Crosses are made between the parents/donors possessing contrasting physiological traits like tissue tolerance, Na^+ exclusion, K^+ uptake and Cl^- exclusion to pool or pyramid the genes governing or contributing for salinity tolerance into one agronomically superior background with an objective to attain overall tolerance.

Table 7.5. Salt tolerance mechanism based grouping of rice genotypes.

Group1 Tissue tolerance to Na^+	Group2 Na^+ excluder	Group 3 K^+ miners	Group 4 Low Cl Uptake	Group 5 Low Na^+/K^+ ratio	Group 6 Early vigor
SR26B	CSR1	Hathwan	HKR128	PR108, Achhi, CSR1	CSR-92-5
CSR21	PAC-831	Swarndhan	Jaya	CSR19, ADT36	89-H1-1-3
CSR26	CSR8	Achhi, CSR23	CSR10	IR4630-22-2-5-1-3	91-H2-6-B-2
CSR27	CSR10	IR36	PR108	CSR18, SLR1214	92-H5-4
Pokkali	CSR13	IR4630-20-22-	CSR19	CSR10, IR42	KR1-24
IR4630-22	CSR19	2-5-1-3, CSR21			CSR11
-2-1-3	CSR20	GR11			CSR1
TCCP266-2	CSR22	SR26B, CSR1			CSR1
-49-2B-3	CSR23	CSR10, CSR11			CSR27
GR11	CSR24	CSR20, Panvel23			

Germplasm Assemblage and Evaluation for Salt Tolerance

In the Indian context, Singh *et al.* (2004) and Singh and Mishra (2004) have thoroughly reviewed the progress and future thrust areas of harnessing salt tolerance in enhancing rice productivity. We have endeavoured probably for the first time to compile and document the voluminous work of almost four decades on collection, evaluation and utilization of rice genetic resources for salt tolerance carried out by CSSRI and its regional stations through Institute's Annual Reports. A detailed perusal of the above documents revealed that various rice materials have been extensively collected, characterized and evaluated under diverse salt stress situations every year as shown in Tables 7.6, 7.7, 7.8 and 7.9. The tested stress conditions of sodicity and salinity were mainly in the states of Haryana

(CSSRI, Karnal, Gudha, Mundalana, Sarswati, Kaithal farms as well as precisely controlled microplots facility at Karnal), West Bengal (Canning Town), Uttar Pradesh (Shivri Farm, Lucknow) and Gujarat (Bhal region). The materials were collected, developed and tested mainly for sodicity and inland salinity in Haryana and Uttar Pradesh mostly during *kharif* season. The germplasm collection and breeding efforts in West Bengal aimed at to tackle coastal salinity accompanied with variable water depths during both *kharif* and *rabi* seasons whereas in Gujarat, material was mainly tested for adaptation to black vertisols. Since, grain yield is the economically important and ultimate target trait; we report here identification of mostly those lines which showed yield superiority under salt stress.

Right from its inception, CSSRI Karnal collected and evaluated more than 11000 germplasm lines consisting of introductions, landraces, improved type material developed in India and at International Rice Research Institute, Philippines. The material was screened for salt tolerance through diverse sodicity and salinity conditions in Karnal, Gudha, Mundalana, Sarswati and Kaithal farms as well as at state-of-the-art precisely controlled sodic and saline microplot facility at Karnal. This massive work was also facilitated by different national and international programmes from time to time. Similarly, more than 1700 genotypes were evaluated across sodic soils in the state of Uttar Pradesh. This resulted in the development and release of many salt tolerant rice varieties mostly suited to sodic and inland saline soils of the country (Table 7.10). Similarly, about 9000 germplasm lines were evaluated from the beginning at Canning Town where material was tested mostly during both *kharif* and *rabi* seasons. This work resulted in the development of suitable salt tolerant varieties for coastal saline areas. CSSRI, Karnal also provided technical trainings and back up to rice breeders of India regarding soil salinity parameters, germplasm handling and material exchange from time to time through different national and international network programs. The compilation of figures since beginning indicates that CSSRI assembled and evaluated more than 23000 germplasm lines for salt tolerance across highly diverse sodic and saline soils of the country. In all, the huge germplasm collection, systematic evaluation through standard protocols and germplasm exchange coupled with technical support are perceived to have gone a long way in the development of many salt tolerant varieties in the country (Tables 7.11 and 7.12).

Utilization of CSSRI Rice Genotypes in Breeding

Genetic studies

Inheritance studies

The inheritance of sodicity tolerance was inferred from the phenotypes observed in F_1 and bulked F_3 generation of three crosses between two tolerant

Table 7.6. Rice germplasm assembled and evaluated for salt tolerance in Haryana (CSSRI, Karnal, Gudha, Mundalana, Sarswati Farms).

Year	No. of germ-plasm	Place of testing	Type of soil	Promising salt tolerant germplasm identified
1970	21	Karnal	Saline and sodic	Damodar (CSR1), MC1, Dasal (CSR2), Basmati 217, SR 26B, NP 130, BC 5, P 221, Jhona 351
1971	1117	Karnal	Sodic (pH -9.5-9.8)	Jaya, IR8-68, Pusa 2-21, Padma, Hybrid mutant 452, Hm 474 , Jhona-20
1972	109	Karnal	Saline and sodic	Vijaya, CR-44-35, Jaya (4)*, IET-1136, IR-8-68,IR-8, Palman-579, C-579, HM 474, Sabarmati, IET-1991, IET 1991, IET 1451, IET 2295, Improved Sabarmati, IET 2868, IET 1991, Ratna
1973	130	Karnal	Sodic (pH 8.6-8.8)	Pusa 2-21(2), Ratna, RP 107-11, Jaya (2), Pusa 4-1-11,CR 110-174, C7306, C24263, RP 6B-11, RP 84-39-1, IET 2508, Bala, Cauvery, Satya, Anupama, Mu1, Ratna, RPW 6-12, RP 176-6, Vijaya, IR 8, RP 5-59, RP 5-46, RP 5-32, Pankaj
1974	45	Karnal	Sodic (pH 8.6 and 8.8)	RPW 6-12, RP 176-6, Vijaya, RP 5-59, RP 5-46, RP 5-32, Pankaj
1975	1685	Karnal	Sodic (pH 9.5 to 10.3)	IR-2153-43, IR 2153-26
1976	332	Karnal	Sodic (pH 7.5 to 9.8)	Pusa 2-21, RP6-51-31-6, CR142-3-8, CR110-170, Pusa4-1-11, RP79-5, PR 106, Pusa33-30-18-3, RP79-9, RPCB2-B-84, CR142-3-2, Mut 1, RP516-31-4, CR44-35, Jaya, RR260-750-3, IR-8, CR129-118, C-8054, RP6-516-31-6, CR115-76, 27092, TNAU7892, MTU6368, TNAU13613, Ratna, RP-967-4-7-2-1, RP-967-4-7-3, RP-967-4-7-4
1977	393	Karnal	Sodic	RP 6-516-31-6, RP 79-5, RP 79-9, CR 110-174, P 4-1-11, CR 142-3-2, CR44-35, CR 142-3-8, UPR 82-1-7-1-2-1-1, CR 141-6058-1-235,

				IET 6225, R 155-355, CR 222-MW 10, CR 216-HK-32-162, CR 141-5036-239, IET 6226, RR 1158-109-1, CR 141-7080-225, RP 1158-91-1, CR 146-7055-225, RP 4, R34-2480, IR442-2-58, HPN 59, IR 2068-65-3, SFC III, IR 1632-93-2-2, RR 967-65-4-3-6, RP 919-189-3-1, RR 825-24-7-5, CR 139-1049, UPR 96-26-2-1-2-1, RP 1021-34-2-1, CR 139-1061, PAU 1-608A, RPW 6-13, PR 106, CAUVERY, MTU 6225, CR 146-5432-208, C 3810, R101-2608, Sona, RP 107-11, RP31-49-2, RP 967-7-4-2-1, C321 or 10-266, M 242, M 117, M 432, M 114, H 62, H 33, M 23, M 152
1979	435	Karnal	Sodic (pH 9.0-9.8)	RJR 113-251, Triveni, BG 34-8, BG 33-2, IET 4089, TNAU 2686, TNAU 7893, RP 872-1-2, MTU 6368, IR 1632-93-2-2, IR 2063-185-2-2, Pusa 37-6-2-3, RP 1021-34-2-1, CR 138-1040, RP 6-51-31-1-8, RP 6- 516-32-2-3-1, C 8054, IR 2153-43-1-5-4, CR 221-W 1014-83, Jaya, CR 22-2685, SAR 28, SAR 43, CSR1, CSR3, UPRM 202, HAU 5-162-3 , Pusa 167
		Gudha	Sodic (pH 10.2-10.4)	TR 17, CSR 1, R 3, CSR 3.
1980	600	Karnal	Partially reclaimed sodic	BG 33-2, RJR 113-251, TNAU 7893, PAU 41-306-1-4-PR422, Nong Nghiep 75-5, Pae Ikulaku, Pae Bone, IR 5908-40-3-2-2, IR 1632-93-2-2, RP 6-516-34-1-8, RP 6-516-32-2-3, IR 14753-49-2, RP 967-11-1-3-9, H3-54, R 1913, IR 13415-46-2, B2360-6-9-2-6-MR-2, CR 235-25, SAR 43, Nonasal, CR 235-11, CR22-2685, CSR 3, BR 40-39-1-3, IR 9575, Ratna, HAU 5-298-2, P150-9-4-1 , HAU 5-162-3.
		Gudha	High sodic (pH sodic	SAR 43, SAR 28, SAR 41, CR 22-2735, M 242.
1981	560	Karnal	Partially reclaimed sodic (pH 9 - 9.5) and saline (ECe-5.0-15)	HXS 53-107, RJR 113-251, SXS 55-301, RP 872-1-2, BG 33-2, Sinank, PAU 125-149-2, IR 8608-125-3-3, IR 9224-225-2-3-3-2, IR 9093-195-1, IR 1632-93-2-2, H 3-54, RP 6-516-32-2-3-1, C8054, B2360-6-9-2-6, M242, M117, SAR41, SAR43, CSR1 (7), entry no. 29, 95, 207, 210,211,28,38,39,40,41,42, RP 974-146-5-3, IR 2035-290-2-1-1,IET 47418, BAS 140, RP 967-4-7-4, C321,P150-21-1, Jhona 349, CSR 3-(6), Ookaghama(4), Oorgondon, IR 4595-4-1-15-5, Jaya(2), Pokkali(4), SAR 26 B, IR 2053-436-1-2, IR 43(2)

1982	43	Karnal	Sodic pH (9.0-10.0)	IR 9224-225-2-3-3-2, IR 8608-125-3-3, RJR 113-251, IR 1632-93-2-2, MTU 6368,IR 3511-39-3-3, IR 5853-162-1-2, SAR 43, M152, IR 10198-66-2, IR 10206-29-2, CSR1, IR 2035-290-2-1-1
	353	Gudha	Sodic	SAR 43, CSR3, M 152, IR 10198-66-2, IR 10206-29-2, CSR 1
1983	41	Karnal	Sodic	Culture no. 25 (IR 8 X H18), Culture no.12 (IR 8 x TKM6)
		Gudha	High sodic (pH-10.0)	CSR 1, RP 1163-191-6-3-2-1, PNL 5-1-30
1984	430	Gudha	High Sodic pH-9.5-10.0and saline (ECe-5.0-10.0dS/m)	IR 19743-25-2-2-31, IR 22105-7-5-2, IR 25865-124-3-3, IR 29708-41-2-2-3, KK 8-7-B , Basmati Mwea, Jaya (2), TNAU 17069, CSR 5,CSR 1 (3), RP-872-1-2, CSR2 (3), H62, CSR3, SAR43, CR 235-11,CR 235-25, CSR3, Pokkali
1985	246	Karnal	Partially reclaimed sodic (pH<9.5) and saline (ECe 5-10dS/m)	Si-pi 692033, Suweon 294 , IR 22107-14-2-1, P 3299-3, 80-H,3-12, 80-H, Jaya, TNAU17069, CSR 5, CSR 1(3), CSR2, CSR3, CR 235-25, M40-432-11-117, CR 235-11
		Gudha	Sodic (pH-9.8-10.0)	81-Hyb.57-7, 81-Hyb.21-4, 81-Hyb.46-6, M1-2-1, M7-14-6, M13-15-6, M14-16-1, M22-18-5
1986	246	Karnal	Sodic (pH-9.5)	80-H-3-11, 81-H-8-42-14-15, 80-H5-76-64, 80-H4-24-36, Suweon, CSR5, Sipi 692033, IR29725-21-1-3-2, IR 25925-84-3-2
		Gudha	Sodic (pH-10)	81-H57-7, 1-H21-2-4, 81-H29-3-5, 1-H25-3-7, 81-H50-7-1, CSR1 mutant M2-1-1
1987	296	Gudha and Karnal	Sodic (pH-7.5-10)	IR 8192-200-3-3-1-1, H3-14 (4), Jaya (2), Taduken (4), MCM-1(2), IR4462-22A-2-10 (2), PAU-21-88-5, IR4462-22A-2-10, IR8192-200-3-3-1-1, CSR1 (2), 1507, 1508, 1530, 1543, 1554, CSR 5, 1509 , 1510, CSR 1 mutant, CSR 5, 1401,1415, 1402, 1407, 60202, IR 36, 60208, 60212, 60309, 60310, 60308, IET 9354, Suweon, CSR 5, Si-pi 692033, IR 29725-21-1-3-2 , IR 25925-84-3-2, IR 31787-85-3-3-3-2, IR 36, PDR 76-D10-D8-D1, M2-5, M2-32, M2-2, M2-29, M2-23, IR 35295-

				82-2-1-3-2, BG 850, Taichung Sen Glutinous, IR 39422-19-3-3-3-3, X 73-3-9, IR 13419-35-1, IR 13524-21-2-3-3-2-2, IR 8
1988	350	Gudha	Reclaimed sodic (pH 8.92-10.07)	1627,1653, 1004,706, IR 8, CSR 9, CSR 13, Jaya, IR 31375-3-3-1-1, IR 31375-3-3-2-2-3, IR 31662-47-2-1-1, IR 9884-54-3-IE-P1-1
		Mundlana	Saline (ECe 5.4-15.05 dS/m)	BC 3, PR 106, BC 1, CSR 10
		Karnal	Sodic and saline (ECe 4.2-9.2 dS/m)	IR 47449-B-B-B-9-B-B, IR 48251-B-B-B-1-1-B-B, IR50129-B-B-9-B-B, IR 51499-B-B-30-B-B, IR 51499-B-B-38-B-B, 1601, 1630, 1603, 1626, 1610, 1654
1989	138	Gudha	Sodic (pH 9.56-10.3)	RP 2597-8-239, RP 2507-14-221, IR 31375-3-3-2-3, IR 31375-3-3-1-1, IR 33552-12-1-2-1, Nona Bokra, CSR 13, CSR 1, CR221-130
			Sodic (pH9.2-10.31)	IR 30864, RP 2269-424-298-98(2), HKR 533-20, HKR 502-37, HKR 490-65, P 837, RP 1810-413-349(2), 80-H5-76-64, RP 1603-390-310-88, 8 J-H8-42-14-1, Hyb 22-713, Hyb 22-706, NDR 637, 81-H57-7
		Mundlana	Sodic and saline	AC 6534-1, IR 1500-AC-1-7, IR 37257-41-3-2-3, IR 31662-34-3-3-2, IR 31362-8-3-2-2, CSR 12, IR 4595-4-1-13, CSR 13
1990	289	Gudha	Sodic(pH 9.5-10.5) and saline (ECe 10 dS/m)	CSR11, BG35-2, RP 1603-39-310-88, RP 2597-3-233, Vikas, CSR10, IR 52731-B-B-B-3-B, IR54745-2-6-16-B-B, TCCP-266-2-49-B-B-3, IR 53655-B-B-B-4-B, IR 53655-B-B-B-5-B, IR 5931-110-1, BC 3
		Mundlana	Sodic (pH 9.2-9.7) and saline (ECe 3.0-9.0 dS/m)	BG 35-2, P837, B5322-B-PN-2-MS-3-NG-2, CSR11, IR4595-4-1-13, 80-H5-76-64, IR 43610-8-8-7-8, RP 2597-3-223, IR 16294-C59-1-30, CSR 10, IR 51500-AC-11-1, TCCP266-B-B-B-13-1-3,IR53655-B-B-B-5-B-B, IR53655-B-B-B-7-B, IR52731-B-B-B-17-B
		Karnal	Saline (5-10 dS/m) and sodic (pH 9.7-10.0)	IR 31361-8-3-2-2, CSR10, CSR13, CSR12, CSR 9, 131-5-8,7906-TR11-1-1 , 80023-TR-166-2-1-4, IR 53655-B-B-B-5-B, IR 52731-B-B-B-17-B (3), IR 53655-B-B-B-14-B, 53655-B-B-B-15-B (2), 53655-B-B-B-3-B

1991	226	Gudha	Sodic (pH~9.2-10.4)	IR 4595-4-1-13 (3), IR 22083-105-3-3-3-3 (2), CSR 10 (4), CSR 19 (3), IR 32809-26-3-3, RP 2597-18-237, RP 2597-3-223,CSR 17, IR 51500-AC-17, IR 31375-3-3-2-2-3, AC 6534-4 , IR 52710-B-B-4-B-B-4, CSR 11, CSR 9, IR 55237-B-B-B-3-2, IR 55237-B-B-B-11-1, IR 51491-AC5-4, IR 53655-B-B-B-6-3, , 52717-B-B-4-B-B-1-3, IR 45247-B-B-2-B-B-1-2, IR 42290-B-B-B-4-B-B-2, IR 52713-B-B-2-B-B-H, IR 49737-B-B-B-28-B-B-2, IR 1471-B-B-2-B-B-B-1, IR 47538-B-B-B-9-B-B-B-1, AC 6534-4, IR 52713-B-B-2-B-B-H, IR 7538-B-B-B-9-B-B-B-1.
		Karnal	Saline (ECe~7dS/m) and sodic (pH 9.2-10.2)	AC 6534-4, IR 52713-B-B-2-B-B-H, IR 47546-B-B-B-1-B-B-B-1 (2), IR 54447-B-B-B-6, IR 47538-B-B-B-9-B-B-B-1, NHT 1, NHT 5, NHT 7, NHT10, NHT 5 , NHT 9.
1992	189	Gudha	Sodic (pH 9.7-10.6)	IR 47538-B-B-B-9-B-B-B-1,IR52713-B-B-2-B-B-H, IR 48065-B-B-B-6-B-B-1, IR 48065-B-B-6-B-B-3, IR 47605-B-B-B-B-6-B-B-1-H, IR 31361-8-3-2-2-2, IR 5931-110-1 , CSR 10.
54447-3B-23-			Sodic (pH 9.7-10.43)	IR 51471-B-B-2-B-B-B-2, G129175,TCCP266-2-49-B-B-3, Deepak, IR46329-SRN-34-3-1-2, LD 183-4, LD 183-7, IR 51471-2B-2-2B-1-1, IR 55179-3B-11-3, IR 52713-2B-8-2B-1-2, IR 57304-4B-4-2, IR 3B-10-2, IR 52724-EB-6-2B-1-1, IR 60485-EB-12-2, IR 55233-3, IR 26916-ES.
		Karnal	Saline (ECe 8.10dS/m)	IR 57304-4B-4-2, IR 60483-2B-10-3, IR 59428-3B-11-2, IR 57048-4B-2-2,IR 60483-2B-17-2, IR 51417-2B-2-2B-1-1, IR 47547-3B-26-26-2B-1 ,IR 52713-2B-8-2B-1-2.
			Saline and Sodic	IR 51417-2B-2-2B-1-1, IR 57304-4B-4-2, IR 52713-2B-8-2B-1-2, IR 55179-3B-11-3 , IR 54447-3B-10-2.
1993	218	Saraswati, Karnal	Saline and sodicMicroplots (pH 10.4)	IR 36, CSR 1, IR 51471-2B-2-3B-1(2), IR 52713-2B-2-2B-H, IR40678-2B-20-3B-4, CSR20, CSR18 (2), CSR19, CSR 87- IR2, CSR 13, CSR 88-IR-1, CSR 87 IR-1 , RP 2597-18-250.
		Karnal	Saline (ECe8-9dS/m)	CSR 87 IR-2, CSR 19, NDRK 5020, CSR 87 IR 1, RP 2597-18-250, CSR 13, CSR 10.

		Saraswati, Karnal	High Sodic (pH 10-10.2)	IET13762,13764,13766,13768,13919,13412, 13765, 13767, 13769, 13917, 13920,CSR 89IR-7, CSR 89IR-8, CSR 89IR-9, CSR 90IR-2,CSR 88IR-11,CSR 88IR-12, CSR 88IR13, CSR 88IR-14,CSR 89IR-11,CSR 89IR-10, CSR 89IR-12, CSR 89IR-13, CSR 88IR-15, CSR 89IR-14, CSR 89IR-15LD 183-3, CSR10(2), LD 183-4, IR 61249-3B-19-2, IR 61250-38-7-1, IR 61247-3B-8-2, IR 63295-AC202, IR 62692-3B-5, IR58440-3B-11-3-3,IR 63295-AC210-4, IR 63295-AC211-3, IR 58440-3B-19-2-1, IR 61228-3B-10-3, IR 61240-3B-10-2, IR 47437-2B-8-2B-3, IR 47605-4B-6-2B-1-H, IR 52713-2B-2-2B-H, IR 31361-8-3-2-2, IR-12, CSR 90IR-2, CSR 89IR-10.
1994	238	Kaithal	Sodic and saline	CSR 89IR-5, CSR 89IR-2,CSR 88IR-6, CSR 88IR-5,92959, CSR 90IR-2, CSR 89IR-10, CSR 88IR-10, CSR 89IR-12, IR 55178-3B-3-2, IR61249-3B-19-2.
		Saraswati	Sodic (pH 10.1-10.4)	CSR 88IR-6, IET No.14545,14538,14536,14540, AT83-1329, IR51471-2B-2-2B-1-1, IR 61311-3B-2,IR 63295-AC202.
		Karnal	Saline and sodic	CSR 11,CSR 18, IR 54465-3B-4-3, IR 5931-110-1, PNL 7-22-48, BW311-7, S818-10-2, IR 63295-AC 210-4C.
1995	146	Kaithal, Saraswati and Karnal	Saline (ECe 5.5-9.7dS/m)and sodic (pH 9.3-10)	88-H3-4-7-1,88-H5-1-1-1, 88-H5-1-1-5, 88-H5-1-1-4, IR 45427-2B-4-2B-H-H-2, IR 49737-3B-26-2B-1-2, IR 49737-3B-26-2B-2-4, IR 45427-2B-4-2B-H-H-1, IR 52736-2B-18-2B-2-1, IR 45427-2B-4-2B-H-H-2, CSR 88IR-4, CSR 88IR-7 (2),CSR 89IR-6, CSR 89IR-2 (4), CSR 89IR-4 (3), CSR 88IR-6(2), PNL 3-5 3-7-1-1, CSR 90IR-6, CSR 90IR-5, CSR 89IR-16 (3), CSR 90IR-5, CSR 90IR-6,CSR 90IR- 4 (2), CSR 89IR-17, IR 61919-38-15-3, IR 61919-38-18-3, IR 64419-38-12-2, IR 64181-3B-12-3, IR 64168-3B-2-1, CSR 1, Pokkali (2),CSR 2,CSR 15, CSR 26(2), CSR 27,CSR 19, BR 4-10 (3), Pokkali, CSR 10 (3), CSR 1,CSR 5 (2), CSR 24, CSR 23, CSR 9(2), CSR 14, CSR 24, CSR 12.
1996	83	Karnal	Sodic (pH 9.59-10.4) and saline (7.9 dS/m)	88-H5-1-1-3-1, 88-H5-2-B-6, 88-H5-1-2-1-4, 88-H5-1-B-3, CSR 89IR-12, CSR 89IR-11, CSR 89IR-14, CSR 88 IR-14, CSR 89IR-9, CSR 89IR-12, CSR 89IR-10, CSR 88 IR-12, CSR 89IR-2, IR 58511-4B-12-3, IR 64419-3B-4-3, IR 58443-6B-10-3, IR 64196-3B-16-3, IR65192-4B-13-2, IR65197-3B-20-2-2.

1997	277	Karnal	Sodic (pH 9.5-10.5) and saline (ECe 8-10 dS/m)	IR 64161-3B-5-1, IR64170-5B-15-3, IR 64168-3B-3-3, IR 61249-3B-19-2, IR 63307-4B-24-2, IR 64197-3B-15-2, IR 62692-3B-5, IR 63295-AC 202, IR 55178-3B-3-2, NLR 33654, UPR 1683-9-1-1, TNAU 92094, UPR 1536-4-1, HKR 94-4, SIR 1295, MTU 1009, IR 71656-5R-B-12P, IR 67570-3B-4-1-3, IR67574-2B-5-1-3, IR67574-2B-17-1-1, IR 67574-2B-4-18-1-2, IR66404-4B-14-1-1, IR66404-4B-21-1-3, IR65775-4B-1-1-2, IR 65775-4B-20-1-3, IR65884-4B-14-1-2, IR65862-4B-16-1-3, IR 65852-4B-25-1-1, IR 6851-4B-23-1-2, 3R-196-1, 3R-230-1, 3R-116, 3R-149, 3R-170, 3R-201, 3R-5, HybridS- IR 72827H, IR 72835H, IR 72834H, Maintainer- PMS 10 B, Restorers-IR 65516-77-2-1-15R, SPRLR 75005-16-3-IR
1998	292	Karnal	Saline and sodic	IR 55233-3B-23-3, IR60485-2B-12-2, CSR89IR-21, CSR 89IR-20, CSR 88 IR-21, IR 26916-ES, IR 55233-3B-23-3, IR 57304-4B-4-2, IR 60483-2B-17-2, IR 60485-2B-12-2, CSR 89IR-22, IR 55233-3B-23-3, IR 57304-4B-4-2, CSR 89IR-21, CSR 89IR-23, IR 54447-3B-10-2, (CSR 13/Panvel 2/IR 36)-6, (CSR 13/Panvel 2/IR 36)-2, 92-H51-4, 88-H3-4-17-2, 88H1-4-B-1, (Panvel 2/CSR 13)-31, CSR 96IR-2, CSR 96IR-3, CSR 95IR-1, CSR 89IR-23, Pusa 2504-1-3-1, Pusa 2504-1-26, IR 67574-2B-18-1-2, IR 66404-4B-12-1-2, IR 66401-2B-6-1-3, IR 65775-4B-10-1-2, IR 65858-4B-13-1-2, IR65199-4B-19-1-1, IR 66065-3B-6-1-2, CSR 88IR-11, CSR 88IR-14, CSR 88IR-15, BR 5331-93-2-8-2, BR 5333-34-4-6, BTS-25
1999	132	Kaithal	Sodic (pH 9.5-9.9)	IR 57304-4B-4-2, CSR 89IR-22, CSR 88IR-19, CSR 88IR-21
2000	307	Kaithal, Karnal	Sodic (pH 9.65-9.83)	CSR 93IR-4, CSR88IR-20, B90-15, NDURK 5042, CN 1413, CSR 92IR-3, CSR 92IR-5(3), CSR 92IR-3, CN 1413, CSR 92IR-4,CSR 92IR-3,(CSR 13/Panvel 2//IR 36)-6-B-2 (96259), (CSR18/Panvel 12)-4,IR 67574-2B-17-1-1, IR 66401-2B-6-1-3, IR 65858-4B-14-1 , IR 65199-4B-6-1-1 , IR 65192-3B-21-1-1 , IR 70868-B-P-30-3 ,BR 5331-93-2-8-4, BTS-11-13, CSR 88IR-12(2), IR63731-1-1-1-3-3-2,BW 267-3(2), CSR 88IR-13,CSR 89IR-13,CSR 89IR-8,IR 5931-110-1, IR 72049-B-R-23-1-1, IR 71895-3R-99-1,IR 69587-3R-P-4-1, IR 69588-4R-P-3-3, IR 70865-B-P-12-2, IR 70865-B-P-20-3, IR 70868-B-P-29-2, IR70868-B-P-30-3, IR 70869-B-9-11-2, IR 70870-B-P-25-2, IR 69587-3R-P-4-

				1, IR 69588-4R-P-5-1, IR 70869-B-P-13-2, IR 70870-B-P-6-3
2001	243	Kaithal, Karnal and Bhaini Majra	Sodic soil (pH 9.58-9.70) and saline (ECe 6-8 dS/m)	CN1413, CSR 10, CSR 92IR-4, CSR 93IR-2, IR 67574-2B 17-1-1, IR 66401-2B 6-1-3, (Bas 370/CSR 10)-4-17-2-3, (Pusa33/CSR 13//CSR 12)-24-l, IR 65192-3B 21-1-1, IR 67574-2B 17-1-1, IR 66404-4B 14-1-1, (Pusa33/CSR 13// CSR 12)-24-l, CSR 89 IR -8 , IR 70868-B-P-30-3 , Jaya, IR 70868-B-P-30-3, IR 72046-B-R-10-1-1, 72046-B-R-15-2-1, IR 72398-B-2R-l-1, IR 69587-3R-P-4-1, IR 71907-3R, IR 71910-3R-32-1, IR 72046-B-R, 15-2-1, IR 71895-3R-99-1, IR 63731-1-1-3-3-2, 93128, CSR10, CSR11, CSR16 , CSR31, CSR 23, 92063, NDRK 5044, Tilakchandan, 96416, CSR 13, Pant Dhan 12, CSR 24, CSR 32, 96228, CSR 19, Naren Usar Dhan 1
2002	86	Kaithal, Karnal	Saline (ECe 6-8 dS/m)and sodic (pH 9.6- 9.7)	IR 6294-CS9-1-30, 92-H51-4, CSR 92IR-4(2), 91-H2-6-B-2, CSR 97IR-2, CSR 97IR-3, CSAR 515-1, TOX3723-103-2-3-1, TOX 3055-82-1-3-2, NDRK 5075, NDRK 5078, IR71910-3R-32-1, IR71895-3R-99-1, IR71907-3R (2), CSR 89IR-8 (2), IR 71895-3R-99-1(2), IR 71910-13R-32, IR 71910-3R, CSR-2K-262, CSR-2K-255, CSR-2K-219, CSR-2K-364, CSR-2K-360, CSR-2K-359, CSR-2K-285 , CSR-2K-269, Entry no. IRRI-2002-3, 6,11, 13, 19
2003	204	Kaithal, Karnal	Sodic and saline	91-H2-6-B-2, Kalanamak, CSR-88 IR-5, KAU-CIRJ-9,CSR 89IR-8, KAU-CIRJ-9, KAU-CIRJ-3, CSR 88IR-5, CSR 2KIR-1, CSR-2K262, CSR-2K219(2), CSR-2K359, CSR-2K 269, IRRI-2002-76, CSR-2K364, CSR-2K360, 96259, CSR-2K-255, CSR-2K -240, CSR-2K -239, CSR-2K -232, CSR-2K -236, CSR-2K -219, CSR-2K -228, CSR-2K -242, CSR-2K -248, CSR-2K -260, IR 72593-B-3-2-2-2-B, IR 75000-59-1-1-B, IR 75000-69-2-1-B, ADTRH1,CORH 2, KRH 1, PHB 1, CNRH 3
2004	208	Karnal	Sodic and saline	91-H2-6-B-2, CSR 89IR-8(2), KAU-CIRJ-3, 91-H2-6-B-2, NDRK 5078, CSR-2K-262(2), CSAR 538-1, NDR 5097(2), CSR-2K-219(3), NDURH-7, NDRK 5088, PNL-1-8-5-17-2, MPH 5401, RH 204, CORH-2, CSR-02K-240, CSR-02K-228, CSR-2K-236, CSR-2K-239, CSR-2K-260, CSR-2K-234, CSR-2K-271, CSR-2K-249, CSR-2K-539, CSR-2K-201, CSR-2K-238, CSR-2K-539,CSR-2K-259, IRRI-2002-11,

				CSR-2K-234, IR 51500-AC 11-1, IR 58427-5B-15, IR 64197-3B-14-2, IR 64419-3B-1-1, IR 64419-3B-3-2, IR 64419-3B-4-3, BW267-3
2005	271	Karnal	Sodic (pH 9.4-9.8) and saline (ECe 6.31-8.1 dS/m)	CSR-2K-219, PNL1-8-5-17-2, NDRK5088 ,CSR 89IR-8, CSR 23, CSRL-04-2366, CSR 13, CSR-2K-239, CSR 23, CSR-2K-228(2), CSR-2K-239, CSR 23, Hybrid Surichi, IR 64, Hybrid DRRH-20, Hybrid DRRH-1, CSRL-04-2366, CSR-2K-232(2), CSR-2K-242(2), CSR-2K-255, CSR-2K-259(2), CSR-2K-349B, IRRI-04-67, IRRI-04-71, IRRI-04-67, CSR-2K-245, IRRI-04-71, IR61919-3B-14-3, IR51500-AC11-1, IR63311-B-3R, IR64197-3B-14, CSR 23, IR63731-1-1-4, IR65192-4B-3-2, IR65195-3B-6-3, IR67075-2B-2-2, BW 267-3(2), IR61925-3B-1-1, IR65185-3B-8, CSR 36 , IR70868-B-P-8
2006	179	Karnal	Sodic (pH 9.5-9.9) and saline (ECe 6.0-11.0 dS/m)	CSR-2K-219 (2), CSR-2K-228, PNL-1-8-5-17-2, CSRL-01-IR97, CR 2092-96-2 , CSRL-01-IR97, CSR23, CSAR-518, NDRK-50001, NDRK-50004, DRRH-28, JRH-8, HRI-148, CSR-2K-232(2), CSR27, CSR-2K-228 , CSR36(9), CSR-2K-228, CPWF 05-15, CSR 90IR-2(3), IR-71895-3R-60-3-1, CSR23, PNL-2-34-5-18, IR69337-AC-2-2-2, IR71907-2-1-2B, IR65847-3B-19-2, PNL-2-34-5-18, IR69337-AC-2-2-2
2007	365	Karnal	Sodic (pH~9.5-9.9) and saline (ECe 12 dS/m)	CSR-2K-219(2), CSR 92063, CSRL-01-IR75, CSR-2K-255(2), CPWF-05-15, CSR-2K-232(2), CSR-2K-228, RIL-06-125, RIL-06-81, RIL-06-51, CSR-2K-242, RIL-06-81, RIL-06-94, RIL-06-103, IRSSTN-06-2718

* The figures within parentheses in the last column indicate the number of trials in which the genotypes were found better salt tolerant.

Table 7.7. Rice germplasm collected and evaluated for salt tolerance in Lucknow, Uttar Pradesh (CSSRI, RRS, Shivri Farm).

Year lines	No. of testing	Place of Stress	Type of Salt	Promising Salt tolerant Germplasm Identified
1996	28	Shivri	Sodic (pH9.95-10.2)	IR 63307-4B-24-2 ,IR64197-3B-15-2 ,IR64168-3B-3-3, 88-H5-1-1-3-1, 88-H5-2-B-6, 88-H5-1-2-1-4, 8-H5-1-B-3.
1997	84	Shivri	Sodic (pH9.59-10.5)	3R-196-1, 3R-230-1, 3R-116, 3R-149, 3R-170, 3R-201, 3R-5, (CSR 13/ Panvel 2/IR 36)- 88H1-4-B-1,(Panvel 2/CSR 13)-312, 92-H51-4, 88-H3-4-17-2.
1998	67	Shivri	Sodic	CSR 96IR-2, CSR 96IR-3, CSR 95IR-1, CSR 89IR-23
1999	132	Shivri	Saline and sodic (pH 9.6-9.9)	CSR 93IR-2, NDURH3, NDURK 5042, IR 68652-3B-81,IR 68652-3B-11-2, IR68652-3B-22-3,IR68651-3B-11-2, IR70870-B-P-10-1, IR 68654-3B-6-1, IR68657—3B-19-3,IR70865-B-P-1-3, IR 70868-B-P-15-1 , IR 70868-B-P-20-2.
2001	339	Shivri	Sodic (pH 9.5-10)	CSR 93IR-2, CN1413, CSR 89IR-23, 91-H2-6-B-2, 92-H51-4, 89-H1-1-3, CSR 24, CSR 32, 92063, 96228, NDRK 5044, 93102, 96416, CSR 29, CSR 21, CSR 23, IR 64, Tilakchandan, CSR 13, Pant Dhan 12, CSR 19, Usar dhan 1, CSR 30, Super-basmati.
2002	251	Shivri	Sodic (pH 9.6)	CSR 97IR-1, CSR 97IR-2, CSR 97IR-3, CSAR515-1, 91-H2-6-B-2, 89-H1-1-3, CSR 92IR-4, CSR 92IR-3, IR71893-3R, IR63731-1-1-1-3-3-2, IR71892-3R, CSR 10, 92-H51-4, CN 1413, NDRK 5075, NDRK 5078.
2003	112	Shivri	Sodic (pH 9.6-9.8)	CSR 13, 91-H2-6-B-2, NDRK 5058, Narendar usar dhan 3, CSAR 515-1, CSR 13, NDRK 5076, NDR 9830149, NDRK 5083, CN 1266-154, CSR-2K-IR-1.
2004	604	Shivri	Sodic (pH 9.4-9.7)	CSR 89IR-8, KAU-CIRJ 9, CSR88IR-5, 91-H2 6B -2, CSR-2K-219,

				PNL 1-8-5-17-2, PNL5-8-17-21, CSR-2K-262, MPH 5401, PHB 71, PRH 122, 6444, DRRH-1, Bas 97 Rajsai, Bas 113-B-A-, Basmati Kota, Bhudawal, IR 63731-1-1-3-3-2, 2366, IR 719073R, IR 71991.
2005	61	Shivri, Mataria and Dhaura	Sodic & saline (pH 9.2-11.5) (ECe 11dS/m)	IR 70023-4B-R-12-3-1, IR 64419-3B3-2, IR 51491-AC10, CSR-2K-219, CSR-2K-262, CSR 89IR-8.
2006	16	Shivri, Mataria and Dhaura	Sodic (pH 9.5-9.9)	CSR36 (2), CSR23 (2)
2007	60	Shivri	Sodic (pH~9.8)	DRRH 52, CSR-2K-219, NDRK 50002, PNL-3-12-8-7-2, CSRL-01-IR97, DRRH 43, CSR-2K-228.

* The figures within parentheses in the last column indicate the number of trials in which the genotypes were found better salt tolerant.

Table 7.8. Germplasm collected and evaluated for coastal salinity tolerance in West Bengal (CSSRI, RRS, Canning Town).

Year	No. of germ-plasm	Coastal salinity	Promising Salt tolerant Germplasm Identified
1972	8	Saline (ECe5.5-11.8)	Damodar (CSR1), Dasal (CSR2), Getu (CSR3)
1975	8	ECe 8.7-14.7 dS/m	Mut-I, Mut-II
1976	76	ECe 5.8-17.4 dS/m	IR2070-435-4-5-5, Damodar, IR 36, IR4-11, Nona Bokara, IR2035-290-2-1-1, RP-6-516-34-1-8, Nonasal, Tongil, B 58b-Mr-105-2.
1977	28	ECe 6.1-14.6 dS/m	SI 61, SI62, Getu Mut-2685, Getu, Mut-2735
1979	361	ECe 8.4-16.2 dS/m	Cult No. 45, cult No. 501, Select.Nos. 129-2, 202-2, Cult No. 5, Cult No. 44, Cult No. 17-2-2, Cult No. 207-2-1, 131-1-2, Select.Nos. 438-1, Select.Nos. 402-1, 451-2 , 448-1.
1980	257	ECe 6.0-11.5 dS/m	Cult No. 2, Cult No. 3, Cult No.7, Cult No.8, Cult No. 14-2, 16-1, Cult No. 53, Cult No.57, Cult No.100, Select.Nos.202-2-2, Select.Nos. 402-1, Select.Nos. 438-1, SAR-41, SAR-43, SAR-28, CSR1, C-3, CR-22-2685, CR-235-25, TNL-4-50, IR-4432-28-5, IR-4630-22-2-5-1-3.
1981	114	ECe 6.3-8.9 dS/m	C3(CSR 1/IR 24), C-13-4, C-14-2, IR 4432-28-5, IR 2307-247-2-2-3, IR-9884-54-3, IR 19660-23-2-1, CST 14-2, RB (C_3).
1982	229	ECe 6.4-16.9 dS/m	CSR 1/IR 24, Jaya/IR 2071-88, CSR 1/IR 2153-43, CSR 1/IR 20, Kalamota, SR 26 B, IET 7903, IET 6695, MI52, C100, Pateni, Patini.
1983	295	ECe 4.5-11.5 dS/m	1R 8/SR 26 B, CSR 1/IR 21 53-43, CSR1/IR24, Jaya xIR 207 1-88, SR 26 B, Kalamota, IET 6905, NC 487/77, NC 491, IET 7904, IET 7903, IET 7907, IR 4422-6-2, IR 4422-6-2-3-1, IR 4630-22-2-17, C 14-1, C2, IR 4630-22-2-5-1-3, 16296-CS-96-14-1, 16298-CS-98-5-2,16298-CS-B-18-2, Valuki, Velki
1984	473	ECe 2.2-10.2 dS/m	3-4(CSR 1/IR 24), 14-2-5(CSR 1/IR2153-43), 158-1-1, NC 491, SR 26 B, NC487/77, SR 26 B, IR 21567-9-2-2-2-1-3, NC 493, CR 292-7056, CR 320-9, Velki,

			NDR 502, CST 7-1, C-2, C I00-1,IR 4630-22-2-5-1-3, IR 9736-16-1-2, 16287-CS-10-1-17, 16287-CS-18-1-2-3, 16294-CS-19-1-24, C-23-2-1, Cult 4-4, PNL 32-10-1, IR 26916-13-1-1, PNL 28-23, IR 2307-247-2-2-3, Gopalbhog.
1985	417	ECe 4.2-7.7 dS/m	C 300 BD-50, TCA 72, TCA 148-3, Jalaplaban, SPR 7297-432-1-1, IR 21074-53-P1-B1-108,CN 574-550-16-3, CN 580-478-7, CN 578-886-52-3,CN 574-550-16-1, CSR 6, CST 7-1, PNL 2-7-10, NDR 503, CR 235-77, 16287-10-1-17, 16294-9-1-30, Pankaj x KMS, Pankaj x SR 26B, Velki, Valuki.
1987	268	ECe 5.0-10.5 dS/m	CAC 615, C200BS-6-23, SR 26 B, C200BD-25-16, Talmugur, IR 16294-9-1-30, C99BD-9-26, Velki, Nonasal mutant, SAR 41/Jaya, Valuki, CAC 615, Nonabokra, Bhurarata, Entry No.31, 45, 40, 7.
1988	764	ECe 4.0-15.0 dS/m	IR 16294 CS-9-1-30, IET 10673, IR 16290-CS-20-1-8, CST 100-1,CAC 615, CSR 6, SR 26 B, C304-25-20, C300-8D-50-11, CSR 6, Sadamota, C99BD-9-43,IET 10353, IET 10674, Nonasal mutant (IET10678), IET 10027, IET9018, IET10173, IET 10109, IET 10097, IET 11198, IET 11197, IET11195, IET 11283, IET 10560, IET 11274, IET11320, Entry No.90, 44, 99, 101,118, IR 32307-107-3-2-2.
1989	775	ECe 5.0-11.0 dS/m	CSRCS 44-1, CSRCS 11-4-4-4-1,CSRCS 21-1, CSRCS 24-1, IE5 10673, IR 16294CS9-1-30, SR26B, C200BS-6-23, IR 16290CS 20-1-1, CST 100-1, Cult 90, Cult 30,CSRCS 5-16-1, Cult 20,Cult 40, Culture 99, C300-BD-50-11, C99-BD-9-43,CSRCD 11-1-1, CSRCS 7-10-2-0-1, CSRCS7-10-4-2-1,Canning 7, Canning 8, C200-BD-25-16, IR 16294-CS-9-1-30, CR644, C200BS-6-23, CR 221-130, CR 719-13, RP 2597-8-239, B30638-CK-8-MD-14-2-1,BR11, Nona Bokra, Cult No.23, Cult No. 4, No. 25, No. 27, No. 30, No. 26, No. 5, No. 14, No.7 , No. 9, No.18, No.23.
1990	646	ECe 6.0-10.8 dS/m	CST 100-1,16290- CS-20-1-8, Cult 5-2-3-4, Cult 11-3-2, 7-1-4, 44-1, C300BD 50-11, C99BD-9-26, Cult 5-0-2, SR 26B, 7-2-5, 2-25-3, 8-B2-2, Talmugur No. 14-2, 5-13-3-8, 5-13-3-5, 12-1-3, 12-8-5-1, IR 16294 CS 9-1-30, C200BD-25-16, CSRC(S)44-1,CSRC(S)5-2-2-5, Pusa NR 580-6, CR 915-1-14-21, IET No. 12089, IET No. 13068, Hamilton, CSRC(S)5-2-3-3.
1991	697	ECe 4.5-9.5 dS/m	CSRC(S)5-2-2-5,B 3063B-CK-MD-14-2-1, IR51485-B-B-20-B-B, CSRC(S)7-10-4-2-1, Canning 7, CSRC(S) 5-2-3-1-1, CSRC(S) 5-3-0, CSRC(S) 7-2-0, CSRC(S) 7-10-7-2, CSRC(S) 11-5-0-2, CSRC(S) 5-13-3-5, CSRC(S) 14-1-8-0, CSRC(S) 18-

			24-1, CSRC(S) 20-43-3, C300-BD-50-11, CSRC(D) 5-2-3-1, C304-25-20, CSRC(D) 5-0-2, CSRC(D) 2-7-2, CSRC(D) 13-11-5, CSRC(D) 5-13-7, CSRC(S) 5-2-2-5, IR 51485-B-B-20-B-B, CSRC(S) 18-24-1, CSRC(S) 20-44-1, RP 1603-390-310-88, Cheriveriuppu, Canning CSR 4, CSRC(S) 15-3-2, CSRC(S) 19-30-6, CSRC(S) 16-15-1, IR 53653-B-B-B-7-B, IR 53655-B-B-B-3-B, IR 50184-B-B-B-18-B-B-1, IR 51337, NC 493, CAC 615, IR 51485-AC1-6, CN845-40-5, CR 626-26-2-3, CR 683-12-3, C340-22-5, C99-BD-9-43,C340-22-17, CAC 706,RAU 3013-4-2-3-1, IR 28453-14-2-1-2.
1992	503	ECe 4-12.5 dS/m	CST 100-1, IR 16294-CS-9-1-30, CSRC(S)5-2-2-5, CST 7-1, CSRC (S)/11-0-8-1, CSRC(S)/2-9-4-1, C99-BD-9-26, C300-B-50-11,C410-9-24, 13-12-2-1, P-G/13-5-6-1, CSRC(S)5-2-2-5, CSRC(S)14-1-B-0,IR 43610-B-B-7-B-B, CSRC(S)5-2-3-3, SR-P/5-3-0,P-Ne/2-2-8, P-SR/7-10-7-2,BW307-6, IR 60485-2B-12-2, IR 52724-2B-6-2B-1-1, IR 60483-2B-17-2, CR 580-17-1, NDR 482, C340-22-5, C99-BD-9-43, Sabita, C304-25-20, SR-P/5-13-7, IR 30864, RP 2524-638,CSRC(S)/7-10-2-0-1.
1993	595	ECe 4.5-7.8 dS/m	IR 16294-CS-9-1-30, IR 16290-CS-20-1-8,CST 100-1, CSRC(S)/5-2-2-5, CSRC(S)/11-3-0, CSRC(S)/12-10-1,SR 26B, CSRC(S)20-43-3, CSRC(S)/11-5-0-2, CSRC(S)/12-11-1-1, C300-BD50-11, C404-32-16, CSRC(D)8-2-0 CSRC(D)/13-12-2-2, CSRC(D)/13-23-4, IET 12858, CSRC(S)/24-1, IET 13422, CSRC(S)/15-5-2, RP 2269-424-398-88, CSRC(S)/7-1-4, CSRC(S)5-2-3-3, CN 921-KGR 1-2-29, Rewa 12-165-3, UPLR3, Pusa 677-50-103-2-9, IET 13876, IET 13670, IET 13860, IR 46319-PM1-36-2-2-1, IR46448-SRN-2-1-1-3, IR 61249-3B-19-2, IR 58440-3B-11-3.
1994	562	ECe 4.7-9.7 dS/m	IR 16290-CS-20-1-8, CST 100-1, CSRC(S)7-13-2-0, CSRC(S)7-1-4, Najani Selection, CSRC(S)20-43-3, CSRC(S)2-9-4-1, C300BD50-11, C99BD9-43, CSRC(D)2-25-2, CSRC(D)5-13-7, IET 14199, CSRC(S)7-1-4, IET 14200 or CSRC(S)20-43-3, CSRC(S)7-B-4-0-1, CSRC(S)7-B-4-0-2, CSRC(S)19-32—1, TTB 150-33-3-1, NDR 1025-2, NDR 1025, IET 14494, 1171-3-2, NDR 40013, KAURM7, IET 11901, IET 13120, NDR 4008.
1995	647	ECe 4.7-7.8 dS/m	IR 55177-3B-9-2, IR 55178-B-B-B-25-1, B 3063B-CK-MD-14-2-1, IR 16294, CS9-1-30, CSRC(S)2-2-8, CSRC(S)11-3-0, CSRC(S)20-43-7, CSRC(S)20-43-3, CSRC(S)11-5-0-2, CSRC(S)2-2-1-3, C300 BD 50-11, CSRC(D)7-2-5,CSRC(D)2-

			7-8, IET 13422, CSRC(S)11-5-0-2, CSRC(S)2-1-7, CSRC(S)5-2-3-1-1, RP 3786-1467-1788, IET 13520, IET 13949 , IET 14920 , IET 14901, CSR 90IR-6, PLA 19934, NDR 4062, NDR 954142, IR 63311-B-3R-B-24-3, IR 63307-4B-24-2,IR 43945-17-3-1-2-1.
1996	165	ECe 4.5-7.5 dS/m	IR 30864, CSRC(S)50-2-1-1-4-B, Cheriviruppu, IR 32429-47-3-2-2, CSRC(S)50-2-1-2-B, IET 12089, IR 46452-B-B-3-B-B, CSRC(S)8-B-2-2, CSRC(S)12-10-1.
2002	553	ECe 4.5.-7.5 dS/m	CSRC(S)24-B-B-N-1-B, CSRC(S)39-B-1-2, CSRC(S)5-2-2-5, IET 10673, IR 16294, CS-9-1-30, CSRC(S) 22-2-4-B-1-1, CSRC(S) 21-1-1-B-1-1, CSRC(S)14-1-4-0, CSRC(S)20-44,IR IR 50363-61-1-2-2, BG 731, CSRC(S)7-10-7-2, CSRC (S)7-9-2-2, CSRC(S)7-1-4, CSRC(S)11-3-0, C300BD50-11, DWCT 156-0-0-0-2-0, CSRC(D)7-12-1, BSRC(O)2-22-6, CSRC(O)7-5-2, CSRC(S)39-B-1-B-2, RP 2525-67, CSRC(S) 24-8-B-B-1, CST 7-1,CSRC(S)32-B-B-B-3-B, Canning 7, CSAR 516-1, NDRK 5077, IET 11353, IR 16294, CS9-1S, SR26B, KJTRH-3, IET 17657, CSRC(S)32-5-B-B-1, CR 2022-1, CN 1163-14-7, NDGR 268, CN 1165-5-17, or 1901-6-25.
2006	462	ECe 5.0-7.4 dS/m	CSRC(S) 50-2-1-1-4-0, CSRC(S)30-10-B-B-1,CSRC(S) 21-2-5-B-1-1, CSRC(S)22-2-B-1-1, CSRC(S) 21-2-7-B-1-1, CSRC(S)5-13-3-1, CSRC(D) 13-19-1, CSRC (D)13-16-9, CSRC(D)4-3-0, CSRC(D)4-10-0, CSRC(S)21-2-5-B-1-1, CSRC(S)7-1-4, IR-16294CS9-1, CSRC(S) 32-B-B-1, CSRC(S)47-7-B-B-1, CN-1347, IET 18702, NDR-5097, IET 19685, IET 19667, IET 20118, IET 20146, IET 19212, IET 19108, IET 20056, CSRC(D) 11-4-3, C-300BD-50-11.
2007	273	ECe 5.5-7.6 dS/m	CSRC (S) 50-2-1-2-B, IR 52280-117-1-1-3, CSRC (S) 47-7-B-B-1, RP-44-17-1, IR 52, CSRC(S)5-13-3-1, CSRC(S)19-32-1, IET 10673, BR 11, CSRC(S)22-2-B-2-1-1-9, CSRC(D)2-22-8, CSRC(D)12-8-5-2, C300BD50-11, CSRC(D)11-4-4, Canning-7, CSRC(S)32-B-B-1, CSRC(S)36-B-B-2-B, NDR 9830/31, CR 661-236-1-3, CSRC(S)7-1-4, NDR 9930013.

Table 7.9. Germplasm evaluated for tolerance to saline vertisols in Bhal region of Gujarat (CSSRI, RRS, Anand).

Year	No. of germ-plasm	Soil Salinity	Promising Salt tolerant Germplasm Identified
1992	164	(ECe~6-10dS/m)pH 8.8 to 9.5	IET 13413, IET 13101,IET 12853, IET 13418, IR 55179-3B-11-3, CSR 19,C200-BS-6-23,CSR-C(S)14-1-8-0,Talmagur, CSRC(S)3-2, CSR 89IR-15, CSR (S)H-520-2,CSR 90IR-1, SR 26 B, Nona Bokra, IR 52717-B-B-4-B-B-1-3, IR 53655-B-B-B-6-3.
1993	118	ECe 9-11 dS/mpH 8.2-8.5	CSRC(S)-44-1,IR43610-8-B-7-B,IR 51485-B-B-20-B-B, CSR 87IR-1, CSR (S)-15-1, CSRC(S)-3-2, Entries 2443,2444, 2448, IR 58440-3B-17-3-3, IR 58440-3B-11-3-3, IR 59445-B-4-2-2 , IR 61249-3B-19-2.
1994	236	ECe 10-13dS/mpH 8.8-9.0	IET 13417,IET 13559, IET 13763, IET 13764, IET 13765,13767, IET 14518, IET 14544, IET 14550, IET 14551, IET 14555, IET 14568, Vikas, PNL-35-23-15-2-24-1,PNL 7-22-49, RP 2270-2-232,RP 2271-433-321, RP 2271-435-326-119, Nona Bokra, Pokkali, CSR 11, IR 63731-1-1-1-4-2-3, IR 63731-1-1-4-3-2, IR 29,TCCP 6533-8-4-2-AC204-2, CSR 3, CSR 8, CSR 10 , SLR 51214.
1995	603	ECe 6-12 dS/mpH 8.8-9.1	CSRC(S) 5-13-3-5, PNL 18-17-2, NDRK 5023, NDRK 5031, VRS1, VRS2, VRS3, NDRK-5012, CSR 90IR-6, CSR 88IR-16, PNL 258 -1-1, PNL 22-46-11-1, NDRK-5032, NDRK 5034, NDRK 5040, RPP-Ext-33-1-1-2, VRS1, VRS2, VRS3, CSR 89IR-16, CSR 90IR-6.
1996	216	ECe 5-8 dS/mpH 8.8-9.0	VRS1,VRS2,NDRK 5031,CSR 88IR-3, VRS 7, VRS 8, VRS1,VRS2,VRS3, VRS4,VRS5,VRSS6, VRS7, VRS8,VRS12 IR51491-AC10, IR61919-3B-24-3, IR 64197-3B-15-2, IR 65192-4B-13-2, IR 40931-33-1-3-2,IR 551777-38-9-2, IR 55179-38-11-3 , IR 63731-1-1-1-1-3.
1997	71	ECe 6-10 dS/mpH 8.7-9.3	VRS1,VRS2,VRS3,VRS8,VRM-1,VRM-3,VRM-4, VRS-5,VRS-6,VRS-12,VRS-13,VRS-14, VRS15, VRS4, VRS10, VRS8, VRM11, VRM12.
1998	126	ECe 4-10 dS/m	VRS2,VRM4,VRS12,VRM11,VRM12,VR mutant-3,VR mutant 5,VR mutant 6, VRS-15, VRS1,VRS25, VR mutant 11,VRM 5,VRM-6,VRM12 , VR mutant 25.
1999	26	ECe 4-9.5 dS/m	VRS1, VRM1, VRM11, VRM31.
2001	56	ECe 6-13.5 dS/m	VRS12, CN1413, NDRK5038, NDRK5064, CSR 91IR-5, VRS1,VRS2, VRS3, VRS12, VRS19, VRM1, VRM5, VRM11, VRM15, VRM25, VRM 31.

parents (CSR 10 and CSR 11) and two sensitive parents (Basmati 370 and Pakistani Basmati) under artificially created sodic soils (pH 9.8) in lysimeters. All the F_3 plants based on IRRI scoring method at adult stage showed continuous variation suggesting that sodicity tolerance is controlled by polygenic trait acting additively in most of the cases along with interactions between the alleles at some loci (Singh *et al.,* 2001). A similar inheritance study for salinity was conducted involving the same populations. Results indicated the role of few major genes along with numerous minor genes involved for salinity tolerance. It was also inferred that salinity tolerance trait is polygenic in nature and lacks maternal influence (Mishra *et al.,* 1998).

Association studies

Initial studies revealed that grain yield/plant in rice was positively and significantly associated with number of grains/panicle, fertile tillers, panicle weight, total tillers/plant and days to 50% flowering and was negatively associated with percent spikelet sterility. Path coefficient analysis showed that the direct effect of panicle weight, fertile tillers and straw yield was maximum in normal and saline soils whereas fertile tillers, number of grains/panicle and panicle weight exhibited maximum direct effect in alkali soils. Panicle weight was not significantly correlated with grain yield though it had maximum direct effect on yield in normal soils. There was lack of significant correlation between salinity score at vegetative stage and score at reproductive stage as well as grain yield.

Another association study comprising of 15 F_1 populations confirmed these findings. Grain yield was correlated positively with filled grains/plant, filled grains/panicle, panicle weight, K content and 100-grain weight. It had strong negative correlation with percent spikelet sterility and reproductive stage salinity tolerance score which is highly desirable (Mishra, 1994). Gupta (1999) also studied the relationship between eight traits (plant height, days to 50% flowering, EBT/P, number of filled grains/plant, grain yield/plant, Na^+, K^+ and Na/K ratio) under the salt stress environments. The role of Na and K becomes apparent from the study as the grain yield correlated negatively with Na content as well as Na/K ratio while it had a positive correlation with K content under alkali as well as saline soil conditions.

Gene action and heritability

Estimates of genetic parameters following Hayman's method showed significant additive and non-additive gene action for salinity tolerance score, K and Na/K ratio in rice in 5 x 5 F_1 diallel (Mishra *et al.,* 1990). Narrow-sense heritability estimates were 0.65 for salinity tolerance score, 0.55 for K absorption and 0.41 for Na/K ratio. The average dominance was within the range of incomplete dominance. Study suggested the involvement of one group of genes for salinity tolerance score and Na/K ratio and two groups for K uptake. Recessive alleles were more concentrated than dominant genes in the tolerant varieties CSR 10 and CSR 1 for reproductive stage salinity tolerance and K uptake. Wr-Vr graphic analysis suggested the involvement of both major and minor genes for all the studied traits. Narrow sense heritability (h^2) revealed the presence of its high magnitude for Na/K ratio for both alkali (0.658) and saline (0.630) soils and high for grain yield in alkali (0.547) and in saline (0.569) soils. Spikelet sterility (%) and number of filled grains/panicle showed high heritability in both stress environments. Thus the preponderance of additive gene action with inheritance of the traits studied is evident

Combining ability analysis

Combining ability analysis by Griffing's method confirmed the significance of both additive and non-additive effects and the former showing greater importance in the inheritance of the traits for salinity. The varieties CSR 1 and CSR10 were found good combiner for all the three traits examined. Their usefulness is suggested for rice breeding programmes aiming to enhance salt tolerance in rice. High heterotic effects found in some crosses suggest the potential of hybrid rice for salt-affected soils (Mishra, 1991). CSR 10 was thus used as a donor in Indian-IRRI collaborative programme on salinity tolerance not only in India but also at IRRI (Table 7.12).

Stability studies

Based on the research conducted at CSSRI, Karnal for phenotypic stability of the 15 genotypes for eight morphological and physiological characters over varying stress environements, Gupta (1999) reported varieties CSR 1, CSR 10, CSR 11, CSR 13 and CSR 21 as the most stable genotypes for grain yield productivity in salt stress soils. These varieties were also found stable in a previous study (Mishra, 1994). The best stable varieties with dual characteristics of Na exclusion and maintaining lower Na/K ratio were found to be CSR 10, CSR 21 and CSR 11, CSR 13 and CSR 1. The genotype CSR 881R-12 was found the most stable genotype and was also highly adaptable over all the environments for most of the traits including grain yield (Mishra *et al.,* 2000).

Heterosis breeding

For breaking the yield ceilings, exploitation of hybrid vigour in rice appears quite promising as exemplified by its success in the favourable environments. Hybrids are also expected to perform better in salinity due to genetic phenomenon of homoeostasis. In the study undertaken at CSSRI, significant heterosis over mid and better parent was observed for almost all the characters studied (Gupta, 1999). Out of 15 F_1 crosses, Pokkali x IR 28 (79.87), CSR 10 x IR 28 (67.18), CSR 13 x IR 28 (54.58) and CSR 1 x IR 28 (48.56) indicated the positive and significant heterotic response over mid parent and a cross Pokkali x IR 28 (49.31) over better parent in sodic soil while the crosses CSR 13 x IR 28 (35.17) and CSR 10 x CSR 13 (26.72) exhibited better heterotic effects over mid-parent and only one cross CSR 10 x CSR 13 (24.53) over the better parent in saline soil.

With an objective to identify suitable salt tolerant maintainers and restorers for WA cytoplasm, a total of 21 CSR varieties were test crossed with IR58025 A (Gautam and Singh, 2004). The lines CSR12, CSR13, CSR14, CSR 19, CSR 22, CSR 23, CSR 24 and CSR 26 proved to be restorers. On the other hand, CSR 2, CSR 5, CSR 10, CSR 11 and CSR 16 turned out to be maintainers which could be converted through back-crossing to generate salt tolerant CMS lines. As regards salt tolerance of parental lines of hybrids, we found recently that IR58025A the widely used CMS line is quite sensitive to sodic soils. Recently, through a systematic study involving 24 parental lines across normal, moderate saline (ECe~6dS/m) and highly saline (ECe~12 dS/m) micro-plot environments conducted up to reproductive stage revealed that among restorers, IR60997-16-2-3-2-2R, BR827-35and BCW-56 were found quite tolerant. Similarly among maintainers, IR80155B and IR80559B were found tolerant to salinity.

Use of F_1 anther culture

F_1 anther culture has twin advantage of speed and efficiency and due to this it has become most effective tool of the plant breeders to attain homzygosity of the recombinants within shortest possible time. The technique also offers the opportunity to screen haploid material at the stage of tissue culture. Anther culture derivatives generated at International Rice Research Institute (IRRI), Philippines and evaluated under ICAR-IRRI collaborative network in India by CSSRI led to the identification of promising rice varieties IR 51500-AC-17, IR 511485-AC-1 and AC 6534-4 for salinity, AC 6533-3 for sodicity and AC 6534-1 for dual tolerance (Singh *et al.* 1992; Singh and Mishra, 1995). IR 51500-AC-17 and AC 6534-1 have been named as CSR 21 (IET 13558) and CSR 28 (IET 13560), respectively which were recommended for minikit trials after national network evaluation. Presently the product of this technique CSR28 derived from anther culture of F_1 between IR42 and IR4630-22-2-5-1-3 is being as salt tolerant check for global evaluation of International Rice Soil Stress Tolerant Nursery (IRSSTN).

Molecular mapping and marker-assisted selection

The complete deciphering of rice genome sequence, the first in any cereal crop, has opened up new vistas in its genetic improvement through molecular marker approaches (IRGSP, 2005). This genomic information is extremely useful and is being used for mapping genomic regions controlling agronomically important traits including salt tolerance with the help of molecular markers. This is likely to enhance breeding efficiency through marker assisted selection.

Ammar *et al.* (2009) used F_2 and F_3 mapping population derived from CSR 27 x MI 48 cross to map QTLs associated with 17 salt tolerance parameters at seedling and reproductive stages in variety CSR 27. They constructed a framework linkage map using 79 SSR and EST markers distributed across 12 rice chromosomes at an average interval of 20.7 cM. One major QTL for multilple salt tolerance parameters was discovered on chromosome 8 and three other major QTLs for Cl ion concentration were detected in CSR 27 variety. We are currently fine-mapping these tolerance attributes through genetically stabilized and immortal recombinant inbred lines (RILs) derived from the above cross. Besides, CSRRI, Karnal has also developed other mapping populations (RILs) through single seed descent method from other crosses CSR 27 x CSR 11, CSR 11 x MI 48, IR 4630 x CSR 11 and Pokkali x IR 28. With an objective to map important salt tolerance traits on the rice genome, these are being studied through phenotyping and genotyping (using SSR and SNP markers) under the prestigious ICAR Network Project on Transgenics in Crops-Functional Genomics component.

Molecular markers are also now capable of detecting the precise inter-varietal similarity. Among different molecular markers, simple sequence repeat (SSR) markers are efficient as well as cost effective to detect precise genetic similarities in rice. Therefore, among evolved basmati varieties, Basmati CSR 30 has shown highest molecular genetic intimacy of 96 % with traditional basmati types (Basmati 370, Dehraduni Basmati, Ranbir Basmati, Basmati 386 and Taraori Basmati) through dendrogram derived from UPGMA cluster analysis using SSR markers (Nagaraju *et al.*, 2002). Neeraja *et al.* (2008) studied the genome wide linkage disequilibrium in a set of 32 rice genotypes including salt tolerant and sensitive genotypes with 64 SSR markers to find genomic regions showing consistency with salt tolerance. They confirmed that most of the markers (44%) with significant LD values were found in the chromosomal regions of reported QTLs for salt tolerance in rice.

Mineral uptake studies

The mineral elements like P and Zn are in low availability in salt affected soils and thus varietal differences for their uptake were studied. The studies conducted at CSSRI by Rao (2000) revealed distinct varietal differences for the P uptake under salt stress. Phosphorus content increased in shoot in different rice genotypes at higher level of salt stress over normal. Maximum P content was observed in CSR 13 (0.234%) and low in IR 28 (0.107%) at ECe 5 dS/m, high P in CSR 10 (0.3%) and in CSR 18 (0.167%) at ECe 10. Further Bas 370 had the least P content (0.113-0.117%) and CSR 10 had the highest P (0.326-0.467 %) at pH 9.5 and 9.8 respectively. The promising genotypes with high P content were CSR 1, CSR 10, CSR 11 and CSR 13 while IR 28 and Bas 370 were found as sensitive genotypes with low P content.

Regarding Zn, salt tolerant rice varieties like BR 4-10, Pokkali, CSR 10, CSR 11, IR 28 and Jaya showed high amounts of zinc in their shoot whereas the sensitive Bas 370, MI 48 and Pusa Basmati 1, exhibited the lowest amount of zinc content. The maximum increase in zinc content over normal was noticed in GR 4 at ECe 5 dS/m and minimum in CSR 22, BR 4-10, IR 28, Panvel 1, IR 36 and CSR 10 (Rao, 2000). It has also been reported that all the sodicity tolerant genotypes are not Zn efficient and vice versa though some of the genotypes have dual tolerance for sodicity and Zn deficiency (Qadar, 2002). High Zn : P ratio was observed in tolerant and low Zn : P ratio in sensitive genotypes. CSR 10 had the highest Zn : P ratio at all levels of stress and Bas 370 showed that lowest Zn : P ratio under stress of both levels of salinity and higher alkalinity. This showed that the better salt tolerance of rice could be related to higher Zn : P ratio (Rao, 2000).

Bio-chemical studies

Proline content

Proline content is reported to increase with increase in salinity and alkalinity in rice. Highest proline content was observed in a highly salt tolerant rice genotype CSR 11 (1.74% mg/g) and least in salt sensitive rice genotype MI 48 (1.16 mg/g) based on overall mean across all the levels of stress compared to other genotypes (Rao, 2000).

Sugar content

The plant encountering salt stress often shows reduction in starch and total carbohydrates and increase in reducing sugars. Limited supply of essential metabolites like carbohydrates retard plant growth under sub-lethal salt stress. Significant decrease in the sucrose content of leaves was observed in rice genotypes under both salinity and alkalinity stresses. Nevertheless, relatively less reduction was observed in tolerant genotypes like CSR 10, CSR 11, CSR 13, CSR 18, CSR 22, CSR 29 and Pokkali and high in sensitive genotypes like Bas 370, MI 48 and Jaya (Rao, 2000).

Protein and enzymes

Enzyme which plays an important role in membrane lipid metabolism is esterase governed by *Est2* locus. Preliminary studies with varieties involved in a diallel revealed very encouraging results. Salt tolerant donors i.e. Nona Bokra, Pokkali, CSR 10 and CSR 13 showed $Est2^1$ isozyme pattern: sensitive genotypes i.e. IR 28, MI 48, and Basmati 370 showed $Est2^2$ pattern while a tolerant genotype CSR 1 showed *Est2* pattern. When these genotypes were grown in varying levels of salinity stresses, the patterns remained the same except in CSR 1 which changed into $Est2^1$ (Mishra, 1994).

Development and Popularization of Salt Tolerant Varieties

Right from its inception CSSRI, Karnal along with its Regional Stations worked in a focused manner in collection, conservation, characterization, evaluation and breeding of salt tolerant rice genetic resources through various national and international networks. Consequently, different salt tolerant varieties were developed and disseminated in salinity prone areas of the country (Table 7.10). Similarly, CSSRI, Regional Research Station, Canning also developed salt tolerant rice varieties like CST 7-1, Sumati, Bhutnath, CSRC(S)7-1-4 suited for the coastal saline areas of the eastern coast. The salinity breeding programmes of other locations in different states as steered and motivated by success stories of CSSRI also identified and developed salt tolerant rice varieties (Table 7.11). The increased food production of the salt affected fields previously having low or zero productivity not only improved economy and well being of the poor farmers/owners of the field but also increased employment avenues for the local people during the crop season and above all contributed to the country's granaries.

CSR Varieties in IRRI Crosses

Besides CSSRI, Karnal, many State Agricultural Universities (SAUs) and other central Institutes have also developed location-specific salt tolerant varieties suiting to local needs (Table 7.11). These varieties when found salt tolerant and adaptable in a wider region comprising of more than one state were released by the Central Variety Release Committee (CVRC). The varieties possessing superior adaptability in a state were released by the respective State Variety Release Committees. The examples of IRRI crosses involving CSR varieties are shown in Table 7.12. This might have resulted in the development of salt tolerant varieties in other countries as well. In India, CSSRI developed varieties like CSR 10, CSR 13, CSR 27, CST 7-1, etc. are included as standard salt tolerant check varieties in the national trials like Alkaline and Inland Saline Tolerant Variety Trial (AL&ISTVT) and Coastal Saline Tolerant Variety Trial (CSTVT) every year. These are also used as checks in the preliminary National Saline and Alkaline Stress Nursery (NSASN). Any newly developed salt tolerant material is adjudged and promoted using these checks as reference varieties, though these check varieties also need to be replaced with new salt tolerant and high yielding varieties like CSR 23 and CSR 36. Similarly, International Rice Soil Stress Tolerant Nursery (IRSSTN) coordinated by IRRI, Philippines involves CSR 28 and CSR 90-IR2 as standard checks in the global evaluation for salt tolerance.

Impact of Salt Tolerant Rice Varieties

Salt tolerant rice varieties like CSR 10, CSR 13, CSR 23, CSR 27, CSR 30 and CSR 36 are widely grown in the salt affected soils of different states and are in great demand by farmers. CSR30 is the first salt tolerant basmati variety which has gained vast popularity even in normal soils of basmati growing areas. Though it is difficult to realize the direct impact of a variety its adoption, popularity and impact could be measured through its breeder/certified seed production chain. To quantify the role/impact of salt tolerant rice varieties, seed distributed by CSSRI, Karnal to various seed multiplying agencies is being taken as a case study. For the period since beginning up to 2004 and 2005-2009, CSSRI, Karnal has supplied 109.3 and 208 quintals respectively of breeder seed of these salt tolerant varieties to different seed multiplication agencies. Besides, CSSRI has also produced and supplied 2300 quintal labeled/certified seed to different organizations of which major share was of UPBSN (UP Bhumi Sudhar Nigam), Lucknow.

The multiplication ratio of breeder seed to certified seed is about 1:200 in rice (breeder to foundation and foundation to certified seed chain is not included in this case study, otherwise it would be much more). If we extrapolate the figures for rice, then 317 quintals of breeder seed would have produced minimum of 63400 quintals of certified seed in one production season. In addition to this, CSSRI supplied 2300 quintals of certified seed. Therefore, the total certified seed quantity would be about 65700 quintals in one season. This seed for sowing purpose is always much costlier than the food grain of rice for consumption purpose so nobody would preferably consume this by purchasing the costlier seed. Even with the conservative estimates, if we consider only 65,000 quintals of total certified seed in the production chain, this might have been cropped in more than 162,500 ha area producing more than 4.80 million quintals paddy (taking average production of only 30 q/ha). This additional food grain particularly from salt affected soils must be worth several million rupees besides employment and income generation. Indian farmers usually do not replace the seed every season and retain it for at least a few years before seed replacement. It is only one season scenario to estimate the output and worth of seed. Hence, one can imagine what would be the scenario after a few seasons. This is the example of the seed distributed by only one institute i.e. CSSRI, Karnal. Many seed multiplication agencies are continuously engaged in bulk seed production and supply to UPBSN, Lucknow; other agencies also supply the seed directly to the farmers. Considering this scenario, the worth and use of salt tolerant rice varieties is much more than what we have extrapolated above. However, further spread of these varieties in the niche target areas can be achieved by creating more awareness by the related extension agencies.

Another major notable impact of using salt tolerant rice varieties is bringing back the unproductive barren or salt affected lands into the fold of productive normal lands. Therefore, faster conversion of degraded lands into normal ones with the use of resilient varieties can ensure more area and thus increased food productivity from these reclaimed lands.

Table 7.10. CVRC released salt tolerant rice varieties developed by CSSRI, Karnal.

Name of variety	CSR10	CSR13	CSR23	CSR27	CSR30	CSR36
Parentage	M40-431-24-114/Jaya	CSR1/Bas. 370//CSR5	IR64//IR 4630-22-2-5-1-3/IP964 -45-2-2	NONA BOKRA/ IR565-33-2	BR4-10/Pak. Basmati	CSR13/ Panvel 2// IR36
IET No.	10349	10348	13769	13765	14720	17340
Year of release	1989	1998	2004	1998	2001	2005
Plant height(cm)	85	115	115	115	155	110
Maturity days	120	145	130	125	155	140
Tolerance limits·						
Salinity (ECe:dS/m)	<11.0	<9.0	<10.0	<10.0	<7.0	<11.0
Sodicity (pH_2)	<10.2	<10.0	<9.9	<9.9	<9.5	<10.0

Yield (t/ha)	>6.0	>6.0	>6.5	>6.5	>3.0	>6.5
Non stressSalt stress	>3.0	>3.0	>4.0	>4.0	>2.0	>4.0
Grain type	Short bold	Long slender	Long slender	Long slender	Basmati type	Long slender
Recommended ecology	Acts as biological amendment for highly deteriorated sodic and inland saline soils.	Sodic and inland saline soils of UP, Haryana, Gujarat & Maharastra.	Sodic soils of Haryana, UP and coastal saline areas of Maharastra. Gujarat, Tamil Nadu, Kerala and West Bengal.	Sodic and coastal saline soils of India	First salt tolerant basmati variety developed for sodic areas of UP, Haryana and Punjab.	Sodic soils of Haryana, U.P. and Pondi-cherry

Indirect Impact

The knowledge emanating from the basic as well as applied research technique at CSSRI are being utilized by several labs within the country and abroad. Screening techniques in artificially created environments are standardized and methodology to breed high yielding salt tolerant varieties in different crop plants is tested and adopted in most of the stress breeding and physiology labs. Database for grouping of donors for varying physiological mechanisms is not available anywhere on a large scale; hence it is being used as the reference data by many laboratories and breeding programmes. Because of this background, a large number of national and international collaborators have collaborated with CSSRI on stress tolerance studies for better understanding of the physiological processes under salt stress and their manifestations. CSSRI, Karnal has worked as the coordinating center for ICAR-IRRI collaborative network projects in which Karnal (Haryana), Canning Town (West Bengal), Panvel (Maharastra), Anand (Gujarat), Vytilla (Kerala), Machilipatnam (AP), Tiruchirapalli (Tamil Nadu), Motto (Orissa), Kanpur (UP), Faizabad (UP), Chinsurah (West Bengal) and Port Blair (Andaman & Nicobar Islands) acted as network centers. The development of salinity breeding research infra-structure like sodic and saline farms, environmentally controlled glass house, sodic and saline microplots, transgenic green house, marker-assisted selection lab etc. all in one place could be a model for other salinity workers.

Table 7.11. Other salt tolerant rice varieties released in India.

Name of variety	Parentage	Salt tolerance limit pH_2	ECe (dSm^{-1})	Stress adaptation (soils)	Year	Days to flowering	Grain type
Varieties released by Central Variety Release Committee							
CST 7-1	CSR1/IR24	-	8	Coastal saline	1991	115	LB
Lunishree	Mutant of Nonasal	-	8	Coastal saline	1992	115	LS
Sumati	Pankaj / NC 678	-	8	Coastal saline soils	2004	105	SB
Bhutnath	SR26B / Pankaj	-	8	Coastal saline soils	2004	150	SB
Jarava (B90-15) IET 15420	B32-Sel-4/ spontanea4//B296	-	8	Coastal Saline soils	2005	150	SB
CSRC(S) 7-1-04	Pankaj / SR26B	-	9	Coastal saline soils	2008	115	LB
State Released Varieties							
Haryana/Punjab							
CSR5 (Vikas)	TKM6/IR8	9.7	7	Alkaline	1979	95	MS
Kerala							
Vyttila 1	Chotupokkali Sel.	4.5-9.7	9	Coastal saline	—	90	MB
Vyttila 2	Cheruviruppu Sel.	4.5-9.7	8	Coastal saline	1980	95	LB
Vyttila 3	Vyttila 1/TN 1	4.5-9.9	9	Coastal saline	1987	85	MB
Vyttila 4	Chethivirippan/ IR4630-22-2-17	4.5-9.7	10	Coastal saline	1993	95	LB
Vyttila 5	Mutant of Mahsuri	4.5-9.5	8	Coastal saline	1996	95	MB
Vyttila 6	Cheruviruppu / IR 5 //Jaya	4.5-9.7	9	Coastal saline	2004	85	MB

Maharashtra							
Panvel 1	IR8/BR4-10	9.7	8	Coastal saline	1984	95	SB
Panvel 2	BR4-10/IR8	9.8	9	Coastal saline	1988	90	LS
Panvel 3	CSR1 / Pankaj	9.8	9	Coastal saline	2000	100	LB
Andhra Pradesh							
MCM 1 SB	CO 18/Kuthir	-	8	Coastal saline			100
West Bengal							
Mohan (CSR 4) SB	Mutant of IR 8	-	7	Coastal saline		1981	115
Karnataka							
CSR 22	IR 64 // IR 4630-22-2-5-1-3 / IR 9764-45-2-2	9.9	8	Inland saline	2006	105	MS
Orissa							
SR 26B	Selection from Kalambank	9.7	8	Coastal saline	1988	115	LB
Sonamani	Velki mutant/ Mahsuri	-	8	Coastal saline	1996	-	-
TamilNadu							
PVR-1	MTU-1/ SR 26B	9.4	4	Coastal saline	1964	115	SB
CO 43	IR20/CSR2	9.7	8	Alkaline & coastal saline	1982	110	MS
ASD 16	ADT 31/CO 39	9.6	7	Coastal saline Alkaline & coastal saline	1986	95	SB

TRY 1	IR578-172-2-2 / BR 1-2-B-19	10	10	Alkaline & coastal saline	1995	110	SB
TRY 2	IET 6238/IR 36	9.9	8		2001	85	LS
Gujarat							
SLR 51214	Vijay/PTB-21	-	8	Inland & coastal saline	1982	100	LS
Dandi (IET 14906)	Panvel 2/IET 8320	-	9	Inland & coastal saline	2003	95	LS -
UP							
Usar dhan 1	Jaya/CSR3	9.9	8	Alkaline	1985	100	SB
Narendra Usar dhan 2	Selection from IR 2058 derivative	9.8	8	Alkaline	1998	95	LB
Narendra Usar dhan 3	Selection from IR46330 derivative	9.9	8	Alkaline	2000	100	LS

Table 7.12. Examples of IRRI crosses involving CSR lines.

Cross	Female Parent	Male Parent
IR 16658	IR 2070-464-1-3	CSR 1
IR 16659	IR 2070-464-1-3	CSR 2
IR 16660	IR 2070-464-1-3	CSR 3
IR 16661	IR 2070-464-1-3	CSR 4
IR 34879	Kalimonch DWF mutant	CSR 1
IR 35155	BR 4-10	CSR 1
IR 37178	IR 36 (MS)	CSR 1
IR 38722	MAHSURI	CSR 1
IR 38781	RD 19	CSR 1
IR 38782	RD19	CSR 2
IR 39828	IR 36	CSR 1
IR 65170	IR 54	CSR 10
IR 65173	IR 68	CSR 10
IR 65176	IR 32429-47-3-2	CSR 10
IR 65177	BG 731-2	CSR 10
IR 65190	IR 8192-200-3-3-1-1	CSR 10
IR 65198	IR 10206-29-2-1	CSR 10
IR 65203	Kuatikputih	CSR 1
IR 65804	Moroberekan	CSR 1
IR 65810	GIZA 171	CSR 10
IR 65815	GIZA 172	CSR 10
IR 65820	GIZA 181	CSR 10
IR 65825	GZ 2175	CSR 10
IR 66059	CSR 10	IR 54/CSR 10
IR 66060	CSR 10	IR 32429-47-3-2-2/CSR 10
IR 66061	CSR 10	BG 731-2/CSR 10
IR 66062	CSR 10	CSR 10/IR 58
IR 66063	CSR 10	CSR 10/IR 10198-66-2

IR 66064	CSR 10	IR 8192-200-3-3-1-1/CSR10
IR 66383	GIZA 171	CSR 1
IR 66386	GIZA 172	CSR 1
IR 66388	GZ2175	CSR 1

Conclusions

Since its inception, CSSRI and its regional stations assembled and evaluated more than 23000 germplasm lines for salt tolerance across highly diverse sodic and saline soils of the country. CSSRI, Karnal also led the country wide breeding program on the development of salt tolerant rice with participation of different centres through national and international programs. The huge germplasm collection, systematic evaluation through standard protocols, germplasm exchange coupled with technical support have also steered the development of many salt tolerant varieties in the country. Already the seed of these salt tolerant rice varieties is disseminated every year in the target salt affected areas of the country through breeder seed indent from Department of Agriculture and Cooperation, Ministry of Agriculture, Govt. of India. Salt tolerance being a complex trait both genetically and physiologically, several germplasm donors having different physiological mechanisms have been identified and used in recombination breeding which resulted in the development of better salt tolerant varieties. Besides, the salt tolerant varieties developed by CSSRI have been widely used to understand the physiological, genetic and molecular bases of salt tolerance which are likely to result in the development of future high yielding salt tolerant rice varieties with better efficiency.

References:

Ammar MM, Pandit A, Singh, RK, Sameena S, Chauhan MS, Singh AK, Sharma PC , Gaikwad K, Sharma TR, Mohapatra T, Singh NK. 2009. Mapping of QTLs controlling Na^+, K^+ and Cl^- ion concentrations in salt tolerant *indica* rice variety CSR 27. *J Plant Biochem and Biotechnol* 18(2):139-150.

Bay ND, Mishra DP, Gupta RK. 1992. Mechanism of salt tolerance in rice in relation to sodium, potassium and polyamine content. *Indian J Agric Biochem* 5(1&2):51-55.

Blumwald E, Grover A. 2006. Salt tolerance. Pages 206-224 in Nigel Halford (ed) Plant Biotechnology. John Wiley and Sons Ltd.: 206-224.

CSSRI. (1997). *Vision 2020: CSSRI Perspective Plan.* CSSRI, Karnal. 95 pp.

CSSRI. (2000). *Final Report of CSSRI-UPLDC Collaborative Research Project 1995-2000.* CSSRI, Karnal. 63 pp.

Dagar JC, Sharma HB, Shukla YK. 2001. Raised and sunken bed techniques for agro forestry on alkali soils of northwest India. *Land Degradation and Development* 12:107-118.

Datta KK, Mishra B, Singh RK, Singh KN, Singh YP. 1996. Technological options for reclamation of salt affected soils under resource constraint environment in Uttar Pradesh. *Agricultural Situation in India* 53(1):17-21.

Fischer RA, Maurer R.1978. Drought resistance in spring wheat cultivars. I. Grain yield responses. *Aust J Agric Res* 29:897-912.

Flowers TJ. 2004. Improving crop salt tolerance. *Journal of Experimental Botany* 55:307-319.

Flowers TJ. 2006. Preface. *J Exp Bot* 57:iv.

Gautam RK, Singh RK. 2004. Identification of salt tolerant varieties as restorers and maintainers for cytoplasmic-genic male sterility for developing salt tolerant rice hybrids. Pages 109-110 in Proc International Symposium on *Rice: From Green Revolution to Gene Revolution* held at Hyderabad during Oct. 4-6.

Gupta D. 1999. *Genetics of salt tolerance and ionic uptake in rice* (*Oryza sativa* L). PhD Thesis, BRA Univesity, Agra, India.

Gupta SK, Sharma SK. 1990. Response of crops to high exchangeable sodium percentage. *Irrigation Science* 11:173-179.

IRGSP (International Rice Genome Sequencing Project). 2005. The map-base sequence of the rice genome. *Nature* 436:793-800.

Ismail AM, Heuer S, Thomson MJ, Wissuwa M. 2007. Genetic and genomic approaches to develop rice germplasm for problem soils. *Plant Mol Biol* 65: 547-570.

Maas EV, Hoffman GJ. 1977. Crop salt tolerance – Current assessment. *ASCEJ Irrig Drain Div* 103:115-34.

Mishra B. 1991. Combining ability and heterosis for yield and yield components related to reproductive stage salinity and sodicity tolerance in rice *Oryza sativa* L. Page 761 in *Rice Genetics II*. International Rice Research Institute, Manila. Philippines.

Mishra B. 1994. Breeding for salt tolerance in crops. Pages 226-259 in Rao *et al.* (eds) *Salinity Management for Sustainable Agriculture - 25 Years of Research at CSSRI*. Central Soil Salinity Research Institute, Karnal, India.

Mishra B. 1996. *Highlights of Research on Crops and Varieties for Salt-affected Soils*. Technical Bulletin. CSSRI, Karnal. 28 pp.

Mishra B, Akbar M, Seshu DV. 1990. Genetic studies on salinity tolerance in rice towards better productivity in salt-affected soils. *Rice Research Seminar*, July 12, 1990. International Rice Research Institute, Philippines.

Mishra B, Singh RK, Jetly Vandna. 1998. Inheritance pattern of salinity tolerance in rice. *J Genet & Breed* (Rome).52:325-331.

Mishra B, Singh RK, Jetly Vandna. 2000. Performance of the rice genotypes under varying soil stresses. *Oryza* 37(3):194-197.

Mishra B, Singh R K, Senadhira D. 2000. Recent advances and future strategies for breeding salt tolerant rice varieties. Pages 275-284 in Peng S and Hardy B (eds) *Rice Research for Food Security and Poverty Alleviation.* Proceedings of the International Rice Research Conference held at IRRI. Philippines during March 31-April 3. 2000. IRRI. Philippines.

Nagaraju J, Kathirvel M, Ramesh KR, Siddiq EA, Hasnain, Seyed E. 2002. Genetic analysis of traditional and evolved Basmati and non-Basmati rice varieties by using fluorescence-based ISSR-PCR and SSR markers. *Proceedings of National Academy of Sciences (*USA) 99(20):5836-5841.

Neeraja CN, Mishra B, Rao Srinivasa K, Singh RK, Padmavathi G , Shenoy VV. 2008. Linkage disequilibrium in salt tolerant genotypes of rice (*Oryza sativa* L). *J Plant Biochem and Biotechnol* 17(1):65-68.

Pitman MG, Lauchli A. 2002. Global impact of salinity and agricultural ecosystems. Pages 3-20 in Lauchli A, Luttge U (eds) *Salinity: Environment-Plants-Molecules.* Kluwer Academic Publishers, Dordrecht.

Qadar A. 1995. Potassium and sodium contents of shoot and laminae of rice cultivars and their sodicity tolerance. *J Plant Nutrition* 18:2281-90.

Qadar A. 2002. Selecting rice genotypes tolerant to Zn deficiency and sodicity stresses. 1. Differences in Zn, Fe, Mn, Cu, P concentrations and P/Zn ratio in leaves. *J Plant Nutr* 25 (3):457-473.

Rao S. 2000. *Adaptation of rices to salinity.* PhD. Thesis. Kurukshetra University, Kurkshetra, Haryana, India.

Sharma RC, Rao BRM, Saxena RK. 2004. Salt affected soils in India – Current assessment. In Advances in sodic land reclamation. International Conference on *Sustainable Management of Sodic Lands* organized during February 9-14, 2004, Lucknow, India

Singh RK, Mishra B, Chauhan MS, Yeo AR, flowers SA, Flowers TJ. 2002. Solution culture for screening rice varieties for sodicity tolerance. *J Agri Sci* (Cambridge, UK). 139:327-333.

Singh RK, Mishra B, Flowers TJ. Yeo AR. 2000. Screening rice genotypes for salinity tolerance based on physiological traits. In Proceedings of the Indo-UK (CSSRI-Sussex) workshop on *Salt-Affected Soils and Breeding for Salt Resistant Crops* held at CSSRI, Karnal from October 12-14. 27 pp.

Singh RK, Mishra B, Jetly V. 2001. Segregations for alkalinity tolerance in three rice crosses. *SABRAO J* 33(1):31-34.

Singh RK, Mishra B, Senadhira D. 1992. Promising salt tolerant F_1 anther culture derivatives (ACDs). *IRRN* 17(1):17

Singh RK, Mishra B. 1997. Stable genotypes of rice for sodic soils. *Indian J Genet* 57(4):431-438.

Singh RK, Mishra B. 2004. Role of Central Soil Salinity Research Institute in genetic improvement of rice in India. Pages 189-242 in Sharma SD and U. Prasad Rao U (eds) *Genetic Improvement of Rice Varieties of India.* Today and Tomorrow Printers & Publishers, New Delhi.

Singh RK, Singh KN, Mishra B, Sharma SK, Tyagi NK. 2004. Harnessing plant salt tolerance for overcoming sodicity constraints: An Indian experience. Pages 81-120 in *Advances in Sodic Land Reclamation.* Concept Paper for the International Conference on *Sustainable Management of Sodic*

Soils held at Lucknow, India from 9-14th February, 2004.

Singh RK. 1996. *Developing the Screening Techniques for Salt Tolerance in Rice*. Report of Research Attachment at Salinity Research Lab, School of Biological Sciences. Univ of Sussex. UK from 1st Nov 95 to 31st July. 1996. 23 pp.

Singh RP, Vasisht AK, Mathur VC. 2003. *Quantitative Assessment of Economic Losses of Degraded Land in India*. Div of Agric Economics, IARI and Advance Publishing Concept, New Delhi. 51 pp.

Yeo AR, Flowers TJ. 1982. Accumulation and localization of sodium ions within the shoots of rice varieties differing in salinity resistance. *Physiol Plant* 56:343-348.

Yeo AR, Flowers TJ. 1983. Varietal differences in the toxicity of sodium ions in rice leaves. *Physiologia Plantarum* 59:189-195.

Yeo AR, Flowers TJ. 1986. Salinity resistance in rice (*Oryza sativa* L) and a pyramiding approach to breeding varieties for saline soils. *Australian J of Plant Physiology* 13:161-173.

Yeo AR, Yeo ME, Flowers SA, Flowers TJ. 1990. Screening of rice (*Oryza sativa* L) genotypes for physiological characters contributing to salinity resistance and their relationship to overall performance. *Theor Appl Genet* 79:377-384.

Yoshida S, Forna DA, Kock JH, Gomez KA. 1972. *Laboratory Manual for Physiological Studies of Rice*. International Rice Research Institute. Manila. 34 pp.

Genetic Resources of Rice in India : 279-321
Editor : S. D. Sharma
Today & Tomorrow's Printers and Publishers, New Delhi - 110 002 (India)

Chapter 8

Collection, Conservation, Evaluation and Utilisation of Rice Genetic Resources at the Vivekanand Parvatiya Krishi Anusandhan Sansthan for the North West Hill Region

P.K. Agrawal, J.P. Aditya, J. Stanley, R.K. Sharma, P. Kumar and J.C. Bhatt

Introduction

Rice is one of the major staple food crops of the hill regions of India. The total area under hill rice in India is about 1.8-2.0 million ha, out of which 0.62 m ha is under North-Western Himalaya, producing about 1.23 million tonnes of rice. The productivity of rice in this hill region was 2,013 kg/ha in 2006-07 (Fertilizer Statistics, 2008) whereas the average national productivity was 2,131 kg/ha.

From agricultural point of view, the north-western hills can broadly be classified into four sub-zones *viz*., lower hills (below 700 m amsl), mid hills (701-1500 m amsl), high hills (1501-2200 m amsl) and very high hills (above 2200 m amsl). Rice in this region was the least beneficiary of the green revolution mainly because of majority of the area (90%) being under rainfed upland condition. In this region, rice is grown as spring (March end/April to September) and June sown (June to September end) under rainfed upland conditions. Rainfed rice is subjected to abiotic (drought, cold and low light) and biotic (blast, brown leaf spot, stem borer) stresses at different stages of crop growth. Therefore, enhancement in productivity and production can be realized by giving emphasis to short duration varieties which can escape various types of stresses in such unfavourable ecosystem. In transplanted rice emphasis was put on resistance/tolerance to blast and low temperature along with high grain yield. Spring rice, which has got maximum area needs cold and drought tolerance at vegetative stage together with higher grain yield, while June sown rice needs the cold tolerance at the reproductive phase.

The Vivekanand Pravatiya Krishi Anusandhan Sansthan (VPKAS) at Almora was initially established by Prof Boshi Sen in 1924 at Calcutta and was named by him as Vivekanand Laboratory. It was shifted to Almora in 1936. He worked mainly on food and fibre crops. This laboratory was taken over by the Government of Uttar Pradesh in 1959 and was mandated to work mainly on rice. The Indian Council of Agriculture Research (ICAR) took it over from Government of Uttar Pradesh in 1974 and renamed it as Vivekanand Pravatiya Krishi Anusandhan Sansthan (VPKAS). The mandate of VPKAS, Almora is to cater to the needs of the NW Himalayan region including the hills of Uttarakhand state where this Sansthan (Institute) is located. This article focuses primarily to the collection, conservation, evaluation and utilization of rice germplasm for the hilly region of the northwestern hill region of India in general but more specifically on Uttarakhand as the Institute (Sansthan) is situated at Almora in Uttarakhand.

The state of Uttarakhand comprises mostly of the hills and the plains of the foot hills of the Himalayas. It is situated between 77°34' to 81°2' East longitude and 28°4' to 31°27' North latitude. The state has subtropical climate in the valleys to temperate on higher slopes. The annual rainfall of Uttarakhand is around 1100 mm. The state has a forest cover of 64.6%. Earlier it was a part of the Uttar Pradesh state of India. It was carved out as a separate state from the state of Uttar Pradesh in the year 2000. This state has another agricultural institute, namely, and Govind Ballav Pant University of Agriculture and Technology (GBPUA&T), Pant Nagar, established in the year 1960, committed to the agricultural research and extension. The state primarily grows rice under three situations: i) irrigated, ii) rainfed upland spring rice (sown in the month of March-April and harvested in September) and iii) rainfed June sown (June to September end) conditions. The area under irrigated rice is 0.2 m ha where as the area under spring rice is 0.08 m ha and rainfed June sown is 0.02 m ha.

Rigorous efforts have been made by VPKAS, Almora over last four decades to develop and identify superior genotypes of rice in terms of grain yield, resistance/tolerance to prevailing biotic/abiotic stresses, quality traits and development of production technologies for their suitability under rainfed upland (spring and summer sowing) and irrigated transplanted conditions. However, the limited success of breeding rice cultivars for the subterranean regions of the Himalayan region is due to the peculiar topography of the hilly region, low temperature, long photoperiod and susceptibility of the varieties to blast disease. Shattering of grain is another disadvantage found commonly in hill areas.

Early Efforts

Prior to 1947, Chinese varieties were first introduced in the Kashmir valley, possibly due to reasons of geographical proximity or contiguity and have been found suitable and so extensively cultivated. The most notable of these introductions is Ch 1039 which was a leading variety in the Kashmir valley. Other introductions from China were Ch 27, Ch 47, Ch 962, Ch 971 and Ch 972. In the early phase of crop improvement in the NW hill regions, because of the common incidence of blast, special attention was paid to incorporate resistance against this disease into the local varieties from the resistant donors like 'Zenith', 'Dawn', 'Tetep', 'RP 79-9' and IR 160, and also to combine cold tolerance as well. Those donors were crossed with many adopted lines like AC 71, Begum, Siga, Madew, Larbond, Khush and Ishikari-Shirogi. Many promising lines were derived from different combinations of crosses *viz.* Ch 988 x IR 20, AC 71 x IR 160, Ch 1039 x IR 127, K71 x IR 127, and AC 71 x IR 532 (Table 8.1). The promising lines were simultanesoly tested in a coordinated trial at Palampur, Joginderpur, Nagrota, Almora, Kalingpong, Wangbal and Shillong (Gangadharan, 1985). Few of the promising materials from them *viz.*, K 84 and K 78-13 were distributed to the farmers. After the establishment of CRRI, Cuttack, many exotic germplasm from China, Japan, Taiwan and Russia were evaluated for their agronomic potentials in this region. Notable among the Chinese introductions were Ch 4, Ch 45, Ch 55, Ch 62 and Ch 63. Ch 45 proved to be a good yielder combined with earliness and resistant against *Helminthosporium* (Anonymous, 1950-51; Ganguli and Padmanabhan, 1959).

Table 8.1. The local varieties and introduced plant type varieties used in hybridization for the hilly regions of India.

Local Varieties		Plant type varieties	
1.	Ch 1039	1.	IR 8
2.	Ch 988	2.	Jaya
3.	Ch 972	3.	IR 532-1-33
4.	Ch 1007	4.	IR 20
5.	AC 71	5.	IR 661-1-140-3
6.	Larbond	6.	IR 127-80-1-10
7.	Siga	7.	IR 537-1-10-388
8.	Madew	8.	IR 442-2-58
9.	Ch 971	9.	IR 160-2-7-2-1-1
10.	Shinei		
11.	Ishikarishirogi		

12.	Shirogi
13.	K 71
14.	Port Brer
15.	KH 6119
16.	KH 10320
17.	K 78-13

Source: Gangadharan, 1985

Evaluation against Biotic and Abiotic Stresses

Evaluation for biotic stresses

The major biotic stresses for this region include many diseases and insect-pests. The major diseases included blast, brown spot, false smut and grain discoloration. Similarly stem borer and leaf folders were the major insect-pests.

Evaluation of rice germplasm against diseases

Screening germplasm against blast: North-Western Himalayas in general and Almora (Hawalbag) in particular is a hot spot for the incidence of blast. The diversity and evolution of blast is very dynamic in this region. Hence breeding for resistance against blast is a major objective for the hilly region of this Himalayan State. In last more than 30 years, more than 24,933 rice genotypes with diverse origin were evaluated at VPKAS, Almora for resistance against blast under the blast nursery from which 647 genotypes were found promising (Table 8.2). Those included genotypes from hills and plains of India, from the International Rice Research Institute, Philippines (IRRI) and other Asian and non-Asian countries through the IRRI. Many of those genotypes were evaluated over years. Many genotypes *viz*., Tetep, Columbia II, Zenith, Ta-poo-cho-z, Diwani and Mlyang 47 were found resistant over years.

Besides the incidences of blast, other important diseases prevalent in the hills of Uttarakhand are brown spot, leaf smut and grain discoloration. More than 7346 genotypes for brown spot, 3402 genotypes for false smut and 1265 genotypes for grain discolouration were evaluated at VPKAS. Among them 247 genotypes against brown spot, 129 genotypes against false smut and 27 genotypes against grain discoloration were found promising (Table 8.3). Genotypes like Tetep and Diwani were found consistently resistance against brown spot over years.

Table 8.2. Evaluation of rice genotypes for resistance against blast.

Year	Genotypes screened		Promising entries
1979	1602	**12**	(Raminad strain-3, P 1231129 (Wagwag), CR 987233263, RP 10571585, RP104571433, RP 10573514, KMP 39, CN 540, RP 1017151, CR 987218265, Adil, IR 5908263)
1980	1450	**28**	(Raminad strain-3, P 1231129 (Wagwag), HPU 2208, k 433, HPU 2202, HPU 800, Zenith, VL 8, Fukini Shiki, PI 4, Dissi Hatif, IR 4547625, Camponi SML, Colombia II, Kusabue, RP 10573511, Tetep, Milyang 47, Toride 1, IR 327333925, IR 59318111, IR 5908263, IR 9202211, IR 590884233, JG 24, KT 311, B 2266b-CW1116521, B 2266b-CW1923)
1982	1700	**28**	(Raminad Str. 3, Lai-Kan-Tao, Fukunishiki, IR 17 488232, IR 17525278112, IR 21015196313, ITA 25, Jinheung, Suweon 300, Suweon 304, Ta-poo-cho-z, Tetep, 16890, 17189, 5287, 54-BC-6-B, RP 1785 75322, Pusa 437 113, CR 366135, NDR 110, RP 105727711, IR 52, TNAU 17005, HPU 5048-11p-411-B-23, Dissihatif (73127), Camponi SML, Columbia II, IR 9202211)
1983	1178	**26**	(Raminad Str. 3, Caloro, H 1152011, INTAN, IR 131497132, Ta-poo-cho-z, 17189, IET 9370 (BAU 14722), IET9371 (BAU 152647), IET 9478 (IR 183483633), IET 9477 (IR18201543223), IET 9480 (IR 20824122), IET 9519 (VRS 163223), IET 9520 (VRS 24212), IET 9521 (VRS 28641), IET 9129 (MEH 116551), IET 9155 (UPR 83150), IET 9157 (UPR 83168), IEV 9158 (UPR 83169), VL 8, Dissihatif (73127), Camponi SML, Tetep, Columbia II, Ta-poo-cho-z, IR9202211)
1984	2027	**20**	(BG 3674, IRAT 109, IRAT 13, IRAT134, IRAT 144, IRAT 147, IR 43, ITA 142, Salumpikit, Rasi, Tetep, Ta-poo-cho-z, Milyang 46, Milyang 47, Camponi SML, Columbia II, VL 8, Dissihatif (73127), IR 9202211, IR 5908263)
1985	408	**27**	(VL 8, VRS 163, VL 24, RAU 729201, IRAT 144, IRAT 13, VL 16, VL 20, Kalinpong 1, ITA 186, HPU 2208, Columbia I, Zenith, Tetep, Ta-poo-cho-z, Milyang 46, Milyang 47, Camponi SML, Columbia II, Dissihatif (73127), INIAP 415, IRAT 134, IRAT 144, IRAT 146, IR 6115, NDR 102, NDR 119)
1987	345	**12**	(VL 8, Zenith, Tetep, Ta-poo-cho-z, Milyang 46, Milyang 47, Columbia II, Disshatiff (73217), Zenith, WAG-WAG, Tadukan, Columbia I)
1988	325	**9**	(Tetep, TA-poo-cho-z, Milyang 46, Camponi SML, Columbia II/ VL 8, IR9202-21-1, Dissi-hatif (73217), IR 1544-238-2-3, C 46-15)
1989	403	**13**	(Tetep, Milyang 46, VL 8, C 46-15, IR 920-21-1, IR 5908-

		26-3, IR 9202-21-3-2, SR 7272-115-4-4, VL 16, BG 367-4, IR 39357-133-3-2-2-2, IRGC 101089, IRGC 101141)
1990	744	**10** (IR 13484-66-2-3-1, Suweon 301, Suweon 303, ITA 132, SR 7272-80-5-8, VRS 163-2-1-4-1, Tetep,VL 8, Dissi-hatif (73127), Camponi SML, IR 9202-21-1)
1991	22	**7** (IR 36, Rasi, VR504-4, Himalaya 741, IET 11064, Ananda, Tulasi)
1993	545	**18** (IR 5908-84-2-2, IR 5873-9-1, IR 57924-8, MENG-LI-WANG-GU (Acc 66985), VL 88974, B 2266 bCw-11-23-4, IR 48725-B-B-11-2, JINBU-9, Suweon 235, TSIPALANDAMBO 650, VL 90-1550, IET Nos. 12606, 12783, 12801, 12870, 12888, 13376, 14218)
1994	527	**20** (SR 3049-58-1-5-1, E 220944, E 220949, E 220976, E 220935, E 220948, E 220985, E 221004, E 221029, E 221078, RP 1451-2439-1225, RP 2533-748-7-18, VL Dhan 163, VL 895193, CN 776-46-B1, IET Nos. 13181, 13183, 13488, 12887, 13689)
1995	726	**35** (IR 578926, Milyang 46, Milyang 104, RAU PT 100972, Suweon 330, Suweon 376, TM 4310, YR 6488-ACP-36, HR 6419-B-56-1-2, IR 8/ENG/SAC, IET Nos. 14008 (HKR 101/MR 603-303), 14018 (IR 28/IR8866-30-3-2-4-2), 14019 (IET 6288/ UPR 231-26-1-2), 14063 (OR 131-5-154/IR 13429-186), 14105 (IR 50/Patnai 23), IET 14804(CR 314-5-10-NC), IET 14828 (Swarna /CR 294-28-1), IET 14224 (IR 54725A/IR 64-R), IET 13766(Nona Bakra/IR 5657332), IET 14370 (Hansraj/ IR/36/ N22), VL 901683, VL 88971, VL 891167, VL 891177, VL 895187, VL 895001, VL 895034, VL 901619, VL 911762, VL 901623, VL 901721, VL 911754, VR 5055, VL 911990, VL 911926)
1996	606	**16** (Boda 271, CT 6514-31-1, E 890744, HPR 978, IR 5908-40-3-2-2, IR 9202-10-2-2-11, IRAT 250, IRAT 254, IRRI 387, Jimbu 12, Milyang 105, Secanado Brazil, Denaulaut Tawar, VSR 286-4-1, Speed 70, Nawyang 4)
1997	765	**22** (Diwani, IR 480-44-157-1-1-3-3, IR 1544-238-2-3, IR 4547-5-3-6, IR 21912-131-3-3, IR 31892-100-3-3-3-3, IRAT 125, Milyang 47, Suweon 303, VL 91-5827, VL 89-5038, VL 91-1990, CSR 88-IR-4, HKR 95-3, IRAT 161, IR 57893-60, OR 1144-114, VL 91-1754, Milyang 46, RAUPT 100-97-2, Suweon 330, VL 89-1194)
1998	2200	**14** (Diwani, IR 4547-5-3-6, PPM 105/IR 9224-117-2-3-3, SPTLR 82078-PIG-B3-24-1-1, SR 8102-140-5, VL 89-5038, VL 89-9419, VL 91-1990, BL 245, CSR 88-IR-4, CSR 89-IR-2, OR 1144-114, LACASSINE, VL 91-1754)
1999	750	**23** (HEXT 4, IR 57893-08, WAB 32-60, SPR 1064, PL 11/SPS-1, HR 4856-1-1-1-1-2, INRC Nos. 15851, 15874, 15876, 16053,

			16066, 16084, 16211,16323, 16337, 16338, 16339, 16638, 17277, 14418, 14558, 14661, 15262)
2000	382	**13**	(INRC nos 541, 542, 861, 1993, 17425, 17426, 17427, 17450, 17455, 17458, 17520, 17553, 17571.)
2001	509	**16**	(VL 95-6399, VL 97-3682, V 3611, VL 3693, VL 4438, VL 7044, IET 16862, IET 16480, VSR 8, VSR 28, VL 4049, VL 4015, XIANG 3510, VL 88971, TO 3107-56-1-22, HPR 1149)
2002	700	**42**	(VL 3964, VL 4049, VL 7045, VL 7059, VL 88971, VL 895038, VL 911762, VL 926041, VL 936052, VL 956394, VL 956502, VL 956394, VL 956502, VL 966821, VL 973677, BL 122, BL 245, VL 4422, VL 4442, VL 956394, VL 973682, RCPL 138, HPR 2086, HPR 2198, SRN 21, SRN 81, SRN 82, SRN 83, SRN 121, SRN 122, SRN 2282, VL 959632, VL 969588, VL 969807, VL 9948, VL 7203, VL 7230, VL 7272, VL 7276, VL 7278, VL 7314, VL 7323)
2003	350	**41**	(A 57, BL 122, BL 245, VL 4040, VL 4015, VL 88971, Tetep, VL 891194, VL891619, VL 926041, VL 891190, VL 895038, VL 936052, VL 956394, VL 956502, VL 966821, VL 963678, VL 973861, SLG 12-Sel., VL 10073, VL 10075, VL 4569, VL 4855, VL 4860, VL 4946, VL 7458, VL 7504, VHC 13, VL 3959, VL 4045, VL 973861, VL 4803, VL 4653, VL 4569, VSR 21, VSR 81 VSR 82, VSR 122, VL 6446, RR 4342, NDR 1054)
2004	4947	**50**	(AD 9408, AS89044, B 6926260-TB-1, BAREN, BG 3677, BLI, Columbia 1, CT 6516641, Diwani, E890715, E 890744, IR 4547336, IR 46292222, IRAT 258, Kehwa 4, Mlyang 47, Milyang 105, Suweon 303, Suweon 351, TKY 1014, TKY 1024, VL 88971, VL 891193, VL 911190, VL 10036, VL 10039, RCPL 139, RCPL 1351, VL 7363, VL 7316, VL 7632, VL 2000- 30018, VL 2000-30019, VL 2000-30027, VL 2000-30167, VL 4769, VL 4930, VL4932, VLUPR 2586722, VL 4455, VL 7072, IR 7822119623, IR 7822119682, IR 7822119656, IR 7822119699, IR 782211987, VL 5001, RR 2862, VL 6394, VL 6446)
2005	520	**69**	(VL 30019, VL10036, VL 7626, VL 30029, VL 30027, VL 30018, VL 7371, VL7316, VL 10039, VL 4860, VL7504, VL 88971, VL 895038, VL 956502, VL 973861, VL 973678, VL 4569, VL 7077, IR 7822422215, IR 7822119682, IR 7822119695, IR 782242224, IR 782219623, IR 7822422298, VL 4040, VL 10382, VL 10411, VL10409, VL7701, Vl7702, VL 7705 VL 7707, VL7738, VL30246, VL 30278, VL 30303, VL 30336, VL30685, VL 30686, VL 30687, VL 30688, HPR 2130, VL 7364, VL 7174, VL 7222, HPR 2529, VL 7316, HPR 2322, IR 7822119695BB, IR 7822119656BB, IR 7822119454BB, IRBLTA2-PI, IR 7822122215BB, IR7822119682BB, IRBLTAK1, IR 78222207503B, IR 7822422298BB, NDR 10951, VL 7630, VL 7505, IR 78908307RE5, IR 789053110-

		2-RE3, VL 7630, VL 3288, DDR 105, IR 7738435365B, IR7822119695, IR 7822119699, Vandana/ WAB 5650)
2006	282	**41** (VL 30685, VL 30686, VL 30688, VL 7674, VL 7702, VL 7705, VL 7802, VL 30511, VL 30560, VL.30590, VL 30389, VL 306876, ARBNCH 1, ARBNCH 21, VL 4040, VL 7504, VL 30018, VL 30018, VL 30029, VL 30336, VL 88971, VL 891190, VL 895038, VL 956502, VL 973861, VL 7316, VL 10411, VL 30425, IR 7822419695, IR 7822119656BB, BR 47262B613, C 1944242, CAN 7227, HPR 1156, IRAT 13, IRAT 144, IRAT 170, VL 901619, VL 936052, VL 943030, VL 953400)
2007	600	**20** (VL 7742, VL 30661, VL 30758, VL 30751, VL 31320, VL 31325, VL 31326, VL 30719, VL 30779, VL 30721, VL 30693, VL 7591, VL 30919, VL 30424, VL 30240, VL 30921, VL 30922, VL 6309, VL 30278, VL 30294)
2008	300	**15** (VL 7849, VL 7853, VL 31289, VL 31291, VL 31293, VL 31451, VL 31287, VL 31296, VL 10591, VL 10554, VL 10536, VL 10919, VL 30917, VL 31329, VL 31334)
Total	**24, 933**	**647**

Source: Annual Reports (1982-83 to 2008-09), VPKAS, Almora

Table 8.3a. Evaluation of rice germplasm against brown spot.

Year	Total Entries/ genotypes screened	Promising entries
1979	80	**9** (CH 1007, CR 11576, CR 126421, K 279, Pratao, Thankote-Marsi, IR 5908263, and B 2266b-CW 1123431)
1980	110	None found resistatant
1982	450	**2** (TNAU 1481, NDR 1)
1983	500	**10** (IET 9519 (VRS 163223), IET 9521 (VRS 28641), HKR 115, OR 437 8022, NDR 102, IR 282246632, CR 4054, VRS 163223, VRS 28641, Rasi)
1984	709	**1** (Rasi)
1989	79	**7** (IET 10888, IET 108889, IET 104002, IET 9985, IET 11006, IET 11164, IET 8087)
1990	510	**11** (Tetep, Ta-poo-choo-z, AD 9408, IR 5533-13-1-1, IR 4547-6-2-5, IR 5908-21-2-3, ITA 230, C 1158-7, C 1321-9, IR 21015-72-3-3-3-1, IR 32429-47-3-2-2)
1992	474	**4** (VR 212, VR 379, VL 90-1685, VL 90-1628)
1993	474	**8** (VL 92-9275, VL 899034, VSR 68, VSR 70, VSR 116, B-448-E-3-5 RI, IR 20897-B-6, IR 40097-6-3-2-1)

1995	171	**3**	(Tetep, AD 9408, Speed 70)
1996	70	**4**	(Milyang 9, Milyang 104, Suweon 235, Suweon 280)
1998	Not Available	**13**	(Diwani, IR 4547-5-3-6, PPM 105/IR 9224-117-2-3-3, SPTLR 82078-PIG-B3-24-1-1, SR 8102-140-5, VL 89-5038, VL 89-9419, VL 91-1990, BL 245, CSR 88-IR-4, CSR 89-IR-2, OR 1144-114, LACASSINE, VL 91-1754)
1999	206	**14**	(VL 95-6394, VL 95-6502, VL 96-6821, VL 97-3677, VL 97-3657, VL 97-3660, VL 97-3680, VL 97-3741, VL 97-3747, VL 95-6394, VL 96-3633, Kalinga- III, VL 95-3326, IRAT 330, ITA 121, IR 49692-44-2-3-3and Milyang 105)
2001	509	**4**	(XIANG 3510, VL 88971, TO 3107-56-1-22, HPR 1149)
2002	538	**19**	(VL 3964, VL 4049, VL 7045, VL 7059, VL 88971, VL 895038, VL 911762, VL 926041, VL936052, VL 956394, VL 956502, VL 956394, VL 956502, VL 966821, VL 973677, BL 122, BL 245, RR 15990, CH 45)
2003	319	**25**	(A 57, BL 122, BL 245, VL 4040, VL 4015, VL 88971, Tetep, VL 891194, VL891619, VL 926041, VL 891190, VL 895038, VL 936052, VL 956394, VL 956502, VL 966821, VL 963678, VL 973861, SLG 12-Sel., VL 9455, VL 9588, VL 10034, RR 36336, NDR 10544, Ch 45)
2004	351	**25**	(AD 9408, AS89044, B 6926260-TB-1, BAREN, BG 3677, BLI, Columbia 1, CT 6516641, Diwani, E890715, E 890744, IR 4547336, IR 46292222, IRAT 258, Kehwa 4, Mlyang 47, Milyang 105, Suweon 303, Suweon 351, TKY 1014, TKY 1024, VL 88971, VL 891193, VL 911190, RR 166645, CH 45)
2005	110	**33**	(VL 30019, VL10036, VL 7626, VL 30029, VL 30027, VL 30018, VL 7371, VL7316, VL 10039, VL 4860, VL7504, VL 88971, VL 895038, VL 956502, VL 973861, VL 973678, VL 4569, VL 7077, IR 7822422215, IR 7822119682, IR 7822119695, IR 782242224, IR 782219623, IR 7822422298, VL 4040, NDR 10951, VL 7630, VL 7505, IR 78908307RE5, IR 789053110-2-RE3, VL 7630, VL 3288, DDR 105)
2006	282	**27**	(VL 4040, VL 7504, VL 30018, VL 30018, VL 30029, VL 30336, VL 88971, VL 891190, VL 895038, VL 956502, VL 973861, VL 7316, VL 10411, VL 30425, IR 7822419695, IR 7822119656BB, BR 47262B613, C 1944242, CAN 7227, HPR 1156, IRAT 13, IRAT 144, IRAT 170, VL 901619, VL 936052, VL 943030, VL 953400)
2007	600	**12**	(VL 30919, VL 30424, VL 30240, VL 30921, VL 30922, VL 6309, VL 30278, VL 30294, VL 30584, VL 30278, CH 45, RR 159-60, RR 168-1)
2008	300	**15**	(VL 7849, VL 7853, VL 31289, VL 31291, VL 31293, VL 31451, VL 31287, VL 31296, VL 10591, VL 10554, VL 10536, VL 10919, VL 30917, VL 31329, VL 31334)
Total	**7346**	**247**	

Source: Annual Reports (1982-83 to 2008-09), VPKAS, Almora

Table 8.3b. Evaluation of rice germplasm against false smut and grain discoloration.

Year	Diseases	Total Entries/ genotypes screened	Promising entries
1999	False smut & grain discoloration	218	**6** (HEXT 4, IR 57893-08, WAB 32-60, SPR 1064, PL 11/SPS-1, HR 4856-1-1-1-1-2)
2001	False smut & grain discoloration	509	**4** (XIANG 3510, VL 88971, TO 3107-56-1-22, HPR 1149)
2002	False smut & grain discoloration	538	**17** (VL 3964, VL 4049, VL 7045, VL 7059, VL 88971, VL 895038, VL 911762, VL 926041, VL936052, VL 956394, VL 956502, VL 956394, VL 956502, VL 966821, VL 973677, BL 122, BL 245)
2004	False smut	400	**33** (AD 9408, AS89044, B 6926260-TB-1, BAREN, BG 3677, BLI, Columbia 1, CT 6516641, Diwani, E890715, E 890744, IR 4547336, IR 46292222, IRAT 258, Kehwa 4, Miyang 47, Milyang 105, Suweon 303, Suweon 351, TKY 1014, TKY 1024, VL 88971, VL 891193, VL 911190, VL 4946, VL 7504, VL 4040, VL 891619, SRSN 8, SRSN 12, VL 4569, SLG 12, VSR 8)
2003	False smut	400	**19** (A 57, BL 122, BL 245, VL 4040, VL 4015, VL 88971, Tetep, VL 891194, VL891619, VL 926041, VL 891190, VL 895038, VL 936052, VL 956394, VL 956502, VL 966821, VL 963678, VL 973861, SLG 12-Sel.)
2005	False smut	95	**26** (VL 30019, VL10036, VL 7626, VL 30029, VL 30027, VL 30018, VL 7371, VL7316, VL 10039, VL 4860, VL7504, VL 88971, VL 895038, VL 956502, VL 973861, VL 973678, VL 4569, VL 7077, IR 7822422215, IR 7822119682, IR 7822119695, IR 782242224, IR 782219623, IR 7822422298, VL 4040, VL 4569)
2006	False Smut	282	**34** (VL 4040, VL 7504, VL 30018, VL 30018, VL 30029, VL 30336, VL 88971, VL 891190, VL 895038, VL 956502, VL 973861, VL 7316, VL 10411, VL 30425, IR 7822419695, IR 7822119656BB, BR 47262B613, C 1944242, CAN 7227, HPR 1156, IRAT 13, IRAT 144, IRAT 170, VL 901619, VL 936052, VL 943030, VL 953400, VL 4569, VL 4930, VL 7316, VL 7504, VL 10036, VL 20246, VL 30219)
2007	False smut	600	**3** (VL 30821, VL 4569, Vl 31328)

2008	False smut	300	**6** (VL 4569,VL 30736, VL 31290, VL 31453, VL 31454, VL 31455)
Total	**False smut**	**3402**	**129**
Total	**Grain discoloration**	**1265**	**50**

Source: Annual Reports (1982-83 to 2008-09),VPKAS, Almora

Screening of germplasm against insect pests

The major insect-pests for the hilly region of Uttarakhand are stem borer and leaf folder. More than 4765 genotypes were screened against stem borer from 1979 to 2008 at VPKAS, Almora out of which 288 were found promising. Similarly 3706 genotypes were screened against leaf folder and 206 genotypes were found promising (Table 8.4). Besides this many genotypes over years were found promising with resistance/tolerance to multiple diseases and insect-pests (Table 8.5) which can be used as donors in the breeding programme.

Table 8.4. Evaluation of rice germplasm for resistance against insect pests.

Year	Biotic stress	Total Entries/ genotypes screened	Promising entries
1979	Stem borer and leaf folder	47	**6** (IR 5867452, IR 586745232, IR 8866303, IR 8866483, IR 9202561, KH 7084)
1980	Stem borer and leaf folder	113	**13** (IR 59082121, IR 590884233, VR 503, IR 8866303, IR 9202561, VRS 16322, VRS28641, VRS 570, T 1432, Ptb 12, Ptb 19, Chemparam Pandi, ARC 14539 G)
1982	Stem borer and leaf folder	147	**8** (IR 156852223, IR 15429268121, IR 9129-K 4, YR 2379792, K 443106, Norin 18, IR 978211121, SR 3055129323)
1983	Stem borer and leaf folder	145	**6** (IR 156852223, IR 1563683, SR 3044783, VL 20, VRS 23022, AS 688)
1984	Stem borer	244	**18** (Suweon 300, IR 92022513, IR 197432522, IR 22623-R-R-4-2, RP 1442232, K 33979, IR 156852223, VL 20, IR 132408223231, IR 43, IR 9209471162, TKM 9, IR 1354056321, IR 1531443233, IR 40, IR 46, RP 21991011, IET 7254)
	leaf folder	210	**5** (RP 21993849562, RP 219941253035, RP 2199102141910, RP 21991011, IET 7254)

1985	Stem borer	268	**16**	(IR 4619571121, RP 2415811121B, K 451131433, IR 19743262231, Cheolwen 32, IF 56, HPU 804, IR 912926333232, UPR 83169, IR 1973552321, IR198193123, Kaohsiung sen yu 252, RP 20683223, TNAU LFR 831311, RP 825247, IR 31868642333)
	Leaf folder	121	**8**	(IR 4619571121, RP 2415811121B, K 451131433, IR 19743262231, IR 43, VRS 1632223, K 45131444, HPU 891)
1987	Stem borer	84	**10**	(Pusa150, K 451-131-4-3-3, Wasteronochi, HPU 800, IR 24632-34-2, IRAT 134, VL 32, VL 29, RP 2418-52-5, HPU 936)
	Leaf folder	84	**7**	(Pusa150, K 451-131-4-3-3, Chealwen 32, RP 2151-33-2, IR 31868-64-2-3-3-3, VL 16, IRAT 133)
1988	Stem Borer	165	**12**	(IR 32429-115-3-2-6, K451-131-4-3-3, RP 2418-52-94, IRAT 125, IR 21912-131-3-3, BG 367-9, IR 28156-84-3-3-2, IR 28154-101-3-2, IR 29692-65-2-3, IR 35366-40-3-3-2-2, IR 66 (IR 32307) 107-3-2-2, IR 13427-45-3-1-2-2-2)
	Leaf folder	119	**6**	(IR 32429-115-3-2-6, IR 28154-101-3-2, RP 2533-23-9-9, RP 2333-252-30, Suweon 235, VR 381)
1989	Stem borer	80	**7**	(IR 1974-25-2-2, IR 28154-101-3-2, IR 1416-1128-5-8, IRAT 125, RP 1674-590-39-1, RNR 772, AD 9408)
	Leaf folder	110	**5**	(VRS 242-1-2, VR 212, RP 2297-11-97, HPU 800, RP 2632-58-7-9)
1990	Stem borer	261	**22**	(IR 28154-101-3-2, NR 10045-20-3-2, CITANDUY, IR 29692-94-2-1-3, RP 1674-590-39-14, Columbia II, IR 18350-93-2, RP 2633-29-5-7, RP 2633-45-12-9, RP 2633-108-7-5, RP 2668-5, WGL 47969, CR 309-268, RP 2151-40-1-72, Pusa 510, Pusa 515, Pallavi, VL 88965, VL 88981, VL 881045, VL 881041, VL 88-964)
	Leaf folder	479	**6**	(IR 35366-40-3-3-2-2, RP 2235-78-14, PTB 12, Pusa 510, Pusa 515, Pallavi)
1991-1994	Stem borer and leaf folder	111	**12**	(IR 28154-101-3-2, EWAN Nos. HR 3990-1-1-5, HR5824-B-3-2-3, IR 52287-153-1-1-2, IR 52280-179-2-1-2, IR 50376-30-1-1-2-2, IR 28154-101-3-2, IR 21912-131-3-3, IR 25924-92-1-3, RNR 772, VR 696, HP 4-845)
1995	Leaf folder	48	**1**	(IRI 023)
1996	Stem borer	606	**16**	(Boda 271, CT 6514-31-1, E 890744, HPR 978, IR 5908-40-3-2-2, IR 9202-10-2-2-11, IRAT 250, IRAT 254, IRRI 387, Jimbu 12, Milyang 105, Secanado Brazil, Denaulaut Tawar, VSR 286-4-1, Speed 70, Nawyang 4)

1997	Stem borer	340	**18**	(Diwani, IR 480-44-157-1-1-3-3, IR 1544-238-2-3, IR 4547-5-3-6, IR 21912-131-3-3, IR 31892-100-3-3-3-3, IRAT 125, Milyang 47, Suweon 303, VL 91-5827, VL 89-5038, VL 91-1990, CSR 88-IR-4, HKR 95-3, IRAT 161, IR 57893-60, OR 1144-114, VL 91-1754)
1998	Stem borer	412	**13**	(Diwani, IR 4547-5-3-6, PPM 105/IR 9224-117-2-3-3, SPTLR 82078-PIG-B3-24-1-1, SR 8102-140-5, VL 89-5038, VL 91-1990, BL 245, CSR 88-IR-4, CSR 89-IR-2, OR 1144-114, LACASSINE and VL 91-1754)
	Leaf folder	412	**22**	(L-82, L-106, P-125, P-370, P-818, P-1379, VHC-182, VHC-228, VHC-540, VHC-654, VHC-696, VHC-1224, VHC-1229, VSR-53, VSR-106, VSR-171, VSR-176, VSR- 183, VSR-191, VSR-200, VSR-229 and VSR-239)
1999	Stem borer	182	**6**	(HEXT 4, IR 57893-08, WAB 32-60, SPR 1064, PL 11/SPS-1and HR 4856-1-1-1-1-2)
2000	Stem borer & leaf folder	238	**4**	(VL 97-3652, VL 97-3653, VL 97-3656 and VL 96-3538)
2001	Stem borer	366	**18**	(VL 953677, VL 963660, VL 974005, VL 974013, VL 3967, VL 4038, VL 4040, VL 4042, VL 4045, VL 4013, VL 963538, VL 973652, VL 973653, VL 973656, XIANG 3510, VL 88971, TO 3107-56-1-22, HPR 1149)
	Leaf folder	366	**14**	(VL 953677, VL 963660, VL 974005, VL 974013, VL 3967, VL 4038, VL 4040, VL 4042, VL 4045, VL 4013, VL 963538, VL 973652, VL 973653, VL 973656)
2002	Stem borer and leaf folder	225	**24**	(VSR 8, VL 891190, VL 953689, VL 3964, VL 4015, CB 98094, CB 97053, HRI 126, JGL 1799, JGL 2671, JGL 3828, GR 3, HKR 46, Tarori basmati, VL 973682, VL 973831, VL 956394, VL 4043, VL 984009, HPR 2061, VL 4092, HPR 2060, UPR 243733, CH 988)
2003	Stem borer and leaf folder	318	**12**	(VL 973741, VL 8, VL 963680, Taraori Basmati, IR 64, ORS 2412, ASD 16, VL 4653, VL 973882, VL3831, Guarani, IR 28228281332)
2004	Stem borer and leaf folder	86	**14**	(WGL 32621, DRRH 20, EXPH 664, KJTRH 3, VL 4930, VL 30218, VL 4637, VL4455, V:L7072, VL 779, VL 7220, VL 7314, VL Dhan 61, VL Dhan 221)
2005	Stem borer and leaf folder	102	**10**	(SKL 761091012, RP 4511257, VL 30183, VL 7364, VL 7222, VL 7316, VL 30032, VL 30087, VL 81, VL 61)
2006	Stem borer and leaf folder	41	**2**	(VL 30278 and VL 30249)
2007	Stem borer and leaf folder	94	**12**	(VL 30919, VL 30655, VL 30920, VL 30921, JKRH 2000, RP 4516-38, RP 4518-26, RP 4621-1842, RP 4621-1845, RP 4510-156, RP 4511-234, RP 4511-254)

2008	Stem borer and leaf folder	90	**9**	(VL 31327, VL 31329, UPR 2992-17-3-2, HPR 2596, RP 2421, Vivek Dhan 154, Benibhog, RP 4687-52-1-1181, IR 64)
Total	**Stem borer**	**4765**	**288**	
	Leaf folder	**3706**	**206**	

Source: Annual Reports (1982-83 to 2008-09), VPKAS, Almora

Table 8.5. Identification of rice genotypes with multiple resistances against diseases.

S. No.	Year	Lines with multiple resistance	Diseases/ insect pest
1.	1979	RP104571433, RP 10573514, KMP 39, CN 540, RP 1017151, CR 987218265	Blast and leaf scald
2.	1986	RP 1015 27 701, HRR 118, P1988 AV 1254 154 32, RP 2151 221 44, CR 294548, RP 215116641, RP 2151, RP 178142.	Blast, brown spot, leaf scald and sheath rot
3.	1988	IR 15579-166, IR 23325-R-R-B-722, B4185E-CW-135154492, B 4219C-MR-4, LA 110, P 135613 M-218, ChiemBau	Blast and sheath rot
4.	1989	Tetep, Milyang 46, VL 8, C 4615, IR 920211, IR 5908-263, IR 9202-2132, SR 7272-11544, VL 16	Brown spot, sheath rot, leaf scald and false smut
5.	1990	RR 217-1, TNAUBPHR 831305	Brown spot and blast
6.	1991	IET 11612, IET 11223, IET 11995, IET 12592, IET 12567, IET 12569, IET 13212, HPR 869, IR 13427, IET 12785, IET 11581, IET 11063, UPR 82-17, CNA 762260, K39/9292.	Blast and brown spot
7.	1992	VRS 286-4-1, HPU-800, TKY 1014, Suweon TKY 1024, IR 4547-5-3-6, IRI 370,	Blast, brown spot, sheath rot, brown spot, leaf scald, grain discoloration and false smut
8.	1993	IR 5908-84-2-2, IR 5873-9-1, IR 57924-8, MENG-LI-WANG-GU (Acc 66985), VL 88974, B 2266 bCw-11-23-4, IR 48725-B-B-11-2, JINBU-9, Suweon 235, TSIPALANDAMBO 650, VL	Blast, brown spot, sheath rot, brown spot, leaf scald, grain discoloration and false smut

		90-1550	
9.	1994	SR 3049-58-1-5-1, E 220944, E 220949, E 220976, E 220935, E 220948, E 220985, E 221004, E 221029, E 221078, RP 1451-2439 -1225, RP 2533-748-7-18, VL Dhan 163, VL 895193, CN 776-46-B1	Blast, brown spot, sheath rot, brown spot, leaf scald, grain discoloration and false smut
10	1995	IR 578926, Milyang 46, Milyang 104,RAU PT 100972, Suweon 330, Suweon 376, TM 4310, YR 6488-ACP-36, HR 6419-B-56-1-2, IR 8/ ENG/SAC	Blast, brown spot, sheath rot, leaf scald, false smut, grain discoloration and stem borer.
11.	1996	Boda 271, CT 6514-31-1, E 890744, HPR 978, IR 5908-40-3-2-2, IR 9202-10-2-2-11, IRAT 250, IRAT 254, IRRI 387, Jimbu 12, Milyang 105, Secanado Brazil, Denaulaut Tawar, VSR 286-4-1, Speed 70, Nawyang 4, P 125, INRC 507, INRC 518, INRC 624, INRC 642, INRC 735, INRC 1307	Blast, sheath rot, leaf scald, false smut and stem borer.
		VR 4225, VR 431 3, VR 549-2, VL 89 9173, VL 89 9416, VL 939317, VL 939419, VL 939325, VL 5044	Leaf blast, sheath rot and leaf scald
12.	1997	Diwani, IR 480-44-157-1-1-3-3, IR 1544-238-2-3, IR 4547-5-3-6, IR 21912-131-3-3, IR 31892-100-3-3-3-3, IRAT 125, Milyang 47, Suweon 303, VL 91-5827, VL 89-5038, VL 91-1990, CSR 88-IR-4, HKR 95-3, IRAT 161, IR 57893-60, OR 1144-114, VL 91-1754	Blast, sheath rot, leaf scald, false smut and stem borer.
		VR 4353, VL 885038, VL 89 5001, VL 896050, VL 901619, VL 901738	Blast, sheath rot, leaf scald, false smut and stem borer.
13.	1998	Diwani, IR 4547-5-3-6, PPM 105/IR 9224-117-2-3-3, SPTLR 82078-PIG-B3-24-1-1, SR 8102-140-5, VL 89-5038, VL 89-9419, VL 91-1990, BL 245, CSR 88-IR-4, CSR 89-IR-2, OR 1144-114, LACASSINE, VL 91-1754	Blast, brown spot, sheath rot, leaf scald, false smut and stem borer.
		BL 245, CSR 88-IR-4, CSR 89-IR-2,	Blast, sheath rot, leaf scald, and

		OR 1144-114, LACASSINE, VL 91-1754	false smut.
14.	1999	HEXT 4, IR 57893-08, WAB 32-60, SPR 1064, PL 11/SPS-1, HR 4856-1-1-1-1-2	Blast, sheath rot, leaf scald, false smut and stem borer.
		VL 97-3677, VL 97-3657, VL 97 -3660, VL 97-3680, VL 97-3741, VL 97-3747	Blast, sheath rot, leaf scald, and false smut.
15.	2000	WAB 9624, WAB 9921, XIANGZI 3150, VL 88971, VL 93 3548, VL 973747, VL 932881, TOX 310756122, TOX 718124, TOX 955, 20B2102, PRR 2, PL8/SPS3, VL 911754, F 501, F 6610, CNA 6710, CT 7242163332P, CTB 249272M1P, HPR 1149, BL 221, BL 222, Tadukan, Fukunishiki, Ishikari shiroae, Shimokita, Yamatenishiki, Jinheng, SR 133632821	Blast, brown spot, sheath rot, leaf scald, false smut and stem borer.
16.	2001	XIANG 3510, VL 88971, TO 3107-56-1-22, HPR 1149	Blast, brown spot, sheath rot, leaf scald, false smut and stem borer.
17.	2002	VL 3964, VL 4049, VL 7045, VL 7059, VL 88971, VL 895038, VL 911762, VL 926041, VL936052, VL 956394, VL 956502, VL 956394, VL 956502, VL 966821, VL 973677, BL 122, BL 245	Blast, sheath rot, leaf scald, false smut and stem borer.
		VL 7203, VL 7230, VL 7272, VL 7276, VL 7278, VL 7314, VL 7323, VL 959632, VL 969588, VL 969807, VL 9948	Blast, sheath rot, leaf scald, and false smut.
18.	2003	A 57, BL 122, BL 245, VL 4040, VL 4015, VL 88971, Tetep, VL 891194, VL891619, VL 926041, VL 891190, VL 895038, VL 936052, VL 956394, VL 956502, VL 966821, VL 963678, VL 973861, SLG 12-Sel., VL 3959, VL 4045, VL 973861, VL 4803, VL 4653, VL 4569	Blast, brown spot, sheath rot, leaf scald and false smut
		VL 10073, VL 10075	Blast, sheath rot, leaf scald, and false smut.
19.	2004	AD 9408, AS89044, B 6926260-TB-1, BAREN, BG 3677, BLI,	Blast, brown spot, sheath rot, leaf scald, grain discoloration and false

		Columbia 1, CT 6516641, Diwani , E890715, E 890744, IR 4547336, IR 46292222, IRAT 258, Kehwa 4, Mlyang 47, Milyang 105, Suweon 303, Suweon 351, TKY 1014, TKY 1024, VL 88971, VL 891193, VL 911190	smut
20	2005	VL 30019, VL10036, VL 7626, VL 30029, VL 30027, VL 30018, VL 7371, VL7316, VL 10039, VL 4860, VL7504, VL 88971, VL 895038, VL 956502, VL 973861, VL 973678, VL 4569, VL 7077, IR 7822422215, IR 7822119682, IR 7822119695, IR 782242224, IR 782219623, IR 7822422298, VL 4040	Blast, brown spot, sheath rot, leaf scald, grain discoloration and false smut
		NDR 10951, VL 7630, VL 7505, IR 78908307RE5, IR 789053110-2-RE3, VL 7630, VL 3288, DDR 105	Blast and brown spot
21.	2006	VL 4040, VL 7504, VL 30018, VL 30018, VL 30029, VL30336, VL 88971, VL891190, VL895038, VL 956502, VL 973861, VL 7316, VL 10411, VL 30425, IR 7822419695, IR 7822119656BB, BR 47262B613, C 1944242, CAN 7227, HPR 1156, IRAT 13, IRAT 144, IRAT 170, VL 901619, VL 936052, VL 943030, VL 953400	Blast, brown spot, sheath rot, leaf scald, grain discoloration and false smut
22.	2007	VOIP 3102	Sheath rot, bacterial blight, and rice tungro

Evaluation of rice germplasm against abiotic stresses

Evaluation for cold tolerance

More than 6537 genotypes of rice were evaluated from 1975 to 2008 at VPKAS, Almora for cold tolerance, out of which 271 were found promising (Table 8.6) (Anon. 1975-76 to 2008-2009). Many genotypes *viz*., L 62-2A, L 62 G and Khonorullo were found to be tolerant to cold in more than one season. Besides this, many basic studies were conducted for cold tolerance at VPKAS, Almora over years which are briefly described here.

Table 8.6. Evaluation of rice genotypes for cold tolerance at VPKAS, Almora.

Year	No. of genotypes evaluated		Tolerance genotypes
1975	462	**19**	(IR 28, IR 2061-294-3-3-2, IR 3487-8, IR 3941-1, IR 3941-8, IR 3941-10, IR 3941-29, IR 3941-33, IR 3942-45, IR 3941-46, IR 394-68, IR 3941-77, Jamula, K 42-55B-3, K 78-13, Kulu, Sirandel Merah, Pankait 117-2-3, Pankait 447-3)
1976	99	**13**	(IR 30, IR 2053-521-1-1, IR 2061-214-3-3-32, IR 3941-CR-126-42, IR 3941-21-CR 126-42, IR 394-35-CR-126-42, JP5, IR 1846-295-1-1, KH 863, K 17-BK-41-1, K 178, IR 3941-9-1A, IR 3941-77-1A)
1977	380	**10**	(Bozu, China 1039, Jodo, Jayanak, Kabjilongtson, Kendzo, Eiko K 28, K 78-13, L1, L12)
1978	198	**6**	(IR 5865-32-5, IR 5867-45-2, IR 5867-50-3-2-1, IR 7167-33-2-4, IR 8866-30-2, MPR-1)
1979	239	**5**	(Jodo, HP 46, IR 1846-296-3, IR 2053-521-1-1, K 427)
1980	245	**8**	(Fuzi 102, Matsumae, Milyang 46, IR 8845-10-2-1-1, IR 92202-23-21-3, IR 9699-K1, IR 13045, Suweon 289)
1981	112	**7**	(IR 19746-28-2-2-2, IR 9758-K2, T 1668, taichung 176, AC 3828, IR 9202-23-2-1-3, K 42-86-8-1)
1983	276	**3**	(IR 9202-6-1-1, PI 160776, Tainung-Sen-Yu)
	30	**7**	(L 62-2A, L 62 G, Khonorullo, Heug-Jodo, Muzudo and K 78-28)
	16	**5**	(L 62 G, L 62-2A, UPRH-299, L-12 1A and Khonorullo)
1985	184	**7**	(NR 10045-20-3-2, NR 10073-167-3-1-1, NR 10073-167-3-1-2, IR 14632-65-2, IR 7167-33-2-3-3-1, IR 9202-23-2-1-3, IR 9202-6-1-1)
1986	88	**8**	(Bir-Ze-Goo, JKAU(K) 449-126-1, JKAU 452-21-4, Ta-Tao, Shun, Tan Wing-Yaangh, IR 288777-8145-3-1-3, Chung-Cheng)
	30	**6**	(VL 502, L 12-1A, Barmi, VHC 1253, KH 100-1A, VL 206)
1987	63	**5**	(IR 13045-104-1, IR 13168-143-1, L4-110, SR 6459-286-1-4-3-4-4, IRAT 346)
	107	**13**	(L 62-2A, L 12-1A, L60, L 1058, VL 8, VL 206, VRS 53, Europa, Salivira, VHC 1253, L 62G, VRS 106, VRS 129)
1988	78	**7**	(HPU 741, IR 15579-135, IR 15579-166, IR 23325-R-R-

			B-7-2-2, B 4313 E-Kn-4-C-C-1, SR 6495-286-4-3-4, RP 1848-216-3-2-1)
	83	**8**	(NR 10073-167-3-1-2, VR 211, Zir-Tan-Tsar, VSR-48, VSR 106, VSR 52, VSR 53, VSR 33)
	280	**3**	(NA)
1989	58	**6**	(AI Yeh, IR 15569, IR 18482, B 4185, Akiyudaka, IR 37358)
	290	**8**	(NA)
1990	79	**9**	(NA)
1990	290	**15**	(C727312)
1991	106	**6**	(IR 18502-PLP 4-3-3-3-1-1, IR 20654-R-R-R-15-1-8, SPTC R 82078-PTG-83-24-1-1, SPTC R 82078-PTG- 88-2165-7, IR 47686-6-6-2-2-1, Meng-Li-Wang-GU)
	291	**20**	(VR 9071)
1992	82	**7**	(NA)
1993	73	**4**	(SPTRL 81399-PTG-9-2-2, SR 11451-T-199, Suweon 347, Suweon 361)
	2000	**10**	(Khonorullo, L-62-G, VL 9071, Barkat, UPRH 299, VL 9170, Europa, VL 9077, VL 9060, L 62-2A)
1995	136	**18**	(NA)
1992-95	60	**10**	(Khonorullo, L 62-G, VL 9071, Barkat, UPRH 299, VL 9170, Europa, VL 9077, VL 9060 and L 62-2A)
1999	15	**5**	(Barkat, VL 94-3027, VL 94-3030, VL 94-3143, VL 95-3336)
2001	69	**12**	(IR 57080, IR 57301, IR 62728, IR 70306, Hawan Ghaezo, IR 71121, IR 61673, IR 67068, IR 67087, Yunlen 2, IR 59624)
2005	18	**1**	(VL 30061)
Total	**6537**	**271**	

Basic studies on the physiological aspects of cold tolerance at VPKAS, Almora: Besides screening rice germplasm against cold tolerance, VPKAS, Almora undertook many experiments to understand the different aspects of cold tolerance among rice genotypes under the hill conditions (Annon. 1980-81 to 2008-09). Some of them are discussed below.

In 1980, 146 genotypes were evaluated at VPKAS for their ability to germinate at low temperatures under field conditions. Seeding was done on 10th

March *i.e.*, about a fortnight earlier than the normal seeding date for spring sown rice, when the air temperature averaged about 16°C at the maximum and 4°C at the minimum. Soil temperatures recorded at 5 cm soil depth were 4.6°C and 23.7°C respectively in the morning and the afternoon. The results showed that several of the local types had much better germination under low temperatures. These included L1-1A, L1-2A, L3-1A, L12-1, L 12-1A, L 12-2A, L 60, L 61, L 62-1A, L 62-2A, L 62A, L 62B, L 62C, L 62F, L 62G, L100A, L100B, L105A, L105B, VL191 and KH 176-5-9, which started germinating within 15 days and completed more than 70% germination within 20 days of sowing. Several other promising cultures were also identified which completed more than 70% germination within 25 days. Similarly one hundred and thirty five genotypes representing a wide spectrum of variation for duration and origin were evaluated in 1981 at low as well as high temperatures under natural field conditions for their germination ability and early growth development. The nursery was seeded on 26th February for studying the effects of low temperature and the data were recorded 18th day onwards when temperature ranged between 14.7 to 20 and 4.5 to 7.4 °C with L 62 C having maximum germination of 50% followed by IR 1846-306-28. Germination was recorded within a week in 60 genotypes. Rest of the cultures either did not germinate or showed extremely poor germination. The soil temperature remained below 18.2/7.8°C during this week. Fifty more genotypes germinated only after 1st of April when temperature increased above 19/9°C. It indicated that genotypes germinated during the month of March seem to posses low temperature tolerance for the germination. The visual scoring of the seedling growth was also done and it was observed that Jodo followed by L 62b, L 62F, Daegoldo, Gang weondo, IR 80-102, Bermi, L 62-A and L 60 showed good seedling vigour at those low temperatures. Some genotypes like L 12-1A, L 105-B and VHC 1253 which had poor seedling growth in the initial stages, showed very good recovery by the end of May. The same set of cultures was also sown in field at comparatively higher temperatures 27.5-31 / 15.5-18.7°C *i.e.*, on 2nd June, 1981. It was observed that the seedling growth at the temperatures 27.5-31 / 15.5-18.8°C was quite satisfactory for the genotypes VHC 1283, L 62B, L 62C, L 12-1A, L 100A, L 62G, Bermi, L 53, L 61, Majhera-3 and L 44. It indicated that L 62-B, Bermi, VHC 1283 and L 12-1A have good adaptability under varied range of environments. In another experiment conducted in 1981, genotypes L1-2A, VL 237, L100, followed by L 62A, L 62G, L 100-A, L 62C, IR 1846-306-2A, K 113-2, Fuzi-102, Daegaldo, Khonorullo, Hawang Haedo, Eiko, Keug-d, Heug-Jodo, VR6, L3-1A-1 and UPR 103-D-6-1were found possessing good vigour at low temperatures. However, Khatiyari Local, L 105-B and L 12-1A showed poor vigour during early stages of seedling growth when low temperatures were prevalent but showed very good recovery at higher temperatures and produced vigorous plants.

Effect of temperature on grain filling duration and spikelet fertility: A set of one hundred genotypes representing varying flowering duration were evaluated at VPKAS in 1981 for the duration from anthesis to grain maturity phase, and spikelet sterility expression under transplanted condition. The duration to anthesis was found ranging from 60 to 132 days whereas it was 27 to 59 days for grain filling. This indicates that the major part of plant life *i.e.*, 69 to 72% contributes for the development of vegetative parts while 28 to 31% contributes towards the production of grain yield.

Effect of temperature on seed water imbibitions capacity: In the year 1981, thirty genotypes were evaluated for their capacity for water imbibitions at 11, 13 and 15°C temperatures. It was observed that the varieties which germinated at 11°C showed more water imbibitions at 11°C, compare to 13 and 15°C temperatures. This technique can be used for quick screening of the genotypes for cold tolerance.

Effect of low temperature on germination and seedling vigour: Thirty genotypes were sown under natural field conditions at Almora in 1983 and 1984. Seven genotypes, namely L 62-2A, L 62 G, Khonorullo, Heug-Jodo, Muzudo and K 78-28 continued to show better seed germination (over 75%) at 11°C field temperature and possess low temperature tolerance for the germination. Rest of the genotypes germinated after temperature reached 15°C. It clearly showed that marked variations exist among genotypes for their germinating ability at low temperatures which can be utilized for the development of new stains.

Spikelet sterility expression in rice under cold: Plants of 16 genotypes with varying degree of tolerance to low temperature germination were grown in pots at Almora till the initiation of booting stage in 1983. One set of pots was transferred to Mirtola (7000') while the other was retained at Almora. The data on ear emergence, leaf colour and sterility per cent were recorded. Genotypes, namely, L 62 G, L 62-2A, UPRH-299, L-12 1A and Khonorullo were found to be highly tolerant to low temperature during the germination as well as grain filling phases.

Effect of chemicals for cold tolerance in rice: Many experiments were conducted at VPKAS from 1988 to 2002 (Anon. 1988-89 to 2002-03) in order to find effects of different chemicals on tolerance of rice genotypes to cold stress. Those included GA_3, Jalshakti, KNO_3, kinetin, ascorbic acid, NaCl, caprylic acid, Farmore, ABA, salicylic acid, tannic acid, thiourea, urea and Alar. Two separate experiments were conducted in 1989 for inducing cold tolerance at germination and reproductive stages through the external application of chemicals. Rice varieties Lee Goo Thao and VL 221 were used for cold tolerance induction at germination and reproductive stages respectively. Among the 18 chemicals of different concentrations, GA_3 (100 ppm), Jalshakti (1g/kg seed), KNO_3 (1%) and kinetin (50 ppm) could enhance period of germination by one month under cold conditions (12°C) as compared to untreated dry seeds. The variety, VL 221 responded differently to chemicals when it was sprayed along with them in different concentrations at tillering and anthesis stages. Spraying of crop at both the stages with ABA (15 ppm), etheral (2000 ppm), NaCl (0.2%) and cycocel (100 ppm) produced higher yields through higher panicle fertility and better grain filling even though the temperature was low (16°C) at the reproductive phase. In another experiment conducted in 1990, seeds of rice variety Lee Goo Thao, highly susceptible to cold temperature, were treated with different chemicals to induce germination under cold condition. Treated seeds were put for germination in the cabinet maintained at 12°C. GA_3 (100 ppm) and kinetin (20 ppm) could induce 10% and 33% germination, respectively which otherwise did not germinate under the same temperature. Alar (1500 ppm) could, however, induce as high as 70% germination in 13 days. This experiment was further repeated in 1991 with five selected chemicals with varying concentrations and was observed that GA_3 (125 ppm) and ascorbic acid (150 ppm) initiated germination within 16 and 18 days, respectively, while the seeds kept in pure water could not show germination even up to one month at 12°C.

Several experiments were conducted at VPKAS for the use of chemicals as spray for inducing tolerance against cold. From the experiments conducted from 1992-1995, foliar sprays of NaCl, GA_3, Caprylic acid and Farmore at tillering and booting stages were found to increase the yield in cold susceptible rice genotypes *viz.*, VL 206 and CTH-1. Significantly higher yield was obtained from Farmore application. With similar experiment conducted in 2001 using foliar spray of ascorbic acid (100 ppm), ABA (10 ppm), salicylic acid (100 ppm), tannic acid (100 ppm), Thiourea (0.5%) and Urea (1.5%) induced the activities of nitrate reductase, peroxidase and the level of soluble proteins over the control during grain filling phase in rice under low temperature condition. Besides, these treatments significantly enhanced the grain yield and improved the cold tolerance efficiency by more than 19 per cent compared to the control. Similarly from the experiments conducted in 2002, it was observed that the foliar spray of ascorbic acid (100 ppm), ABA (10 ppm), salicylic acid (100 ppm), tannic acid (100 ppm), Alar (100 ppm) and Thiourea (0.5%) induced the activities of nitrate reductase, peroxidase, the level of photosynthetic pigments and soluble protein over the control during grain filling phase in rice under low temperature condition. Besides, this aforesaid treatment significantly enhanced grain yield and improved the cold tolerance efficiency by more than 17 per cent in comparison to the control.

Association of morpho-physiological characters with cold tolerance: From many experiments conducted between 1992 to 1995 at VPKAS, it was found that longer coleoptiles (1-1.3 cm), broader and more foliages with chlorophyll content in the range of 2.6 to 3.1 mg/g fresh weight at 21 days after germination and higher NR activity had the positive association with cold tolerance. In another experiment conducted in 1995 using 14 rice genotypes, it was noted that longer coleoptiles, small, wider and more number of leaves with higher chlorophyll content, NR activity and proline content had the positive association with cold tolerance.

Photosynthetic response of rice genotypes under low temperature condition: To understand the photosynthetic basis of grain yield under low temperature during reproductive phase, 15 genotypes were evaluated at VPKAS in 1999 at two dates of sowing, May 26 and June 26, to expose the material to low temperature condition at flowering. The minimum and maximum temperature ranged between 6.0 - 20.5°C and 21.0 - 30.0°C. One month old seedlings were planted in well puddled field. Canopy photosynthesis under both the plantings had the positive association with total dry matter production. Besides, the rates of photosynthesis (r=0.526*) and canopy photosynthesis (r=0.762*) showed significantly positive correlation with grain yield under normal planting and poor association under low temperature condition, *i.e.*, late planting. Rate of photosynthesis and crop growth recorded during the vegetative phase under both the sowing dates were found almost similar. Harvest index exhibited significant positive association with grain yield (r=0.716**) under low temperature condition only, which in turn, suggests that the low temperature reduced the partitioning of photo assimilates resulting into poor grain yield. Five genotypes, *viz.*, Barkat, VL 94-3027, VL 94-3030, VL 94-3143 and VL 95-3336 emerged as cold tolerant genotypes. The promising genotypes for different photosynthetic attributes are listed in Table 8.7. In another experiment conducted in 2000, a set of fifteen genotypes was grown at two dates of sowing: May 25 (normal) and June 25 (Late). The minimum and mean temperature during grain filling phase in late planting were <10°C and <18°C, respectively. Canopy photosynthesis exhibited significant positive correlation with total dry matter under normal (r=0.713**) and late (r=0.674**) planting. Rate of photosynthesis recorded at grain filling (r=0.611*) and total mean canopy photosynthesis (r=0.713**) estimated at different physiological stages showed significant positive relationship with grain yield under normal planting rate of both the photosynthesis and mean canopy whereas under low temperature condition due to late planting, photosynthesis was poorly associated with grain yield. Rate of crop growth and photosynthesis recorded during vegetative phase under both the plantings were found almost similar. Under low temperature conditions, reduction in rates of photosynthesis, transpiration, stomatal conductance, photosynthetic pigment and nitrate reductase activity at grain filling stage, and harvest index and test weight were observed. Moreover, harvest index showed significant, positive association with grain yield (r= 0.729**) under low temperature condition only, which indicates that the low temperature reduced the partitioning of photo assimilates resulting in poor yield. Cold tolerance efficiency (%) exhibited negative association with grain sterility (r=-0.776**) and positive association with anthocynins contents (r=0.572*), proline content (r=0.742**) and nitrate reductase activity (r=0.818**). The promising genotypes for different photosynthetic attributes under low temperature conditions are listed in Table 8.8.

Again in 2001, a field experiment with 15 genotypes of rice was conducted to assess the photosynthetic basis of yield under low temperature in hills (Anon. 2001-02). Rates of canopy photosynthesis at vegetative stage under both plantings were found almost similar. However, photosynthesis rates recorded at vegetative stage were higher under the late planting. A significant reduction in rates of photosynthesis, canopy photosynthesis and level of photosynthetic pigments was observed at grain filling stage under the late planting due to low temperature, below the critical limit. Reduction in the grain filling rate was also noted under late planting due to low temperature. Furthermore, harvest index under low temperature condition exhibited the significant positive association with grain yield (r=0.843**) and cold tolerant efficiency which in turn indicted that low temperature reduced the partitioning of photo assimilates which resulted in to poor grain yield. Three genotypes, namely, VL 94-3288, VL 98-3894 and VL 98-3895 were identified as cold tolerant (Table 8.10). A significant relationship of grain sterility (r=-0.961**), HI (0.886**), panicle weight (r=0.812**) and test weight (r=0.586*) with cold tolerance efficiency suggested the importance of screening rice genotypes for cold tolerance. The promising genotypes for different photosynthetic attributes under low temperature are listed in Table 8.9.

In order to understand the effect of low temperature on the losses in photosynthesis, an experiment was conducted at VPKAS in 2002 using three rice varieties namely VL Dhan 206, Vivek Dhan 82 and Thapachini and different grain filling periods were assessed. The grain filling phase was exposed to different regimes of low temperature stress by altering sowing of rice on three dates *i.e.*, normal nursery sowing on 25May and 15 and 30 days late to normal sowing. Rate of leaf photosynthesis and the level of photosynthetic pigments were reduced as crop experienced the low temperature stress during grain filling period. VL Dhan 206 experienced the highest photosynthetic loss. Nitrate reductase activity also showed the similar trends. However, proline content showed the opposite trend to leaf photosynthesis. Besides this, membrane thermo stability showed the improvement in respect to low temperature when the crop was grown under late sowing / low temperature condition. Regression analysis between leaf photosynthesis versus stomatal conductance, transpiration rate, photosynthetic pigments (Chl a, Chl b, Total Chl, Carotenoid), nitrate reductase activity and grain filling rate showed the linear positive association while negative association with proline content. Such regression equations may be used to assess the impact of low temperature stress of leaf photosynthesis in rice (Table 8.10).

Table 8.7. Promising rice genotypes for photosynthetic attributes under cold condition.

Characters	Promising entries under low temperature conditions
High photosynthetic rate	Barkat, VL 91-1754, VL 91-1998, VL 94-3143
High canopy photosynthesis	VL 91- 1754, VL 91-1998, VL 91-3149, VL 94-3030, VL 94-3143
High photosynthetic rate and high canopy photosynthesis	VL 91- 1754, VL 91- 1998, VL 91-3143
High photosynthetic rate, high canopy photosynthesis and high yield	VL 94-3143

Table 8.8. Different photosynthetic attributes of rice genotypes under cold condition.

Characters	Promising entries under low temperature conditions
High photosynthetic rate	VL 91- 1754, VL 94-3143, VL 94-3148
High canopy photosynthesis	VL 91-1754, VL 94-3143, VL 94-3148
High canopy photosynthesis and high yield	VL 91-2877, VL 94-3143, VL-94-3148
High photosynthetic rate, high canopy photosynthesis and high yield	VL 94-3143, VL 94-3148
Cold tolerant genotypes	VL 91-1923, VL 91-1990

Table 8.9. Different photosynthetic attributes of rice genotypes under cold condition.

Characters	Promising entries under low temperature conditions
High photosynthetic rate	VL 94-3288, VL 96-3610, VL 97-3653
High canopy photosynthesis	VL 93-2767, VL 94-3288, VL 96-3610, VL 97-3653, VL 97- 3861, VL 98-3895
High grain yield	VL 93-2767, VL 94-3288, VL 97-3653, VL 97-3657, VL 98-3894, VL 98-3895

High photosynthetic rate, high canopy photosynthesis and high grain yield	VL 94-3288, VL 97-3653
Promising cold tolerant genotypes	VL 94-3288, VL 98-3894, VL 98-3895

Table 8.10. Linear regression equations between leaf photosynthesis rate (Pn) and other physiological traits in rice.

Traits	Linear regression equations	R^2
Pn vs. stomatal conductance	$Y=3.947 + 0.0017 X^2$	0.836
Pn vs. transpiration rate	$Y=-6.667 + 3.170 X^2$	0.620
Pn vs. nitrate reductase	$Y=-2.657 + 8.595 X^2$	0.885
Pn vs. proline	$Y= 22.276 - 0.336 X^2$	0.820
Pn vs. chlorophyll a	$Y= 0.092 + 10.903 X^2$	0.930
Pn vs. chlorophyll b	$Y=-0.041 + 45.759 X^2$	0.865
Pn vs. total chlorophyll	$Y=1.1006 + 7.847 X^2$	0.870
Pn vs. total carotenoids	$Y=-0.812 + 39.099 X^2$	0.803
Pn vs. grain filling rate	$Y=-0.578 + 16.876 X^2$	0.799

Evaluation of rice genotypes for drought tolerance

Over last 25 years (1978 to 2003), more than 1027 rice genotypes were evaluated for drought tolerance from which 78 were found to be promising against drought (Table 8.11, Anon. 1978-79 to 2003-04). Besides this many basic studies related to drought tolerance were taken up at VPKAS, Almora and GBPIED, Koshi, Katarmal which are described here briefly.

Table 8.11. Evaluation rice genotypes for drought tolerance.

Year	No. of genotypes evaluated	Tolerant genotypes
1978	190	7 (N22, CTG 1516, IRAT 109, IRAT 111, Dular, DV 110)
1979	118	1 (IAC-25)
1981	104	3 (Salumkit, ITA 117, ITA 164)
1983	207	6 (IRAT 104, DR 92, ARC 10372, Dourado, Precoce, NDR 220)
1985	117	5 (GNA 762260, DD 961, HAU 741, IRAT-112, BG 367-7)
1986	15	2 (VRS 163, Khonorullo)

1987	**33**	5 (AVS 463, IAC 463, IAC 118/81, IAC 233/79, IRAT 112)
1989	**25**	3 (BG 621-1, ITA 257, IRAT 112)
1990	**51**	8
1991	**81**	10
1995	**55**	14
1998	**5**	3 (Vandana, VL Dhan 163, VR 435-3)
2001	**16**	3 (VL 88-5038, VL 88-5001, Vivek Dhan 62)
2003	**10**	1 (VL 6394)
Total	**1027**	78

Effect of moisture stress on seed germination: Effect of moisture stress on seed germination can be evaluated under laboratory condition by many ways. One of the reliable methods is to create osmotic concentration by use of polyethylene glycol (PEG-6000). A large number of rice genotypes was evaluated at VPKAS, Almora using PEG from 1984 to 2004. Fifteen genotypes were evaluated in 1984 for their germinating ability at five simulated moisture stress conditions *i.e.*, 0, -4, -8, -12 and -16 bars, created by PEG at 25°C. None of the genotypes showed germination at 16 bars. However, genotypes *viz.*, Daegaldo, Khonorullo, Muzudo and Jodo showed some germination at -12 bars after 7 days of putting seeds in the solutions. These genotypes also showed longest root length as well as shoot length at -12 bars except Daegaldo which produces the longest roots and more roots at -8 bars. Again in 1985, 80 genotypes were evaluated by germinating under different osmotic concentrations created by PEG 6000. It was found that none of the genotypes germinated at -16 bars. Eight genotypes *viz.*, Khonorullo, Muzudo, Judo, Kaegaldo, Heug Jodo, IRAT-102, K 333 and Eiko showed germination at -12 bars. Out of these, only Khonorullo recorded the maximum germination and longest root length at -12 bars (Table 8.12).

Likewise in 1986, 30 genotypes were evaluated under different osmotic potentials (PEG 6000). It was found that none of the genotypes germinated at -16 bars. Eight genotypes however, showed germination at -12 bars. Khonorullo recorded the maximum germination and root length at -12 bars. Maximum root number and seedling length were observed in Judo (Table 8.13). Again in 1998 a set of 33 genotypes were evaluated following the same procedure. Out of 33 genotypes only seven genotypes namely HPR 1156, VL 3682, VL 3831, VL 6041, VL 6050, VL 6394 and VL 6704 proved their drought tolerance efficiency. Similarly from the experiments done in 2003, amongst the 129 genotypes, only two genotypes namely, VL-7385 and VL-7392 proved their drought tolerance efficiency (>80%)

during germination. From another experiment conducted in 2004, amongst the 78 genotypes screened for moisture stress only two genotypes namely, VL-7620 and VL-30032 proved their drought tolerance efficiency during germination.

Table 8.12. Germinating ability of rice genotypes under various osmotic concentrations.

Genotypes	Germination %			Root length (mm)			Seeedling length (mm)		
Moisture stress (bars)	0	-8	-12	0	-8	-12	0	-8	-12
Khonorullo	100	88	65	150	80	50	100	10	6
Muzudo	100	100	38	95	80	35	100	25	0
Judo	100	56	12	110	100	25	110	40	0
Daegaldo	100	88	30	100	100	35	100	35	0
Heug Jodo	96	96	30	100	100	20	90	40	0
IRAT 102	100	88	24	130	65	20	110	5	0
K 333	92	92	20	130	100	30	130	18	12
Eiko	100	84	8	130	80	0	150	30	0

Table 8.13. Germination ability of rice genotypes under various osmotic concentrations.

Genotypes	Germination %			Root number			Root length (mm)			Seedling length (mm)		
Moisture stress (bars)	0	-8	-12	0	-8	-12	0	-8	-12	0	-8	-12
Khonorullo	100	90	65	5	2	1.6	160	90	55	105	10	0
Muzudo	100	98	40	8	4	1.2	95	78	37	107	22	0
Judo	98	60	10	8	4	2.0	115	100	30	110	40	0
Daegaldo	95	90	30	7	4	1.0	85	110	25	115	30	0
Heug Jodo	90	90	35	9	4	1.5	95	95	18	85	30	0
IRAT 102	98	90	25	6	2	0.0	140	50	20	100	8	0
K 333	92	92	25	10	3	1.0	140	100	30	135	20	15
Eiko	100	85	10	8	5	0.0	155	90	0	155	30	0
VL 206	100	92	0	7	4	0.0	110	80	0	120	10	0
Thapachini	100	95	0	8	3	0.0	160	110	0	140	30	0
L 62 G	100	70	0	9	4	0.0	130	75	0	110	10	0
KH 100-1A	100	10	0	7	1	0.0	130	0	0	120	0	0

Working at the GB Pant Institute for Environment and Development, Koshi Katarmal, Agnihotri *et al.* (2007) investigated the effect of moisture stress amongst 30 landraces of rice caused by the treatment with polyethylene glycol for seed germination and seedling growth using three concentrations of osmoticum (-0.40, -0.50 and -0.75 MPa). They observed that, decreasing the water potential from -0.40 to -0.75 MPa resulted in decreased percentage of germination and seedling growth in most of the landraces. Overall Tilansi showed higher germination (83%) even at the concentration of -0.75 MPa whereas only 8% germination was recorded in Syaudhan. More than 80% germination was observed in 40% landraces at -0.40 MPa and 40% germination or less was observed at -0.75 MPa in 23% of the total landraces examined. The per cent and rate of germination was consistently and significantly reduced in all the landraces whenever the osmotic potential was reduced. In control seedlings, the difference in root length over initial values ranged between 0.07 cm (Dudhikapkoti) and 2.37 cm (Kaladur) while that of seedling (shoot) length ranged from 0.60 (Dehradooni basmati) to 4.59 cm (Kaladur). When a critical osmotic potential (-0.75 MPa) was applied, very limited root length increment took place and ranged between 0.02 and 1.66 cm in Anjani and Syaudhan, respectively. A similar trend was also found for seedling growth at the same osmotic potential (-0.75 MPa) and the increment in shoot length ranged from 0.01 cm (Bindudhan, Kantolia and Thapachini) to 2.67 cm in Saunji. The studies revealed that a few landraces of rice from those still being cultivated in the hilly region of Kumuan Himalaya may prove useful for introgression of genes for osmotic stress tolerance landraces in future breeding programmes.

Chlorophyll stability index and nitrate-reductase activity in relation to dry matter production and drought tolerance: The studies related to chlorophyll stability index and nitrate reductase activity were carried out at VPKAS in relation to dry matter production and drought tolerance. A set of 30 genotypes were evaluated in 1985 for morpho-physiological characters like leaf area, total dry matter, tillers number, ear head characters, yield and nitrate-reductase activity at different stages of plant development and the chlorophyll stability index were taken. Genotypes VL 16, IRAT 102, Jodo, Muzudo and Thapachini were found possessing higher nitrate-reductase activity and total dry matter at active tillering stage. This indicated that higher activity of nitrate-reductase is associated with higher total dry matter production. Total chlorophyll contents were observed higher in IRAT-102, UPRH 211, Khonorullo, L 1058, VHC 1253, VL 16, K 78-13, JC 99 and Muzudo while VRS 163, VL 502, VL 206, L 62G, Khonorullo and VL 16 showed higher chlorophyll index, which indicated that these genotypes have capacity to tolerate moisture stress during growth and development. Similarly in another experiment conducted in 1986, a set of 30 genotypes representing different maturity groups was studied. The nitrate-reductase activity was measured in top leaves at different stages of the plant development. Genotypes VL 16, IRAT 102, Jodo, Muzudo, Daegaldo, VL 8, Hawang Haedo, Thapachini, K 78 -13, L 62-2A, IR 26 and Gangweondo showed markedly higher activity at the vegetative stage (60 days after sowing). All these genotypes (except VL 8) also showed relatively more dry matter accumulation at this stage which indicates that higher NRA is associated with higher dry matter production. Three genotypes, *viz.*, VL 8, K 78-13 and Daegaldo, showed consistently higher NRA till middle of grain filling; however, their yield levels were low due to late maturity and effect of cold. Genotypes which showed less damage because of cold during grain filling and produced higher yields were VL 502, L 12-1A, Bermi, VHC 1253, KH 100-1A and VL 206. Thus, cold tolerance and higher activity of nitrate- reductase are essential parameters for obtaining higher yield levels in rice.

Chlorophyll stability index and water holding capacity of leaves in relation to drought tolerance: In an experiment conducted in 1986 at VPKAS, flag leaves of 30 genotypes were used for studying chlorophyll stability index and % loss of water. Among them six genotypes namely, VRS 163, Khonorullo, VLK 39, Jodo, IR 1846-3064 and L 62G showed over 70% chlorophyll stability and less than 25% moisture loss (Table 8.14).

Table 8.14. Chlorophyll stability index and loss of water in promising rice genotypes.

Genotypes	Chlorophyll stability index %	% loss of water		
		2 hrs	6 hrs	24 hrs
VRS 163	75	21.6	56.7	62.2
Khonorullo	74	18.2	58.4	62.3
VLK 39	73	17.3	51.9	59.6
Jodo	73	20.9	50.7	62.7
IR 1846-3064	72	18.2	47.0	60.6
L 62G	71	25.9	63.0	66.7
L 62-2A	70	27.1	58.3	58.3
Hawang Heado	70	25.0	56.8	63.6
Gangweando	70	18.7	52.1	62.5
Daegaldo	70	26.8	61.0	61.0
VL 206	68	27.4	60.0	62.0
IR 26	63	22.5	50.0	62.5

LR 1A	60	27.3	61.5	64.4
Barmi	60	10.4	55.2	61.2
IRAT 102	58	19.4	56.7	58.8

Drought tolerance in relation to proline accumulation: Proline accumulation is an indication of drought tolerance in many crop genotypes. In 1984 and 1985, fifteen genotypes were studied in pots under glass house conditions (for two years). The proline contents were estimated in the flag leaves under irrigated and stressed conditions. It was observed that VRS 163 and UPRH 211 showed higher proline contents under stressed conditions. All the genotypes possessed similar degree of stress and showed good recovery in the plants.

Basic Studies Related to the Physiological Aspects of Grain Yield

Effect of yield related traits on grain yield

Over last 30 years, many rice genotypes have been studied at VPKAS for their yield related traits. Some of the genotypes found promising for panicle traits were: Angi Kafotsy, H 278-51-1-1, Hexi-15, H 13-4-1-1-2, H 13-4-1-2-1, IR 67087-AC 5-7 and for early duration were: VR 116-1-2, 212, 549-4, 395, VL 92-6020, VL 93-6052, 6300, 6288, 6286, VL 94-3250, K 969-2-1-28, K681-3-2, VL 92-6036, L 56 and SR 12192-T-8. Few genotypes *viz.*, HS 361, Ch 1039, K 680-22-1, EC 401-1595 and EC 401-5079 were observed to take less than 80 days to flower (Anon. 1996-97 to 2008-09).

An experiment was conducted in 1978 under rainfed conditions with six rice genotypes *viz.*, VL 206, Bala (late), KH 863, KH 6199, IR 781-KH 7086 and IR 667-KH 811 to study the effects of different growth parameters on grain yield. Observations were recorded at regular intervals of 7 days on total dry matter production, leaf area index and number of tillers per plant, time taken to flowering and maturity, plant height at harvest, grain weight and number per spikes, grain yield, percentage of effective tillers, distribution index and harvest index. Data on crop growth rate and leaf area index are presented in Table 8.15 (a & b). Dry matter production at different growth stages of growth differs significantly amongst the varieties. VL 206, KH 6199 and KH 863 showed maximum dry matter production. In the case of VL 206 and IR 781-KH 7086, the growth rate showed a marked increase after anthesis, while for IR 667-KH 811 and KH 863 the maximum growth rates were observed at the time of flowering. All the varieties showed a continuous increase in leaf area index up to 90 days.

Table 8.15a. Growth rate among the rice genotypes (rate/sq meter/week).

Varieties	Days after sowing					
	55	62	69	76	83	90
VL 206	0.182	0.114	0.165	0.172	0.100	0.305
Bala	0.165	0.068	0.083	0.186	0.274	0.185
KH 6199	0.129	0.073	0.050	0.338	0.258	0.237
IR 781-KH 7086	0.084	0.090	0.063	0.348	0.178	0.109
IR 667-KH 811	0.070	0.138	0.131	0.205	0.120	0.299
KH 863	0.075	0.174	0.240	0.298	0.170	0.690
Mean	0.118	0.107	0.105	0.260	0.183	0.217
C.D. at 5%	0.063	0.063	N.S.	0.093	0.117	0.129

Table 8.15b. Leaf Area index among the rice genotypes.

Varieties	Days after sowing						
	48	55	62	69	76	83	90
VL 206	1.48	2.79	4.50	6.94	6.95	7.54	8.72
Bala	1.44	1.80	5.03	5.95	5.95	8.27	9.69
KH 6199	1.15	1.53	3.44	4.60	4.60	6.68	8.79
IR 781-KH 7086	1.62	2.33	4.91	6.34	6.34	6.88	7.34
IR 667-KH 811	1.90	1.89	3.59	4.41	4.41	5.24	7.58
KH 863	1.21	2.16	4.42	4.66	4.66	5.21	7.96
Mean	1.46	2.12	4.32	5.49	5.49	6.63	8.39
C.D. at 5%	N.S	N.S	N.S.	1.86	N.S.	2.09	N.S.

In another experiment, a set of 100 genotypes belonging to different maturity groups were evaluated in 1981 under transplanted condition using 40 days old seedlings. Data were recorded on days to anthesis, period for grain filling and percent spikelet fertility. The duration for anthesis was found to range from 79 days in Shinei to 123 days in Gogo puthi and Hung-80 and, for grain filling duration from 28 days in Shoanan to 48 days in IR 26 and Taichung 176. It was found that out of the total duration of rice plant, 71 to 74 days can be attributed to the formation and development of plant vegetative parts and only 26 to 29% to grain production. The genotypes at the Almora location (1600 m altitude) flowered from last week of August to 1st week of October. These data indicate that large variation exists among the cultures for grain filling duration (Anon. 1981-82). Similarly Bist *et al.* (2007) collected 13 landraces of rice from Nankoshi microwatershed in Almora district of Uttarakhand and evaluated the diversity among them. They found variations for traits related to grain and straw yield among the collected genotypes.

Phasic developmental studies in rice

A preliminary experiment was conducted at VPKAS in 1978 using eleven rice genotypes to study the length of duration of different developmental stages in rice. The germination took place within five to seven days of sowing. Gaza 14 had the earliest spike initiation (43 days after germination) and Saca took the longest (68 days after germination). Traits like time taken for spike development, period between termination of spike development and anthesis differed greatly among the genotypes. A similar experiment was conducted in 1980 to collect preliminary information on the duration of various developmental stages within a set of five rice genotypes. Observations were recorded on two components, days to panicle initiation (days from germination to panicle initiation) and days from panicle initiation to heading (Table 8.16). Differences among cultures for both these components were highly significant. The average contribution of the duration from germination to spike initiation was 50.5 days. It was markedly higher than the duration from floral initiation to earhead emergence (37.2 days). The situation was similar in all the varieties (Anon. 1978-79).

Table 8.16. Duration of the developmental stages among the rice genotypes.

Sl. No.	Genotypes	Days to panicle initiation (GI)	Days from panicle to earing (IE)	Total days to flowering (GI +IE)
1	VL 8	58.8	46.8	102.6
2	VLK 39	48.0	37.5	83.5
3	VL 501	44.0	40.8	84.8
4	VL 502	49.5	36.3	85.8
5	JC 99	41.5	34.8	75.3
6	RPKN 2	58.3	40.5	98.8
7	Jodo	47.3	34.3	81.6
8	Norin-18	57.5	39.0	96.5
9	Reemi	40.5	45.3	85.8

10	Bermi	51.5	27.3	78.8
11	Himdhan	57.3	38.5	95.8
12	Thapachini	54.3	32.3	86.6
13	Khatiyali local	50.8	35.5	86.3
14	L-K-1	53.5	34.0	87.5
15	L 56	48.3	37.5	85.8
	Mean	50.54	37.23	87.7
	C. D. at 5%	1.02	2.16	

A set of 15 genotypes representing two different maturity groups was studied during 1983 at Almora (Anon. 1983-84). Six distinct developmental phases were identified as seedling to germination (SG phase), germination to spike primordial /initiation (GI phase), spike primodial initiation to last spikelet formation (IT phase), ear emergence to anthesis (EA phase) and anthesis to grain maturity (AM phase). Different character expressions of these genotypes are presented in Table 8.17. The SG phase did not show significant difference among the genotypes. However differences for other characters among the genotypes were significant. The GI phase was obtained to be the longest and ranged from 40 days in Reemi to 58 days in RPKN-2, followed by AM phase, 39 days in L-12-1A to 56 days in RPKN-2. The shortest duration was recorded in EA phase which ranged from 2 day in Bermi and L 56, to 5 days in RPKN-2. All the genotypes suffered from low temperatures prevalent during grain filling and expressed high spikelet sterility due to delayed flowering which was caused by longer durations of all the phases (Table 8.17). Total dry matter production (TDM), taken at ear emergence stage was the lowest in Reemi (2.26g/plant) and highest in Khatyari local (9.52 g/ plant). Genotypes which had longer GI phase duration, in general, recorded higher TDM. The unproductive tillers remained the lowest in short duration genotypes like Reemi and Jodo and the highest in RPKN-2 while in the late varieties the productive tillers were more in VL 501 and Bermi. Main ear grain weight and total plant yield were recorded maximum in Khatyari local followed by L 12-1A. The overall expression of these characters indicated the longer duration of IT phase with around 50 days duration of GI phase had favourable impact on grain yield in rice.

Table 8.17. Expression of different physiological characters among rice genotypes.

Variety	GI Phase (days)	IT phase (days)	TE phase (days)	EA phase (days)	AM phase (days)	Un-productive tillers/ plant	Productive tillers/ plant	Main ear grain wt (g)	Grain yield/ plant (g)	DM at earing (g/ plant)
VL 8	56.0	10.2	39.0	3.0	50.5	6.51	2.50	0.68	0.68	7.16
VLK 39	45.0	11.0	25.0	1.1	44.0	1.78	5.00	2.50	7.15	5.56
VL 501	41.0	11.4	30.0	1.3	50.0	5.90	1.90	1.90	5.20	5,31
VL 502	51.0	10.9	24.0	1.2	44.8	2.10	2.18	2.18	6.50	6.34
JC 99	43.0	11.0	23.2	5.0	55.6	3.10	2.95	2.95	6.25	3.44
RPKN-2	58.0	9.5	42.0	5.2	56.0	8.62	0.15	0.15	0.15	7.96
Jodo	47.0	9.9	22.0	2.0	43.0	0.92	3.20	3.20	8.95	3.54
Norin 18	57.7	11.0	18.7	3.3	51.6	6.95	0.50	0.50	0.50	6.36
Remi	40.0	10.0	30.0	1.4	44.0	0.71	2.55	2.55	8.60	2.26
Berml	47.7	8.0	20.0	1.0	49.0	0.68	2.15	2.15	8.50	5.50
Himdhan	58.0	9.7	31.0	3.7	44.0	5.12	2.00	2.00	7.90	7.98
Thapachini	52.0	12.0	24.0	3.4	43.0	4.10	1.40	1,40	8.52	8.52
Khatyari Local	51.0	12.0	23.0	2.2	44.0	1.80	3.52	3.52	10.20	9.52
L 12-1A	50.0	12.2	20.8	2.6	38.0	2.12	3.25	3.25	9.85	9.31
L-56	48.0	10.9	23.0	1.0	42.1	2.70	2.35	2.35	9.00	5.21
CD at 5%	2.9	1.3	2.4	0.6	3.4	1.64	0.98	0.98	1.91	1.26

SG phase: Seedling to germination; GI: Germination to spike primordial /initiation; IT phase: Spike primordial initiation to last spikelet formation; EA phase: Ear emergence to anthesis; AM phase: Anthesis to grain maturity.

Studies on grain sterility and other related traits

Grain sterility in rice

Grain sterility is one of the very serious problems in rice. The magnitude of sterility varies from variety to variety. Several factors have been suggested as the possible reasons for grain sterility and include very high or low temperature, long photoperiod or possibly limitation of photosynthates. With this objective in view, exploratory studies were conducted in 1978 and three experiments were conducted at VPKAS to study the (i) effects of different dates of sowing D_1 (12th June) and D_2 (22nd June); (ii) the effect of photoperiod normal day photoperiod (ND) *vs* short day photoperiod (SD, eight hours light) and (iii) the effect of population density on grain sterility. Though there were wide differences between the sowing dates and photoperiodic treatments, the overall effect of the two treatments did not fall in any pattern. The change in the sowing date (from D_1 to D_2) or photoperiod (ND to SD) caused an increase in grain weight per spike.

Testing photosensitivity of local strains

A set of 85 short duration local collections was evaluated in 1980 for their response to photosensitivity. The nursery was planted in two sets, one under extended photoperiod and the other under normal photoperiodic situation (Annon. 1980-81). The extended photoperiod was created by lighting incandescent bulbs throughout the night with the light intensity of about 3000 lux at the farthest point from the source of light. With the extended photoperiod, six of the cultures failed to flower while twelve flowered very late but failed to set seeds. Several others showed varying degrees of enhancement in the duration of flowering. Only a few cultures showed stability in flowering under both the photoperiodic situations. The photoinsensitive type included genotypes *viz.*, VHC 240, VHC 243, VHC 265, VHC 685, VHC 1121, VHC 1141, L 1-1A and China 1039.

Screening of paddy genotypes for tolerance to Fe deficiency

In 1977, a set of twenty genotypes of paddy were screened for tolerance to Fe deficiency *in vitro* by the use of excised embryo. It was observed that some of the genotypes (VL 203 and VL 224) did not germinate in the absence of Fe while genotypes like Kalinga I, VTE-8, IET-3359 showed better growth. Maximum height, root number and number of leaves were recorded in Kalinga I under Fe.

Evaluation of Rice Genotypes for Quality Traits

Quality for rice includes the shape, size and chemical compositions of rice grains. The ratio of length and breadth of rice grain also influences its acceptability and cost, hence contributes significantly to the quality parameters. From 1983 to 2008 many rice genotypes have been evaluated for different quality traits (Anon. 1983-84 to 2008-09). Thapachini is a local cultivar suitable to the hills of the Himalayan states. In 1983, a set of four varieties of rice *viz*., VL 16, VL 18, VL 8 and Thapachini were evaluated for their quality parameters. VL 18 recorded maximum grain length (6.10 mm), L/B ratio (2.84), elongation (8.0 mm) in addition to the highest protein content (8.22%). It also had medium grain shape and seemed to possess intermediate amylase content as the rice cooked dry and puffy and retained soft structure on cooking. Thus, VL 18 was found most suitable for cooking and eating qualities. VL 8 with short and bold grains not only recorded the minimum L/B ratio but also lowest protein (6.13%) and did not confer good cooking quality (Table 8.18). Similarly in 1987, three genotypes namely, VL 501, VL Dhan 161 and Bala were evaluated for grain quality characteristics. It was found that VL Dhan 163 recorded more protein content (9.30%) than VL 501 (9.1%) and Bala (8.98%). The starch gelatinization temperature in VL Dhan 163 was, however, low as compared to VL 501 and Bala, which show intermediate gelatinization temperature.

Table 8.18. Cooking characteristics of selected rice genotypes.

Variety	Length (mm)	Breadth (mm)	L/B raio	Size and shape of the grain	Time for cooking (mins.)	Ratio of expansion (cooked to raw rice)	Elongation (mm)	1.7% KOH test Scores	Gelatinization temp.	Protein (%)
VL 16	5.75	2.70	2.13	medium	30	2.92	7.5	3	Intermediate	7.01
VL 18	6.10	2.15	2.84	Medium	35	3.43	8.0	6/7	low	8.22
VL 8	4.50	3.00	1.50	Short bold	35	3.13	6.0	6/7	low	6.13
Thapachini	5.60	2.60	2.15	medium	25	3.53	7.0	4/5	intermediate	7.34

In another experiment conducted in 2006, 18 genotypes were evaluated for their protein, P, K, Fe, Zn, Cu and Mn content. The protein content varied from 5.32% in VLD 61 to 9.34% in VLD 221. P and K content varied between 0.21- 0.34 and 0.29- 0.49%. Highest K was observed in VLD 62 and VLD 208 respectively. Iron and Zinc content ranged between 24.1- 40.1 and 7.2-15.0 ppm. Highest Fe and Zn were observed in VLD 82 and VLD 208 respectively (Table 8.19, Anon. 2006-07).

Table 8.19. Micronutrient and pasting properties of rice varieties.

Variety	Protein (%)	Ash (%)	P (%)	K (%)	Fe (ppm)	Zn (ppm)	Cu (ppm)	Mn (ppm)
VLD 8	6.38	1.28	0.317	0.34	36.3	13.07	1.24	6.47
VLD 16	9.22	0.78	0.214	0.305	33	9.68	1.54	6.97
VLD 61	5.32	0.80	0.275	0.312	24.06	11.05	2.14	5.64
VLD 62	7.68	-	0.34	0.293	37.99	9.64	1.62	7.29
VLD 65	5.47	0.83	0.282	0.295	24.67	7.45	1.18	4.27
VLD 81	8.63	0.88	0.284	0.3	32.61	10.96	1.87	7.24
VLD 82	7.42	0.50	0.269	0.303	40.06	11.07	2.52	6.16
VLD 85	9.23	0.78	0.225	0.373	30.24	11.96	1.88	5.55
VLD 86	6.71	-	0.248	0.315	28.46	14.19	2.83	6.23
VLD 84	5.36	1.15	0.271	0.307	35.96	8.84	1.47	9.65
VLD 154	8.35	-	0.242	0.300	29.95	9.72	1.16	5.82
VLD 207	8.14	0.75	0.244	0.315	29.13	13.46	1.94	10.45
VLD 208	8.27	1.00	0.265	0.485	25.98	15	4.88	31.49
VLD 209	6.94	0.50	0.298	0.297	24.16	9.94	1.39	2.91
VLD 221	9.34	0.93	0.256	0.32	25.71	7.23	1.86	6.37
VL 30424	7.97	0.88	0.284	0.287	35.07	9.54	1.82	10.44
VL 30425	7.19	0.45	0.271	0.335	30.1	8.94	1.02	4.80
VL 3831	7.01	0.90	0.258	0.297	32.56	8.09	1.27	3.06

A set of 17 rice genotypes (Table 8.20) were evaluated in 2006 for pasting properties and found that the rice genotypes widely varied for the pasting properties of starch when RVA analysis was done. Peak viscosity varied from 44.67- 220.17. Peak viscosity indicates the water binding capacity of starch. VLD 8 and VLD 61 have lowest and highest water binding capacity respectively. Setback value of VLD 65 indicates firmer and drier cooked texture of whole rice (Anon. 2006-07). Besides this, wide variation was observed in most of the physicochemical properties (Table 8.21). The apparent amylose content ranged from 17.4 - 26.3%. Length and width also showed variation across the varieties. Maximum and minimum length was found in VL 30424 (7.85mm) and VLD 8 (4.87mm). Similarly wide variation was found for 100 grain weight, ranging between 1.41- 2.29 g.

Table 8.20. Pasting properties of rice varieties.

Variety	Peak viscosity	Hot paste viscosity (RVU)	Cool paste viscosity	Setback
VLD 8	44.7	45.6	114.3	69.7
VLD 16	149.8	151.7	318.8	169.1
VLD 61	220.2	191.5	402.3	182.1
VLD 62	179.8	181.3	369.9	190.1
VLD 65	166.2	167.0	412.3	246.1
VLD 81	179.1	179.1	364.3	185.2
VLD 82	193.6	152.8	372.1	178.5
VLD 85	58.3	58.6	122.0	63.8
VLD 86	154.1	154.6	335.7	181.6
VLD 154	139.2	140.8	321.6	182.4
VLD 207	75.4	75.3	159.5	84.1
VLD 208	85.3	85.8	180.8	95.5
VLD 209	114.1	94.3	192.7	78.6

Table 8.21. Physico-chemical properties of rice varieties.

Variety Amylose content (%)	100 grain wt (g)		length (mm)	Width (mm)
VLD 8	2.09	4.87	-	17.40
VLD 16	1.98	6.45	2.57	22.04
VLD 61	2.29	7.37	2.59	25.16
VLD 62	1.86	6.12	2.52	20.64
VLD 65	1.80	6.81	2.09	20.77
VLD 81	2.11	7.47	2.30	22.03
VLD 82	2.14	6.46	2.51	21.24
VLD 84	2.04	5.78	2.58	21.43
VLD 85	1.41	6.15	2.34	20.38
VLD 86	2.04	5.97	2.61	19.75
VLD 207	1.64	5.65	2.45	23.44
VLD 208	1.60	6.07	2.51	23.66

VLD 209	1.70	5.76	2.44	26.25
VLD 221	1.69	6.30	2.39	19.67
VL 30424	2.21	7.45	2.13	20.72
VL 30425	2.22	7.85	2.13	19.56
VL 3831	1.51	5.43	2.23	20.94

Rice Germplasm Collected and Used for the Development of Cultivars

VPKAS, Almora has a total collection of rice germplasm of 1767 genotypes which are maintained in the gene bank of VPKAS, Almora. Out of this 789 genotypes are also maintained at the long term storage at the NBPGR, New Delhi. It includes 116 exotic collections. Among the genotypes maintained at VPKAS, the days to 50% flowering ranged from 68 days (KHH 154) to 128 days (KHH 330). Similarly the range for panicle length was 10.7 cm (AKS 63) to 33.83 cm (VOHP-CR-3121) and for plant height was 55.33 cm (KJ-5) to 168.67 cm (VOHP-CR-3120).

Out of this collection, many genotypes were released as cultivars and utilized in breeding programme as donors. A total of 17 rice genotypes have been released for the hills of Uttarakhand from VPKAS (Table 8.22). Those include genotypes suitable for rainfed spring sown, rainfed June sown and irrigated ecosystem. Many of the local and Chinese lines were found to be good combiners resulting in varieties suitable to this region. Even if VL Dhan 221 is a cultivar suitable for the rainfed June sown condition, the crosses involving this genotypes resulted in cultivars suitable for spring sown (VL Dhan 209) and irrigated (VL Dhan 82, VL Dhan 85, VL Dhan 86) conditions.

Table 8.22. Varieties released from VPKAS for the hills of Uttarakhand state.

Sl. No.	Name	Parents	Year	Released by	Ecosystem
1.	VL Dhan 206	Bamni (local)	1983	SVRC	Rainfed Upland Spring sown
2.	VL Dhan 207	VL Dhan 206 x Annada	2006	SVRC	Rainfed Upland Spring sown
3.	VL Dhan 208	VR 410-19 x VR-212	2006	SVRC	Rainfed Upland Spring sown
4.	VL Dhan 209	Himdhan x K-39 x VL Dhan 221	2006	SVRC	Rainfed Upland Spring sown
5.	VL Dhan 163	IR 2053 x Kn-16-767 x	1987	SVRC	Rainfed Upland

		Ch-1039			June sown
6.	VL Dhan 221	IR 2053-521-1-1-1 x Ch-1039	1992	CVRC	Rainfed Upland June sown
7.	Vivek Dhan 154	VL Dhan 221 x VL- 24	2006	SVRC	Rainfed Upland June sown
8.	VL 8*	Selection from Kaoshuing 22	1969	SVRC	Irrigated late
9.	VL Dhan 16*	JP-5 x YRL-1	1984	SVRC	Irrigated Medium
10.	VL Dhan 61	Jaya x Ta-boo-choz	1997	CVRC	Irrigated Medium
11.	Vivek Dhan 62	China-4 x BG-367-4	2000	CVRC	Irrigated Medium
12.	VL Dhan 65	Himalaya-2 x VL Dhan 16	2006	SVRC	Irrigated Medium
13.	VLK Dhan 39*	Ch-1039 x IR 580-19-2-3-2	1980	SVRC	Irrigated Early
14.	VL Dhan 81	China 988 x HPU-741	1999	CVRC	Irrigated Early
15.	Vivek Dhan 82	VL Dhan 221 x UPR-82-1-7	2001	CVRC	Irrigated Early
16.	VL Dhan 85	HPU-799 x VL Dhan 221	2006	SVRC	Irrigated Early
17.	VL Dhan 86	VL Dhan 16 x VL Dhan 221	2006	CVRC	Irrigated Early

Utilization of Donors in Molecular Breeding

Pyramiding blast resistance genes for durable resistance in elite cultivars VL Dhan 206 and Basmati 370

Sridhar *et al.* (1999) screened a number of isogenic lines with genes in different combinations for resistance against blast. Those included *Pi* 1, *Pi* 2, *Pi* 4 and *Pi* 9 (Table 8.23). Among them *Pi* 1, *Pi* 2 and *Pi* 9 were found effective against blast at Almora. Since linked markers are available for those genes, marker assisted selection using DNA markers is a viable approach for pyramiding two or more genes for durable resistance. Based on this result and so many years of work at VPKAS, Almora, we chose two genes, *Pi* 2 and *Pi* 9 for pyramiding them under two different backgrounds, VL Dhan 206 and Basmati 370. Both these resistance genes have been mapped and site specific markers are available for both the genes. The donor used for *Pi* 2 was C101A51 and the donor for the *Pi* 9 was *O. minuta* derivative (WHD-IS-75-1-127). Both the primers (RG 64 for *Pi* 2 and P 28 for *Pi* 9) were found polymorphic to the recipient parents VL Dhan 206 and Basmati 370. The PCR product amplified by the primer RG 64 need to be restricted using Hae III in order to get the polymorphism since it generates a cleaved amplified polymorphic sequence (CAPS). Six populations with different cross combinations (Table 8.24) were used for the marker assisted selection. The rice plants with both these gene are nearly fixed and were found promising. Many of these lines score less than 3 in a scale of 0 to 9, under field conditions. These lines are available for the multilocational testing (VPKAS Annual Report, 2008-09).

Table 8.23. Reaction of near isogenic lines at different locations of India.

Lines/ Cultivars	Resistant gene(s)	Locations					
		Cuttack	Jagdal-pur	Ambik-apur	Almora	Jorhat	Suttur
O. minuta derivative (WHD-IS-75-1-127)	*Pi* 9(t)	1	3	1	4	1	4
C101 LAC	*Pi* 1(t)	5	2	1	5	1	1
C101 A51	*Pi* 2(t)	5	3	2	4	3	4
C101 PKT	*Pi* 4a(t) (*Pi*-ta)	5	5	5	4	4	5
C102 PKT	*Pi* 4a(t) (*Pi*-ta)	4	4	5	5	3	4
C104 PKT	*Pi* 3(t)	5	5	5	5	5	5
BL 122	*Pi* 1(t), *Pi* 2(t)	1	3	1	2	1	4
BL 142	*Pi* 1(t), *Pi* 4(t)	2	5	1	4	5	1
BL 245	*Pi* 2(t), *Pi* 4 (t)	1	2	1	1	3	1
Karuna	Nd	5	5	5	5	3	5
HR 12	nd	5	nt	5	5	Nt	nt
Co 39	*Pi*-a (Co 39)	5	5	5	5	2	5
Kalinga III	nd	5	4	5	5	2	4
Vandana	nd	5	4	5	5	3	5

Source: Sridhar *et al.* (1999), nd = not detected. nt = not tested, ng = not germinated, scale 1 to 5

Table 8.24. Populations used for MAS of *Pi* 2 and *Pi* 9.

S. No.	Crosses
1.	(VL 206 x *O. minuta*) x (VL 206 x C101A51)
2.	VL 206 x (VL 206 x C101A51) x *O. minuta*
3.	(VL 206 x C101A51) x (VL 206 x *O. minuta*)
4.	Bas 370 x (Bas 370 x *O. minuta*) x (Bas 370 x C101A51)

5.	Bas 370 x (Bas 370 x C101A51) x *O. minuta*
6.	Bas 370 x (Bas 370 x C101A51) x (Bas 370 x *O. minuta*)

References:

Agnihotri RK, Palni LMS, Pandey DK. 2007. Germination and seedling growth under moisture stress: Screening of landraces of rice *(Oryza sativa* L.) from Kumaun Region of Indian Central Himalaya. *Journal of Plant Biology* 34:21-27.

Anonymous. 1950-51. CRRI Annual Report, Cuttack, India.

Anonymous. 1982-83 to 2009-09. VPKAS Annual Reports, Almora, India.

Bist V, Singh B, Rao KS, Maikhuri RK and Nautiyal AR. 2007, Genetic divergence of paddy landraces in Nanakosi micro-watershed of Uttarakhand Himalaya. *Journal of Tropical Agriculture* 45:48-50.

Fertilizer Statistics. 2008.

Gangadharan G. 1985. Breeding. Pages 73-109 in Padmanabhan, SY (ed) *Rice Research in India.* Indian Council of Agricultural Research, New Delhi.

Ganguli D, Padmanabhan SY (1959) Helminthosporium Diseases of Rice III. Breeding resistant varieties - Selection of resistant varieties. *Indian Phytopath* 12: 99-110

Sridhar R, Singh UD, Agrawal PK, Reddy JN, Sanger RBS, Bhatt JC, Rathaiah Y, Rao KVSRK. 1999. Usefulness of blast resistance genes and their combination in different blast endemic loacation in India. *Intern Rice Res Notes* 24(2): 22-24.

VPKAS. 1982-83 to 2008-2009. Annual Report, 2008-09. Vivekanand Parvatiya Krishi Anusandhan Sansthan, Almora.

Genetic Resources of Rice in India : 323-346
Editor : S. D. Sharma
Today & Tomorrow's Printers and Publishers, New Delhi - 110 002 (India)

Chapter 9

Exploration, Collection, Conservation and Utilization of Wild Rices for the Improvement of Cultivated Rice

S.D. Sharma and B.C. Patra

There are archaeological evidences that wild rices were harvested and consumed by the people of middle Gangetic plains in Neolithic times much before rice was domesticated. Wild rices were used in rituals by the Aryans since the late Vedic period. According to Ashvaghosha (a poet of first century AD) and Kalidasa (a poet of fourth century AD), wild rices were considered as food for ascetics. Wild rices are harvested even today by the tribal populations of Orissa and Chhattisgarh and sold in the market (Mehra, 2005; Sattar *et al.,* 2010). On days of fasting, many people consume wild rice and call it god-given rice *deobhat.*

The modern scientific approach to study wild rices in India started with the arrival of British plant explores in the nineteenth century. These plant explorers started collecting the specimens of wild rices from different parts of the country and depositing them either in the Indian herbaria (at Calcutta and Saharanpur, the latter was later shifted to Dehradun) or sent them to Kew, UK. Specimens of wild rices collected in India in the nineteenth century and deposited in the Indian herbaria were studied by the present authors (Sharma, 1964; Patra, 2008). It reveals that the three main wild species of rice in India had already been collected by plant collectors before 1850 (Table 9.1). The number of specimens of wild rices collected from different parts of India and deposited in herbaria increased progressively as we move forward in the time scale.

Table 9.1. Specimens of wild rices collected by plant collectors in India in the nineteenth century as available in the Indian herbaria.

Sl.No.	Species	Name of collector(s)	Collectors No.	Period of collection	Locality	Herbarium ACC No.	Remarks
1	*O. rufipogon*	Wallich	1030	Oct 1815		DD	*Frequens ad margins aqauarum stagnanteum. O. sativa spontanea*
2	*O. granulata*	W. Wight		Aug 1835	Peninsular India	CAL	
3	*O. officinalis*	W. Wight	3026	Dec 1849	Ganjam, Orissa	CAL	
4	*O. granulata*	S. Kurz		Oct 1868	Little Ranjit	CAL	
5	*O. granulata*	C.B. Clarke 1876	24861	Nov 18,	Parasnath, Bihar	CAL	
6	*O. granulata*	C.B. Clarke 1883	33832	Oct 11,	Hazaribagh, Bihar	CAL	
7	*O. nivara*	J.F. Duthie	6788	Nov 5, 1887	Silori, Chanda Dist, Maharashtra	DD	
8	*O. nivara*	A.E. Lowrie	9975	Oct 16, 1889	Chanda, Maharashtra	DD	
9	*O. nivara*	J.F. Duthie	9976	Dec 3, 1889	Wardha, Maharashtra	DD	
10	*O. rufipogon*	J.F. Duthie	9978	Dec 5, 1889	Bajrighat Tank, Chanda Dist, Maharashtra	DD	
11	*O. officinalis*	J.F. Duthie	9977	Dec 12, 1889	Chanda Dist, Maharashtra	DD	

12	*Porteresia coarctata*	R. Ellis X1		June 30, 1891	Chhar Gang, Sundarbans	CAL
13	*Porteresia coarctata*	R. Ellis X3		Jul 11, 1891 mouth, Sundarbans	Kaga Boga River, 3 miles from its	CAL
14	*Porteresia coarctata*	R. Ellis X4		Jul 20,	Sundarbans 1891	CAL
15 Dap	*O. rufipogon*	Radha Ballav		5.9.1891	Sundarbans	CAL 15
16 1894	*O. rufipogon*	J.D. Nusker	1132	Oct 12,	Raniganj, Bihar	CAL 16
17	*O. nivara*	Mokim	1381	Oct 1894	Gaya, Bihar	CAL, DD
18	*Porteresia coarctata*	J.D. Nusker	1318	Sept 11, 1894	Malta	DD BSI-Poona
19	*O. nivara*	G. King Central India			Portye tank,	CAL
20	*O. nivara*	G. Watt	7881		Bihar	DD
21	*O. granulata*	J.D. Hooker			Sikkim	CAL

NB: sn = without any collector's number, nd = without any date, CAL= available at Central National Herbarium, Calcutta, DD = available at Forest Research Institute Herbarium, Dehradun.

These collections helped the flora writers to delimit the species of *Oryza*, describe them and enumerate the species of *Oryza*. The earliest such efforts were done by Roxburgh (1832). Describing the wild rices (*O. nivara*) of Bhandara district of Maharashtra, Watt (1891:626) says:

> *It springs up of itself in swamps and other marshy places and is collected by the lower classes who may be seen sweeping with a piece of stick, this self-sown rice into their basket. The gathering season is rather early, as the constant supply of water quickens the growth. This* deodhan *is bartered to dealers and other well-to-do classes for regularly grown rice. Sometimes a tithe of* deodhan *sweeping is made over its collector to the village landlords. Their superiors are glad to get hold of it, as it comes into use on the Risshepunchum* [Rishi Panchami] *(in August) and Shirawunderwadsee* [Shravan Dwadashi] *celebrated 11 days after the Risshepunchum. ——. In the times of Mahratta dynasty,* deodhan *was specially distributed from the government storehouse* (ambarkhana).

After Watt (1891), Hooker (1897) has described the wild *Oryza* species of India. Subsequently, regional flora writers have described the wild species of *Oryza* available in their respective compendiums (Prain, 1903; Haines, 1928; Gamble, 1934).

Our present knowledge about the species of *Oryza*, their inter-relationships, evaluation of their characters and scope of utilization of different species in the improvement of cultivated rice is due to the collective efforts of scientists of many countries of the world. Though Indian scientists have made their own contributions in this field, their efforts cannot be viewed in isolation. A general view of the situation especially with regard to their taxonomic delimitation, species relationships and evolutionary trends has been presented by Sharma (2003, 2005).

In so far as the study of wild rices is considered, the period from 1891 up to the present day may roughly be split into three phases. The first phase started in 1891 when Watt (1891) enumerated Indian species of wild rices in his *Dictionary of Economic Products of India.* In this phase, morphology, geographical distribution and taxonomy of the species were studied and finally monographs on genus *Oryza* were published by Roschevics (1931) and Chevalier (1932). The period from 1933 to 1963 marks the second phase when cytogenetics and genome relationship of species were studied mainly by the Japanese scientist Toshitaro Morinaga and his associates. The Indian rice scientists made their own contribution in these fields. This phase culminated with a symposium on Genetics and Cytogenetics of Rice organized in 1963 at IRRI where the work done up to that time was reviewed and genomic symbols were standardized (IRRI, 1964). The third phase i.e. the phase of exploration, collection, conservation, evaluation of wild rices for their desirable traits and utilization started in 1964 and has continued up to the present times. However, it must be clearly understood that taxonomic and biosystematic studies of *Oryza* species have continued up to the present time and some evaluation and utilization of wild rices was done even prior to 1963.

Based on the studies made so far, the following species of *Oryza* are available in India:

O. sativa L. is the Asian cultivated rice. It is widely cultivated in the sub-continent. This species has enormous morphological variation and has also ecotypic differentiation. There is another cultivated species of rice (*O. glaberrima*) that is confined to West Africa only.

O. nivara Sharma *et* Shastry is an annual wild species. It grows in seasonal pools and roadside ditches in the Deccan plateau and also in the middle Gangetic plains. It was earlier known as *O. fatua* or as *O. sativa* var. *fatua* by flora writers. Some rice biosystematists do not recognize it as a separate species; according to them, it is a form of *O. rufipogon* only.

O. rufipogon Griff. is a perennial species. It grows in swamps and on edges of ponds in coastal plains and in lower valleys of Ganga and Brahmaputra. It has been recognized as a distinct species only since 1950s. It was, however, initially identified as *O. perennis.*

O. officinalis Wall. *et* Watt, is comparatively a tall plant growing in forests shades of eastern and northeastern India and also in Western Ghats.

O. malampuzhaensis Krishn. and Chandr. was collected from Kerala and Tamil Nadu only in 1958. Some taxonomists consider it as a subspecies of *O. officinalis* only.

O. granulata Nees *et* Arn. *ex* Watt is a small plant that grows in forest shades especially in Eastern India. *Oryza indandamanica* that has been reported from Andaman Islands is a form of *O. granulata* only.

Porteresia coarctata (Roxb.) Tateoka grows on coastal regions in tidal swamps. It was earlier treated as a species of genus *Oryza* and was called *Oryza coarctata* Roxb.

The cultivated rice *O. sativa* and the two wild species *O. nivara* and *O. rufipogon* are closely related. Besides, the cultivated rice *O. sativa* hybridizes in nature with *O. nivara* and also with *O. rufipogon.* These natural hybrids and their intergrades are generally known as *O. sativa* f. *spontanea.* Since these species and the various intergrades of their hybrids are difficult to be distinguished in natural populations, they are sometimes collectively grouped as *rufipogon-nivara-spontanea* complex. Table 9.2 provides the chromosome numbers, genomes, ecology and geographical distribution of species of Oryza. This chapter will deal mainly with exploration, collection, conservation, evaluation and utilization of wild rices as genetic resources for the improvement of cultivated rice mainly in an Indian perspective. However, the international scenario will also be presented if it has a direct bearing on our understanding of the species relationships among wild rices and the scope of their utilization in the genetic improvement of rice in India.

Table 9.2 Species of genus Oryza, their chromosome number, genome, ecology and geographical distribution.

Species	Chromosome no. (2n)	Genome	Ecology	Geographical distribution
O. alta Swallen	48	CCDD	Partial shade	Central and south America
O. barthii A. Chevalier	24	AA	Open habitat	Tropical Africa
O. brachyantha A. Chev. *et* Roehr.	24	FF	Open habitat	West and Central Africa
O. eichingeri A. Peter	24	CC	Partial shade	Equatorial Africa
O. glaberrima Steud.	24	AA	Open habitat	Tropical west Africa
O. glumaepatula Steud.	24	AA	Open habitat	Central and south America
O. grandiglumis (Doell) Prod.	48	CCDD	Partial shade	Central and south America
O. granulata Nees *et* Arn. *ex* Watt	24	GG*	Shade	South and Southeast Asia
O. latifolia Desv.	48	CCDD	Partial shade	Central and South America
O. longiglumis Jansen	48	HHJJ*	Partial shade	Southeast Asia
O. longistaminata A. Chev. *et* Roehr.	24	AA	Open habitat	Tropical Africa including Madagascar
O. malampuzhaensis Krishn. *et* Chandr.	48	BBCC	Partial shade	South India

O. meridionalis Ng	24	AA	Open habitat	Australia and Indonesia
O. meyeriana (Zoll. *et* Mor. *ex* Steud.) Baill.	24	GG*	Shade	Southeast Asia
O. minuta J. S. Presl *ex* C. B. Presl	48	BBCC	Partial shade	Philippines
O. nivara Sharma *et* Shastry	24	AA	Open habitat	South and southeast Asia including south China
O. officinalis Wall. *ex* Watt	24	CC	Partial shade	South and southeast Asia including south China
O. punctata Kotschy *ex* Steud.	24	BB	Open habitat	Sudan and Ethiopia
O. ridleyi Hook. f.	48	HHJJ*	Partial shade	Southeast Asia
O. rufipogon Griff.	24	AA	Open habitat	South and southeast Asia, south China
O. sativa L.	24	AA	Open habitat	South and southeast Asia including south China
O. schlechteri Pilger	48	HHKK**	Shade	Papua New Guinea
O. schweinfurthiana Prod.	48	BBCC	Partial shade	Tropical Africa

* Aggarwal *et al.* 1997; **Ge *et al.* 1999

As back as 1921, Roy collected samples of wild rices of *rufipogon-nivara-spontanea* complex from Central Provinces and Berar (mainly from Chhattisgarh) and classified them into 24 groups. Majority of his collections bred true whereas some segregated. He discarded the general view prevailing among the farmers that cultivated rices if left uncultivated turn into wild rices. Bhalerao (1928) studied the wild rices (*O. nivara*) of Kolaba and Ratnagiri districts of Maharashtra and Belgaum and Dharwar districts of Karnataka though he identified them merely as 'wild forms of *O. sativa*'. He described them as annual aquatic growing "in marshy areas, in small pools and ponds and on the margins of big tanks". He mentions about their seed dormancy. He also mentions about their early flowering habit, copious tillering, shedding habit, small panicle, long and very rough awn (10-12.5 cm) and uneven maturity of their panicles. He states that the "seeds seem to require a long period of rest before they can germinate." He also remarked that due to natural crossing, every cultivated variety in the region has its own grain shedding type. It is evident that the above two authors were correctly describing the annual wild species (*O. nivara*) and its natural hybrids with the cultivated rice. Mitra and Ganguly (1932) reported the perennial wild rice (*O. rufipogon*) from Surma valley and have even provided a photograph of the plant (their Fig 1, Plate XXVI).

Cytology of wild species was studied by many workers in India. Interspecific hybrids of Oryza species were also produced and their cytology was studied. Mitra and Ganguly (1932) studied the inheritance of characters in rice hybrids. The first use of a wild species for the genetic improvement of cultivated rice was made at the Paddy Breeding Station, Coimbatore. Here *O. sativa* was crossed with *O. longistaminata* and the drought resistance varieties CO 15 and CO 16 were developed (Ramiah, 1953).

Exotic Oryza spp. *O. longistaminata, O. barthii, O. schweinfurthiana, O. minuta* and *O. latifolia* were introduced into India during 1930s especially to study their relationships with the cultivated rice. The African cultivated species *O. glaberrima* was also introduced (Ramiah, 1953). Crosses were made between many of these wild species and the cytology of their F_1 hybrids was studied (Table 9.3).

Table 9.3. Interspecific hybrids reported by Indian rice scientists during 1930-1950.

S.No	Inerspecific hybrid	Genome hybrid	Pairing behaviour of chromosomes	Authors
1	*O. sativa* x *O. longistaminata* (a)	AA	24 II	Ramiah (1928)
2	*O. sativa* x *O. officinalis*	AC	24 I	Ramiah (1934), Ramanujam (1937)
3	*O. sativa* x *O. glaberrima*	AA	12 II	Ramiah (1938)
4	*O. minuta* x *O. officinalis*	BCC	12 II+12 I	Nandi (1938)
5	*O. sativa* x *O. minuta*	ABC	12 II+ 12 I (b)	Nandi (1938)
6	*O. sativa* x *O. rufipogon (a) var*	AA	24 II (c)	CRRI (unpubl)
7	*O. sativa* x *O. barthii* (a)	AA	24 II (c)	CRRI (unpubl)
8	*O. glaberrima* x *O. barthii*	AA	24 II (c)	CRRI (unpubl)

9	*O. sativa* x *O. rufipogon*	AA	24 II (c)	CRRI (unpubl)
10	*O. sativa* x *O. schweinfurthiana* (d)	ABC	36 I	CRRI (unpubl)

NB: (a) Names of some of the species have been changed in view of the present knowledge. *O. perennis* is now *O. rufipogon, O. breviligulata* is now *O. barthii*
(b) This finding of Nandi (1938) has not been confirmed by other cytologists.
(c) Information added by the present authors.
(d) *O. schweinfurthiana* (syn *O. punctata* 4x) was misidentified as *O. eichingeri* by early scientists.
Source: Ramiah (1953:264), Ghose, Ghatge and Subrahmanyan (1960:155)

After the establishment of Central Rice Research Institute at Cuttack (Orissa) in 1946, S. Sampath assembled a very good collection of wild Oryza species at the Institute. He and his students studied *rufipogon-nivara-spontanea* complex, produced many interspecific hybrids and studied their cytology. Chatterjee (1948) studied the herbarium specimens of Oryza and published a fresh enumeration of species of the genus. The Indian Society of Genetics and Plant Breeding organized a symposium on introduction of crop plants in which Ramiah and Ghose (1951), Sampath and Rao (1951) and Chatterjee (1951) presented papers on the origin of cultivated rices. These scientists recognized the perennial forms of *rufipogon-nivara-spontanea* complex as *O. perennis.* Sampath and Rao (1951) proposed that the cultivated species *O. sativa* has originated from the perennial wild species *O. perennis* and all the intermediate types of this complex are natural hybrids between *O. sativa* and *O. perennis*.

The Indian Agricultural Research Institute (IARI) started its post-graduate courses in 1958. Dr S.V.S. Shastry who had worked earlier at the Central Rice Research Institute and had later on pachytene analysis of *Melilotus* spp with for his Ph.D. dissertation at Wisconsin University, joined the post-graduate school of IARI. He and his students initiated their studies of the rice chromosomes at pachytene stage of meiosis (Shastry, Ranga Rao and Misra, 1960). They also studied the cytology of many interspecific hybrids and reported their behaviour especially at pachytene stage. During 1960 to 1963, he and his group made an excellent collection of wild species *O. rufipogon, O. nivara* and *O. sativa f. spontanea* from Uttar Pradesh, Bihar, Orissa and Chhattisgarh and studied the interspecific relationships of these elements and proposed *O. nivara* as a distinct annual wild species (Sharma and Shastry, 1965). They also collected plants of *O. officinalis* and *O. granulata* from Assam, Tripura, Bihar and Orissa.

The collections made up to 1960s were maintained and shared by the rice scientists mainly for biosystematic and cytological studies and to determine their interrelationships. The need to conserve them or to evaluate and utilize them in rice breeding was not felt because (a) the scientists did not feel that there was any danger of loss of genetic diversity in their natural habitats, and (b) the genetic diversity available within the cultivated rice was not exhausted by then in so far as the genetic improvement of the cultivated rice is concerned. These collections were maintained at research centres mostly for their academic studies and were lost whenever the concerned scientist retired or was transferred. Only in a few cases some of their collections were sent to other research centres.

The decade of 1960s was the period when the agricultural scientists were becoming increasingly aware about the loss of genetic diversity of cultivated and wild rices and also about the need to conserve this biodiversity. The danger of loss of genetic diversity of wild species was much greater because their habitats were being encroached upon and destroyed by the development activities of the growing human population. The spread of high yielding rice varieties in 1960s and growing these varieties under high levels of fertilizers and taking two crops in a year created an ideal setting for the spread of insect pests and diseases which were only minor ones so far. This necessitated looking for genes for resistance beyond the cultivated species. New techniques were by now available to make distant crosses and recombine characters. Besides, the taxonomic, biosystematic and cytogenetic studies already carried out during 1930s to 1960s provided a clearer understanding of the species relationships in the genus. This further encouraged the rice scientists to collect, conserve, evaluate and utilize the wild species in rice breeding.

The International Board of Plant Genetic Resources (IBPGR) was formed by the Consultative Group for International Agricultural Research (CGIAR) at Rome in 1974. This generated a renewed interest in collection, conservation, evaluation and utilization of wild rices. The National Bureau of Plant Genetic Resources (NBPGR) was established at New Delhi in 1976. Thus there was a renewed interest in the collection of wild rices to conserve them. Besides, lack of resistance to grassy stunt virus in any of the available cultivated varieties at IRRI and its availability in a wild species (*O. nivara*) emphasized the importance of collection and conservation of wild rices of India.

Exploration and Collection

After the establishment of the NBPGR and its regional stations, the initiative for collection and conservation of rice germplasm rested with the Bureau. The NBPGR created long term and medium term storage facilities for the safe storage of germplasm, organized and provided logistic support for exploration and collection, trained scientists on various aspects of genetic conservation of crop plants, etc. The role of CRRI was rejuvenation, seed multiplication, characterization, documentation and to maintain a working collection. The role of DRR was to evaluate the rice germplasm at its various coordinated centres.

In early 1980s, CRRI collected the samples of *O. rufipogon* and *O. nivara* from different parts of India. Crosses made among *O. nivara* collections indicated that wider the spatial separation among *nivara* populations, greater was the sterility of their F_1 hybrids (Sharma *et al.*, 2000).

During 1984-89, the NBPGR and CRRI jointly organized exploration and collection of wild species namely *O. nivara, O. rufipogon* and *O. officinalis* from north Orissa, South Bihar, West Bengal, Sikkim, Mizoram, Maharashtra, Karnataka, Goa and Gujarat. A total of 204 accessions were collected in this period. In addition, the CRRI had its own exploration of wild rice, especially for *Porteresia coarctata* from the Bhitarkanika (Tables 9.4 and 9.5). In 1987, additional 75 samples comprising *O. granulata, O. rufipogon, O. nivara, O. officinalis* and *O. sativa* f. *spontanea* were collected from coastal, mid-land and mountainous areas of Kerala.

Table 9.4. Collaborative efforts of CRRI and NBPGR for wild rice collection during 1991 to 1999.

Sl. No.	Month, Year	Collected by	Location	No. of accessions	Materials collected
1.	Nov. 1991	Patnaik (CRRI) & Tomar (NBPGR)	23 districts of eastern Uttar Pradesh (wild rice)	40 accessions	*O. nivara-30, O. rufipogon-8 & O. spontanea-2*
2.	Nov. 1989	NBPGR Base Centre (Malik & Dikshit)	South Bihar, North Orissa, & West Bengal (in collaboration with IRRI)	134 accessions	*O. rufipogon-68, O. nivara-38 O. spontanea-25 O. granulata-1 (Similipal) O. coarctata-2 (Chandipur on Sea, Balasore)*
3.	1999	Marndi (CRRI) & Pani (NBPGR)	Sundargarh, Sambalpur (Orissa) & Raigarh (Chhattisgarh)	35 accessions	*O. nivara, O. rufipogon & O. spontanea*

Table 9.5. Collection of wild/weedy species of rice by CRRI.

Year	Species	No. of collections	Area/State
1984-89	*O. nivara*	204	Orissa,
	O. rufipogon		West Bengal,
	O. officinalis		Maharashtra, Karnataka, Goa, Gujarat
1994-95	*Porteresia coarctata*	9	Bhitarkanika (Orissa)
1999-2003*	*O. nivara*	161	Orissa
	O. rufipogon	151	
	O. sativa f. *spontanea*	6	
1999-2003*	*O. nivara*	78	West Bengal
	O. rufipogon	84	
	O. sativa f. *spontanea*	1	

* Under NATP

Under the NATP project, the scientists of NBPGR and CRRI collected 483 samples of *O. nivara* and *O. rufipogon* from West Bengal and Orissa states which are quite rich in the biodiversity of these two species (Table 9.6.).

As India is rich in the genetic diversity of wild rices, rice scientists of many countries have visited India for exploration and collection of its wild rices. The first such trip was mad by Dr H.I. Oka and his team from National Institute of Genetics, Mishima, Japan in 1957. They explored eastern and southern parts of India. In 1959, Dr H. Kihara and Dr S. Nakao of the same Institute visited India and collected wild rices from Assam. In 1971, Prof Tadayo Watabe of Tottori University of Japan and his team visited India and collected *O. nivara* and *O. rufipogon* from almost all the rice growing regions of India. They also collected samples of rice embedded in the bricks of ancient buildings to check if there has been any shift in the type of rice cultivated in the country during the last several centuries.

International collaborative explorations for wild rice were also undertaken. The CRRI-IRAT-ORSTOM explorations resulted in the collection of 79 accessions of *O. nivara* and *O. officinalis* from western India during 1984. In 1986 French teams from Institut de Recherche Agronomiques Tropicales (IRAT) and Office de Recherche Scientifique et Technique en Outre Mer (ORSTOM) in collaboration with the ICAR made sizeable collection of *O. nivara, O. rufipogon* and *O. officinalis* from Goa, Karnataka, Maharashtra and Gujarat. Additional 75 samples

Table 9.6. Inventory of wild rice germplasm collected under NATP (PB).

S.No	Period of collection	Areas explored	State	Persons involved	Acronym	No. of accessions	
						O. nivara	*O. rufipogon*
1.	16.11.99 to 25.11.99	Khurda, Nayagarh, Phulbani, Boudh, Sonepur, Bolangir, Kalahandi, Rayagada, Gajapati & Ganjam districts	Orissa	B.C. Patra & K. Samsudeen	PS	30	25
2.	16.11.99 to 3.12.99	Sundargarh & Sambalpur districts	Orissa	D. Pani & B.C. Marndi	PB	25	15
3.	16.10.99 to 6.12.99	Cuttack, Khurda, Jagatsinghpur & Puri districts	Orissa	S.R. Dhua	SRD-99	16	9
4.	2.12.99 to 9.12.99	Bhadrak & Balasore districts	Orissa	S.S.C. Patnaik K. Samsudeen & NBPGR	KVPS	1	3
5.	16.12.99 to 23.12.99	Balasore, Bhadrak, Jajpur, Dhenkanal & Cuttack districts	Orissa	P. Nayak & NBPGR	RM-5	1	5
6.	16.12.99 to 23.12.99	Ganjam district	Orissa	B.C. Marndi & NBPGR	RM-6	1	3
7.	2.1.00 to 9.1.00	Bhadrak, Balasore, Mayurbhanj & Keonjhar districts	Orissa	S.S.C Patnaik &B.C. Marndi	CM	2	38
8.	27.1.00 to 29.1.00	Jagatsinghpur district	Orissa	S.R. Dhua, B.C. Patra & K. Samsudeen	DPS	2	1
9.	27.10.00 to 02.11.00	Bhadrak, Balasore, Mayurbhanj, Keonjhar, Sundargarh, Jharsuguda, Sambalpur &	Orissa & WB	B.C. Patra & P.K. Nayak	BP	39	20

		Midnapore districts					
10.	27.10.00 to 04.11.00	Koraput, Malkangiri, Nawarangpur & Kalahandi districts	Orissa	S.S.C. Patnaik & B.C. Marndi	CM	43	12
11.	31.10.00 to 1.11.00	Cuttack, Dhenkanal, Angul, Sambalpur, Jharsuguda, Sundargarh & Bargarh districts	Orissa	S.R. Dhua	SRD-00	19	18
12.	25.11.00 to 03.12.00	Midnapore, Bankura, Purulia, Burdwan, Nadia & Hooghly districts	West Bengal	S.S.C. Patnaik &B.C.Marndi	CM	2	24
13.	10.10.01 to 16.10.01	Purulia, Bankura and Midnapur districts	West Bengal	S.R. Dhua	SRD–01	29	14
14.	4.11.01 to 13.11.01	North and South 24 Parganas districts	West Bengal	R.K. Sahu & B.C. Marndi	RB	16	2
15.	4.11.01 to 13.11.01	Malda, South Dinajpur, North Dinajpur, Darjeeling, Jalpaiguri & Cooch Bihar districts	West Bengal	S.S.C. Patnaik & P.K. Nayak	SP	5	15
16.	9.11.01 to 16.11.01	Hooghly, Nadia, Burdwan, Murshidabad, Birbhum & Malda districts	West Bengal	B.C. Patra	BCPW	19	25
17.	29.11.01	Puri district	Orissa	S.R. Dhua, R.K. Sahu, B.C. Marndi, U. Dhua & P.K. Nayak	SRBP	-	4
Total						250	233
Grand Total						**483**	

comprising *O. granulata, O. rufipogon, O. nivara, O. officinalis, O. spontanea* were collected from coastal, mid-land and mountainous areas of Kerala during 1987.

Realizing the importance of wild rices in rice breeding and the danger of their loss in the their natural habitats, International Rice Research Institute appointed a biosystematist, Dr Duncan A. Vaughan, who could collect wild rices from all over the world and also study the specimens available in all the important herbaria of the world. He visited almost all parts of India and collected samples of all the wild species that grow in India. Dr Vaughan did an excellent job and published an authoritative monograph on wild rices wherein he has provided photographs of the wild species and maps of their geographic distribution (Vaughan, 1994). He also visited all the important herbaria of the world and has listed the exsiccatae of Oryza spp with notes on them.

Conservation

O. nivara is an annual species propagated from seeds; the seeds, however, have dormancy for a long period. *O. rufipogon* is a perennial species; it can be propagated by seeds as well as by its stubbles. *O. officinalis, O. malampuzhaensis and O. granulata* are also perennial species and can be propagated from their seeds as well as through stubbles. However, it is easier to multiply this species by splitting its stubbles unless one needs a large number of plants at one time. The seeds are preferably kept in the medium and long term storage for their longer viability whereas the stubbles are often maintained in pots only. *Porteresia coarctata* is also a perennial species and can be easily propagated through stubbles.

The medium term and long term storage facilities became available in 1980s only. Before this time, they were generally maintained by growing them in pots only. For medium term storage, the seeds are dried up to 10-12% for reducing their moisture content and kept in 3-layered aluminum foil pouches. The outer layer of the pouch is polyester of 12 micron; intermediate layer is aluminum of 12 micron and the innermost layer is polythene of 250 gauges. These aluminum foil pouches are stored in cold module at a regulated temperature of 4° C and 33% relative humidity (RH). The seeds can be safely kept stored in the medium term storage for 8-10 years. These wild rices are also being shared with researchers throughout the country. Another set of these wild rices is conserved in the long-term storage at NBPGR, New Delhi at –20° C with 3-4% RH.

Evaluation

In general, the wild species of Oryza harbour genes for resistance to insect pests and diseases in greater frequency than the cultivated species (Table 9.7). In India, not many species have been screened for their resistance to biotic or abiotic stresses. The African perennial rice *O. longistaminata* was screened at CRRI for resistance to bacterial leaf blight and was found to be immune (Devadath, 1983). Similarly, *Porteresia coarctata* has been screened for tolerance to salinity and has been found tolerant to a high level of salinity at MSSRF (Table 9.8).

Table 9.7. Screening for resistance to insect pests in rice and wild rice.

Insect	Rice accessions tested (no.)	Resistant varieties (%)	Wild rice accessions tested (no.)	Resistant populations (%)[a]
Brown planthopper				
Biotype 1	30,790	0.91	248	48.0
Biotype 2	8,922	1.41	248	39.1
Biotype 3	10,711	1.38	248	39.5
Whitebacked planthopper	46,488	0.84	248	37.7
Green leafhopper	47,944	2.50	228	49.1
Zizag leafhopper	2,370	1.52	208	55.2
Striped stem borer	15,000	0.15	167	4.2
Yellow stem borer	19,961	0.13	181	13.3
Leaffolder	17,914	0.20	264	3.0
White maggot	15,787	0.01	328	2.1

[a]Damage rating of 0.3%

Source: Heinrichs *et al.* (1985), as quoted by Vaughn (1994:13)

Table 9.8. Wild rice accessions that showed resistance to Bacterial Blight pathotypes.

Species	Acc. Nos.	Collectors' No.	IC Nos.
O. nivara	CR-100268	SRD00-7	309815
	CR-100311	BP-13	282947
	CR-100312	BP-14	282948
	CR-100426	RB-26	330651

	CR-100427	SP-1	330652
	CR-100428	SP-2	330653
	CR-100431	SP-5	330656
	CR-100432	SP-6	330657
	CR-100448	BCPW-2	330583
	CR-100450	BCPW-4	330585
	CR-100454	BCPW-8	330589
	CR-100460	BCPW-14	330595
	CR-100463	BCPW-17	330598
	CR-100477	BCPW-31	330612
	CR-100481	BCPW-35	330616
	CR-100482	BCPW-36	330617
	CR-100486	BCPW-40	330621
O. rufipogon	CR-100444	SP-18	330669
	CR-100445	SP-19	330670
	CR-100462	BCPW-16	330597
	CR-100464	BCPW-18	330599
	CR-100466	BCPW-20	330601
	CR-100488	BCPW-42	330623
	CR-100490	BCPW-44	330625

Source: Vikal, Das, Patra, Goel, Sidhu and Singh (2006)

While screening the rice germplasm of cultivated and wild species for drought tolerance, two accessions of *O. nivara* were found to be tolerant to drought along with that of some cultivated ones. It bears significance because the habitat of this wild species is usually roadside seasonal ditches. Hence getting drought tolerance was not expected in this species. However, AC 100374 and AC 100476 were identified as drought tolerant during large scale screening of rice germplasm under field condition (Patra *et al.,* 2008) (Table 9.9).

Table 9.9. Performance of two identified drought tolerant wild rice accessions.

Genetic resources	Species	Drought score ('SES')	Recovery score ('SES')
Vandana (Tolerant check)	*O. sativa* (HYV)	1	1
IR 20 (Susceptible check)	*O. sativa* (HYV)	9	9
IC-330470 (AC-100374)	*O. nivara*	1	1
IC-330611 (AC-100476)	*O. nivara*	1	1

Source: Patra, Dhua, Marndi, Nayak, Swain, Kumar and Singh K (2008)

The grassy stunt virus is transmitted by brown plant hopper (BPH). This disease makes the rice plant look like a clump of grass and the plant does not produce any panicles. The disease was causing a great loss to rice crop in Southeast Asia. At IRRI, 6,723 accessions of cultivated rice and many accessions of wild rices were screened for resistance to this virus and only one collection of *O. nivara* from Basti district of Uttar Pradesh (IRRI Acc 101508) was found to be the only source of resistance (Jena and Khush, 2000).

In China, Prof Yuan Long Ping identified a cytoplasmic male sterile plant of *O. sativa* f. *spontanea* in Hainan Island. With this source of cytoplasmic male sterility, China could develop hybrid rice. This cytoplasmic source is designated as WA (wild abortive) and now has been shared with other countries including India.

The various traits for which the different species have been screened and found tolerant/resistant at IRRI are provided in Table 9.10.

Table 9.10. Useful traits of various species of Oryza.

Species	2n	Genome	Useful traits
Ser. Sativae:			
O. nivara	24	AA	Resistance to grassy stunt virus, blast; drought avoidance
O. rufipogon	24	AA	Resistance to BB, BPH, tungro virus; tolerance to aluminum and soil acidity; source of CMS; elongation ability
O. meridionalis	24	A^mA^m	Elongation ability; drought avoidance
O. glaberrima	24	A^gA^g	Tolerance to drought, acidity and iron toxicity; resistance to RYMV, blast, African gall midge, nematodes; weed competitiveness

O. barthii	24	A^gA^g	Resistance to GLH, BB; drought avoidance
O. longistaminata	24	A^lA^l	Resistance to BB, BPH, stemborer, nematodes; drought avoidance
O. glumaepatula	24	$A^{gp}A^{gp}$	Elongation ability; source of CMS
Ser. Latifoliae:			
O. officinalis	24	CC	Resistance to thrips, WBPH, BB, stem rot
O. rhizomatis	24	CC	Drought avoidance
O. eichingeri	24	CC	Resistance to BPH, WBPH, GLH, yellow mottled virus
O. punctata	24	BB	Resistance to BPH, zigzag leafhopper
O. australiensis	24	EE	Resistance to BPH, drought avoidance
O. minuta	48	BBCC	Resistance to BB, blast, BPH, GLH; tolerance to sheath blight
O. latifolia	48	CCDD	Resistance to BPH, BB; high biomass production
O. alta	48	CCDD	Resistance to striped stemborer; high biomass production
O. grandiglumis	48	CCDD	High biomass production
Ser. Brachyanthae:			
O. brachyantha	24	FF	Resistance to yellow stem borer, leaf folder, whorl maggot; tolerance to lateritic soil, easy threshability
Ser. Meyerianae:			
O. granulata	24	GG	Resistance to BB; adaptation to aerobic soil; shade tolerance
O. meyeriana	24	GG	Adaptation to aerobic soil; shade tolerance
Ser. Ridleyanae:			
O. ridleyi	48	HHJJ	Resistance to blast, BB
O. longiglumis	48	HHJJ	Resistance to stem borer, whorl maggot, blast, BB
Ser. Schlechterianae:			
O. schlechteri	48	HHKK	

Source: Adapted from Singh and Khush (2000) and Brar and Khush (2003).

Utilisation

Generally a breeder avoids making a wide cross for transferring any desirable character from a wild species to the cultivated one if the same desirable character is available within the cultivated species because, in the process, many undesirable characters of the wild rice also get transferred and then the breeder has to make efforts to eliminate these undesirable traits. It is generally the resistance to insect pests and diseases or tolerance to some rare physiological traits that the wild species are screened for and transferred to the cultivated species.

With regard to utilisation of wild species for the genetic improvement of a crop plant, Harlan and De Wet (1971) classified the species into three classes which they called (a) primary gene pool, (b) secondary gene pool, and (c) tertiary gene pool based on their isolation barriers. In the case of *O. sativa*, it is easier to transfer a trait from a wild species which is a diploid (like *O. sativa*) and has the same genome (AA) as that of *O. sativa.* Transferring a trait from a diploid species having a different genome or from a tetraploid species is comparatively difficult. However, B, C and D genomes are considered homoeologous and they show some affinity with AA genome species (Shastry, 1965). It is, therefore, easier to transfer a trait from, say, *O. officinalis* (CC) to *O. sativa* (AA) than transferring a trait from, say, *O. granulata* (GG) or *O. brachyantha* (FF). Transferring the characters from tetraploid (2n=48) wild species to diploid (2n=24, *O. sativa*) involves additional problem of bringing the hybrid to a diploid level retaining mostly the *sativa* chromosomes with a few introgressed characters of the tetraploid wild species.

As mentioned already, the earliest attempt to transfer any trait from a wild Oryza species to *O. sativa* was done at Paddy Breeding Station, Coimbatore. It is claimed that tolerance to drought was transferred from "*O. longistaminata*" to *O. sativa* though the former is a species that grows in deposits of water. The rice varieties developed from this hybridization are said to be CO 15 and CO 16 (Ramiah, 1953). It is suspected that this wild species was *O. rufipogon* only but misidentified as *O. longistaminata* which is an African rhizomatous species. Both these wild species are, however, diploid (2n=24) and have the same genome (AA) as that of *O. sativa.*

In 1965, when some collections of *O. nivara* (2n=24, genome AA) were sent by Dr SVS Shastry from IARI to IRRI, one of them turned out to be the sole source of resistance to grassy stunt virus. At that time IRRI was looking for resistance to grassy stunt virus that was not available in any of its accessions of cultivated rice and several other accessions of wild species. The resistance from this *O. nivara* was soon transferred to the cultivated rice and varieties like

IR28, IR29, IR30, IR32, IR34 and IR36 were developed and released for cultivation in countries affected with this disease (Jena and Khush, 1990).

The species having different genomes or ploidy or both may be treated to form secondary gene pool as transfer of genes from these species is difficult to produce. As Jena and Khush (2002:276) have remarked:

> *Interspecific hybrids between rice and wild species belonging to secondary gene pool are difficult to produce. Very low crossability and degeneration of hybrid embryos during early stages of development are the major constraints of these crosses. These interspecific hybrids are male sterile and embryo rescue is a must to develop F_1 hybrids and backcrossing to recurrent* O. sativa *parent is needed until partially fertile plants with normal chromosome complement (disomics) having 2n=24 or monosomic alien addition lines (MAAL) having 2n+1 = 25 chromosomes are produced. The fertile progenies are selfed to develop alien gene introgression lines and evaluated for transfer of desirable agronomic traits.*

One such instance that has been successfully achieved by IRRI is the transfer of BB, BPH and WBPH resistance from *O. officinalis* to *O. sativa.* Though both these species are diploid, the genome of *O. officinalis* is different (CC) from that of *O. sativa* (AA). *O. sativa* was crossed with *O. officinalis* and the F_1 was backcrossed with *O. sativa* to produce BC_2F_1 disomic progenies that segregated for resistance for BPH and WBPH. The recurrent *sativa* parent was susceptible to BPH whereas the wild species was resistant to BPH. Several introgression lines resistant to BPH biotypes of Philippines, India and Bangladesh were produced. Three of these BPH resistant lines have been released in the Mekong Delta of Vietnam.

Despite such difficulties, IRRI has been successful in transferring many of the useful traits from wild rices to the cultivated rice (Table 9.11). Many of these have not been utilized in developing new rice varieties but IRRI has created an "arsenal" of such pre-breeding material to fight against any outbreak of disease or pest in future because even a single new biotype of insect or a race of a pathogen can cause havoc.

Table 9.11. Agronomically important genes introgressed from wild Oryza species into cultivated rice.

Genes transferred to *O. sativa*	Donor species	Genome	IRRI accession number
Grassy stunt virus resistance	*O. nivara*	AA	101508
Bacterial blight resistance	*O. longistaminata*	AA	—

	O. officinalis	CC	100896
	O. minuta	BBCC	101141
Blast resistance	*O. minuta*	BBCC	101141
BPH resistance	*O. officinalis*	CC	100896
	O. australiensis	EE	100882
WBPH resistance	*O. officinalis*	CC	100896
Cytoplasmic male sterility	*O. sativa* f. *spontanea*	AA	——
Yellow stemborer resistance	*O. brachyantha*	FF	101232
	O. ridleyi	HHJJ	
Sheath blight resistance	*O. minuta*	BBCC	101141

Source: Jena and Khush (2000)

Biotechnology and Other Modern Tools

Porteresia coarctata is a wild rice that grows in the mangrove swamps and has tolerance to salinity. Efforts are going on to transfer its resistance into the cultivated rices and with this objective, F_1 hybrids have been produced (Jena, 1994; Brar *et al.,* 1997). With the availability of new biotechnological tools, transfer of genes from distant genera and species is now possible. In this sense, the spheres of primary, secondary and tertiary gene pool are now greatly enlarged. The role that molecular genetics and biotechnological tools can play in various aspects of conservation of genetic resources has been discussed in Chapter 10 (Role of Molecular Genetics and Biotechnical Tools in Conservation of Genetic Resources of Rice).

References:

Aggarwal RK, Brar DS, Khush GS. 1997. Two new genomes in the Oryza complex identified on the basis of molecular divergence analysis using total genomic DNA hybridization. *Mol Gen Genet* 254:1-12.

Bhalerao SG. 1928. The wild rice (*Oryza sativa*) of the Bombay Presidency. *Poona Agric College Magazine* 20:45-49.

Brar DS, Elloran RM, Talag JD, Abbas F, Khush GS. 1997. Cytogenetic and molecular characterization of an intergeneric hybrid between *Oryza sativa* L and *Porteresia coarctata* (Roxb.) Tateoka. *Rice Genetics Newsletter* 14:14:43-44.

Brar DS, Khush GS. 2003. Utilization of wild species of genus Oryza in rice improvement. Pages 283-309 in Nanda JS (ed) *Rice Breeding and Genetics: Research Priorities and Challenges.* Oxford & IBH, New Delhi.

Chatterjee D. 1948. A modified key and enumeration of species of genus *Oryza* Linn. *Indian J*

Agric Sci 8:185-192.

Chatterjee D. 1951. Note on the origin and distribution of wild and cultivated rices. *Indian J Genet & Plant Breed* 2:18-22.

Chevalier A. 1932. Nouvelle contribution a l'etude systematique des Oryza. *Rev Bot Appl Agri Trop* 12:1014-1032. (In French) [An English translation of the text is available in Nanda JS, Sharma SD (eds) *Monograph on Genus Oryza*. Oxford & IBH, New Delhi].

Devadath S. 1983. A strain of *O. barthii* an African wild rice immune to bacterial blight of rice. *Curr Sci* 52(1):27-28. (The author means *O. longistaminata*, which was earlier misidentified as *O. barthii* by many rice workers).

Gamble JS. 1934 Gramineae. In Fischer, *Flora of the Presidency of Madras*. Part X.

Ge S, Sang T, Lu BR, Hong DY. 1999. Phylogeny of rice genomes with emphasis on origin of allotetroploid species. *PNAS* 96(25):14400-14405.

Ghose RLM, Ghatge MB, Subrahmanyan V. 1960. *Rice in India*. Indian Council of Agricultural Research, New Delhi.

Haines HH. 1921-25. *The Botany of Bihar and Orissa (Part I to VI)*. (Reprint Edition, 1961). Botanical Survey of India, Calcutta.

Harlan JR, DeWet JMJ. 1971. Toward a rational classification of cultivated plants. *Taxon* 20:509-517.

Heinrichs EA, Medrano FG, Rapusas HR 1985. *Genetic Evaluation for Insect Resistance in Rice*. International Rice Research Institute, Manila, Philippines. 356 pp

Hooker JD. 1897. *Flora of British India. Vol VII*. Reeve & Co, London.

Jena KK. 1994. Development of intergeneric hybrid between *O. sativa* and *Porteresia coarctata*. *Rice Genet Newsl* 11:78-79.

Jena KK, Khush GS. 1990. Introgression of genes from *Oryza officinalis* Wall *ex* Watt to cultivated rice, *O. sativa* L. *Theor Appl Genet* 80: 737-745.

Jena KK, Khush GS. 2000. Exploitation of alien species in rice improvement — Opportunities, achievements and future challenges. Pages 271-285 in Nanda JS (ed) *Rice Breeding and Genetics: Research Priorities and Challenges*. Oxford & IBH, New Delhi.

Khush GS, Brar DS. 2001. Rice genetics from Mendel to functional genomics. Pages 3-25 in Khush GS, Brar DS and Hardy DS. (eds). *Rice Genetics IV*. Proceedings of the Fourth International Rice Genetics Symposium, Oct 22-27, 2000, IRRI, Manila.

Mehra KL. 2005. Rice in Indian history and culture. Pages 1-31 in Sharma SD and Nayak BC. (eds) *Rice in Indian Perspective*. Today and Tomorrow's Printers and Publishers, New Delhi.

Mitra SK, Ganguly PM. 1932. Some observations on the characters of wild rice hybrids. *Indian J Agric Sci* 2:271-79.

Patra BC. 2008. Diversity study of rice (*Oryza sativa*) in herbarium. Environment & Ecology 26(1):250-255.

Patra BC, Dhua SR, Marndi BC, Nayak PK, Swain P, Kumar GAK, Singh K. 2008. Exploration, collection, characterization, evaluation and conservation of wild rice germplasm of east India. *Oryza* 45(2):98-102.

Prain D. 1903. *Bengal Plants. Vol II.* NW & Co, Calcutta.

Ramiah K, Ghose RLM. 1951. Origin and distribution of cultivated plants of south Asia. *Indian J Genet & Plant Breed* 11:7-13.

Ramiah K. 1953. *Rice Breeding and Genetics.* Indian Council of Agricultural Research, New Delhi.

Roschevicz RJ. 1931. Critical botanical review of species of rice. *Appl Bot Genet Pl Breed Bull 27(4):3-133.* (In Russian). [An English translation of the text is available in Nanda JS and Sharma SD (eds) *Monograph on Genus Oryza.* Oxford & IBH, New Delhi].

Roxburgh W. 1832 *Flora Indica. Vol II.* Fascimile edition (1976) Bishan Singh Mahendra Pal Singh, Dehradun and Periodical Experts, Delhi, India

Roy SC. 1921. A preliminary classification of wild rices of central Provinces and Berar. *Agric J India* 16:365-380.

Sampath S, Rao MBVN. 1951. Inter-relationship between species in the genus Oryza. *Indian J Genet & Plant Breed* 11:14-17.

Sattar M, Sharma SD, Pokharia, AK. 2010. History of Rice in South Asia (up to 1947) in Sharma SD (ed) *Rice: Origin, Antiquity and History.* Science Publishers Inc, New Hampshire, NH 03748. (In press).

Sharma SD, Tripathy S, Biswal J. 2000. Origin of *O. sativa* and its ecotypes. Pages 349-369 in Nanda JS (ed) *Rice Breeding and Genetics: Research Priorities and Challenges.* Oxford & IBH, New Delhi.

Sharma SD. 1964. *Interspecific Relationships in Genus Oryza.* Ph.D. thesis submitted to Indian Agricultural Research Institute, New Delhi.

Sharma SD. 2003. Species of genus Oryza and their inter-relationships. Pages 73-111 in Nanda JS and Sharma SD (eds) *Monograph on Genus Oryza.* Oxford & IBH, New Delhi.

Sharma SD. 2005. Wild rices and their utilization. Pages 195-216 in Sharma S.D. and Nayak, B.C. (eds) *Rice in Indian Perspective.* Today and Tomorrow Printers and Publishers, New Delhi.

Sharma SD, Shastry SVS. 1965. Taxonomic studies in genus Oryza. III. *O. rufipogon* Griff *sensu stricto* and *O. nivara* Sharma *et* Shasty *nom nov. Indian J Genet & Plant Breed* 25(2):157-165.

Shastry SVS. 1965. Genomic differentiation in the genus Oryza. *Indan J Genet* 26A:258-271.

Shastry SVS. Ranga Rao DR and Misra RN. 1960. Pachytene analysis in Genus Oryza. I. Chromosome morphology in *O. sativa. Indian J Genet & Plant Breed* 20:15-21.

Singh RJ, Khush GS. 2000. Cytogenetics of rice. Pages 287-311 in Nanda JS (ed) *Rice Breeding and Genetics: Research Priorities and Challenges.* Oxford & IBH, New Delhi

Vaughan DA. 1994. *The Wild Relatives of Rice: A Genetic Resources Handbook.* International Rice Research Institute, Manila, Philippines. 137 pp.

Vikal Y, Das A, Patra BC, Goel RK, Sidhu JS, Singh K. 2006. Identification of new sources of bacterial blight (*Xanthomonas oryzae* pv. *oryzae*) resistance in wild Oryza species and *O. glaberrima. Plant Genetic Resources: Characterization and Utilization* 5(2):108-112.

Watt G. 1891. A Dictionary of the Economic Products of India. Vol V. Reprinted in 1972 by Cosmo Publications, Delhi.

Genetic Resources of Rice in India : 347-363
Editor : S. D. Sharma
Today & Tomorrow's Printers and Publishers, New Delhi - 110 002 (India)

Chapter 10

Role of Biotechnology in Exploiting Rice Germplasm

D.R. Pani and L. Behera

Plant genetic resources are the major biological basis of the world food security. Conservation, management and use of germplasm maintained in genebanks pose a number of challenges to the researchers. Germplasm evaluation and value addition to genetic resources are particularly important because they help identify traits and provide the foundation to enhance the use of collections. Common problems include the development of strategies for sampling representative individuals in natural populations, improvement of tools and technology for long-term conservation or high throughput genetic analysis of an increasingly high number of stored accessions. Therefore, the characterization of the accessions maintained in the germplasm collections and the determining the genetic relationship between them is important for the sustainable conservation and increased use of crop genetic resources. Germplasm characterization of plant accessions deposited in genebanks has been limited and is probably a major cause for the limited use of accessions in breeding programs.

The tools of modern biotechnology are being increasingly applied for plant diversity characterization and playing a major role in assisting plant conservation programs. Integrating biotechnology with germplasm conservation programmes is a prerequisite to achieve success in sustainability and to complement the existing technologies. There are five main areas of biotechnology, which can assist rice germplasm conservation and utilization programmes: (a) wide hybridization, (b) cryopreservation, (c) DNA markers technology (d) genomic tools, and (e) transgenics.

Wide Hybridization

Rice productivity is affected by several biotic and abiotic stresses. The genes for some of these stresses like resistance to sheath blight, tungro and yellow stemborer are not available or very limited in cultivated rice germplasm. Hence, there is need to broaden the rice gene pool by introgressing genes for such traits from diverse sources. The wild species of *Oryza* representing AA, BB, CC, BBCC, CCDD, EE, FF, GG and HHJJ genomes and related genera are rich source of desirable traits such as resistance to biotic stresses, tolerance to abiotic stresses, increased biomass production, improved grain qualities, cytoplasmic male sterility, increasing yield, etc. Utilization of wild rices, therefore, has great potential for widening the genetic base of the cultivated rice and increasing the yield potential and the stability of high yielding rice varieties (Brar and Khush, 1997). Transfer of useful genes from the secondary gene pool (AA genome wild species) to *O. sativa* may be accomplished by conventional sexual hybridization method but from tertiary gene pool (non-AA genome wild species) is normally not possible by this method. Several incompatibility barriers such as low crossability and limited recombination between chromosomes of wild and cultivated species limit the transfer of useful genes. Advances in tissue culture have enabled production of a series of wide hybrids, introgression lines and monosomic alien addition lines (MAALs) and molecular marker techniques have enabled to precisely detect the introgression of chromosome segments from wild into cultivated species (Jena and Khush, 1984; Brar *et al.*, 1991; Khush and Brar, 1992). It has been possible to transfer certain traits like resistance to grassy stunt virus, bacterial blight, blast, BPH, WBPH, yellow stemborer, sheath blight, tungro virus, cytoplasmic male sterility, tolerance to acidity, iron and aluminium toxicity, yield enhancing QTLs, etc from wild rices to the cultivated rice through wide hybridization between cultivated rice and several distantly related species, embryo rescue, back crossing and selection process (Table 10.1). Introgression lines and MAALs have been used to map several useful genes/QTls and assign genes to particular chromosome of rice. One of the genes, *Xa-21* introgressed from *O. longistaminata* has been cloned and physically mapped on chromosome 11 of rice. Some of the high yielding breeding lines which have introgressed genes for insect resistance from wild rice are now grown in the farmers' fields in Vietnam (AS996) and in the Philippines (Matatag 9). One of the elite breeding lines (IR73885-1-4-3-2-1-6) carrying the genes for resistance to tungro introgressed from *Oryza rufipogon* in the background of IR64 has been released as a stop-gap variety for cultivation in tungro-prone areas of the Philippines.

Table 10.1. Successful introgression of genes from wild *Oryza* species to cultivated rice.

Genes transferred to *O. sativa*	Donor species	Genome	References
Grassy stunt virus resistance	*O. nivara*	AA	Khush *et al.*, 1977
Bacterial blight resistance	*O. longistaminata*	AA	Khush *et al.*, 1990
	O. officinalis	CC	Jena and Khush, 1990
	O. minuta	BBCC	Amente-Bordeos *et al.*, 1992; Mariam *et al.*, 1996
	O. latifolia	CCDD	Multani *et al.*, 2003
	O. australiensis	EE	Multani *et al.*, 1994
	O. brachyantha	FF	Brar *et al.*, 1996
Blast resistance	*O. minuta*	BBCC	Amente-Bordeos *et al.*, 1992
BPH resistance	*O. officinalis*	CC	Jena and Khush, 1990
	O. australiensis	EE	Multani *et al.*, 1994
	O. latifolia	CCDD	Multani *et al.*, 2003
	O. minuta	BBCC	Brar *et al.*, 1996
WBPH resistance	*O. officinalis*	CC	Jena and Khush, 1990
	O. latifolia	CCDD	Multani *et al.*, 2003
Cytoplasmic male sterility	*O. perennis*	AA	Virmani *et al.*, 1988; Dalmacio *et al.*, 1995; Dalmacio *et al.*, 1996
Yellow stem borer*	*O. brachyantha*	FF	Brar *et al.*, 1991
	O. ridleyi	HHJJ	Brar *et al.*, 1991
Sheath blight resistance*	*O. minuta*	BBCC	Brar *et al.*, 1991
Tungro tolerance*	*O. rufipogon*	AA	Brar *et al.*, 1991
Tolerance to acidity, iron and aluminium toxicity*	*O. rufipogon*	AA	Brar *et al.*, 1991
	O. glaberrima	AA	Brar *et al.*, 1991
QTls for yield and yield components	*O. rufipogon*	AA	Xiao *et al.*, 1996; Septiningsih *et al.*, 2003; Cheema *et al.*, 2008

* Advance backcross progenies are being produced by IRRI

Cryopreservation of Germplasm

Cryopreservation is the most reliable method for long-term storage of plant genetic resources and is becoming a complementary approach to the conventional conservation methods. It is cost effective, requires small storage space and stored materials can be easily exchanged between countries. Cryopreservation involves preservation of germplasm in a dormant state at ultra low temperature, usually in liquid nitrogen (-196 ^{0}C) that can provide the potential for indefinite preservation of biological material (e.g. seeds, pollens, embryos, cell culture) of crop plants and reduce the problems associated with seed regeneration, especially the risk of loss of genes caused by environmental problems (Shibli *et al.,* 2006). Cryopreservation is also used for storage of 'elite' cell lines responsible for the production of secondary metabolites or of cell lines derived from genetic transformation, somatic hybridization, etc. During cryopreservation, cell division and metabolic and biochemical processes are arrested and thus the cells are allowed to retain their properties unchanged (Niino *et al*., 1995). It offers maximum stability of phenotypic and genotypic characteristics of the stored germplasm (Moukadiri *et al.,* 1999).

In rice, cryopreservation of embryogenic cells provides a more efficient means of ensuring a constant supply of competent cells for genetic manipulation. The recovery of embryogenic rice cultures after cryogenic storage has been reported by several groups. Embryogenic callus and suspension cultures from a range of different rice varieties have been cryopreserved, including *indica* and *japonica* varieties (Huang *et al*., 1995; Jain *et al*., 1996; Guangcun *et al.,* 1998; Marassiab *et al.,* 2006). Transgenic rice suspension cultures have also been successfully recovered from cryogenic storage. *In vitro* culture studies have resulted in the creation of genetic variability in rice i.e., somaclonal variants, genetic transformants, somatic hybrids and cybrids, mutations, etc. Cell cultures on repeated sub-culturing undergo genetic erosions and molecular changes such as polyploidy, aneuploidy, gene amplification, translocation, chromosome elimination, mutation, etc, thus leading to the regeneration of variant plants (Bajaj, 1990). Extensive studies on cryopreservation of rice have been conducted by several workers (Sekai and Noshiro, 1975; Gresshoff and Gartner, 1977; Ishikawa and Sekai, 1978; Stanwood, 1980; Stanwood and Bass, 1981; Bajaj and Sala, 1991; Benson and Lynch, 1999; Zhang *et al.* 2000; Marassiab *et al.,* 2006). Cryopreservation of rice as seeds, excised immature embryos, endosperms, anthers/ pollens, callus, cell suspension, protoplasts have been achieved successfully retaining their genetic stability.

DNA Marker Technology

Several DNA marker techniques such as RFLP, RAPD, AFLP, SSR, etc have been developed in recent years. They differ in principle, detection of polymorphism rate, application, cost and time requirement. DNA markers are ubiquitous, discrete, innumerable, non-deleterious and not influenced by environment. DNA marker technology has been extensively used for assessment of genetic variability of breeding lines, cultivars, gene and genome mapping, marker-assisted selection breeding, genome synteny studies, prediction of hybrid performances, management of alien gene introgression, population and ecological studies, monitoring of gene(s) in genetic engineering, identification of somatic hybrids, map based cloning of desired genes, etc.

The genetic maps in rice using different molecular markers have been constructed and efficiently employed in tagging and mapping several useful traits that are extremely important for breeding programs like yield, disease resistance, stress tolerance, seed quality, etc. A large number of both monogenic and polygenic loci for various traits have been identified in rice. Molecular markers tightly linked to genes controlling bacterial blight, blast, gall midge, submergence, etc are now employed in marker-assisted selection for developing high yielding rice varieties with resistance/tolerance to multiple biotypes/pathotypes/stresses (Huang *et al.*, 1997; Hittalmani *et al.*, 2000; Singh *et al.*, 2001; Narayanan *et al.*, 2004; Joseph *et al.*, 2004; Muralidharan *et al.*, 2004; Butaman *et al.*, 2005; Sundaram *et al.*, 2008; Zhou *et al.*, 2009). Apart from mapping and tagging of genes, molecular markers have been used in assisting backcross breeding of major genes and QTLs into elite cultivars from wild or cultivated species and synteny mapping among closely related species (Jena *et al.*, 1992; Frisch *et al.*, 1999; Tian *et al.*, 2006). A major QTL (*Sub1*) explaining about 70% of phenotypic variation in submergence tolerance has been identified and fine-mapped on chromosome 9 in the submergence tolerant cultivar FR13A (Xu *et al.*, 2000). This major QTL has provided the opportunity to apply marker-assisted backcrossing (MAB) to develop submergence tolerant version of rice cultivar Swarna (i.e. Swarna-*Sub1*) that is widely grown in the lowland rainfed ecosystem of India (Neeraja *et al.*, 2007).

Genetic Diversity Analysis of Rice Germplasm

Diversity analysis measures the level of genetic similarity or differences among plant germplasm and plays a vital role in conservation and varietal development. This information can serve as a basis for rational use in conservation of genetic resources since collection, storage and maintenance of germplasm are difficult and time consuming. It is also necessary to understand the genetic variations between the existing cultivars, wild progenitors and a number of wild species that exist in nature. Following domestication, genetic variation in crop plants has continued to become narrower due to continuous selection pressure for specific traits such as yield, thus rendering them more vulnerable to disease and insect epidemics and jeopardizing the potential for sustained genetic improvement over a long period. Thus, it is extremely important to study the genetic composition of the germplasm of existing modern-day cultivars in comparison with their ancestors and related species. This will not only provide information on their phylogenetic relationship but will also offer a chance of finding new and useful genes, as the accessions with most distinct DNA profiles are likely to contain a greater number of novel alleles.

DNA fingerprinting is a well known DNA marker technique for measuring genetic diversity among cultivars, establishing authenticity of strains, individuals, varieties, identification of duplicate accessions, etc. Apart from its forensic applications, it has also evolved as a simple yet elegant technique in solving paternity disputes, confirming quality parameters in plant biotech industries, characterizing crop varieties, micro organisms, protecting breeders right and other applications in plant sciences. With the availability of ultra dense genetic and molecular linkage maps in rice, it is now possible to choose locus specific, highly polymorphic and co-dominant markers like microsatellite to carry out fingerprinting studies in a more structured way. This is not only helps in the accurate estimation of genetic diversity but also provides an opportunity for studying relationship between the rice genotypes. Rice is one of the few crop plants with rich genetic diversity. A vast reservoir of genetic variation exists in the rice germplasm. An extent of genetic diversity in a germplasm collection is of immense importance for selecting diverse genotypes for genetic improvement of the crops. Precise, reliable, fast, cost-effective identification and assessment of genetic diversity of rice cultivars is essential for effective utilization, variety registration, prevention of unauthorized commercial use and protection of breeder's as well farmers' rights. Traditionally used morphological and biochemical markers for estimating genetic diversity and DUS (distinct, uniform and stable) testing are time consuming, expensive and have not been found to be discriminative enough, warranting more precise techniques. Further, these markers are not reliable because many characters of interest have low heritability and are genetically complex. DNA markers such as RFLP, RAPD, microsatellite, ISSR, AFLP, SNP, etc have been extensively used to accurately asses genetic diversity, parentage determination, pedigree analysis, cultivar identification, checking of purity of hybrids and parental lines, etc (Ishii and McCouch, 2000; Blair *et al.*, 2002; Nagaraju *et al.*, 2002; Jain *et al.*, 2004; Nandakumar *et al.*, 2004; Singh *et al.*, 2004; Juneja *et al.*, 2006; Zhang *et al.*, 2009).

Taxa Identification, Phylogeny and Genome Synteny

The taxonomic classification is an essential step to determine whether any germplasm is a part of the primary, secondary or tertiary gene pool of the system concerned. This is especially important in cases where morphological markers can prove to be inaccurate and misleading. A good example of this is the lines Azucena and PR 304 which have been classified as *indica* using morphological characters but clearly revealed their *japonica* nature when analysed by RAPD markers.

The molecular techniques can provide path-breaking information regarding the fine time scale on which closely related species have diverged, genetic variations that are associated with species formation and the pattern of genetic variation within the species. A number of DNA markers are being used extensively for reconstructing phylogenies of various *Oryza* species and also of more distantly related grass species (Ishii and McCouch, 2000; Ni *et al.*, 2002; Ren *et al.*, 2003; Juneja *et al.*, 2006; Barry *et al.*, 2007). Though RFLPs, minisatellites and transposons are useful for carrying out genetic variability analysis, the trend is now shifting towards the use of PCR-based markers like STMS (sequence-tagged microsatellite markers), AFLP (amplified fragment length polymorphisms), ISSR (inter simple sequence repeat), STS (sequence tagged site), SNP(single nucleotide polymorphism), etc which have proved extremely valuable in the analysis of gene pool variation of crops during the process of cultivar development as well as in classification of rice germplasm.

Scientists are making rapid progress toward a comprehensive framework for comparative genomics of the grass family (Poaceae) that will permit comparative studies at new levels of intricacy. Comparison of rice genome with genomes of other cereals and grasses promises to clarify the grass gene set, providing phylogenetic data helpful in resolving what are sometimes striking differences between independent annotations of the same genome sequence (Ouyang *et al.*, 2007; Tanaka *et al.*, 2008). Similar gene content and order of cereals and related grasses, especially those such as *Oryza*, sorghum and *Brachypodium* that have not incurred lineage-specific genome duplications, provide a starting point for accelerating progress in the study and improvement of many additional taxa. Indeed, the vast majority of monocots lack sufficient genomic tools to investigate pertinent problems in agricultural productivity, conservation biology, ecology, population biology and systematic biology. For example, orphan crops, which are collectively planted to 250 million ha year^{-1} and yield US$100 billion per year farm gate value in the developing world, are largely without genomics resources (Naylor *et al.*, 2004; Andrew *et al.*, 2009).

Genomics Tools for Discovery of Novel Genes/Alleles

Creation, evaluation and selection of right combination of genes/alleles either by natural selection or through the efforts of breeders are important in rice improvement programme. The manipulation of a large number of genes is often required for improvement of even the simplest of characteristics. The map-based complete rice genome sequence is now freely available to researchers worldwide, providing the most fundamental tool that will further accelerate efforts to improve rice yield. The complete genetic information of rice will serve as a 'gold mine' for genomic research in rice and other cereal species.

SNP (nucleotide polymorphism information) on multiple cultivars/accessions opens a new dimension of genome sequence comparison. Dense coverage of the rice genome with polymorphic DNA markers is an invaluable tool for cultivar discrimination, marker-assisted selection breeding, positional cloning, identification of novel genes/alleles and a wide range of evolutionary studies. Various genomics tools like DNA markers, EcoTILLING, etc are now used for genetic analysis, speculate about evolutionary processes, transfer of genes from wild rices and discover of novel genes/alleles and single nucleotide polymorphism (SNP) variations using diverse rice germplasm. The International Rice GeneBank Collection (IRGC) at IRRI comprises the largest collection of rice germplasm with more than 102,547 accessions of the Asian cultivated rice (*Oryza sativa*), 1,651 accessions of the African cultivated rice (*Oryza glaberrima*) and 4,508 accessions of 22 related wild relatives. IRRI maintains records of pedigree of all modern rice varieties derived from mating of traditional varieties (International Rice Information System; http://www.iris.irri.org/). The existing diverse germplasm collections allow a detailed characterization of various traits, such as tolerance to biotic and abiotic stresses, yield, nutrition, grain quality, etc and are "gold mines" for analysis of allelic diversity of all rice genes. Latha *et al.* (2004) used a rice calmodulin gene, a gene for encoding a late embryogenesis-associated protein and salt-inducible rice gene for allele mining of stress tolerance on identified accessions of rice and related germplasm. Wang *et al.* (2004) analyzed 905 rice lines that represent natural rice germplasm and identified ethylene responsive binding protein 3(ERF3) which co-localized with drought QTL on chromosome-1. Feltus *et al.* (2004) have aligned whole genome sequences of two rice subspecies, *indica* and *japonica* and analyzed levels and patterns of genetic diversity. After filtering multiple copy and low quality sequence, 408, 898 candidate DNA polymorphisms (SNPs/INDELs) were identified between the two subspecies. Monna *et al.* (2006) searched the genomes of rice cultivars and wild rice accessions for nucleotide polymorphisms and identified 7805 polymorphic loci, including single-nucleotide polymorphisms (SNPs) and insertions/deletions (InDels) in predicted intergenic regions. These markers will be valuable resource and important tool for genetic analysis and map-based cloning of rice gene. Using alternative EcoTILLING protocol, Kadaru *et al.* (2006) successfully identified four new SNPs in a 922-bp DNA region from the *alk* gene associated with cooking quality and an SNP cluster containing a deletion within a 472-bp fragment of the *waxy* gene that encodes for the granule-bound starch synthase enzyme in addition to 4 previously reported SNPs in a collection of 57 diverse rice accessions. McNally *et al.* (2009) used LR-PCR amplicons spanning 11,343 non-overlapping fragments and resequencing microarrays to interrogate 100.1 Mb of the unique fraction of the Nipponbare genome for 20 diverse varieties and landraces. They could identified the distribution of 160,000 nonredundant SNPs and some introgressed regions associated QTLs for traits like leaf area, vascular bundle number, plant height, root number, cooking quality, amylose content, gel consistency and shoot weight.

Transgenics

Transgenic approach is an attractive alternative but complementary to conventional breeding procedure for improving value and performance, production of pharmaceuticals or even addition of new traits like C4 photosynthetic capacity, apomixis, etc to the existing rice cultivars/germplasm. Novel genes from a variety of sources like bacteria, animals and plants can be introduced into rice through several transformation methods such as *Agrobacterium,* biolistic and DNA uptake by protoplast. It provides several advantages over traditional breeding methods for improving resistance to biotic and abiotic stresses, nutritional quality and agronomic performance: (a) quick and precise transfer of desired genes into suitable variety without drastically changing the genetic background, (b) transfer of genes that are not available so far in the same plant species or gene pool and (c) provides access to modified rice gene that give enhanced performance not attainable by mutation and recombination *in vivo*. Enormous progress has been made for introduction of useful genes such as resistance to insects, diseases, abiotic stresses, herbicides and quality in terms of micronutrients, vitamins or amino acids into rice cultivars (Table 10.2). Ye *et al.* (2000) have developed transgenic of *japonica* rice capable of accumulating provitamin A. This was done through modification of rice plant by introducing two genes (*Psy* and *lcy*) from a plant (daffodil) and one gene (*crtI*) from a bacterium *Erwinia uredovora* to produce and accumulate provitamin A in the grain, a trait not found in the nature. Later, Paine *et al.* (2005) from Syngenta group have reported the development of Syngenta Golden Rice 2 (SGR 2) in the background of Kaybonnet variety with carotenoid levels up to 37mg per gram of grain. Further, efforts by IRRI in collaboration with India are underway to introgress the β-carotene loci from GR lines into popular Asian rice varieties, using marker aided backcrossing (MAB) approach. Genetic modification of elite rice cultivars is seen as a simple and less expensive alternative to vitamin supplements and save 125 million children particularly those in developing countries suffer from vitamin A deficiency (VAD), causing blindness and death. Hong *et al.* (2009) developed transgenic rice suspension culture system containing anti-tumor associated glycoprotein 72 (TAG 72) gene. Purified hzAb antibody from cell suspension culture was shown to bind to the recombinant antigen, TAG 72, and to bind specifically to human LS174T colon adenocarcinoma cells expressing the TAG 72 antigen. The binding occurred to the same extent as was seen with animal-derived antibody. The results indicated that the hzAb antibody derived from rice cell suspension culture will have great potential for pharmaceutical applications and add great value to the rice germplasm.

Table 10.2. Important genes transferred to rice

Genes transferred	Traits	References
Bar	Herbicide resistance	Christou *et al.*, 1991
CP-stripe virus	Stripe virus resistance	Hayakawa *et al.*, 1992
Bt- cry1A(b)	Sheath blight, leaf folder resistance	Fujimoto *et al.* 1993
Chi 11 (Chitinase)	Sheath blight resistance	Lin *et al.*, 1995
HVA1	Osmo protectant	Xu *et al.*, 1996
Adh/pdc	Submergence tolerance	Qunmio *et al.*, 1997
Bt- cry1A(b)	Stem borer resistance	Wu *et al.*, 1997
Bt- cry1A(b)	Stem borer resistance	Datta *et al.*, 1998
Xa-21	Bacterial blight resistance	Tu *et al.*, 1998
PR-5 (rice thaumatin – like protein)	Sheath blight resistance	Datta *et al.*, 1999
Ferritin (Soyabean)	Increased iron content	Goto *et al.*, 1999
PEPC (MC)	Photosynthetic capacity	Ku *et al.*, 1999
Bt- cry1A(b)	Resistance to eight lepidopteran species, bacterial blight	Shu *et al.*, 2000
PEP –CK (BBC)	Photosynthetic capacity	Suzuki *et al.*, 2000
Xa-21 and *Bt*	Bacterial blight and stem borer resistance	Tu *et al.*, 2000
Psy, lyc, Ctrl	Beta-carotene biosysnthesis (Pro-Vit-A)	Ye *et al.*, 2000
PPDK (MC)	Photosynthetic capacity	Fukayama *et al.*, 2001

GNA	Brown planthopper resistance	Maqbool *et al.*, 2001
cry1A(b)/cry1A(c)	Resistance against four lepidopteran species	Ye *et al.*, 2001
Xa-21	Bacterial blight resistance	Baisakh *et al.*, 2002
TPSP	Salt and drought tolerance	Garg *et al.*, 2002
CAX1	Abiotic stress tolerance	Kim *et al.*, 2005
Chi 11	Sheath blight resistance	Sridevi *et al.*, 2005
cry1A/cry1Ac	Stem borer resistance	Ho *et al.*, 2006
PEPC (ME)	Photosysnthetic capacity	Bandyopadhyay *et al.*, 2007
hzAb (anti-TAG 72)(anti tumor-associated glycoprotein 72)	Pharmaceutical applications	Hong *et al.*, 2008
OsiSAP8 (stress associated protein gene family)	Salt and drought tolerance during anthesis stage	Kanneganti and Gupta, 2008
OsYSL18 (iron phytosidero phoretransporter)	Translocation of iron in reproductive organs and phloem in joints	Aoyama *et al.*, 2009
OsMT1a (type 1 metallothionein)	Zinc homeostasis and drought tolerance in rice	Yang *et al.*, 2009

References:

Amanta-Bordeos A, Sitch LA, Nelson R, Dalmacio RD, Oliva NP, Aswidinoor H, Leung H. 1992. Transfer of bacterial blight and blast resistance from the tetraploid wild rice *Oryza minuta* to cultivated rice. *Theor Appl Genet* 84:345-354.

Andrew H., Paterson, John E. Bowers, Frank A. Feltus, Haibao Tang, Lifeng Lin, Xiyin Wang. 2009. Comparative Genomics of Grasses Promises a Bountiful Harvest. *Plant Physiology* 149:125–131.

Aoyama T, Kobayashi T, Takahashi M, Nagasaka S, Usuda K, Kakei Y, Ishimaru Y, Nakanishi H, Mori S, Nishizawa NK. 2009. *OsYSL18* is a rice iron (III) –deoxymugineic acid transporter specifically expressed in reproductive organs and phloem of lamina joints. *Plant Mol Biol* 70:681–692.

Baisakh N, Datta K, Oliva N, Ona I, Rao GJN, Mew TW, Datta SK. 2001. Rapid development of homozygous transgenic rice using anther culture harboring rice chitinase gene for enhanced sheath blight resistance. *Plant Biotechnol* 18:101–108.

Bajaj YPS, Sala F. 1991. Cryopreservation of germplasm of rice. Pages 553-571 in Bajaj YPS (ed) *Biotechnology in agriculture and forestry. Vol 14. Rice*. Springer. Berlin Heidelberg, New York.

Bajaj YPS. (ed) 1990. *Biotechnology in agricultura and forestry, Vol II. Somaclonal variation in crop improvement. I.* Springer, Berlin Heidelberg, New work.

Bandyopadhyay A, Datta K, Zhang J, Yang W, Raychaudhuri S, Miyao M, Datta SK. 2007. Enhanced photosynthesis rate in genetically engineered *indica* rice expressing *pepc* gene cloned from maize. *Plant Sci* 172: 1204-1209.

Barry MB, Pham JL, Noyer JL, Billot C, Courtois, Ahmadi N. 2007. Genetic diversity of two cultivated rice species (*O. sativa* and *O. glaberrima*) in Maritime Guinea. Evidence of interspecific recombination. *Euphytica* 154:127-137.

Benson EE, Lynch PT.1999. Cryopreservation of Rice Tissue Cultures Pages 83-94 in Hall, Robert D (ed) *Methods in Molecular Biology: Plant Cell Culture Protocols.* Vol 111. Humana Press.

Blair MW, Hedetale V, McCouch SR. 2002. Fluorescent-labeled microsatellite panels useful for detecting allelic diversity in cultivated rice (*Oryza sativa* L). *Theor Appl Genet* 105:449-457.

Brar DS, Elloran R, Khush GS. 1991. Interspecific hybrids produced through embryo rescue between cultivated and eight wild species of rice. *Rice Genet Newsl* 8:91-93.

Brar DS, Dalmacio R, Elloran R, Aggrawal R, Angeles R, Khush GS. 1996. Gene transfer and molecular characterization of introgression from wild Oryza species into rice. Pages 477-485 in Khush GS (ed) *Rice Genetics III.* International Rice Research Institute, Manila, Philippines.

Brar DS, Khush GS.1997. Alien introgression in rice. *Plant Mol Biol* 35:35–47.

Bustaman M, Refliar Agisimanto D, Lubis E, Suwarno. 2005. Application of molecular markers for the assessment of blast resistance genes, *Pi-1* and *Pi-2*, incorporated into Indonessian rice cultivars. *J Biotechnol Pertanian* 10(2):51-54.

Cheema KK, Bains NS, Mangat GS, Das A, Vikal Y, Brar DS, Khush GS, Singh K. 2008. Development of high yielding IR64 X *Oryza rufipogon* (Griff.) introgression lines and identification of introgressed alien chromosome segments using SSR markers. *Euphytica* 160: 401–409.

Christou P, Ford TL, Kofron M. 1991. Production of transgenic rice plants from agronomically important *indica* and *japonica* rice varieties via electric discharge particle acceleration of exogeneous DNA into immature zygotic embryos. *BioTechnol* 5:957-96.

Dalmacio R, Brar DS, Ishii T, Sitch LA, Virmani SS, Khush GS. 1995. Identification and transfer of a new cytoplasmic male sterility source from *Oryza perennis* into *indica* rice (*O. sativa*). *Euphytica* 82: 221–225.

Dalmacio RD, Brar DS, Virmani SS, Khush GS. 1996: Male sterile line in rice (*Oryza sativa*) developed with *O. glumaepatula* cytoplasm. *Inter Rice Res Notes* 21(1): 22–23.

Datta K, Velazhahan R, Oliva N, Ona I, Mew T, Khush GS, Muthukrishnan S, Datta SK. 1999. Over-expression of the cloned rice thaumatin –like protein (*PR-5*) gene in the transgenic rice plants enhances environment friendly resistance to *Rhizoctonia solani* causing sheath blight disease. *Theor Appl Genet* 98: 1138-1146.

Datta K, Vasquez A, Tu J, Torrizo L, Alam MF, OlivaN, Abrigo E, Khush GS, Datta SK. 1998. Constitutive and tissue specific differential expression of *CryIA(b)* gene in transgenic rice plants conferring enhanced resistance to insect pests. *Theor Appl Genet* 97:20–30.

Feltus FA, Wan J, Schulze SR, Estill JC, Jiang N, Paterson AH. 2004. An SNP resource for rice genetics and breeding based on subspecies *indica* and *japonica* genome alignments. *Genome Res* 14:1812-1819.

Frisch M, Bohn M, Melchinger AE. 1999. Comparison of selection strategies for marker-assisted backcrossing of a gene. *Crop Sci* 39:1295-1301.

Fujimoto H, Itoh K, Yamamoto M, Kyozuko J, Shimamoto K.1993. Insect resistant rice generated by introduction of a modified endotoxin gene of *Bacillus thuringiensis*. *Biotechnol* 11: 1151-1155.

Fukayama H, Tsuchida H, Agarie S. 2001. Significant accumulation of C_4-specific pyruvate orthophosphophate dikinase in a C_3 plant, rice. *Plant Physiol* 127:1136-1146.

Garg AK, Kim Ju-Kon, Owens TG, Ranwala AP, Choi YD, Kochian LV, Wu RJ. 2002. Trehalose accumulation in rice plants confers high tolerance levels to different abiotic stresses. *Procd Natl Acad Sci* (USA) 99:15898-15903.

Goto F, Yoshihara T, Shigemoto N, Toki S, Takaiwa F. 1999. Iron fortification of rice seed by the soybean ferritin gene. *Nature Biotechnol* 17: 282–286.

Gresshoff PM, Gartner E. 1977. Cryopreservation of *Arabidopsis thaliana* and other seeds by storage in liquid nitrogen. *Arabidopsis Inf Serv* 14:12.

Guangcun HE, Lihui S, Lanjie L, Xiaohui Y, Lahong SH ENG , Xiaoling W. 1998. Somatic cell cryopreservation and protoplast regeneration of important disease-resistant wild rice *Oryza meyeriana* Baill. *Science in China* 41(4):393-399.

Hayakawa T, Zhu YF, Itoh K, Kimura Y, Izawa T, Shimamoto K, Toriyama K, Toriyama S. 1992. Genetically engineered rice resistant to rice stripe virus and insect transmitted virus. *Proc Natl Acad Sci* (USA) 89: 9865-9869.

Hittalmani S, Parco A, Mew TV, Ziegler RS & Huang N. 2000. Fine mapping and DNA marker-assisted pyramiding of the three major genes for blast resistance in rice. *Theor Appl Genet* 100:1121-1128.

Ho NH, Baisakh N, Oliva N, Datta K, Frutos R, Datta SK. 2006. Translational fusion hybrid *Bt* genes confer resistance against yellow stem borer in transgenic elite Vietnamese rice (*Oryza sativa* L.) cultivars. *Crop Sci* 46: 781-789.

Hong SY, Lee TS, Kim J, Jung JH, Choi CW, Kim TG, Kwon TH, Jang YS, Yang MS. 2008. Tumor targeting of humanized fragment antibody secreted from transgenic rice cell suspension culture. *Plant Mol Biol* 68:413–422.

Huang CN, Wang JH, Yan QF. 1995. Plant regeneration from rice (*Oryza sativa* L.) embryogenic cells cryopreserved by vitrification. *Plant Cell Rep* 14:730-734.

Huang N, Angeles ER, Domingo J, Magpantay G, Singh S, Zhang G, Kumaravadivel N, Bennet J & Khush GS. 1997. Pyramiding of bacterial blight resistance genes in rice:marker-assisted selection using RFLP and PCR. *Theor Appl Genet* 95:313-320.

Ishii T, McCouch SR. 2000. Microsatellites and microsynteny in chloroplast genomes of *Oryza* and eight other Gramineae species. *Theor Appl Genet* 100:1257-1266.

Ishikawa M, Sekai A. 1978. Freezing avoidance in rice and wheat seeds in relation to water content. *Low Temp Sci* B36:39.

Jain S, Jain RK, McCouch SR. 2004. Genetic analysis of Indian aromatic and quality rice (*Oryza sativa* L) germplasm using panels of fluorescently-labeled microsatellite markers. *Theor Appl Genet* 109:965-977.

Jain S, Jain RK, Wu R. 1996. A simple and efficient procedure for cryopreservation of embrogenic cells of aromatic varieties. *Plant Cell Rep* 15:712-717.

Jena KK, Khush GS. 1984. Embryo rescue of interspecific hybrids and its scope in rice improvement. *Rice Genet Newsl* 1: 133–134.

Jena KK, Khush GS. 1990. Introgression of genes from *Oryza officinalis* Wall. ex Watt to cultivated rice. *O. sativa* L. *Theor Appl Genet 80:*737-745.

Jena KK, Khush GS, Kochert G. 1992. RFLP analysis of rice (*Oryza satva* L) intorogression lines. *Theor Appl Genet* 84*:*608-616.

Joseph M, Gopalakrishnan S, Sharma S, Singh RK, Singh VP, Singh AK, Singh NK, Mohapatra T. 2004. Combining bacterial blight resistance and basmati quality characteristics by phenotypic and molecular marker assisted selection in rice. *Mol Breed* 13*:*377-387.

Juneja S, Das A, Joshi SV, Sharma S, Vikal Y, Patra BC, Bharaj TS, Sidhu JS, Singh K. 2006. *Oryza nivara* Sharma *et* Shastry, the progenitor of *O. sativa* L subspecies *indica* harbours rich genetic diversity as measured by SSR markers. *Curr Sci* 91:1079-1085.

Kadaru SB, Yadav AS, Fjellstrom RG, Oard JH. 2006. Alternative EcoTILLING protocol for rapid, cost-effective single-nucleotide polymorphism discovery and genotyping in rice (*Oryza sativa* L.). *Plant Mol Biol Report* 24: 3–22.

Kanneganti V, Gupta AK. 2008. Overexpression of *OsiSAP8*, a member of stress associated protein (SAP) gene family of rice confers tolerance to salt, drought and cold stress in transgenic tobacco and rice. *Plant Mol Biol* 66:445–462.

Khush GS, Bacalangco E, Ogawa T. 1990. A new gene for resistance to bacterial blight from *O. longistaminata*. *Rice Genetic Newsl* 7*:*121-122.

Khush GS, Ling KC, Aquino RC, Aguierd VM. 1977. Breeding for resistance to grassy stunt in rice. Proc Intern Cong, *Society for Breeding Researches in Asia and Oceania* (*SABRAO*) 6*:*3-9.

Khush GS, Brar DS. 1992. Overcoming the barriers in hybridization. *Theor Appl Genet* 16: 47–61.

Ku MSB, Agarie S, Nomura M, Fukayama H, Tsuchida H, Ono K, Hirose S, Toki S, Miyao M, Matsuoka M. 1999. High-level expression of maize phosphoenolpyruvate carboxylase in transgenic rice plants. *Nature Biotechnol* 17:76-80.

Latha R, Rubia L, Bennett J,. Swaminathan MS. 2004. Allele mining for Stress Tolerance genes in *Oryza* species and related germplasm. *Molecular Biotechnology* 27:101-108.

Lin W, Anurtha CS, Datta K, Potrykus I, Muthukrishnan S, Datta SK. 1995. Genetic engineering of rice for resistance to sheath blight. *BioTechnol* 13: 686-691.

Maqbool SB, Husnain T, Riazuddin S, Masson L, Christou P. 1998. Effective control of yellow stem borer and rice leaf folder in transgenic rice indica varieties Basmati 370 and M7 using the novel d-endotoxin *cryIIA Bacillus thuringiensis* gene. *Mol Breed* 6:1–7.

McNallya KL, Childs KL, Bohnert R, Davidson RM, Zhao K, Ulat VJ, Zeller G, Clark RM, Hoen DR, Bureau TE, Stokowsk R, Ballinger DG, Frazer KA, Cox DR, Padhukasahasram B, Bustamante CD, Weigel D, Mackill DJ, Bruskiewich RM, Ratsch G, Buell CR, Leung H, Leach JE. 2009. Genomewide SNP variation reveals relationships among landraces and modern varieties of rice. *Proc Natl Acad Sci* (USA) 106: 12273–12278.

Marassiab MA, Scocchia A, Gonzaleza AM. 2006. Plant regeneration from rice anthers cryopreserved by an encapsulation/dehydration technique *In Vitro Cellular Develop Biology-Plant* 42:31-36.

Mariam AL, Zakri AH, Mahani MC, Normahm MN. 1996. Interspecific hybridization of cultivated rice, *Oryza sativa* L with the wild rice, *O. minuta* Presl. *Theor Appl Genet* 93:664- 671.

Monna L, Ohta R, Masuda H, Koike A, Minobe Y. 2006. Genome-wide Searching of Single-nucleotide polymorphisms among eight distantly and closely related rice cultivars (*Oryza sativa* L.) and a wild accession (*Oryza rufipogon* Griff). *DNA Research* 13:43–51.

Moukadiri O, Deming J, Conner JE, Cornejo MJ. 1999. Phenotypic characterization of the progenies of rice plants derived fro cryopreserved calli. *Plant Cell Rep* 18:625-632.

Multani DS, Jena KK, Brar DS, Delos Reyes BC, Angeles ER and Khush GS. 1994. Development of monosomic alien addition lines and introgression of genes from *Oryza australiensis* Domin to cultivated rice *O. sativa* L. *Theor Appl Genet* 88:102-109.

Multani DS, Khush GS, delos Reyes BG, Brar DS. 2003. Alien genes introgression and development of monosomic alien addition lines from *Oryza latifolia* Desv. to rice, *Oryza sativa* L. *Theor Appl Genet* 107:395–405.

Muralidharan K, Laha GS, Krishnaveni D, Reddy CS, Reddy JN, Sridhar R. 2004. Field performance of genotypes generated by molecular-assisted breeding to bacterial leaf blight in India. *Rice Genet Newsl* 21:29-31.

Nagaraju J, Kathirvel M, Ramesh Kumar R, Siddiq EA, Hasnain SE. 2002. Genetic analysis of traditional and evolved *Basmati* and non-*Basmati* rice varities by using fluorescence-based ISSR-PCR and SSR markers. *PNAS* 99(9):5836-5841.

Nandakumar N, Singh AK, Sharma RK, Mohapatra T, Prabhu KV, Zaman FU. 2004. Molecular fingerprinting of hybrids and assessment of genetic purity of hybrid seeds in rice using microsatellite markers. *Euphytica* 136: 257-264.

Narayana NN, Baisakh N, Oliva NP, VeraCruz CM, Virmani CM, Gnanamanickam SS, Datta K, Datta SK. 2004. Molecular breeding: Marker-assisted selection combined with biolistic transformation for blast and bacterial blight resistance in *indica* rice (cv. CO 39). *Mol Breed* 14:61-67.

Naylor RL, Falcon WP, Goodman RM, Jahn MM, Sengooba T, Tefera H, Nelson RJ. 2004. Biotechnology in the developing world: A case for increased investments in orphan crops. *Food Policy* 29: 15–44.

Neeraja CN, Maghirang-Rodriguez R, Pamplona A, Heuer S, Collard BCY, Septiningsih EM, Vergara G, Sanchez D, Xu K, Ismail AM, Mackill DJ. 2007. A marker-assisted backcross approach for developing submergence-tolerant rice cultivars. *Theor Appl Genet* 115:767–776.

Ni J, Colowit PM, Mackill DJ. 2002. Evaluation of genetic diversity in rice subspecies using microsatellite markers. *Crop Sci* 42:601-607.

Niino T, Shirata K, Oka S. 1995. Variability of mulberry buds cryopreserved for 5 years at – 135°C. *J Sericulture Sci Japan* 64:370-374.

Ouyang S, ZhuW, Hamilton J, Lin H, CampbellM, Childs K, Thibaud-Nissen F, Malek RL, Lee Y, Zheng L. 2007. The TIGR Rice Genome Annotation Resource: Improvements and new features. *Nucleic Acids Res* 35:D883–D887.

Paine JA, Shipton CA, Chagger S, Howells RM, Kennedy MJ, Vernon G, Wright SY, Hinchliffe E, Adams JL, Silverstone AL and Drake R. 2005. Improving the nutritional value of Golden Rice through increased pro-vitamin A content. *Nature Biotechnology* 23: 482-487.

Quimio C, Torrizo LB, Ellis M, Abrigo EM, Oliva NP, Vasquez AR, Ella ES, Setter TL, Carpena Al, Ito O, Peacock WJ, Dennis L, Datta SK. 1999. Enhancement of submergence tolerance in transgenic rice overproducing pyruvate decarboxylase. *J Plant Physiol* 156:516-521.

Ren F, Lu Bao-Rong, Li S, Huang J, Zhu Y. 2003. A comparative study of genetic relationships among the AA-genome *Oryza* species using RAPD and SSR markers. *Theor Appl Genet* 108:113-120.

Sekai A, Noshiro M. 1975. Some factors containing to the survival of crop seeds cooled to the temperature of liquid nitrogen. In Frankel O, Hawkes JG (eds) *Crop Genetic Resources for Today And Tomorrow*. Univ. Press, Cambridge. 317 pp.

Septiningsih EM, Prasetiyono J, Lubis E, Tai TH, Tjubaryat T, Moeljopawiro S, McCouch SR. 2003. Identification of quantitative trait loci for yield and yield components in an advanced backcross population derived from the *Oryza sativa* variety IR64 and the wild relative *O. rufipogon*. *Theor Appl Genet* 107:1419–1432.

Shibli RA, Shatnawi MA, Subaih W, Ajlouni MM. 2006. *In vitro* conservation and crypreservation of plant genetic resources: A Review. *World J Agril Sciences* 2(4):372-382.

Singh S, Sidhu JS, Huang N, Vikal Y, Li Z, Brar DS, Dhaliwal HS, Khush GS. 2001. Pyramiding three bacterial blight resistance genes (*xa5, xa13 and Xa21*) using marker-assisted selection into *indica* rice cultivar PR106. *Theor Appl Genet* 102:1011-1015.

Singh RK, Sharma RK, Singh AK, Singh VP, Singh NK, Tiwari SP, Mohapatra T. 2004. Suitability of mapped sequence tagged microsatellite site markers for establishing distinctness, uniformity and stability in aromatic rice. *Euphytica* 135:135-143.

Sridevi G, Sabapathi N, Meena P, Nandakumar R, Samiyappan R, Muthukrishnan S, Veluthambi K. 2003. Transgenic indica rice variety Pusa Basmati 1 constitutively expressing a rice chitinase gene exhibits enhanced resistance to *Rhizoctonia solani*. *J Plant Biochem & Biotechnol* 12: 93-101.

Standwood PC, Bass LN. 1981. Seed germplasm preservation using liquid nitrogen. *Seed Sci Technol* 5:699.

Stanwood PC. 1980. Tolerance of crop seeds to cooling and storage in liquid nitrogen (-196 ^{0}C). *J Seed Technol* 5(1):26.

Sundaram RM, Vishnupriya MR, Biradar SK, Laha GS, Reddy GA, Rani NS, Sarma NP, Sonti RV. 2008. Marker-assisted introgression of blight resistance in bacterial blight resistance in Samba Mahsuri, an elite *indica* rice variety. *Euphytica* 160:411-422.

Suzuki S, Murai N, Burnell J, Arai M. 2000. Changes in photosynthetic carbon flow in transgenic rice plants that express C_4-type phosphoenolpyruvate carboxykinase from *Urochloa panicoides*. *Plant Physiol* 124:163-172.

Tanaka T, Antonio BA, Kikuchi S, Matsumoto T, Nagamura Y, Numa H, Sakai H, Wu J, Itoh T, Sasaki T. 2008) .The rice annotation project database (RAP-DB): 2008 update. *Nucleic Acids Res 36*: D1028–D1033.

Tian F, Li DJ, Fu Q, Zhu ZF, Fu YC, Wang XK, Sun CQ. 2006. Construction of introgression lines carrying wild rice (*Oryza rufipogon* Griff.) segments in cultivated rice (*Oryza sativa* L) background and characterization of introgressed segments associated with yield-related traits. *Theor Appl Genet* 112:570-580.

Tu J, Datta K, Alam MF, Fan Y, Khush GS, Datta SK. 1998. Transgenic rice variety IR72 with *Xa21* is resistant to bacterial blight. *Theor Appl Genet* 97: 31-36.

Tu J, Datta K, Khush GS, Zhang Q, Datta SK. 2000. Field performance of *Xa21* transgenic *indica* rice (*Oryza satva* L), IR72. *Theor Appl Genet* 101:15-20.

Virmani, S.S. and Wan, B.H., 1988. Development of CMS lines in hybrid rice breeding. *In*: Hybrid Rice. Int. Rice Res. Inst., Manila, Philippines, p. 103-114.

Wang HH, Naredo MEB, Wu JL, Till BJ, Greene EA, Comai L, Henikoff S, Leung H, McNally KL. 2004. EcoTILLING candidate genes for drought tolerance in rice. Rockefeller Foundation Resilient Crops for Water Limited Environments Workshop, Cuernavaca, Mexico, 24-28th May, poster abstract.

Wu C, Fan Y, Zhang C, Oliva N, Datta SK. 1997. Transgenic fertile *japonica* rice plants expressing a modified *cryIA(b)* gene resistant to yellow stem borer. *Plant Cell Rep* 17:129-132.

Xiao J, Grandillo S, Ahn SN, McCouch SR, Tanksley SD,Yuan L.1996. Genes from wild rice improve yield. *Nature* 384:223–224.

Xu K, Xu X, Ronald PC, Mackill DJ. 2000. A high-resolution linkage map in the vicinity of the rice submergence tolerance locus *Sub1*. *Mol Gen Genet* 263:681–689.

Yang Z, Wu Y, Li Ye, Ling HQ, Chu C. 2009. *OsMT1a*, a type 1 metallothionein, plays the pivotal role in zinc homeostasis and drought tolerance in rice. *Plant Mol Biol* 70:219–229.

Ye X, Al-Babili S, Klöti A, Zhang J, Lucca P, Beyer P, Potrykus I. 2000. Engineering the Provitamin A (**ß-carotene**) biosynthetic pathway into (Carotenoid-Free) rice endosperm. *Science* 287:303-305.

Ye GY, Tu J, Hu C, Datta K, Datta SK. 2001. Transgenic IR72 with fused *Bt* gene *cry1A(b)/cry1A(c)* from *Bacillus thuringiensis* is resistant against four lepidopteran species under field conditions. *Plant Biotechnol* 18:125–133.

Zhang YX, Wang JH, Bian HW, Zhu M. 2001. Pregrowth-desiccation: A simple and efficient procedure for the cryopreservation of rice (*Oryza sativa* L) embryogenic suspension cells. *Cryo Lett* 22:4221–229.

Zhang D, Zhang H, Wang M, Sun J, Qi Y, Wang F, Wei X, Han L, Wang X, Li Z. 2009. Genetic structure and differentiation of *Oryza* L. in China revealed by microsatellites. *Theor Appl Genet* DOI 10.1007/s00122-009-1112-4.

Zhou YL, Xu JL, Zhou SC, Yu j, Xie XW, MR, Sun Y, Zhu LH, Fu BY, Gao YM, Li ZK. 2009. Pyramiding *Xa23* and *Rxo1* for resistance to two bacterial diseases into an elite *indica* rice variety using molecular approaches. *Mol Breed* 23:279–287.

Genetic Resources of Rice in India : 365-389
Editor : S. D. Sharma
Today & Tomorrow's Printers and Publishers, New Delhi - 110 002 (India)

Chapter 11

Regulatory Aspects in Exchange of Rice Germplasm: Indian Perspective

Arjun Lal and Vandana Tyagi

Introduction

Rice remains the most important staple food, feeds roughly half the planet's population and approximately three quarters of a billion of the world's poorest people depend on rice to survive. Contributing about 42 per cent to country's food grain production, it provides 80% of food calorie requirements. The revolutionary feature of rice is its steady, long-term evolution with societal change. India is the largest rice growing country accounting for about one-third of the world acreage under the crop. It is grown in almost all the states of India, covering more than 30 per cent of the total cultivated area.

Rice is indigenous to India. The sub-continent is rich in genetic diversity. In the crop improvement programmes not only indigenous but also exotic germplasm has been used for specific traits. There is still a need to augment diversity of genes that exist in rice gene pool and hence, there is necessity of procuring germplasm from other climatic/geographical regions for increasing the yield, stress management and value addition. Unregulated germplasm exchange activity may, however, result in inadvertent introduction of weeds, insects, nematodes and pathogens exotic to the country and harmful to agriculture. It is, therefore, necessary for a country to have adequate safeguards while introducing seed/ planting material. Plant quarantine provides such safeguards as it regulates the introduction of plants, planting materials, plant products, and soil, living animals etc with a view to prevent the introduction of harmful pests, their establishment and further spread.

Need for the establishment of an organization to regulate plant introduction activities in India was felt as early as 1935 by the Board of Agriculture and Animal Husbandry. Late Dr. B.P. Pal, in his classic paper, 'Search for New Genes' (published in 1937) also emphasized the importance of germplasm augmentation and utilization in crop improvement programmes. The need was reiterated in a meeting of the Indian Society of Genetics and Plant Breeding in 1941, which *inter alia* discussed the subject of economic crops. Dr. B.P. Pal, working at the then Imperial (now Indian) Agricultural Research Institute (IARI), New Delhi had approached the then Imperial (now Indian) Council of Agricultural Research (ICAR) to set up a unit for assembly of global germplasm under phytosanitary conditions in India. The ICAR scheme for 'Plant Introduction' commenced functioning in 1946 in the then Botany Division of IARI under the leadership of Late Dr. Harbhajan Singh as the first 'Operational Scientist'. The scheme was further expanded and strengthened as 'Plant Introduction and Exploration Organization' in the Botany Division in 1956, and later established as a separate 'Division of Plant Introduction' in IARI in 1961 headed by Dr. Harbhajan Singh (Dhillon *et al.,* 2001).

Subsequently, on the recommendations of the 'High Level Committee' constituted by the Government of India in 1970, and the foresight of visionaries like, Dr. B.P. Pal, Dr. M.S. Swaminathan and Dr. A.B. Joshi, the 'Division of Plant Introduction' was upgraded into an independent organisation 'National Bureau of Plant Introduction (NBPI) in 1976 and later renamed as National Bureau of Plant Genetic Resources (NBPGR) in 1977. The units of Entomology, Nematology and Plant Pathology in their respective divisions at IARI carried out the quarantine inspection, treatment and clearance of introduced seed/planting material. Subsequently, the Division of Plant Quarantine was created at NBPGR with the amalgamation of Entomology, Nematology and Plant Pathology Units.

NBPGR operates under ICAR system controlled by the Department of Agricultural Research and Education (DARE) of the Government of India. NBPGR is now playing a very dynamic and key role in overall regulation and management of Plant Genetic Resources (PGR) in India. NBPGR is augmenting the large amount of variability for broadening the genetic base by importing genetic wealth of different crop plants from different ecological areas of the world. These materials have been used all over the country by various institutes in various crop improvement programmes.

Historical Perspectives of Exchange

Genetic diversity is considered as an important criterion while choosing the parents for realizing heterosis and recombination in breeding programmes.

Within a limit, inclusion of more diverse parents in hybridization is believed to increase the chances of obtaining stronger heterosis in hybrids. In order to provide a rational basis for the selection of suitable parental lines, it is important to ascertain the magnitude of diversity among maintainers and elite restorer lines. Some introductions have been used directly as new varieties and became popular commercial cultivars while others are being used as a donors/parents and then released as new varieties after several selection cycles. The introduction of Taichung (Native) 1 from the semi dwarf mutant *Dee-geo-woo–gen* of Taiwan in the 1960s was the most significant introduction in the history of rice breeding in India. This was the single variety adaptable to whole of India and out yielded both in *kharif* and *rabi* seasons. Introduction of *japonica* rice for inter-sub-specific (*indica* x *japonica*) hybridization to recombine high fertilizer responsiveness and non-lodging habit of the *japonica* rice with the grain quality and adaptation of *indica* rice resulted in release of varieties such as Mahsuri and Adt 27.

Exotic germplasm has been utilized in various crop improvement programs enhancing yield parameters, incorporating resistance/tolerance to various biotic and abiotic stresses and value addition. A total of 5,86,991 accessions were imported/ introduced from Belgium, Bhutan, Brazil, Canada, China, Columbia, France, Indonesia, Japan, Kenya, Nepal, New Zealand, Nigeria, Philippines, South Korea, Sri Lanka, Switzerland, Thailand, USA, USSR and Vietnam. Additionally, every year materials for trials/nurseries are received from IRRI, Philippines for multilocation trials. Some important trait-specific introductions included EC121003 for cold tolerance; EC121807-15 and EC307145-61 for blast resistance; EC121816-23 and EC201878-975 for bacterial leaf blight resistance; EC121824-25 for resistance to stem borer; EC121828-53 and EC122935-40 for leaf folder resistance and salinity tolerance; EC199465-66 for sheath blight and stem borer resistance; EC309699-705 for green hopper resistance, EC161320-46 for rice tungro virus resistance, EC268456-931 for soil stresses (salinity, alkalinity, acid sulphate toxicity, phosphate deficiency) and submergence tolerant germplasm and many more accessions in last three decades were augmented from various countries particularly from Philippines. International Rice Research Institute (IRRI), Philippines now maintains, on behalf of humanity, the world's most complete and diverse collection of rice germplasm. Other introductions having broad adaptability gene *CP-SLO-7* from Philippines and multiple disease resistance varieties MR 219, MR 220 from Malaysia were also added to Indian gene pool. *Oryza* species collections and varieties Akibara with good cooking quality and stable yield, Akilikari with high grain yield and Hassa Rumochi are important introductions for rice breeding programme. Short statured upland and waxy variety Hatsukaori , varities Kinpa with good grain quality and tolerant to bacterial blight, Nipponbare and short statured Roimui were introduced from Japan. Variety Barkat (K-78-13) -highly resistant to lodging and frost with cold tolerance developed from a cross between two *japonica* cultivars, (Shin-ei x Ch.971) was released in 1978 for cold and high altitudes in J & K. Other varieties like China -972, 988, 1007 and China - 1039 released during 1975-78, were the secondary selections from mother stocks having good cooking quality. Two direct introductions released as varieties in Indian condition, viz, Intan, from Indonesia and cv. Munal from USA were adapted in hilly areas in Karnataka. Tropical *japonicas*, *indica* and *japonica* varieties of *O. sativa, glaberrima* rice from WARDA; upland varieties, super rice from China and Brazil and other trait-specific varieties need to be introduced further. Some of the rice introductions (non-transgenic and transgenic) made since 2000 are given in Table 11.1.

Regulatory Aspects of Germplasm Introduction

Exchange of plant material on a world wide basis had been mostly carried out without regard to well defined procedures, but several countries have established plant introduction services/organization for channelising,exchange (import and export) of plant materials. The National Plant Introduction Services of most countries have some features in common which facilitate their efficient operation.

- Plant Introduction Services usually forms part of the Department of Agriculture or a major research organization and hence is able to draw upon extensive resources in carrying out its work.
- Plant Introduction Services either has direct responsibility for plant quarantine or works in close collaboration with the quarantine authority.
- Plant Introduction Services has at its disposal good testing facilities in all the major climatic zones of the country concerned.
- Plant Introduction Services maintain records of introduced plants, usually including plants introduced by other agencies and is thus able to ensure, that national requirements are met without unnecessary duplication.
- Plant Introduction Services provide a contact point for the International Collaboration through FAO and other agencies.

By maintaining extensive collection, and by establishing close collaboration with its counter parts abroad, the plant introduction service is able to provide plant material needed by specialists, across the country.

Table 11.1. Some major introductions of exotic rice germplasm through NBPGR for research purposes.

Exotic Collection . (EC) No	Source Country	Year	Specific traits of introduced material	Supplied to
Non-transgenic				
453763-765	USA	2000	Variety Madison, Lemont and Katy, typical long grain, aromatic and light brown pericarp	•NBPGR Regional Station, Cuttack
459717-920	Philippines	2000	Bacterial blight, stem borer and sheath blight resistant lines.	•NBPGR Regional Station, Hyderabad
463280-281	Philippines	2000	Drought tolerant lines	•TNAU, Coimbatore
469595	Philippines Cuttack	2001	Var. CP-SLO-17 with wide compatibility gene	•NBPGR Regional Station,
494242-249517348-401	Philippines	20022003	Salinity tolerant lines	•Pt JL Nehru College of Ag. & Res., Karaikal, Pondicherry
494446	Japan	2002	Var. Suifou fujisaka-5 resistant to rice blast *(Pyricularia grisea)*	•DRR, Hyderabad, •NBPGR Regional Station, Cuttack
4944447 478048-052	Japan Philippines	2002	Var. Mochi - Glutinous rice	•DRR, Hyderabad, •NBPGR Regional Station, Cuttack
526554-561	Philippines	2003	Submergence tolerant lines	•NDUAT, Faizabad, UP
512353-361513198-237	Philippines	2003	New Plant types (NPT) - less tillering, all tillers bearing panicle, more number of grains per panicle and high harvest index	•DRR, Hyderabad
552817	USA	2004	GP KBNT Lpa 1-1 with low phytic acid (39%)	•DRR, Hyderabad

552822	USA	2004	GP GLPA, gold hull, low phytic acid germplasm	•DRR, Hyderabad
550889	USA	2004	Var. GP Aromatic, having good aroma and cooking quality of Basmati 370, early maturing	•DRR, Hyderabad
552825	USA	2004	Var. GP LGRU ef with early flowering	•DRR, Hyderabad
548000-001	Malaysia	2004	Var MR 219, MR 220 having multiple disease resistance	•DRR, Hyderabad
550177	USA	2004	Var. Boliviar having resistance to blast, narrow brown leaf spot (*Cercospora janseana*), long grained, superior parboiling and canning quality, early maturing	•DRR, Hyderabad
550175	USA	2004	Var. Saber having resistant to blast (*Pyricularia grisea*), sheath blight (*Rhizoctonia solani*), brownleaf spot (*Cercospora janseana*), leaf smut (*Entyloma oryzae*) and panicle blight. Long grained, superior parboiling and canning quality and early maturing	•DRR, Hyderabad
546319-340	Philippines	2004	Phosphorus deficiency and Iron toxicity tolerant lines	•IGAU, Raipur
541907	Philippines	2004	High Iron content	•IGAU, Raipur
552826-856	Philippines	2005	CMS, Maintainer, TGMS and Restorer lines	DRR, Hyderabad
572259-267	USA	2006	Early maturing with resistance to blast disease, kernel smut, rice water weevil and rice stalk borer	•DRR, Hyderabad
581124-127	Philippines	2006	QTL lines	•CSSRI, Karnal

583411-453	Philippines	2006	Resistant to Grassy stunt virus	•DRR, Hyderabad
609612	USA	2007	Indica rice genetic stock mutants	•DRR, Hyderabad
634219-220	USA	2008	Mutant germplasm lines having improved resistance to sheath blight, bacterial panicle blight & narrow brown leaf spot	•DRR, Hyderabad
638154-165	Philippines	2009	Lines resistant to blast and tolerant to Rice tungro virus	•Rice Research Stn, Tirur, TN
Transgenic				
497299	Switzerland	2002	Containing Psy bacterial phytoene desaturase (Crtl) and lycopene B-cyclase responsible for synthesis of beta carotene in rice endosperm	•DRR, Hyderabad
499743	IRRI, Philippines	2002	Containing Xa-21 gene responsible for imparting resistance to bacterial blight	•Mahyco Research Foundation, Hyderabad
501369-443	Belgium	2002	Containing *Cry1Ab* gene expressing a protein and imparting resistance to lepitopteran insects; *BAR* gene expresses a protein conferring resistance to glufosinate anmmonium	•M/s Hybrid Rice International, New Delhi
531285-305	Vietnam	2003	Containing phytoene synthase (*Psy*), bacterial phytoene desaturase (*Crtl*) and lycopene beta-cyclase (*lcy*) genes responsible for synthesis of beta carotene in rice endosperm,	•DRR, Hyderabad
541413-437	Germany	2004	Containing single chain Fv fragments genes (*ScFv*) from chicken and mice, antifungal protein (*AFP)-AG* gene from *Aspergillus giganteus*; AFP-RS gene from radish; AFP-Lactoferricin gene from human; AFP-Chitinase from wheat and hygromycin phoshotransferase (hpt) from *Streptomyces hygroscopicus*	•M/s Maharashtra Hybrid Seeds Company Ltd, Mumbai, Delhi

			having resistance to certain bacterial and fungal pathogens and hygromucin resistance.	
560441-446	UK	2005	American long grained cultivar Cocodrie capable of synthesizing ß carotene, containing *Psy,* Crtl gene	•DRR, Hydearbad Division of Genetics, IARI, New Delhi
571843-48	USA	2006	Containing *Psy* (Phytoene synthase), *Crtl* (Phytoene desaturase) genes	•Division of Genetics, IARI, New Delhi
588381	China	2006	Medium maturity and showing enhanced resistance against infestations of rice stem borer and leaf folders expressing Bt fusion gene (*Cry1Ab-1AC*)	•M/s Nath Seeds Ltd., Aurangabad, MS
604745	USA	2007	Expressing 5-enolpyrulshikimate-3 phosphate synthase (EPSPS) gene conferring tolerance against glyphosate	•M/s Maharashtra Hybrid Seeds Company Ltd, Mumbaı
611943-003626367-398	Belgium	2008	Containing *CryAb, Cry1c* and BAR genes	•M/s Bayer Biosciences Pvt. Ltd., New Delhi
EC630492-494	China	2008	Expressing *Cry2A, Cry1c & Cry1Ac* genes for imparting resistance against rice stem borer & rice leaf folder	•Pioneer Overseas Corp, Hyderabad
EC630495-96	IRRI, Philippines	2008	Containing ferritin gene for insect and pest tolerance	•RRS, Chinsurah, West Bengal

In India, there is a single window system for the exchange of small samples of plant germplasm (including transgenics) meant for research, and NBPGR is the nodal institution for that. It regulates the import of seeds/ planting material for research under the provisions of Plant Quarantine (Regulation of Import into India) Order, 2003 of the Destructive Insects and Pests (DIP) Act of 1914. The plant introductions include germplasm, elite strains, improved varieties, genetic stocks and related species in field, horticultural, agri-silvicultural, plantation and industrial crops from various parts of the world. In order to prevent the introduction and spread of pests associated with PGRs and to ensure that only pest-free material is supplied to the indentors, strict plant quarantine protocol is followed including post-entry quarantine (Ram Nath, 1996; Parakh and Gautam, 1999). The imported consignments treated with pesticides are grown and observed in post-entry quarantine facilities (screen houses, glass-houses) for detection of exotic pests. Seed harvested only from pest-free healthy plants is supplied to the indenters (Mathur and Lal, 1996; Khetarpal *et al.*, 2001). NBPGR is empowered by the Ministry of Agriculture to undertake quarantine processing of all germplasm and research material including transgenics under exchange for both public and private sectors.

Destructive Insects and Pests (DIP) Act, 1914

The Government of India in 1914 legislated a comprehensive Destructive Insects and Pests (DIP) Act, 1914 (Act 2 of 1914) to regulate or prohibit the import into India, or from one state to another, of any article likely to carry any pest (fungus, bacteria, nematode, insect, mite, weed etc) that may be destructive to any crop (Mathur and Lal, 1996; Majumdar, 1998; Bhalla *et al.*, 2009).

- It authorizes the Central Government to prohibit or to regulate the import of any article or class of articles into India or any zone/area of its territory.
- It authorizes the Central Government to prohibit or regulate the export from any state or movement from one state or movement to another within the country of plants or planting material
- It authorizes the state Governments to make rules for detection, inspection, disinfection or destruction of any article or class of articles in respect of which Central Government has issued notifications.
- It provides for penalties for person(s) who knowingly contravene the rules and regulations issued under the ACT.
- It also protects the person(s) from any suit or prosecution or other legal proceedings- for anything done in good faith under the Act.

DIP Act passed on 3rd Feb, 1914, has been revised and amended several times. The seed was not originally included in the DIP Act, but in 1984, the Government of India passed the Plants, Fruits and Seeds (Regulation of Import into India) Order of 1984.

Plants, fruits and seeds (Regulation of import into India) Order 1984 & 1989

Under the Plants, Fruits and Seeds (PFS) Order, seed was brought under the purview of the DIP Act and the main features included that no consignment can be imported into India without an Official Import Permit and Official Phytosanitary certificate. Post-entry isolation of specified crops at approved locations was stipulated.

The main features of the order are as follows:

- No consignment shall be imported into India without a valid import permit to be issued by a competent authority.
- No consignment shall be imported into India unless the same is accompanied by a Phytosanitary Certificate issued by an authorized official of the plant quarantine service of the source (exporting) country.
- All consignments shall be imported only through prescribed ports of entry.
- Seed/planting materials requiring isolation growing in detention, shall be grown in approved post-entry quarantine facility

The responsibility to enforce the quarantine regulations basically rested with the Directorate of Plant Protection, Quarantine and Storage headed by the Plant Protection Adviser to the Government of India. The order regulated the import of only 19 crops. Therefore, there was a gap left out having high risk of the introduction of exotic pests and diseases with the plant and plant materials not covered in the Order. Therefore, the need was felt to revise the plant quarantine regulation to safeguard from exotic pests and diseases.

To provide the best planting materials available worldwide to Indian farmers and seed growers to enhance productivity, Government of India through Gazette notification, liberalized the existing rules by announcing New Policy on Seed Development (NPSD) on 16 September 1988. To accommodate the provisions of NPSD, Plants, Fruits and Seeds (Regulation of Import into India) Order, 1989 was promulgated by the Government of India.

New policy on seed development

The Government of India introduced the New Policy on Seed Development (NPSD) in October, 1988 with an objective to make available to the Indian farmers the best genetic materials available anywhere in the world. The policy aimed to ensure that while importing seeds/planting materials, all care is to be taken with regard to plant quarantine requirements/procedures to preclude the entry of exotic pests, diseases and weeds detrimental to Indian Agriculture (Radhey Shyam, 1995).

The policy laid emphasis on two basic aspects firstly, liberalization of imports alongwith streamlining of plant quarantine procedures and secondly, certain incentives to encourage domestic seed industry. The policy specially emphasized on the import of high quality seeds/planting materials, strengthening/ modernization of plant quarantine facilities; effective observance of plant quarantine requirements/ procedures and incentives to the domestic seed industry.

The Agencies eligible to import the seeds/planting material under the Policy were

a) Department of Agriculture/Horticulture of the State Governments, State Agricultural Universities and ICAR

b) Seed producing Indian companies/firms, after registration with National Seeds Corporation

c) National Seeds Corporation and State Seeds Corporation or State Farms Corporation

d) Food processing industrial units

e) Growers of vegetables/flowers registered with the Director of Agriculture/ Horticulture of the State Governments.

As, the significance of plant quarantine increased in view of globalisation and liberalisation in international trade of plants and plant material in the wake of Sanitary and Phytosanitary (SPS) Agreements under WTO, the import of agricultural commodities is presently regulated through the Plant Quarantine (Regulation of Import into India) Order, 2003 issued under sub-section (1) of section 3 of the Destructive Insects and Pests Act, 1914 (2 of 1914) incorporating the provisions of New Policy on Seed Development, 1988. The phytosanitary certification of agricultural commodities being exported is also undertaken through the scheme as per International Plant Protection Convention (IPPC), 1951 to which India is a signatory.

Plant Quarantine (Regulation of Import into India) Order, 2003

Plant Quarantine (Regulation of Import into India) Order, 2003 commonly referred as PQ Order, 2003 is the latest regulation for the purpose of prohibiting and regulating the import into India of agricultural articles and came into force on the 1st of January, 2004 except sub-clause (22) of Clause 3 which came into force from 1st April 2004. The text of this order is available at *www.plantquarantineindia.org* which states the conditions of import into India of seed/planting material. Till June 2009, 14 amendments to the PQ Order, 2003 have been notified and there are five draft amendments, with the revised list of crops under the Schedules VI, VII and VIII. The revised list under Schedules VI and VII include 677 and 286 crops/commodities respectively (Bhalla *et al.*, 2009). The PQ Order, 2003 ensures the incorporation of "Additional/Special Conditions" for import of crops/commodities free from quarantine pests on the basis of pest risk analysis (PRA) following the international guidelines (Table 11.2).

The primary objectives of PQ Order, 2003 are to prevent the introduction and spread of exotic pests that are destructive to crops by regulating/restricting the import of plants/plant products and to facilitate safe global trade in agriculture by assisting the producers and exporters by providing a technically competent and reliable phytosanitary certificate system to meet the requirements of trading partners. The major activities include:

- Inspection of imported agricultural commodities for preventing the introduction of exotic pests and diseases, inimical to Indian Fauna and Flora.
- Inspection of agricultural commodities meant for export as per the requirements of importing countries under International Plant Protection Convention (IPPC).
- Survey and surveillance for detection of exotic pests and diseases already introduced for containing/controlling them by adopting domestic quarantine regulations.
- Undertaking Post Entry Quarantine Inspection in respect of identified planting materials.
- Conducting the Pest Risk Analysis (PRA), to finalise phytosanitary requirements for import of plant/plant products.

The commercial imports of seeds of coarse cereals, pulses, oilseeds and fodder crops and rootstocks/seedlings of fruit plant species for propagation shall only be permitted based on the recommendation of Export-Import [EXIM] Committee of Department of Agriculture and Cooperation except the trial material of 19 crops as specified in Schedule XII of PQ Order, 2003 (*www.plantquarantineindia.org*). The Plant Protection Advisor (PPA) to the

Table 11.2. Quarantine pests of rice as per plant quarantine (Regulation of Import into India) Order 2003.

Schedule	Purpose of Imports	Additional declaration required to be incorporated in Phytosanitary Certificate	Special conditions of import	Institutions Authorized to import
Schedule-V Clause 3 (3) (6) and 10 and 11(3)	Seed for sowing	(i) Freedom from (a) Granary weevil (*Sitophilus granarius*) (b) White tip nematode (*Aphelenchoides besseyi*) (c) Leaf streak (*Pseudomonas avenae*) (d) Sheath brown rot (*Pseudomonas fuscovaginae*) (e) Seeding rot (*Pseudomonas glumae*) (f) Bacterial halo blight (*Pseudomonas syringae* pv. *Oryzae*)	Seed soaking overnight and hot water treatment at 52 °C for 10 min.	Shall only be imported by Director, NBPGR New Delhi/ Director, Directorate of Rice Research, Hyderabad

Schedule	Purpose of Imports	Additional declaration required to be incorporated in Phytosanitary Certificate	Special conditions of import
Schedule-VI Clause 3 (3) (6) and 10 and 11(3)	Grain for consumption	Freedom from (a) Granary weevil (*Sitophilus granarius*)	Fumigation with Methyl Bromide @32 gm/cu m at 28 °C and above under NAP and the treatment to be endorsed on Phytosanitary Certificate or by any other fumigant/ substance in the manner approved by the Plant Protection Adviser for this purpose.

Government of India or other issuing authorities as listed in Schedule X shall issue the import permit for import of seed and plants for sowing or planting and for import of plants/ plant products for consumptions. The import permit shall be issued subject to such conditions and restrictions prescribed in Schedule V, VI and VII. All consignments of seeds and plants for propagation and regulated articles such as live insects, microbial cultures, biocontrol agents and soil shall only be imported into India through regional plant quarantine stations namely Amritsar, Chennai, Kolkata, Mumbai or New Delhi.

Under the PQ Order, 2003 the Government of India has made it obligatory for all the importers including plant breeders and researchers intending to import seed/ plant products, to fulfill the following two mandatory requirements

- Import permit from the issuing authority of Government of India.
- Phytosanitary certificate from the country of origin.

These two documents must accompany every consignment of seed/ plant products imported from abroad.

Issuance of import permit for germplasm

As per PQ Order 2003, Clause 6 (2) Director, NBPGR, New Delhi is authorized to issue import permit for import of plant germplasm/transgenics/ genetically modified organisms for research/experimental purpose by the public/ private organization. The import of germplasm for research is permitted through New Delhi Airport only. NBPGR receives such imported consignment from custom authorities for its quarantine inspection and clearance. The applicant desirous of importing seed/planting material has to apply to the Director, NBPGR on a prescribed application form (*PQ Form 08*) for obtaining the Import permit (IP). The IP issued in form PQ 09 in triplicate is valid for six months and is non-transferable. The application form (PQ 08) and other details (processing fee charges) are available at NBPGR website www.nbpgr.ernet.in

After obtaining import permit the applicant should send it to the concerned supplier that has agreed to supply the required germplasm for use in research with the request that the import permit in duplicate must be enclosed alongwith the consignment of seed/planting material. However, it should also be ensured that the consignment must be addressed to the Director, NBPGR only and the exporter/ supplier of the material should be requested not to treat the seed or planting material with any chemical until and unless asked to do so in the import permit.

Phytosanitary Certificate (PSC)

The second mandatory requirement is that of Phytosanitary Certificate which is to be issued by the official agency of the donor country. Every consignment shall be accompanied by PS (original copy) issued by the Government Official from country of origin in the prescribed format of Food and Agriculture Organization (FAO). Phytosanitary Certificate should specify additional declarations for freedom from specific pests and diseases as specified in the import permit or that the pests specified do not occur in the country or state of origin and be supported by documentary evidence. Phytosanitary Certificate is also issued by the NBPGR for all germplasm material meant for export to foreign countries. It is issued by Plant Quarantine Division, NBPGR, New Delhi or by NBPGR Regional Station at Hyderabad (for ICRISAT mandated crops). A flow chart of the different steps involved in processing of applications for import of plant germplasm/ transgenics/ genetically modified organisms is given in Fig. 11.1.

Gazette of India extra ordinary Part II Section-3 sub section (1) published by authority no. 621, New Delhi has defined Rules for the Manufacture, Use, Import, Export and Storage of Hazardous Microorganism/Genetically engineered organisms or cells made under sections 6,8 and 25 of the Environment (Protection) Act, EPA 1986 (29 of 1986). EPA plays a major role in minimizing the risk from pollutants and contaminants affecting flora and fauna, human and animal health and preserving the environment. In accordance with this Act, all transgenic plants are regulated items and hence, the provisions of PQ Order, 2003 are applicable to import of transgenic seeds as well.

Department of Biotechnology (DBT), Ministry of Science and Technology and the Ministry of Environment and Forests (MoEF) have separate set of prescribed procedures for providing permission for import of transgenics. An indenter who wishes to import transgenics has to submit the proposal through the Institutional Biosafety Committee (IBSC). IBSC recommends the application to Review Committee on Genetic Manipulation (RCGM) which is an authorized agency of the Government of India, and functions under DBT.

RCGM assesses the import indents, and examines the desirability of import of transgenic line, from the biosafety point of view, for research purposes and issues Seed Transfer Clearance Letter (valid for one year). RCGM includes representatives from DBT, Indian Council of Medical Sciences (ICMR), Indian Council of Agricultural Research (ICAR), Council of Scientific and Industrial Research (CSIR) and other experts in their individual capacity. Genetic Engineering Approval Committee (GEAC) functioning under the MoEF regulates the commercial introduction (large-scale use/ experiment) of transgenic material (Tyagi

Submission of application (PQ 08) to Director, NBPGR, New Delhi alongwith prescribed fee and relevant documents*

↓

Scrutiny of application (PQ 08) and issuance of Import Permit in triplicate having validity for six months

↓

Import Permit in duplicate is sent to indentor/ supplier of the material with important instructions to be followed

↓

Consignment got released from customs/ foreign post office or received from custom services Delhi Airport

↓

Consignment is checked for IP and PSC by Exchange Division and sent to Plant Quarantine Division

↓

Material released by PQD is received back by Germplasm Exchange, EC/ ET numbers assigned

↓

Indentor is informed to pay the handling charges and Quarantine fee

↓

Material is dispatched to the indentor

* Relevant documents may include Certificate of R&D recognition from Department of Scientific & Industrial Research (DSIR), Ministry of Science & Technology, Govt. of India. For import of transgenics, clearance letter from DBT and undertakings as per para 4, 5, and 6 must be submitted with PQ 08 form

Fig. 11.1. Steps in Import of Germplasm for Research

et al., 2007).

After getting the technical clearance for import from RCGM, the indentor has to apply to the Director, NBPGR, New Delhi for the issuance of IP in the PQ form 08 enclosing with it the Department of Biotechnology Permit/ Seed Transfer Clearance Letter. The indentor is also required to furnish the information and provide undertakings as per para 4, 5 and 6 of the DBT permission.

Para No. 4

- No transgenic material is permitted for experimentation in open environment without prior authorization from the Government of India.
- Full account of transgenic plants raised from the imported seeds is to be kept in a bound book, which should be available for inspection by the authority in case such a need arises.
- All transgenic materials prescribed by the indenters may be available for inspection, whenever required.
- All the unwanted transgenic materials may be destroyed by burning after the experiments are conducted.
- For any use of the transgenic material for propagation in the open environment, the applicant will make a separate application to RCGM through Institutional Biosafety Committee (IBSC).
- All precautions would be taken to prevent the escape of the genetic material into the open environment and shall follow the Recombinant DNA Safety Guidelines of the Government of India (http://dbtindia.nic.in/uniquepage.asp).
- The applicant shall certify to the NBPGR that to his/ her best of knowledge and belief the material being imported confirm to the description given in the import clearance.
- NBPGR shall retain up to 5% of the transgenic seeds in its facility and keep them under safe custody. NBPGR shall further issue a receipt to that effect to the applicant.
- The seeds retained by the NBPGR shall be used by the Government for future reference.

Para No. 5

- The supplier of the transgenic material shall certify that transgenic material has the genes as has been described in the permission.

- The supplier shall also certify that these transgenic materials do not contain any embryogenesis deactivator gene sequence.

Para No. 6

The transgenic material shall be handled, packaged and transported as specified in "r-DNA safety guidelines and revised guidelines for research in transgenic.

Release of the consignment (transgenic and non-transgenic)

After obtaining import permit the recipient should send both the copies to the concerned official/supplier (abroad) that have agreed to supply the required germplasm for use in research. The supplier should be instructed to send two copies of IP (one pasted outside the seed parcel and other inside the seed packet) and other relevant documents with the consignment that are required for custom clearance. It should be clearly instructed to the sender/ supplier that consignment should be addressed only to the Director, NBPGR, New Delhi. The port of entry of germplasm is New Delhi Airport only.

The imported/ introduced material is released to the indentor and assigned an Exotic Collection (EC) number which is unique national identity for all material imported into India. The number should be quoted in all future references and should remain unchanged.

Documentation of germplasm

All indents for import of germplasm are registered for assigning the case number and then forwarded to the Plant Quarantine (PQ) Division without opening the parcel alongwith duly filled Import Quarantine (IQ) form for detailed quarantine inspection and clearance. A flow chart of the different steps involved in quarantine processing of imported germplasm at NBPGR is given in Fig. 11.2. After clearance from PQ Division, the samples are first arranged taxonomically indicating their genus, species, common name and cultivar name etc. for national accessioning in the national record. Accessioning of Exotic accessions are being done online (http://www.nbpgr.ernet.in/geq) and each introduction/ accession is assigned an EC (Exotic Collection) number (which remains unchanged) and documented alongwith all the detailed information such as

- Import Quarantine (IQ) number and date
- Source country/ country of origin

- Complete postal address of the source country
- Import Permit number and date of issue
- Phytosanitary Certificate number and date
- Biological status/variety name/type/category of the material
- Distribution details
- Indenter of the material
- Whether the material was introduced under MoU/restored/ transgenic?
- Additional information/specific traits/remarks if any.

Regulatory aspects of exports: Access to biological resources of India vis a vis Biological Diversity Act (BDA), 2002

Under the provisions of the Convention on Biological Diversity (CBD), Government of India enacted legislation called the Biological Diversity Act (BDA), 2002 (18 of 2003) and also notified the Biological Diversity Rules, 2004. The objective is to provide for access to biological resources of the country and equitable share in benefits arising out of the use of biological resource together with sustainable use and conservation of biological diversity. As per the Act (Section 3), no person from outside India or a body corporate, association, organization incorporated or registered in India having non -Indian participation in its share capital or management, can access any biological resources or knowledge associated, for research, commercial utilization, bio-prospecting or bio-utilization, without prior approval of National Biodiversity Authority (NBA). Without obtaining prior approval of the NBA no person can apply for any IPR in or outside India for any invention based on any research or information on a biological resource obtained from India, as per Section 6 of BDA, 2002. Also, as per Section 4 of the BDA, without prior approval of the NBA, results of any research relating to any biological resources occurring in, or obtained from India for monetary consideration or otherwise can be transferred to any person who is not a Citizen of India or Citizen of India who is non- resident or a body corporate or organization which is neither registered nor incorporated in India or which has any non- Indian participation in its share capital or management.

However, the provisions of Section 3 and 4 shall not apply to collaborative research projects involving transfer or exchange of biological resources or information relating thereto between institutions including Governments sponsored institutions of India, and such institutions in other countries, if such collaborative projects conform to the policy guidelines issued by the Central Government in

Fig 11.2. QUARANTINE PROCESSING OF IMPORTED GERMPLASM

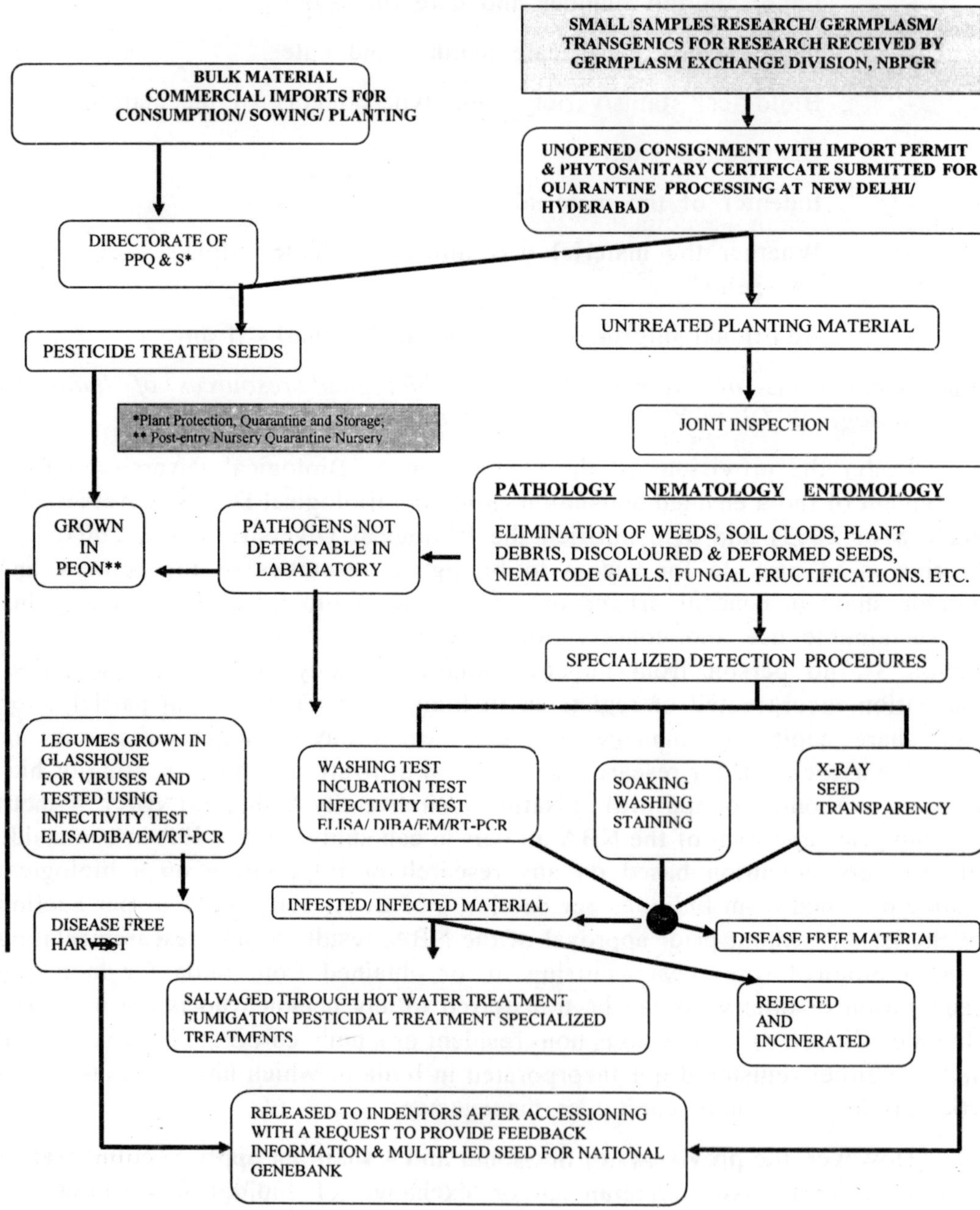

Fig. 11.3. Processing of applications for the export of PGRs for research purposes.

this behalf or approved by the Central Government (Fig. 11.2).

The person who shall be required to take the approval of the National Biodiversity Authority [Section 3 (2)] are the following, namely (a) a person who is not a citizen of India; (b) a citizen of India, who is a non resident as defined in clause (30) of section 2 of the Income Tax Act, 1961; (c) a body corporate, association or organization- (i) not incorporated or registered in India; or (ii) incorporated or registered in India under any law for the time being in force which has any non Indian participation in its share capital or management.

For exchange of Plant Genetic Resources for Food and Agriculture (PGRFA) under the multilateral system (MLS) of the International Treaty on Plant Genetic Resources for Food and Agriculture (ITPGRFA), only those crops that are mentioned in Annex I of the Treaty shall be considered. Presently, Annex I covers 35 food crops and 29 forage genera based on world food security and interdependence of countries for PGRs. This exchange shall be only among the contracting countries (countries signatory to the Treaty) and under the conditions of Standard Material Transfer Agreement (SMTA), approved by the Governing Body of the Treaty (http://www.planttreaty.org/).

Export of PGRs under collaborative research projects, in the light of notification of the Biological Diversity Act 2002 is now routed through the PGR Export Facilitation Committee (PGR-EFC). The PGR-EFC scrutinizes each proposal and submits its recommendation to DARE/ Chairman, NBA for final approval/ clearance. A Flow chart showing different steps in export of germplasm for research purposes is given in Fig. 11.3.

Any supply/exchange/transfer of genetic resources for food and agriculture/ germplasm/genetic material/genetic components for research out from India is done under Material Transfer Agreement (MTA). The MTA lists the mutually agreed conditions for the transfer done between two parties, and has some essential clauses, viz. name(s) and addresses of both parties, name(s) identity and other passport details of the materials to be exchanged, purpose of transfer, cost involved, treatment of derivates after further research IPR(s) generated, acknowledgements and arbitration conditions or dispute settlement etc. (Tiwari, 2006). The MTA specifies the conditions for the transfer of seed/planting materials under different categories mainly

- Within India, not covering persons as described in Section 3 (2) of the Biological Diversity Act, 2002 (18 of 2003) (BDA)
- Within India, wholly or partly covering persons as described in Section 3 (2) of BDA

Fig. 11.3. Processing of applications for the export of PGRs for research purposes.

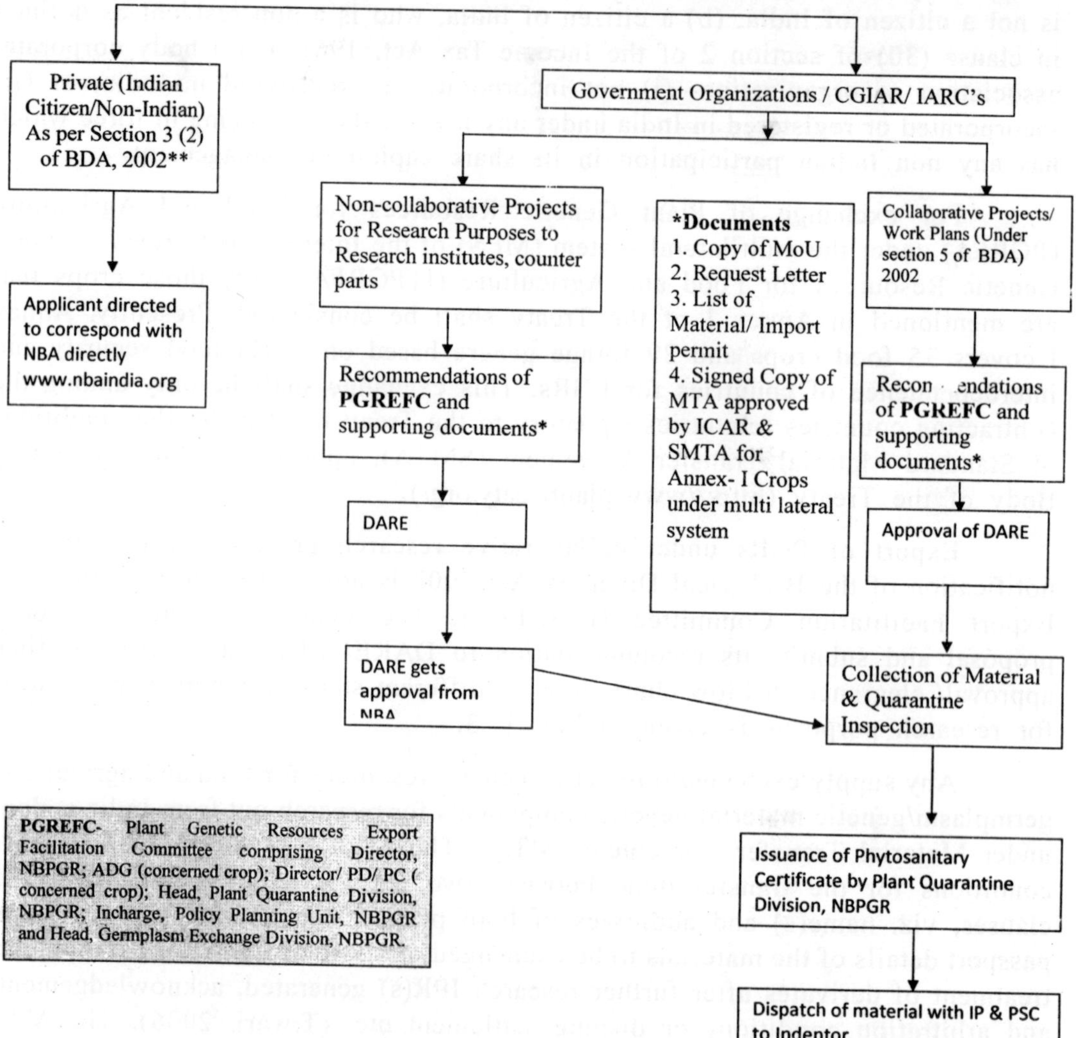

** The persons who shall be required to take the prior approval of the National Biodiversity Authority under Section 3 sub-section (2) are the following. (a) a person who is not a citizen of India; (b) a citizen of India, who is a non-resident as defined in clause (30) of section 2 of the Income-tax Act, 1961; (43 of 1961) (c) a body corporate, association or organization (i) not incorporated or registered in India, (ii) or incorporated or registered in India under any law for the time being in force which has any non-Indian participation in its share capital or management.

Fig 11.2. Quarantine processing of imported germplasm.

- Outside India, with non-members of ITPGRFA, and wholly or partly covering persons as described in Section 3 (2) of BDA

Standard Material Transfer Agreement (SMTA) specifies the conditions for exchange of germplasm between signatories (parties) with members of the ITPGRFA, wholly or partly covering persons as described in Section 3 (2) of BDA.

Commercial imports vis-à-vis EXIM policy

The Govt. of India, Ministry of Commerce and Industry announces Export Import Policy every five years. The Export Import Policy (EXIM Policy) is updated every year on the 31st of March and the modifications, improvements and new schemes are effective w.e.f. 1st April of every year. Chapter 6 of the EXIM policy deals with Export Oriented Units and Chapter 7 deals with units in the Special Economic Zones. Similarly, Govt. of India also releases the Hand Book of Procedures detailing the procedures to be followed in each of the schemes covered in the EXIM Policy. The objectives and salient features of EXIM policy are

- To accelerate the country's transition to a globally oriented vibrant economy with a view to derive maximum benefits from expanding global market opportunities.
- To stimulate sustained economic growth by providing access to essential raw materials intermediates, components, consumables and capital goods required for augmenting production.
- To enhance the technological strength and efficiency of Indian agriculture, industry and services, thereby improving their competitive strength while generating new employment opportunities, and encourage the attainment of internationally accepted standards of quality.
- To provide consumers with good quality products at reasonable prices.

These objectives are achieved through the coordinated efforts of all the departments of the government in general and the Ministry of Commerce and Industry and the Directorate General of Foreign Trade and its network of Regional Offices in particular, with a shared vision and commitment and in the best spirit of facilitation, in the interest of export promotion.

All the proposals for bulk import and export of rice are submitted to EXIM Committee Meeting and decisions are taken or approved on case to case basis and the recommendations are notified in the proceedings of EXIM Committee meetings. These meetings are conducted frequently (monthly/ bimonthly) depending on the number of proposals received. The committee members represent various offices of Seeds Division, Horticulture Division, Plant Protection Division of DAC, ICAR and other organizations and Chaired by Additional Secretary, DAC, Government of India.

Conclusions

The responsibility for import and export of plant genetic resources for research purposes has been delegated by the Ministry of Agriculture to the Indian Council of Agricultural Research (ICAR)/ Department of Agricultural Research and Education (DARE). NBPGR is the nodal agency recognized for PGR exchange within the country as well as outside the country. After enactment of the Biological Diversity Act, Protection of Plant Varieties and Farmers Rights Act (PPVFRA) and ratification of International Treaty on Plant Genetic Resources for Food and Agriculture by India, NBPGR has to follow new rules for exchange of plant genetic resources for food and agriculture. The import of rice germplasm for research purposes and bulk imports of seeds for multiplication or grains for consumption are regulated under the provisions of PQ Order, 2003.

Regulation of access to rice germplasm and information under the new regime in India has to take into account the established institutional mechanism and various acts in force relating to agro- biodiversity. The requests of indenters are to be dealt with depending on the status of requesting party and the conditions for access under different categories. NBPGR/DARE is the nodal agency for the purpose, has to regulate the access based on projects approved by State Government Department or Central Government, or MoU/Work plan and other relevant agreements. Collaborative research projects should clearly state key investigators/institute involved, details of the biological resources/associated knowledge to be exchanged (exported), special status of material if any as per any international agreement (e.g. FAO designated, INGER Nurseries, etc.), intended use for research and quantity for exchange to be limited to experimental purpose only. Conditions of MTA and SMTA also need to be strictly followed and monitored.

References and Useful Websites:

Bhalla S, Chalam V Celia, Arjun Lal, Khetarpal RK. 2009. *Practical Manual on Plant Quarantine.* NBPGR, New Delhi. 204 pp.

Dhillon BS, Pareek SK, Singh BM, Srinivasan K. 2001. NBPGR: Past and present. Pages 1-30 in Dhillon BS, Varaprasad KS, Srinivasan K, Mahendra Singh, Sunil Archak, Umesh Srivasatava and Sharma GD (eds) *National Bureau of Plant Genetic Resources: A Compendium of Achievements.* NBPGR, New Delhi.

Khetarpal RK, Varaprasad KS, Arjun Lal, Agarwal PC, Lal B. 2001. Plant Quarantine of germplasm under exchange. Pages 90-115 in Dhillon BS, Varaprasad KS, Srinivasan K, Mahendra Singh, Sunil Archak, Umesh Srivasatava and Sharma GD (eds) *National Bureau of Plant Genetic Resources: A Compendium of Achievements.* NBPGR, New Delhi, India.

Majumdar A. 1998. Plant quarantine procedures in relation to plant germplasm movement. Pages 174-187 in Gautam PL, Dabas BS, Umesh Srivastava, Duhoon SS (eds) *Plant Germplasm Collections: Principles And Procedures.* NBPGR, New Delhi.

Mathur VK, Arjun Lal. 1996. *Plant Quarantine Activities at NBPGR (1976-1996).* Scientific Monograph No 6, NBPGR, New Delhi. 96 pp.

Ministry of Agriculture (Department of Agriculture and Cooperation). 2003. *Plant Quarantine (Regulation of Import into India) Order 2003 and Its Amendments.* The Gazette of India Extraordinary, Part II Section 3 subsection (ii), No 1037, 2003. Notification dated 18th November, 2003.

Parakh DB, Gautam PL. 1999. Role of quarantine in prevention of diseases of horticultural crops. Pages 668-689 in Varma LR and Sharma RC (eds) *Diseases of Horticultural Crops – Fruits.* Indus Publishing Co, New Delhi.

Radhey Shyam. 1995. New policy on seed development and role of designated inspection authorities. Pages 119-134 in Rana RS, Mathur VK, Kazim M, Arjun Lal, Manju Lata Kapur, Brahmi P (eds) *Exchange & Quarantine of Plant Genetic Resources.* NBPGR, New Delhi.

Ram Nath. 1996. India's plant quarantine system works to thwart potentially devastating crop losses. *Diversity* 12(3):49-50.

Tiwari SP. 2006. *Regulatory and Operational Mechanisms as Related to Agro- Biodiversity.* Second and Revised Edition, NAARM (ICAR) Publication. 216 pp.

Tyagi Vandana, Deep Chand, Singh IP, Singh SP, Ranga SS. 2007. Guidelines for exchange of Transgenic Planting material. Pages 74-81 in Gurinderjit Randhawa, Shashi Bhalla, V Celia Chalam and SK Sharma (eds) *Cartagena Protocol on Biosafety: Decisions to Diagnostics.* NBPGR, New Delhi.

Websites:

http://www.nbpgr.ernet.in (NBPGR)

www.plantquarantine.india.org (PQ Order)

http://dbtindia.nic.in/uniquepage.asp (DBT, India)

http://www.nbaindia.org/contact.htm (NBA, Chennai)

http://www.planttreaty.org/ (ITPGRFA, FAO)

Genetic Resources of Rice in India : 391-409
Editor : S. D. Sharma
Today & Tomorrow's Printers and Publishers, New Delhi - 110 002 (India)

Chapter 12

Rice Germplasm: Conservation and Utilization in Tamil Nadu

M. Subramanian, S. Robin and K. Thiyagarajan

Tamil Nadu, one of the major and important rice growing states, currently cultivates rice, the staple food crop, in about 2.2 million hectares. Rice, the global grain crop is grown in varying agro-climatic and ecological conditions ranging from irrigated wetlands receiving water from canals or tanks (rainfed/riverfed) to less conspicuous rain-fed areas of dry and semi-dry systems of cultivation. In Tamil Nadu, rice soils also differ widely; they are sandy, clayey, acidic, alkaline and saline. Rice in Tamil Nadu is always under the mercy of rivers flowing from the Western Ghats during southwest monsoon season followed by the capricious northeast monsoon. Hence it is greatly affected by natural hazards like drought, flood, high wind, etc. The area under rice in this state is fluctuating; it ranged from 1.5 million hectares in 2002-03 to 2.7 million hectares in 1976-80. The reduction of rice area from 2.7 million hectares to the present 2.2 million hectares is mainly attributed to conversion of rice lands into non-agricultural purposes and crop diversification. Besides, uncertainties caused by erratic monsoon behaviour have also motivated farmers against rice.

Rice cultivation in Tamil Nadu state is most varied in the whole of India as its duration ranges from about 100 days to 240 days in different seasons. There are five major duration groups of rice cultivation in the state depending on seasons of cultivation (Table 12.1).

Among the duration groups, very early (up to 100 days), early (101-120), medium (121-140), long (141-160) and very long duration rice varieties are cultivated in Tamil Nadu. In a normal year, early and medium duration rice varieties cover larger area while long duration varieties also occupy considerable acreage

Table 12.1. Season and duration of rice cropping in Tamil Nadu.

S. No.	Name of the season (months of sowing)	Duration (days)
1.	*Sornavari* (April-May)	Short duration (101-120) Medium duration (121-140)
2.	*Kar* (May-June)	Short duration (101-120)
3.	*Kuruvai* (June-July)	Short duration (101-120)
4.	Early *Samba* (July-Aug)	Long duration (141-160)
5.	*Samba/Thaladi/Pishanam* (Sep-Oct)	Medium duration (121-140)
6.	Late *samba*/Late *Thaladi*/ Late *Pishanam* (Oct-Nov)	Short duration (101-120)
7.	*Navarai* (Nov-Dec)	Short duration (101-120)

in this state.

More than 65 million people living in Tamil Nadu are rice eaters preferring to consume rice three times a day; only occasionally they intake wheat products. Therefore the production of rice in this state should be enhanced and sustained to meet the demand of the ever growing population in this state. Rice productivity in this state before independence was only 805 kg/ha; this has subsequently enhanced to 3.5 t/ha so much so that the state is often ranked either first or second among the rice growing states in India. This current productivity level, however, is not sufficient and has to be increased further at least to 5 t/ha at the earliest possible time to feed the growing population and decrease the dependence on the central pool. The gradual increase in the productivity and production of rice since independence could be possible mainly due to concerted efforts by rice research workers. Their continuous and strenuous research on rice improvement has resulted in the release of many high yielding rice varieties and modern cultivation technologies which have helped to achieve this high level of yield in the state. Nevertheless, the success in the research should primarily be attributed to and could be possible because of the enormous variability gifted by nature to this crop. This variability, however, has not yet been exploited fully.

Rice (*Oryza sativa* L.) belongs to the family *Poaceae* (Graminae) and the tribe *Oryzeae*. The tribe *Oryzeae* consists of 12 genera including the genus *Oryza* which was so named by Linnaeus. The genus has 20 wild and two cultivated species, *viz., O. sativa,* the Asian cultivated species with its two varietal groups, viz., *indica* and *japonica* and the African cultivated rice, *O. glaberrima*. *Oryza* is considered closely related to bamboos (Tzvelev, 1989). They possess specific differences among their traits. Examples of potentially useful traits in genera related to *Oryza* are plants adapted to cold water (*Zizania*), salt water (*Porteresia*) and plants with unisexual spikelets (*Zizania, Luziola* and *Zizaniopsis*). Genus *Oryza* consists of species adapted to a broad range of habitats. Several species grow in shady forests, others in vast stands in deep water swamps. Wild rice can be found in the Himalayan foot hills, Asian river deltas, tropical Caribbean islands, the Amazon basin, the inland swamp lands of Southern and Western Africa as well as in temporary pools of the arid Savannas of the tropics. The wild species of *Oryza* are found almost exclusively within the tropics; cultivated rice, however, is grown as far as 50°N in China and 40°S in Argentina.

Rice is regarded as the first crop cultivated in Asia. Although the first recorded mention of rice originates from China in 2800 BC, Copeland (1924) and Chang (1976) asserted that rice might have originated in South or Southeast Asia including India, China, Thailand and Indonesia. Carbon dating of pottery shards with the imprints of rice grains in Thailand indicated that rice could be dated back to at least 4000 BC (IRRI, 1994). The three oldest archaeological evidences for rice cultivation are from Mahagara in India (6500–4500 BC), Non-Nok-Tha in Thailand (about 6000 BC) and Pengtoushan in China (7150–6250 BC) (Abrol and Gadgil, 1999). Great diversity of rice and linguistic evidences suggest that rice cultivation originated in Southeast Asia which includes Northeast Zone of India. Therefore, Indian subcontinent is considered as one of the centres of origin of rice *Oryza sativa* var. *indica.* This subcontinent is an important centre of genetic diversity where a large number of landraces, indigenous varieties of cultivated rice, forms of wild species and related genera are found.

Watt (1891) and De Candolle (1886) mentioned South India as the centre of origin of rice cultivation owing to richness in diversity of its cultivars. Unfortunately, majority of this genetic wealth has eroded due to neglect and developmental activities of selective breeding especially in the early decades of 20th century which also helped in the collection, preservation and purification of several landraces.

Unlike other crops, rice is gifted with enormous biodiversity spread in many parts of the globe including India. The landraces, indigenous varieties, cultivars, modern varieties, genetic stocks, breeding lines and wild species constitute the major components of the rice genetic diversity. Rice germplasm with wide array of variability is the wealth of the country because development in rice crop ever depends upon the crop genetic diversity. Particularly, the genetic differences scattered among the germplasm are the treasures and are much useful to improve and breed high yielding rice varieties with resistant to biotic and abiotic stresses and grain quality.

Exploration and conservation of the valuable rice genotypes was initiated during 1960s with a view to investigate their habitat, origin, natural variability and also to evaluate their relationship. This approach was very much useful to utilize elite genotypes to improve rice and to breed new varieties.

All the states in India were explored for the landraces of rice, indigenous varieties of cultivated rice, forms of wild species and related genera. Like other states, Tamil Nadu also has traditional varieties, indigenous rice genotypes. A few of them are still under cultivation. In particular upland and lowland rainfed rice are available with wide array of variation in their characteristic features like duration, height, kernel traits, panicle features, medicinal properties and resistance to biotic and abiotic stresses.

Traditional Varieties

Farmers over millennia left behind thousands of locally adapted rice varieties that have evolved under natural and artificial selection forces. The traditional varieties cultivated for millennia were highly adapted to their local conditions. They generally have low yielding ability and do not respond to high input managements. Under high input management, they grow tall with long leaves and more number of unproductive tillers. But most of them have good quality grains with better cooking, nutritive and medicinal properties besides resistance to biotic and abiotic stresses of their region. Most of the landraces that are still cultivated by the farmers are in areas where mechanization and irrigation are not possible or where agro-chemicals are not available or usable. The seeds of the landraces that are already preserved in seedbanks are out of continuous cultivation. On return, stored seeds often lose their chances to adapt to changing environments and agricultural practices. To obviate this lacuna, conservation by farmers is encouraged as it leads to continuous cultivation and selection by farmers and evolution of new types. Traditional rice varieties are genetic wealth of the country as they are the basic resources for further genetic improvement. However these traditional varieties are now slowly becoming extinct after the spread of high yielding varieties.

The short-statured new high yielding varieties performed well when favourable management practices were followed. Therefore, the spread of modern high yielding varieties and adoption of improved management practices like fertilizer and pesticide application, irrigation and mechanization were quick and rapid under high input agriculture. But these modern varieties exhibit narrow genetic differences. For example, most of them have white slender kernels as against local varieties which had various coloured kernels with varied hull thickness, grain weight and length besides possessing resistance to biotic and abiotic stresses. Introduction and spread of high yielding varieties pose major threat to the genetic diversity and cause genetic erosion. The modern varieties have almost completely replaced thousands of traditional landraces that were previously cultivated by farmers. Dwindling of landraces from cultivation is a common phenomenon in Tamil Nadu as in other states.

During 1966-67, only 25% of the total area was under high yielding varieties; the area increased to 57.8% during 1986-87 and to 76.7% by 1996-97 (Ratho, 2004; Saxena *et al.*, 2003). This indicates that in many areas, high yielding modern varieties were adopted by farmers and the cultivation of landraces / traditional varieties declined as high as 85–100%. Some of the traditional rice varieties cultivated in the rice growing areas of Tamil Nadu are provided in Box 12.1. Among these, some of the varieties are still under cultivation in different regions of this state offering their niche adaptation.

Collection of Rice Germplasm

The collection and conservation of rice germplasm are closely linked as the germplasm are utilized for rice improvement. In the colonial India in 1912, two Government Economic Botanists were appointed to work on rice - one at Dhaka (now in Bangladesh) and the other at Coimbatore in Madras province. The Paddy Experiment Station at Coimbatore which is a hundred years old is presently designated as Department of Rice under the Centre for Plant Breeding and Genetics of Tamil Nadu Agricultural University, Coimbatore. The Station had 2000 occasions of rice during 1911-1914 (Rai, 1999). It now possesses the old germplasm collections and relevant records on them. These accessions differ widely in duration, grain quality, rice texture, flavour, fineness, etc. and for morphological characters such as stem colour, seed coat, etc.

The prime task of this centre at the time of its inception was to collect and conserve the native genetic diversity that existed in rice. The old volumes of records (Fig. 12.1) indicate the efforts of germplasm collection undertaken with the help of revenue officials at that time. The preliminary collections included mostly traditional varieties known after their area of adoption and season. The collected germplasm were characterized and documented for their plant and grain morphology. Records on characterization indicated the efforts taken up to study the diverse parameters of phenology, plant morphology and grain characteristics. (Fig. 12.2) The germplasm collection grew as the years passed and presently a total of 2508 accessions are preserved in the local genebank at Coimbatore. The collections have been grouped based on their maturity duration *viz.*, early (< 120

Box 12.1. Traditional Rice Varieties of Tamil Nadu.

Traditional rice varieties of Tamil Nadu		
Anaikomban	*Kuliadichan*	*Sambalai*
Arivikiruvai	*Kullankar*	*Sembuli*
Arupatham samba	*Kuruvai*	*Senthinayagam*
Arupatham Kodai	*Kuruvaikalayan*	*Senkuruvai*
Avasara samba	*Manakkottai*	*Sigappu kuruvai*
Ayyan samba	*Manavari*	*Sivappu sirumani*
Bangaru theegalu	*Maranellu*	*Vellai sirumani*
Boombalai	*Marathondi*	*Sivappu ottadan*
Boon samba	*Mosam*	*Sornavari*
Chandadi	*Mottai kuruvai*	*Tedlla samba vadan*
Chettu Valian	*Mattaikar*	*Tella sanna vadlu*
Chingleput sirumani	*Munginel*	*Thanjavur vadan samba*
Channasamba	*Muthu samba*	*Thattara vellai*
Channavadan samba	*Muthuvellai Naraiyan*	*Thiruchengodu samba*
Chinthamani	*Nellore samba*	*Thiruthuraipoondi kar*
Chitrakali	*North Arcot Vadan samba*	*Thuyamalli*
Chittraikar	*Norungan*	*Uvarkondan*
Erachannavadlu	*Palthondi*	*Vadakkathikar*
Ghanthasala	*Panamarathu samba*	*Vadan samba*
Cheenellu	*Peruthanduvellai samba*	*Valiyan*
Garudan samba	*Periakitchli*	*Valsaramundan*
Gobikar	*Pisini*	*Varamilagi*
Kodaikannan	*Poombalai*	*Varappukudanchan*
Kaiveri samba	*Poongar*	*Vari garudan samba*
Kallurundaikar	*Pon samba*	*Veedivadangan*
Kappasamba	*Pudupatti samba*	*Velam samba*
Karsamba red	*Pulithipatti*	*Vellai gowni*
Kar samba white	*Rajayogam*	*Vellaikar*
Karthigai samba	*Rangoon samba*	*Vellai kattai*
Kattavellai	*Red ottadan*	*Vellai samba*
Kattuvanam	*Red sirumani*	*Vella kuruvai*
Kayama (non scented)	*Sadai samba*	*Vellai ottadan*
Kayama (scented)	*Sanna samba*	*Venginel*
Kesari Kolavali	*Sarapalli*	*Verlumbala*
Konakuruvai	*Seenganni*	*White ottadan*
Korungu samba	*Seevanu samba*	*White sirumani*
Kothanda	*Saradi*	*Yera sanna vadlu*
Kattanur		

days), medium (121-140 days) and late (>141 days). The Coimbatore genebank consists mostly of the traditional and improved cultivars from Tamil Nadu, neighbouring states, rest of India and includes overseas collections (Fig. 12.3). More than 75 per cent of the collections are local varieties (both traditional and improved) whereas the rest of the collections came from 24 different countries. Among them Philippines, followed by China, Japan and West Africa have contributed substantially whereas the rest of the countries have contributed very few rice genotypes.

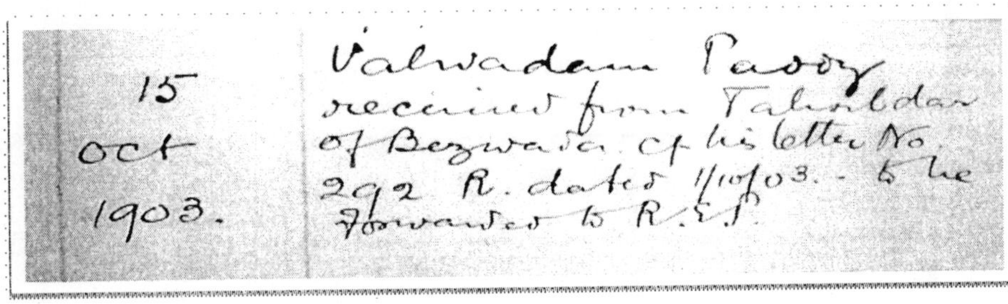

15 Oct 1903. Valwadam Paddy received from Taluqdar of Bezwada cf his letter No. 292 R. dated 1/10/03. - to be forwarded to R.E.S.

Fig 12.1. An old record dating back 1903 maintained at Coimbatore on the traditional crop variety, *Valwadam* collected through the revenue officials

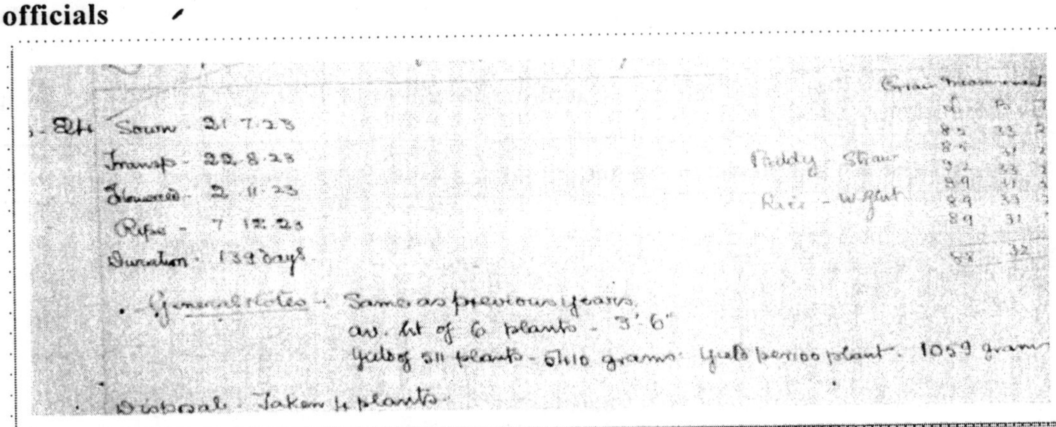

Sown - 21.7.23
Transp. - 22.8.23
Flowered - 2.11.23
Ripe - 7.12.23
Duration - 139 days.
General notes - Same as previous years.
av. ht of 6 plants - 3'6"

Fig 12.2. An old record at Coimbatore showing the germplasm characterization taken up in the early periods.

To develop new rice varieties and management technologies to meet the various ecological situations in Tamil Nadu, rice research work has been initiated at different centres on location specific basis; the rice research stations at these places have also collected rice germplasm. They mostly consist of traditional varieties, new varieties developed, varieties from other states, etc. The rice germplasm maintained at different research stations of Tamil Nadu are furnished in Table 12.2.

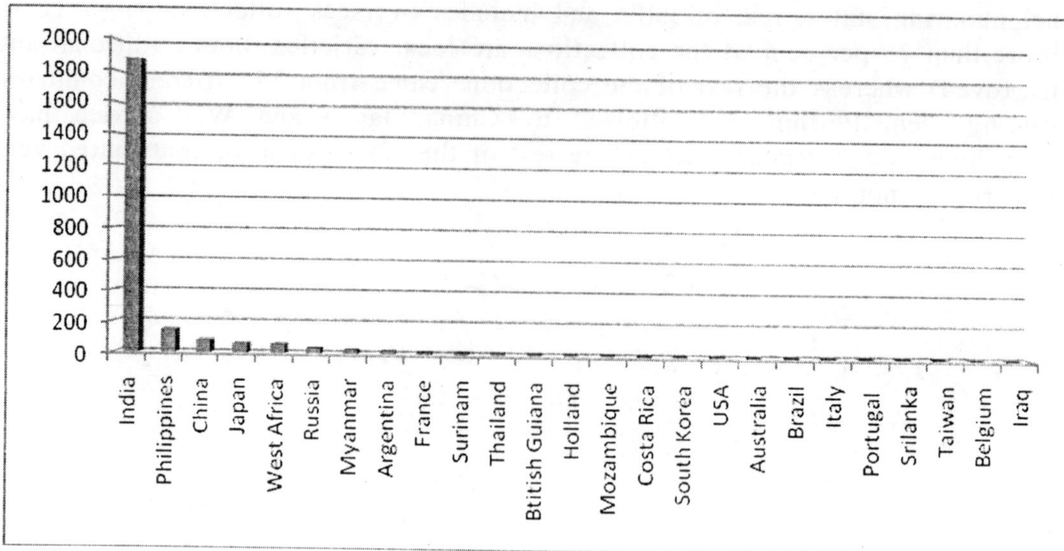

Fig. 12.3 Composition of world collections in Coimbatore genebank

Table 12.2. Rice germplasm collections at different Rice Research Stations in Tamil Nadu.

S.No.	Name of the Research Station	Number of collections
1.	Department of Rice, TNAU	2508
2.	Tamil Nadu Rice Research Institute, Aduthurai	1550
3.	Agricultural Research Station, Paramakudi	10
4.	Agricultural Research Station, Thirupathisaram	10
5.	Rice Research Station, Ambasamudram	25
6.	Rice Research Station, Tirur	243

Germplasm Conservation

Conservation is the management of resources to derive sustainable benefits to meet future needs. 'Germplasm conservation' preserves genetic resources for a longer period of time without loss of viability and also without any need of its frequent rejuvenation and to distribute the required accessions for breeding and other research purposes. At Tamil Nadu, all the collected germplasm are preserved in medium-term storage at 5°C. Approximately, 50 panicles of each accession are collected, threshed, dried to roughly 10% moisture content, sealed and preserved. The germplasm is rejuvenated and characterized after every five years.

To highlight, many germplasm collections available at Tamil Nadu have many good and special traits which are very much useful in the rice improvement programme for the future. For instance, *Karuppu Kavuni,* a local variety, is reported to have medicinal properties useful for the people with diabetics. *Kattuvanam* (TNR 2), *Thiruthuraipoondikar* (TNR 1), Red *Ottadan* (ADT 6) and White *Ottadan* (ADT 7) are established as genotypes with tolerance to deep water conditions. Accessions like *Panamarathur Samba*, *Uvarkondan* and PVR 1 are the inland saline tolerant varieties. TKM 6, GEB 24, *Rangoon samba, Sadai ponni*, Red *Sirumani*, White *Sirumani*, *Karthigai samba, Chingalpattu samba, Konakuruvai* (ADT 16), *Yeera Sannavadlu* and TKM 8 are known for their excellent cooking quality. *Manavari, Kurvaikalayan, Kattisamba, Norungan, Chithraikar, Noothripathu, Varappukodanchan, Poong*ar and *Kallurundaikar* have been identified as good drought tolerant rice genotypes.

Rice germplasm vary in their maturity duration also. Of these, very early (about 100 days) and very long duration (165 – 220 days) groups have played critical role in breeding rice varieties suitable for specific ecosystems and environments. The accessions *Thuyamalli* and *Avasara samba* (80-85 days), *Kuruvai* (90 days) and MDU 5 and PY 2 (100 days) are very early accessions while *Korangu samba*, *Chinna samba* and Red *ottadan* (165 days), *Nellore samba*, *Sadai samba*, *Garudansamba* and White-*sirumani* (170-175 days), *Anaikomban*, *Ayyan-samba*, *Senthinayagam*, *Chinglepattu sirumani* and *Kattuvanam* (180-200 days) and Red *ottadan* and White *ottadan* (220 days) are coming under very long duration groups.

Majority of the germplasm accessions at Coimbatore has been spared to be a part of the genebank at the International Rice Research Institute (IRRI) through NPBGR, New Delhi.

Utilisation of Rice Germplasm

The rice germplasm consisting of landraces, local cultivars, obsolete varieties, special varieties, breeders' lines, improved strains, wild species, related weedy species, etc., have an inherent genetic value because of their adaptation to different farming conditions and resistance to biotic and abiotic stresses. Knowledge of these traits and their adaptation to different farming conditions enhance the value of the conserved germplasm. The use of germplasm in rice improvement could be facilitated by systematic evaluation and documentation of the acquired data.

A total of 184 rice varieties and four hybrids have been released/introduced in Tamil Nadu as indicated in the Table 12.3.

Table 12.3. Rice varieties/hybrids released/introduced in Tamil Nadu.

S.No.	Name of the Station	Rice		Total
		Varieties	Hybrids	
1.	Tamil Nadu Rice Research Institute, Aduthurai	ADT 1 – ADT 48	ADTRH 1	49
2.	Rice Research Station, Coimbatore (Dept. of Rice, TNAU	CO 1 – CO 50	CORH 1 – 3	53
3.	Rice Research Station, Ambasamudram	ASD 1 – ASD 20	-	20
4.	Rice Research Station, Tirurkuppam	TKM 1 – TKM12	-	12
5.	Rice Research Station, Peravurani (Coastal salinity)	PVR1	-	1
6.	Rice Research Station, Thalaignayar (Deep water)	TNR 1 – TNR 2	-	2
7.	Agricultural Research Station, Palur	PLR 1 – PLR 2	-	2
8.	Agricultural Research Station, Paramakudi	PMK 1 – PMK 4	-	4
9.	Agricultural Research Station, Thirupathisaram	TPS 1 – TPS 3	-	3
10.	Regional Research Station, Paiyur	Paiyur 1	-	1
11.	Agricultural College & Research Instt., Madurai	MDU 1 – MDU 5	-	5
12.	Anbil Dharmalingam Agrl. College & Res. Institute, Tiruchirapalli	TRY 1 – TRY 2	-	2
13.	Kamarajar Farm Science Centre of Pondicherry	PY 1 – PY 5	-	5
14.	Annamalai University, Chidambaram	AU 1 – AU 2	-	2

Introductions:			
1.	Tamil Nadu Rice Research Institute, Aduthurai	IR 20, IR 34, IR 50, IR 64, Pusa 33, Pusa 2-21, IET 1722, IET 1444, IET 4786, IET 3257, Ponni (Red), Improved White Ponni NGR 9672 and CR 1009	14
2.	Paddy Research Station, Coimbatore	Bhavani	1
3.	Varieties introduced in Tamil Nadu after 1965	Annapoorna, Tiruveni, Imp. Sabarmati, Jaya, Ratna, IR 5, IR 8, IR 22, IR 26, IR 34, IR 36 and IR 72	12
		Total	**188**

Earlier attempts of rice improvement were mostly concentrated on improving the local rice through selection in Tamil Nadu as was done elsewhere in India. These pure line selections from locally collected germplasm resulted in the release of 65 new varieties by pureline selection and two by hybridization yielding about 10–20% higher than their original population and were adapted to the places where their original populations were cultivated (Table 12.4). Some of the traditional varieties were also utilized in the hybridization programme and new high yielding rice varieties were developed and released for large scale cultivation (Table 12.5).

Table 12.4. Varieties improved through pure line selections from traditional varieties in Tamil Nadu (Subramanian *et al.*, 2007).

S.No.	Traditional variety	Improved varieties by selection	Year of release	Season
	ADUTHURAI (ADT)			
1.	Sivappu (Red) Sirumani	ADT 1	1925	*Samba* / *Thaladi* irrigated.
2.	Vellai (White) Sirumani	ADT 2	1925	*Samba* / *Thaladi* irrigated.
3.	Kuruvai	ADT 3 ADT 4	1925 1925	*Samba* / *Thaladi* irrigated. *Samba* / *Thaladi* irrigated
4.	Nellur samba	ADT 5	1925	*Samba* / *Thaladi* irrigated.
5.	Sivappu (Red) Ottadan	ADT 6	1925	*Samba* irrigated
6.	Vellai (White) Ottadan	ADT 7	1925	*Samba* irrigated

7.	Poongar	ADT 9	1933	*Kar / Kuruvai* irrigated.
8.	Kurangu samba	ADT 10	1933	*Samba / Thaladi* irrigated.
9.	Nellore Samba	ADT 11	1933	*Samba* irrigated
10.	Chitrai kali	ADT 12	1936	*Samba* irrigated
11.	Chinna samba	ADT 13	1936	*Samba* irrigated
12.	Vellaikar	ADT 14	1937	*Kar/Kuruvai* irrigated.
13.	Senguruvai	ADT 15	1937	*Kar/Kuruvai* irrigated.
14.	Kona kuruvai	ADT 16	1938	*Kar/Kuruvai* irrigated.
15.	Muthu samba	ADT 17	1938	*Samba* irrigated
16.	Vellai kuruvai	ADT 18	1939	*Kuruvai* irrigated
17.	Sarapalli	ADT 19	1939	*Kuruvai* irrigated
18.	Thanjavur vadan samba	ADT 21	1939	*Samba* irrigated
19.	North Arcot vadan samba	ADT 22	1942	*Samba* irrigated
20.	Kullankar	ADT 23	1949	*Sornavari/Navarai* irrigated.
21.	Boon samba	ADT 24	1951	*Samba/Thaladi* irrig.
22.	Red Mottai kuruvai	ADT 26	1964	*Kuruvai* irrigated
	COIMBATORE (CO)			
23.	Kitchli samba	GEB 24	1921	*Samba* irrigated
24.	Peria kitchili	CO 1	1921	*Samba* irrigated
25.	Boombalai	CO 2	1923	*Samba* irrigated
26.	Vellai samba	CO 3	1923	*Samba* irrigated
27.	Anaikomban	CO 4 & CO 8	1923 & 1931	*Samba* irrigated
28.	Chinna samba	CO 5	1926	*Samba* irrigated
29.	Sadai samba	CO 6 CO 7	1926	*Samba* irrigated
			1926	*Samba* irrigated
30.	Tinnelveli Anaikomban	CO 8	1932	*Samba* irrigated
31.	Kar samba red	CO 9	1932	*Kar/kuruvai* irrigated
32.	Gobi kar	CO 10	1932	*Kar/navarai* irrigated
33.	Ayyan samba	CO 11	1933	*Samba* irrigated
34.	Senthinayagam	CO 12	1933	*Samba* irrigated
35.	Arupatham kodai	CO 13	1940	*Kar/Kuruvai* irrigated

36.	Chinna Vadan samba	CO 17	1941	*Kar/Kuruvai* irrigated
37.	Vellaikar	CO 18	1941	*Kar/navarai* irrigated
38.	Chengalpattu sirumani	CO 19	1941	*Samba* irrigated
39.	Tella sanna vadlu	CO 20	1941	*Kar* irrigated
40.	Arupatham samba	CO 21	1942	*Kar/ kuruvai* irrigated
41.	Manavari	CO 22	1942	*Kar* rainfed
42.	Rangoon samba	CO 23	1944	*Kar/Navarai* irrigated
43.	Pudupatti samba	CO 27	1952	*Samba* irrigated
44.	Bangaru theegalu	CO 28	1952	*Samba* irrigated
45.	Thiruchengodu samba	CO 32	1964	*Samba* irrigated
	AMBASAMUDRAM (ASD)			
46.	Kar samba Red	ASD 1	1943	*Kar* irrigated
47.	Kar samba White	ASD 2	1944	*Kar* irrigated
48.	Veedhivadangan	ASD 3	1944	*Pishanam* irrigated
49.	Kuruvaikalayan	ASD 4	1944	*Pishanam* irrigated
50.	Karthigai samba	ASD 5	1945	*Pishanam/samba* irrig.
51.	Anaikomban	ASD 7	1945	*Pishanam/samba* irrig.
52.	Thuyamalli	ASD 8	1951	*Kar* irrigated
53.	Avasara samba	ASD 9	1951	*Kar* irrigated
54.	Kolavalai	ASD 10	1952	*Pishanam* irrigated
55.	Arivikiravai	ASD 13	1965	*Pishanam* irrigated
	TIRURKUPPAM (TKM)			
56.	Pisini	TKM 1	1950	*Samba* rainfed
57.	Sembalai	TKM 2	1950	*Samba* rainfed
58.	Sornavari	TKM 3	1950	*Sornavari* rainfed
59.	Yera Sannavadlu	TKM 4	1950	*Navarai* irrigated
60.	Manakattai	TKM 5	1952	*Navarai* irrigated
61.	Kullankar	TKM 7	1973	*Sornavari/Navarai* irrig.
	THALAIGNAYAR (TNR)			
62.	Thiruthuraipoondi kar	TNR 1	1964	*Samba* deep water
63.	Kattuvanam	TNR 2	1972	*Samba* deep water

	PALUR (PLR)			
64.	Garudan samba	PLR 1	1940	*Samba* irrigated
65.	Chitraikali	PLR 2	1942	*Navarai* irrigated
	PAIYUR (PAIYUR)			
65.	Sadai Ponni	Paiyur 1	1982	*Samba* irrigated

Table 12.5. Rice varieties developed by hybridization involving traditional varieties.

Rice Parents	New Rice varieties
Mola Kolukulu / ADT 2	ADT 8
IR 8 / Katti samba	TPS 1

In addition to high yielding potential, some traditional varieties possessed resistance to pests and diseases, abiotic stresses and quality grains with good cooking qualities. Among them GEB 24 and TKM 6 need special mention, which were identified as good donors for grain quality and multiple resistance respectively. Utilizing these two varieties many high yielding varieties were developed in India, at IRRI, in Sri Lanka, Thailand, etc. GEB 24, a pure line from *Kitchili samba,* was particularly known for its superior grain quality characterized by high amylose content and soft gel consistency that was extensively used in hybridization. TKM 6, the multiple resistant rice variety was developed by crossing GEB 24 / CO 18 (a pure line selection from *Vellaikar*) and was used in many hybridization programmes and many improved varieties have been developed subsequently (Fig. 12.4 and Fig. 12.5). Cauvery (CO 35) yet another high yielding rice variety released from this state has been utilized by other rice research stations in India and new rice varieties like *Manohar*, *Ugadhan*, *Aditya* and *Vagadhan* were released. Similarly, the wonder rice IR 8 and the high yielding dwarf variety T(N)1 introduced into Tamil Nadu were used in the crossing programme with local varieties to develop and release new high yielding varieties (Table 12.6).

Table 12.6. Rice varieties developed through hybridization involving IR 8, T(N)1 and traditional varieties.

S.No.	Parents	New variety
1.	IR 8 / Cul. 340	ADT 31 and ADT 33
2.	IR 8 / Rod Jolele	ADT 34
3.	IR 8 / ADT 27 (Norin 8 x GEB 24)	CO 33

4.	IR 8 / CO 32 (Thiruchengode samba)	CO 36
5.	IR 8 / CO 25 [CO 4 (Anai komban) / CO 10 (Gobikan)]	CO 40 (Rajarajan)
6.	ADT 31 / Ratna / ASD 8 / IR 8	ASD 17
7.	TKM 7 (Kullankar) / IR 8	TKM 9
8.	IR 8 / Chitraikar	MDU 1
9.	CO 25 [CO 4 (Anaikomban) / CO 10 (Gobikar)]	IR 8
10.	T(N) 1 / CO 29 [CO 13 (Arupatham-kodai) / CO 4 (Anaikomban)]	CO 34 & CO 37
11.	T(N) 1 / TKM 6 [GEB 24 (Kitchli samba) / CO 18 (Vellaikar)]	CO 35
12.	T(N)1 / ASD 1 (Kar samba Red)	ASD 14
13.	TKM 6 [GEB 24 (Kitchili samba) / CO 18 (Vellaikar)] / T(N) 1	TKM 8

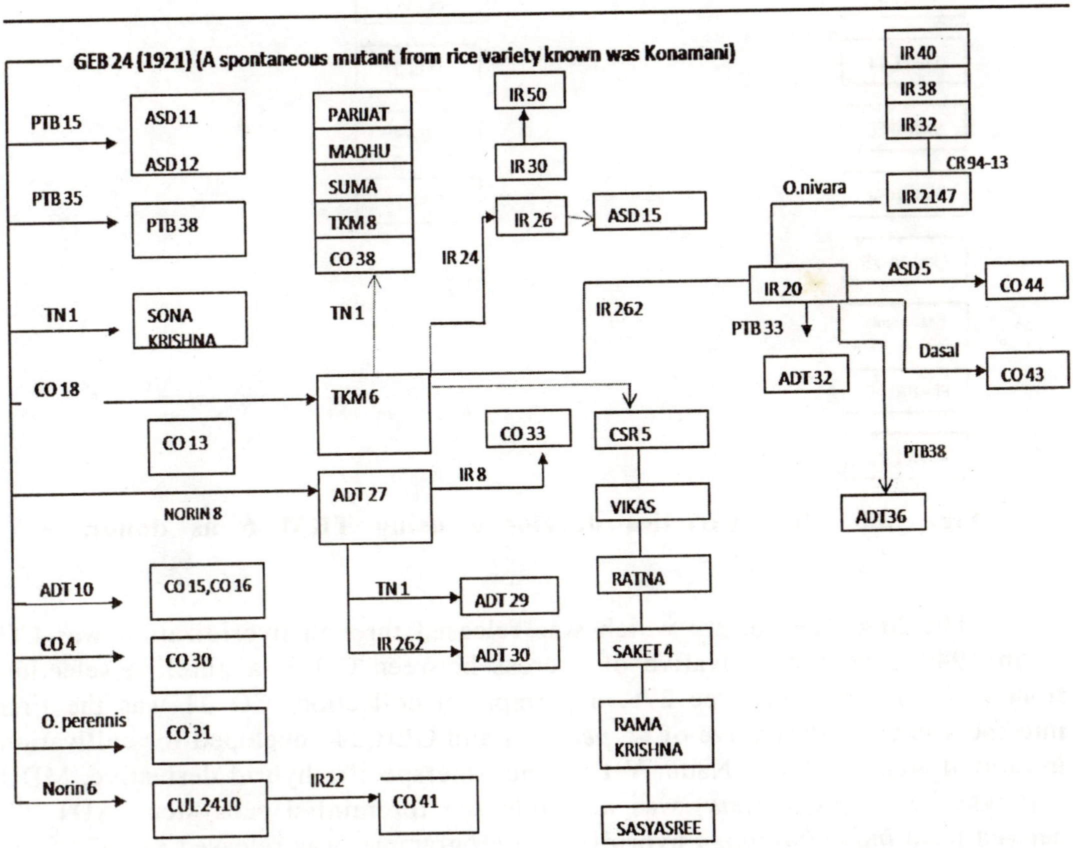

Fig. 12.4. Rice varieties developed using GEB 24 as donor.

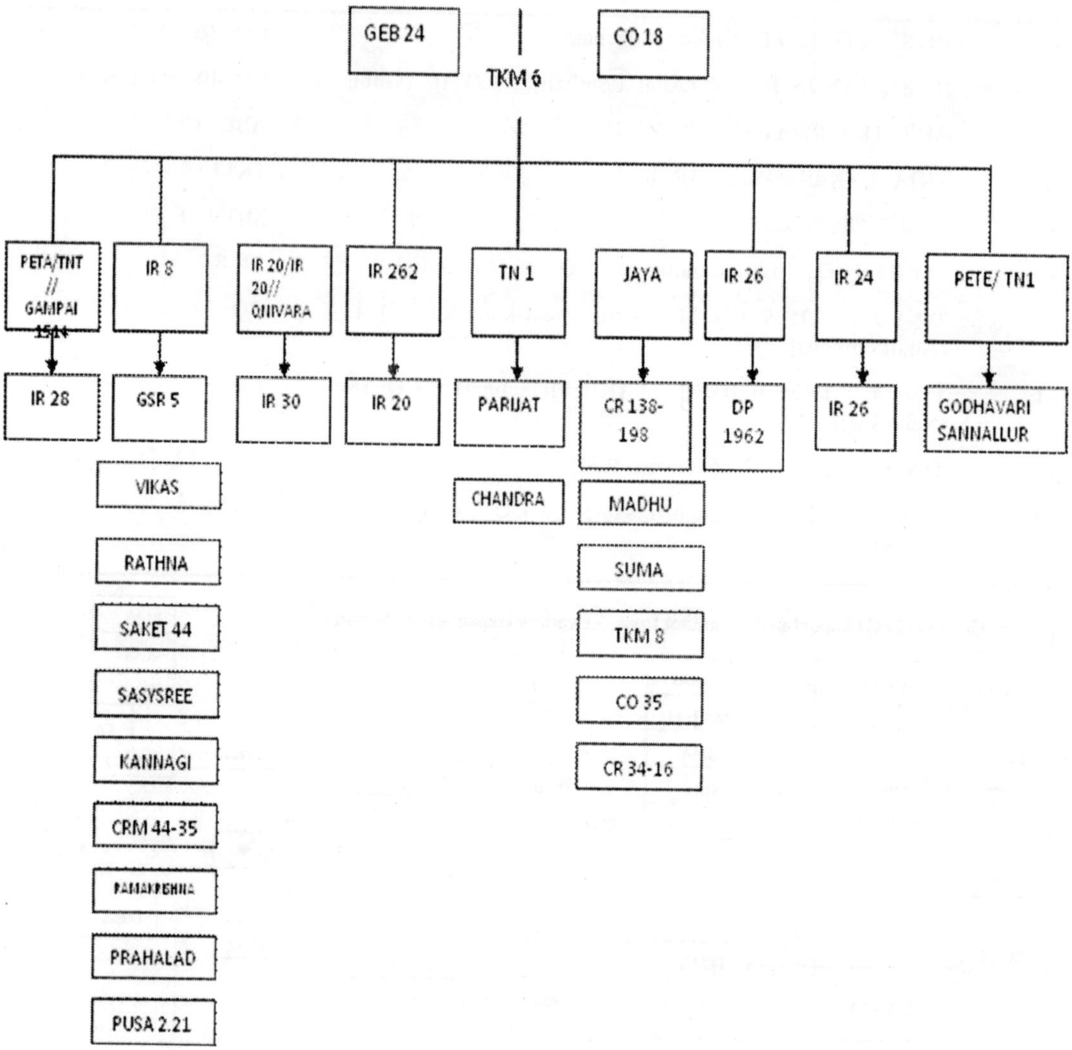

Fig. 12.5. Rice varieties developed using TKM 6 as donor.

The first rice variety which was released through hybridization was CO 14 in 1940. It was a derivative of a cross between CO 3, a pureline selection from *Vellai samba* and Type 275, a germplasm collection. CO 31 was the first interspecific cross derivative of *O. perennis* and GEB 24 developed for cultivation in rainfed areas of Tamil Nadu. Yet another interspecific hybrid derivative, MDU 5 (Pokkali / *O. glaberrima*) was also released for rainfed ecosystem. ADT 27, derived from *indica/japonica* hybridization programme, was released in 1978 from the cross between exotic Norin 8 and GEB 24. A red rice variety TKM 9 was released in 1978 from the cross between TKM 7 and IR 8; it has been popular for a long period and is still under cultivation in major areas of Tamil Nadu and Kerala both under semi-dry and wet ecosystem. A total of 15 new crop varieties was released directly from the cross involving traditional cultivars or their PLS as one of the immediate parents. These cultivars carried forward the useful traits further through second generation hybridization. The new cultivars coming out of hybridization programmes substantially improved rice yield in this state.

Varietal improvement of rice started with the release of GEB 24 and till date 184 varieties and 4 hybrids have been released. GEB 24 at the time of its release recorded 3.5 t/ha. The average productivity of the subsequent cultivars developed through pure line breeding also stood around that yield level during the initial phase of the breeding. However, several unique cultivars were released during that period with fine grain, pest and disease resistance, flood tolerance, salt and drought resistance. The yield level, however, stagnated around 3.5 t/ha even after resorting to hybridization involving traditional rice varieties as both the parents. The yield level jumped to 5 t/ha after the introduction of fertilizer responsive semi-dwarf varieties into the hybridization programme and it often touched 7 t/ha. The landraces in continuation with the semi-dwarf parents offered several new varieties with high yield and good adaptability. Thus, the landraces of rice grown and cultivated in Tamil Nadu have played a significant role in the rice culture of the state by contributing substantially through their genetic diversity and in developing superior new rice cultivars with long standing adaptability.

Documentation

The Rice Breeding Centres such as Department of Rice, TNAU, Coimbatore and Tamil Nadu Rice Research Station, Aduthurai and the other stations have already described all the rice germplasm maintained at their centres. The Tamil Nadu Agricultural University is now giving importance to the maintenance of germplasm of all crops and documentation of all the accessions maintained; it has started a new department named as Department of Plant Genetic Resources.

Conclusion

- Exploration, collection and conservation of rice should be further strengthened in Tamil Nadu as many of the hills and forest areas are yet to be explored for the wild species as well as landraces.
- The Tamil Nadu Agricultural University, Regional Agricultural Colleges and Research Centres are to be given the responsibility of collection and

conservation of germplasm.

- Conservation of germplasm still remains an important and challenging task.
- *In situ* conservation of landraces through on-farm conservation in farmers' fields is more effective because it maintains allelic diversity, is very accessible and provides high security. It should be made a viable option for farmers by providing them incentives and rewards.
- Supply of quality seed at affordable cost, procurement of seed at reasonable price and providing simple and effective seed storage facilities are important needs for successful seed conservation by farmers.
- The farmers should be often consulted to solve their problems and constraints encountered through technical backstopping. This is a novel way not only to conserve gene pools but also to restrict genetic erosion of valuable germplasm.
- Knowledgeable well-trained scientists should be placed in charge of all genebanks.
- Research on seed technology needs be strengthened to intensify the study on quality, viability, dormancy and storability of rice seeds to raise healthy plants even for germplasm.
- Storage facilities at all conserving centres should be improved with latest and modern infrastructure facilities at par with international level to store the seeds for long term without loss of their viability and erosion of valuable genotypes.
- Free exchange of breeding materials among scientists, particularly the donors for desirable agronomic characters, pest and disease resistant characters and quality traits will be very useful and, for this, a network programme should be established.
- Use of biotechnological tools like *in vitro* techniques have to be strengthened and practised after imparting suitable training to the scientists to maintain and conserve the germplasm collections.
- Documentation of the details of genetic diversity of rice germplasm and their characteristics deserves immediate attention and the available information on rice germplasm should be made available and accessible to all rice breeders at national level.
- The Indian Council of Agricultural Research (ICAR) should provide financial and logistic support for collection and conservation to save the

valuable regional rice germplasm from erosion.

- Policy on intellectual property right (IPR) should be well-documented and implemented to protect the property rights for the rice germplasm and to facilitate free exchange of genetic materials on mutual understanding and benefit sharing to promote breeding desirable rice plants by the rice growing states in India.

References:

Abrol, Yash P, Gadgil Sulochana (eds). 1999. *Rice in a Variable Climate*. APC Publications, New Delhi.

De Candolle. 1886. *Origin of Cultivated Plants.* Reprint Edition (1967). Hafner, New York NY.

Chang, Te-Tzu. 1976. The origin, evolution, cultivation, dissemination, and diversification of Asian and African rices. *Euphytica* 25:435–441.

Copeland, Edwin Bingham. 1924. *Rice*. Macmillan, London.

IRRI (International Rice Research Institute). 1994. *Rice Almanac* (2d ed). International Rice Research Institute, Manila.

Rai, M. 1999. Rice germplasm evaluation and enhancement in India: Issues, status, options and future plan of action. In *Proceedings of the International Symposium on Rice Germplasm Evaluation and Enhancement*, Publ. No. 195:83-91, University of Arkansas.

Ratho SN. 2004. Role of Central Rice Research Institute in genetic improvement of rice varieties of India. Pages 39-99 in Sharma SD and Nayak BC (eds) *Genetic Improvement of Rice Varieties of India (Part I)*. Today and Tomorrow's Printers and Publishers, New Delhi.

Saxena SK, Chandar SB, Ghosh R, Sinha N, Jain, Gupta, Anil K. 2003. Costs of conservation of agrobiodiversity in India. Pages 137-174 in Detlef Virchow (ed) *Efficient Conservation of Crops Genetic Diversity: Theoretical Approaches and Empirical Studies*. Springer Verlag, Berlin.

Subramanian M, Mannual WW. 1998. *Varietal Description of Rice* (*Platinum Jubilee Release*). Tamil Nadu Rice Research Institute, TNAU, Aduthurai. 80 pp.

Subramanian M, Neelakantapillai K, Thirumani S and Kannaiyan S. 2007. *Biodiversity of Rice, Its Conservation and Utilization*. National Biodiversity Authority, Chennai, Tamil Nadu, India. 92 pp.

Tzvelev NN. 1989. The system of grasses (Poaceeae) and their evolution. Bot Rev 55:141-207.

Watt G. (1891). *Dictionary of Economic Products of India*. Vol V. (Reprint edition). Periodical Experts, New Delhi.

Genetic Resources of Rice in India : 411-429
Editor : S. D. Sharma
Today & Tomorrow's Printers and Publishers, New Delhi - 110 002 (India)

Chapter 13

Rice Germplasm of Kerala

Leena Kumary S.

Introduction

The cultivated rices of the world belong to two species, *O. sativa* and *O. glaberrima*. Commonly known as Asian cultivated rice, *O. sativa* is the species cultivated throughout the rice growing areas of the world and it is in south and southeast Asia that both *O. sativa* and its wild relatives occur in great profusion and variability. *O. glaberrima,* the African rice, is generally known from tropical west Africa and has also been reported from Guyana and El Salvador in Central America (Nayar, 1973). Based on phytogeographical evidences, Nair *et al.* (1964) supported the consideration by earlier workers that peninsular India and more particularly the Malabar coast is the centre of origin of rice. The occurrence of as many as five wild taxa, the presence of high varietal diversity and also the wide occurrence of several dominant genes in local rice varieties, such as red pericarp colour, awning, bold grains and black husk colour, all support the above consideration.

The Asian cultivated rice exhibits great amount of variation with respect to sterility relationships between geographical races (*indica, japonica* and *javanica*), morphology (spikelet size, size of glume, colour of glume, colour of pericarp) and grain characters (glutinous and non-glutinous and also scented and non-scented rices). Varietal groups corresponding to ecotypes or ecological races each adapted to different conditions of photoperiod (photosensitive and photo-insensitive) and land type (upland, lowland and deep water) may also be distinguished.

Rice Genetic Resources in Kerala

Rice covers a vast array of ecological niches in Kerala. Kerala lies entirely with in the tropics and is located between North latitudes 8°18' to 12°48' and East longitude 74°52' to 77°22'. Based on topography, Kerala is classified into four distinct regions viz., high ranges, high lands, midlands and lowlands (Anonymous, 2000). The high ranges lie 750 m above MSL along the Western Ghats constituting the entire Waynad and Idukki districts and eastern parts of other districts bordering the Western Ghats. The region experiences temperature ranging from 19° C - 30° C and is dominated by plantations of tea, coffee and cardamom. Rice is grown in the broad and extensive valley bottoms in Wayanad. The hill tracts on the western side of the Westen Ghats are classified as the highlands which lie between 75 and 750 m above MSL. Covered with forests and small streams, the soil of the highlands are generally forest loam with wide variation in depth and organic matter contents. Rice varieties grown in these area have such plasticity that they are conditioned to grow under a wide spectrum of moisture regime. The midland plains (7.5 – 75 m above MSL) sandwich between the highlands and the lowlands and extend to the entire length of the state. Undulating terrains intersected by numerous rivers and rivulets, wetland systems form a major feature of this area. The lowlands bordering the Arabian Sea is a strip along the coast which lie from below MSL up to 7.5 m above MSL. Characterised by marine land forms with swamps and lagoons, several places are liable to be flooded. Extensive paddy fields intercepted with plantations of coconut is the major feature of this region. Annual rainfall of the state ranges from 2203 mm to 3584 mm with an average of 3228 mm.

High rainfall coupled with undulating topography has given rise to different moisture regimes resulting in a wide range of micro environments. Taking into consideration the variation in resource endowments, topography, soil, abiotic factors and seasonal differences, rice grown in the state is grouped under eight different agro-ecosystems. The natural and artificial selections operating in these areas over long periods of time have resulted in a large number of traditional varieties suited to each region and possessing special traits like resistance to biotic and abiotic stresses, ability to survive extreme agro-edaphic situations, quality attributes like medicinal value, aroma and suitability to special purposes which constitute an invaluable reservoir of genes that are used by plant breeders for development of superior crop varieties. The full spectrum of genetic resources of rice in Kerala consists of wild species, natural hybrids between cultigens and wild relatives, primitive cultivars or land races, commercial types, pureline selections of farmers varieties, elite varieties of hybrid origin and other breeding materials including mutants as well as hybrid derivatives.

Distribution of Wild Species of Rice in Kerala

Of the twenty wild species reported in the genus *Oryza*, more than five species viz., *O. nivara, O. rufipogon, O. officinalis,, O. granulata* and *O. sativa* f. *spontanea* are found in Kerala (Vaughan and Muralidharan, 1989). Krishnaswamy and Chandrasekharan (1958) reported the occurrence of a tetraploid form of Oryza in the Emurbhagavati hills near Malampuzha reservoir in Palakkad district and named it as *O. malampuzhaensis*. Later workers considered this species as a tetraploid form of *O. officinalis* viz., *O. officinalis* ssp. *malampuzhaensis* (Tateoka, 1963). *Oryza sativa* f. *spontanea,* the closest relative of the cultivated rice is of common occurrence in Kerala. It is considered to be a hybrid between *O. sativa* and *O. nivara* or *O. rufipogon.* Table 13.1 lists the wild spp. of Oryza found in Kerala as reported by Vaughan, 1994.

Table 13.1. Wild species of *Oryza* found in Kerala.

Sl. No.	Species	Ploidy level	Genome Group
1	*O. nivara*	Diploid (2n=24)	AA
2	*O. rufipogon*	Diploid (2n=24)	AA
3	*O. officinalis*	Diploid (2n=24)	CC
4	*O. granulata*	Diploid (2n=24)	Unknown
5	*O. malampuzhaensis*	Tetraploid (2n=48)	BBCC

Source: Vaughan (1994)

A wide array of useful traits could be found in most of these wild species. *O. officinalis* has been a source of resistance to yellow stem borer, brown plant hopper, green leaf hopper, white backed hopper and zigzag leaf hopper (Heinrichs *et. al.,* 1985). *O. rufipogon* has genes for elongation ability, salinity tolerance and resistance to bacterial blight (IRRI, 1990; Ikeda *et al.,* 1990) and has been identified as a source for cytoplasmic male sterility (Virmani and Shinjyo, 1988). Rice improvement programmes world over have benefited substantially by transferring useful traits from wild rices to cultivated rices.

But the phenotypic diversity of the cultivated rices far exceeds that of the wild species because of the ease with which humans, over the millennia, used phenotypic characters as criteria for selection. Farmers have made selections in the Asian rice for its adaptation to grow in areas well outside the range of wild progenitors at both high and low altitudes. Thus the cultigens have a greater range of genetic variation for many traits and continued productivity of the rice crop depends largely on harnessing this genetic variation within the species

It is assumed that near about two thousand traditional varieties which were well suited for different agro-climatic situations and seasons of rice cultivation were cultivated in Kerala. They include varieties possessing genes for resistance against biotic and abiotic stresses, varieties which were locally used as medicine or as ingredients in medical preparations, and scented varieties which differ from the scented "Basmati" rice with respect to growth habits as well as physico-chemical properties of the grains (Table 13.2).

Table 13.2. Landraces of Kerala possessing specific traits.

Trait	Traditional varieties possessing the trait
Resistance to BPH	*Athikkiraya, Arikkirai, Chenninayakan, Chemben, Cheriyachitteni, Chitteni Chemparampandi, , Eravapandy, Eswaramangalam, Kayama, Karivennel, Karutha-vellathan, Kavunginpoothala, Kochuvithu, Kodiyan, Parakulam chitteni, Thavalakkannan, Thekkan, Vadakkan chitteni, Valiyachampan, Valsarachampavu, Vellathilcheera, Veluthacheera, Vellachenipan.*
Resistance to Gall Midge	*Athikkiraya, Aruvakkari, Arikkirai,Chennellu, Chitteni, Eravapandy, Kattamodan, Kodiyan, Parambuvattan, Parakulam, Thavalakkannan,Vadakkan chitteni, Vellathilcheera, Veluthacheera.*
Resistance to Leaf folder	*Athikkiraya, Arikkirai, Chempan, Champav, Chitteni, Vellathilcheera*
Resistance to Stemborer	*Athikkiraya, Eravapandy, Kavunginpoothala, Thekkanchitteni, Thekkancheera, Valsarachampavu, Kulaperuvala*
Resistance to WBPH	*Chemben, Cheriyachitteni, Chemparampandi, Chitteni, Thekkan.*
Resistance to GLH	*Athikkiraya,Eravapandy, Kodiyan, Ponnaryan, Thavalakkannan, Thekkan, Vattan, Veluthacheera.*
Resistance to Blast	*Karuthamodan, Kattamodan, Karivennel, Kodiyan, Parambuvattan, Thadukan, Thondy, Palthondy, Vadakkanchitteni*
Resistance to Sheath blight	*Eravapandy, Thondy, Palthondy, Vellari*
Resistance to Foot rot	*Parambuvattan*
Resistance to BLB	*Karuthamodan, Thekkanchitteni*
Resistance to Drought	*Chuvannamodan, Cheriyaryan, Modan, Karuthamodan,*

	Kattamodan, Kuttimodan, Parambuvattan, Parapilarppan, Karanellu, Thonnuran.
Resistance to Water-logging and Flood	*Aryan, Chenkayama, Champavu, Kayama, Kavunginpoothala, Kodiyan, Kuttadan, Thavalakkannan Vellakkoli, Vayalithuva, Vellathilkolappala*
Resistance to Salinity	*Anakodan, Chettiviruppu, Cheruvirippu, Choottupokkali, Kuthiruvithu, Orkayama, Orumundakan, Pokkali, Virippu*
Medicinal Value	*Chennellu, Erumakkari Kalamappari, Karimkuruva, Karuthachembavu, Kavunginpoothala, Kunjinellu, Njavara, Neduvali, Narikari, Perunellu, Poovali, Tanavala, Ulimkathi, Valanellu, Varakan,,Velvali*
Cooking Quality	*Thekkan, Vattan, Kavunginpoothala*
Aroma	*Jeerakasala, Gandhasala, Kayama, Kothampalarikkayama, Neycheera, Pookkulathari, Kunjikkayama and Rajakayama.*

Resistance to Pests and Diseases

Many of the traditional rice varieties of Kerala offer a pool of resistant genes against insect pests like brown plant hopper, green leaf hopper, gall midge, leaf folder and stem borer (Kalode *et al.*, 1977) and diseases like sheath blight, sheath rot, foot rot, blast, etc. that cause considerable damage to the rice crop (Leena Kumary *et al.*, 1998). Among them varieties like *Thekkan, Eravappanndy, Thekkancheera, Veluthacheera, Karivennel, Kochuvithu, Orpandy, Veluthacheera, Orumundakan, Arikkirai,* etc. have been used in the resistant breeding programmes world over. Many of these varieties are proven to be multiple resistant to biotic and abiotic stresses and can form the building blocks in many rice improvement programmes since they act as donors of valuable genes.

Drought Resistance

Rice varieties like *Kattamodan, Karuthamodan, Chuvannamodan, Arimodan, Karimodan, Parambuvattan, Karanellu,* etc. are used in the uplands of Palakkadu (known as *modan* lands) for direct seeding while varieties like *Kochuvithu, Vykatharyan, Karavala, Champavu, Parapilarppan,* etc. are used in *Onattukara*. These varieties possess resistance to drought and diseases like blast and are adapted to the low light intensity experienced during the *kharif* season. *Karanellu* also known as *Erinellu* grown by the tribal people of Attappadi is a highly weed competitive, long duration, tall variety with deep root system.

Resistance to Salinity, Water-logging and Flood

The rice varieties grown in the Pokkali tract (extending over the districts of Ernakulam and Alappuzha) during *kharif* season are photo-insensitive, have a duration ranging from 120 to 125 days and plant height of over 125 cm and have red kernelled bold grains are tolerant to water-logging, salinity, and acidity. High genetic diversity exists within the Pokkali varieties viz., *Pallippuram Pokkali, Kuzhippalli Pokkali* and *Vettikkal Pokkali* collected from different regions of this tract. Other saline resistant varieties of the tract include *Cheruviruppu, Chettiviruppu, Kuruka* and *Anakodan*. Varieties like *Orumundakna, Eravapandy, Orpandy* and *Orkayama* are photosensitive and are grown during the second crop season. *Kuthiruvithu, Kochumundon and Mundon* are the saline tolerant varieties of the *Kaipad* system of Kannur District. Traditional varieties viz., *Mundakan, Karamundakan, Karimundakan, Vellamundakan, Athikkirazhimundakan* and *Oarumundakan* are photosensitive tall *indica* varieties possessing tolerance to salinity and are popularly grown during the *Mundakan* (second crop) season in the Karthikappally Taluk of Kollam District which comes under the *Onattukara* ecosystem where saline water inundation is of common occurrence. Varieties *Vellakkoli, Karuthakoli, Kodiyan,* etc. are flood tolerant varieties suited to the deep submerged areas of Ponnai Kole while *Vayalithuva* and *Vellathilkolappala* are flood tolerant varieties of South Kerala.

Medicinal Value

Though much of the wide variability that was available in the medicinal rices of Kerala is now lost, some rice varieties are still grown by farmers for their medicinal properties. These include *Njavara, Chennellu, Kunjinellu, Erumakkari, Karuthachembavu* and *Kavunginpoothala* (Leena Kumary, 2004). The medicinal properties of these rice varieties have to be validated by clinical studies but traditional knowledge associated with these varieties substantiate their use either as medicine or as ingredients in medicinal preparations. *Njavara,* a unique medicinal rice variety from Kerala, deserves special mention in this regard. This variety is known as *Shashtikam* in Sanskrit due to its extra short duration, coming to harvest within 60-70 days. Indigenous medicinal preparation using *Njavara* along with *Kurunthotti* (*Sida*) rejuvenates the muscles and nerves. Two types of *Njavara* have been identified, the white glumed and black glumed, both of which are used in Ayurvedic treatments. *Chennellu* and *Kunjinellu* are varieties indigenous to North Kerala. One type of *Chennellu* with bright red grains, grown as an upland variety in parts of Kannur district, is used in treatment of diarrhoea and vomiting. Another type of *Chennellu* with straw coloured grains is grown in wetlands in Wayanad district. *Kunjinellu* is a variant of red coloured *Chennellu* with small red grains and is given to patients recovering from jaundice. *Erumakkari* and *Karuthachambavu* are the traditional rice varieties indigenous to South Kerala. *Karuthachambavu* has black grains, blackish red kernels and black endosperm. The gruel made from the parched grains of *Karuthachambavu* is used to treat nausea, vomiting and stomach pains. *Erumakkari* was used for treatment of cough. *Annoori*, a wild species of rice is used by the Kani tribes for treatment of small pox. *Kavunginpoothala* indigenous to Palakkad District is given to diabetic patients to reduce discomfort (Sujithkumar, 1999).

Aroma

Gandhasala and *Jeerakasala* are the predominant scented varieties of Wyanad District of Kerala which comes under the high ranges at an elevation of about 750 m above MSL. These varieties differ from the famous "Basmati" rice with respect to growth habits as well as physico-chemical properties of the grains. They are tall, long duration varieites and are weakly photosensitive. They posses long thin culms, produce long panicles with large number of small well filled grains. The grains are short to medium in size (not long slender like Basmati) but have very good elongation on cooking. *Neycheera, Kothampalanikayama, Pookkulathari, Kayama, Kunjikkayama* and *Rajakayama* are the scented rice varieties grown in the high lands which lie between 75 and 750 m above MSL on the western side of the Westen Ghats (Leena Kumary *et al.*, 2003).

Population explosion and its chain reactions resulting in increased pressure on land for food and shelter leave no choice than to compete for limited land resources causing denudation of forests and erosion of primitive and wild rice species. This alarming situation has warranted collection, evaluation and preservation of the rich genetic resources with utmost urgency.

Collection and Conservation of Rice Genetic Resources in Kerala

Systematic management of the rice genetic resources in Kerala has received adequate attention particularly in the post green revolution phase and has gained remarkable momentum in recent years. Collection and conservation of genetic diversity being one of the main mandates of the various rice research stations under the Kerala Agricultural University, independent collections were made by individual stations from their respective localities and maintained by them. The major rice research stations involved in rice germplasm collection of Kerala are

1. Regional Agricultural Research Station, Pattambi
2. Regional Agricultural Research Station, Ambalavayal

3. Onattukara Regional Agricultural Research Station, Kayamkulam
4. Rice Research Station, Moncompu
5. Rice Research Station, Vytilla

Around 1000 rice accessions collected from different parts of Kerala were transferred to IRRI for conservation in the Global Gene Bank during 1976. During 1999, the activity on collection and conservation of genetic diversity of rice in Kerala was strenghthened under the National Agricultural Technology Progect on "Sustainable Management of Plant Biodiversity". Under this programme, fifteen exploration trips were conducted covering the various agro-ecological regions of Kerala State to collect the landraces/traditional cultivars and wild and weedy relatives of rice available in these areas. One hundred and eighty nine landraces including many rare and almost extinct varieties were collected which have been conserved under Long Term Storage (LTS) in the National Gene Bank, New Delhi, Medium Term Storage (MTS) at Regional Station of the NBPGR, Thrissur and also maintained in the field genebanks of the rice research stations under Kerala Agricultural University. Fig.13.1. shows the diagrammatic representation of collection, conservation and use of rice genetic resources by the various institutions in Kerala.

Characterisation of the Rice Germplasm of Kerala

The local landraces grown in different agro-climatic regions of Kerala differ for a range of characters including crop duration, plant height, tillering, pigmentation of various plant parts, panicle characters, grain characters as well as grain and straw production. There are photo-insensitive varieties as well as photosensitive varieties maturing in 60-260 days. Tremendous variation exists for grain characters viz., grain color, grain shape, grain size, kernel color, aroma, puffing, popping, flattening, cooking and eating qualities. They reflect varietal preferences for different traditional food preparations. Grain color itself varies from straw to gold to brown tinges to complete brown and to black. The grain size varies from round to short bold to long bold to long slender with long bold red kernelled grains being mostly preferred by the local farmers. In order to assess the variability available for each of these characters in the rice germplasm of Kerala, the landraces collected through various exploration programmes were characterised for the following quantitative and qualitative characters as per the descriptor list on rice prepared by IRRI-IBPGR:

(1) Seedling height, (2) Leaf length, (3) Leaf width, (4) Blade pubescence, (5) Blade color, (6) Basal leaf sheath color, (7) Leaf Angle, (8) Flag leaf

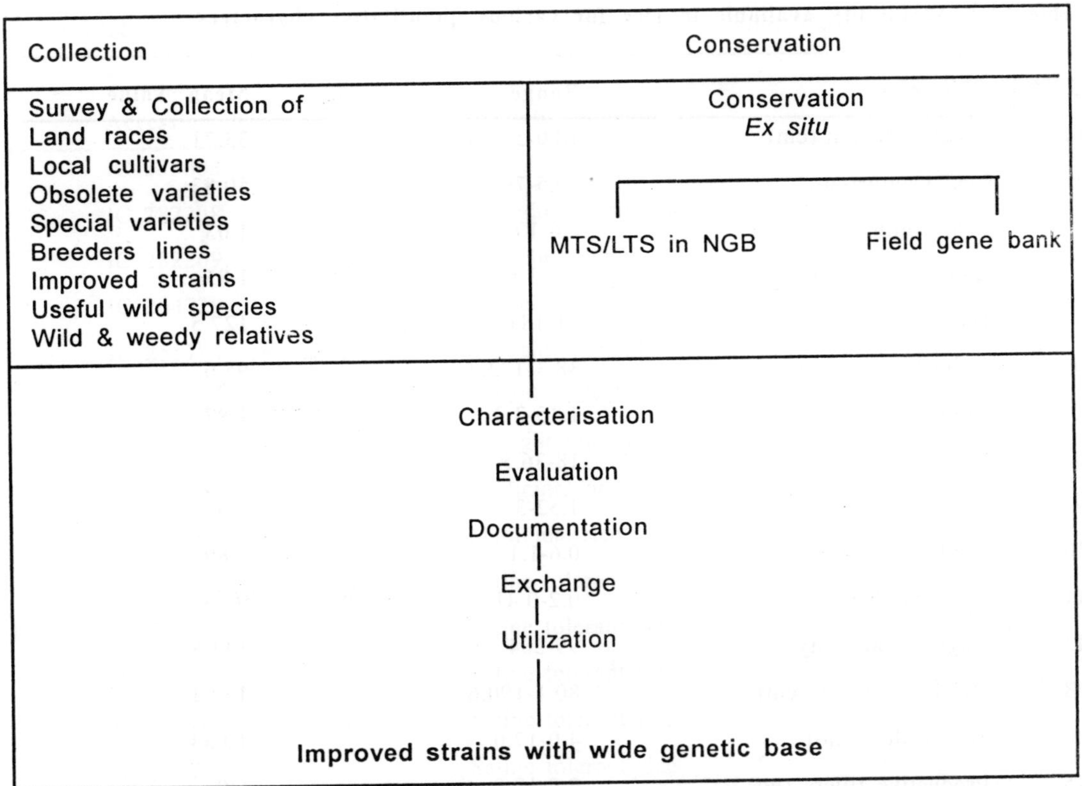

Fig. 13.1. Collection, conservation and use of rice genetic resources in Kerala.

angle, (9) Ligule length, (10) Ligule color, (11) Ligule shape, (12) Collar colour, (13) Auricle color, (14) Days to heading, (15) Culm length, (16) Culm angle, (17) Culm diameter, (18) Internode colour, (19) Culm strength, (20) Panicle length, (21) Panicle type, (22) Secondary branching of panicle, (23) Panicle exsertion, (24) Panicle axis, (25) Grain shattering, (26) Threshability, (27) Grain awning, (28) Awn color, (29) Apiculus color, (30) Stigma color, (31) Lemma and palea color, (32) Lemma and palea pubescence, (33) Sterile lemma color, (34) Sterile lemma length, (35) Spikelet sterility, (36) 100-grain weight, (37) Seed length, (38) Seed width, (39) Seed coat color, (40) Scent, (41) Leaf senescence, (42) Maturity (days), (43) Height of plant, (44) Total tillers, (45) Productive tillers, (46) Grain yield per plant, (47) Straw yield per plant, (48) No of spikelets per panicle, (49) No. of grains per panicle.

Rich diversity was noticed in all the characters studied except in the case of characters like stem diameter, scent, sterile lemma colour, leaf blade pubescence, etc. Table 13.3 depicts the variability available in the crop for various quantitative characters.

Table 13.3. Variability available in rice for various quantitative characters.

Sl. No	Character	Range	Mean Value
1	Seedling height (cm)	18.0-51.16	33.71
2	Leaf length (cm)	30.5-79.4	50.83
3	Leaf width (cm)	0.6-1.6	1.03
4	Ligule length (cm)	0.5-3.5	1.85
5	Days to 50% flowering	71-184	127.5
6	Culm length (cm)	38.3-172.2	95.6
7	Culm girth (cm)	1.15-3.2	1.87
8	Panicle length (cm)	18-36.5	25.5
9	100 grain weight (g)	1.52-3.65	2.67
10	Seed length (cm)	0.6-1.1	0.89
11	Seed width (cm)	0.2-0.41	0.36
12	Days to maturity	99-204	133.8
13	Height of plant (cm)	80.1-190.6	132.1
14	Total tillers (no)	4.0-17.0	10.34
15	Productive tillers (no)	2.0-10.0	6.0
16	Grain yield per plant (g)	1.0-28.1	10.6
17	Straw yield per plant (g)	3.5-110.2	45.28
18	No. of grains per panicle	52-405	143.3

Evaluation of Rice Genetic Resources in Kerala

Genetic evaluation of the landraces for specific characters was taken up with utmost priority in Kerala from the early period of rice improvement. The landraces collected through various exploration programmes were maintained in the field genebanks, characterised for quantitative and qualitative characters and evaluated for resistance to biotic and abiotic stresses and quality attributes and promising donors with special features have been identified. National accessioning of the germplasm collected and maintained at different rice research stations has also been done. Table 13.4 gives the details of promising accessions from the rice germplasm collection of Kerala

Table 13.4. Promising accessions from the rice germplasm collection of Kerala.

Accession No.	Drought tolerant rice
IC 255803	Drought tolerant rice
IC 282208	Drought tolerant rice
IC 282209	Drought tolerant rice
IC 264729	Drought tolerant rice
IC 255784	Drought tolerant rice
IC 255782	Drought tolerant rice
IC 282225	Drought tolerant rice
IC 255787	Medicinal rice
IC 349691	Medicinal rice
IC 255767	Saline tolerant rice
IC 255761	Saline tolerant rice
IC 255762	Saline tolerant rice
IC 255764	Saline tolerant rice
IC 255765	Saline tolerant rice
IC 264730	Flood and saline tolerant rice
IC 264731	Flood and saline tolerant rice
IC 265194	Deep water rice
IC 265196	Deep water rice
IC 265197	Deep water rice
IC 265183	Aromatic rice
IC 343379	Aromatic rice
IC 349677	Aromatic rice

Utilization of Genetic Diversity in Crop Improvement

One hundred years have passed since research programmes for rice improvement were initiated in Kerala in 1905 at the Agricultural Research Station, Thalipparamba during the British regime to face the problems of low production and food scarcity. Scientific breeding for evolution of improved varieties were initiated in late nineteen twenties when Agricultural Research Station at Pattambi was established in 1927. With a view to meeting the needs of the time, breeding efforts were made to stabilize and increase rice production by incorporating various useful characters including photo-insensitivity, early maturity, lodging resistance, tolerance to biotic and abiotic stresses and high yield into the native rice varieties of good adaptability. As a result, improved cultivars have been developed from many native ones and further achievements have been made in breeding using these improved cultivars and exotic materials.

Improvement of Tall *indicas* through Selection

In late twenties and early thirties of twentieth century, traditional rice varieties that were popular in Kerala were improved by pure line selection and released by various research stations within the State (Table 13.5). These earlier attempts for rice improvement before the advent of green revolution were largely concentrated on improving the local varieties through selection. More than half of the famous 'PTB' varieties released from Regional Agricultural Research Station, Pattambi (thirty four), the first three 'MO' (MO-1, MO-2, MO-3) varieties from Rice Research Station, Moncompu, the saline tolerant selections from Rice Research Station, Vytilla have all been developed either through mass selection or pureline selection of traditional varieties which were popular in the respective regions. This increased the average yield of the varieties to nearly two fold (1.5 t/ha to 3.0 t/ha). These varieties were widely cultivated within and outside the state because of their increased yield compared to their parents. Bred for yield improvement over their local progenitors, these varieties combine high yield as well as the special traits of their parents like resistance to pests and diseases, quality attributes, physiological attributes, adaptability to different ecological situations and systems of cultivation.

Table 13.5. Improvement of traditional tall *indicas* by various Rice Research Stations in Kerala.

Research Station	Varieties developed
Regional Agricultural Research Station, Pattambi	PTB 1 to PTB 34
Rice Research Station, Moncompu	MO 1 to MO 3
Rice Research Station, Vytilla	Vytilla 1 and Vytilla 2
Regional Agricultural Research Station, Kayamkulam	KTR 1, KTR 2, KTR3, UR 19,UR 10, Kylm 4
Agricultural Research Station, Adoor	Adr 1 to Adr 10

PTB 10 is a variety resistant to BPH, stem borer and gall midge. It is a very short duration variety and is utilised in rice breeding throughout the world for hybridisation programmes It has contributed to the evolution of many of the HYVs released in Kerala. This unique variety is also known to possess the gene for better translocation efficiency and photosynthetic efficiency (Natarajan *et al.*, 1996). PTB 15 and PTB 16, two flood tolerant rice varieties derived from the local variety *Kavunginpoothala* are still popular in Kerala and Tamil Nadu due to high consumer acceptance for the white grains which are also good for making pearl rice. PTB 18 (*Eravapandy*) and PTB 21 (*Thekkan*) are multiple resistant varieties, PTB 7 (*Parampuvattan*) is resistant to foot rot and gall midge, PTB 18, PTB 21 and PTB 33 (*Arikkirai*) show resistance to BPH, PTB 27 (*Kodiyan*), PTB 28 (*Kattamoden*) and PTB 32 (*Aruvakari*) are gall midge resistant varieties, PTB12 (*Chitteni*) and PTB 29 (*Karuthamodan*) show resistance to bacterial blight and PTB 28 (*Kattamodan*), PTB 29 (*Karuthamodan*), PTB 30 (*Chuvannamodan*) and PTB 42 (*Suvarnamodan*) are upland varieties.

Varieties Developed through Hybridization and Selection

Hybridization between the native germplasm and exotic varieties to combine the high yield of exotic varieties with the desirable qualities of native rices was carried out in Kerala Agricultural University right from the sixties. A remarkable breakthrough in the field of rice improvement was the release of Annapoorna (PTB 35), the first high yielding short duration dwarf variety released in India in 1966 through hybridization of PTB 10 with T(N)1. This brought about a great revolution in total rice production scenario of the state with an average yield of nearly 5 t/ha. Subsequently several rice varieties were developed by incorporating the semi-dwarfing gene into the genetic background of the traditional cultivars of Kerala. Introduction of varieties like T(N)1 and IR 8 which were high yielding but susceptible to the pests and diseases of tropical region in the sixties especially when grown with increased use of chemical fertilisers resulted in an outbreak of pests and diseases. Breeding programmes emphasising the need for development of varieties with built-in resistance to pests and diseases were undertaken and crosses were made between outstanding varieties and resistant donors which resulted in high yielding and multiple resistant rice varieties (Table 13.6).

Table 13.6. Hybrid derivatives developed in Kerala using native germplasm.

Sl No	Variety	Parentage	Native germplasm involved
1	Annapoorna	T(N) 1 / PTB 10	*Thekkancheera*
2	Rohini	PTB 10 / IR 8	*Thekkancheera*
3	Aswathy	PTB 10 / DGWG	*Thekkancheera*

4	Triveni	Annapoorna / PTB 15	*Thekkancheera, Kavunginpoothala*
5	Jyothi	PTB 10 / IR 8	*Thekkancheera*
6	Sabari	IR 8/2 / Annapoorna	*Thekkancheera*
7	Bharathy	PTB 10 / IR8	*Thekkancheera*
8	Swarnaprabha	Bhavani / Triveni	*Thekkancheera, Kavunginpoothala*
9	MattaTriveni	Reselection from Triveni	*Thekkancheera., Kavunginpoothala*
10	Jayathi	Triveni / IR 2061	*Thekkancheera., Kavunginpoothala*
11	Nila	Triveni / Vellathil Kolappala // Co.25	*Thekkancheera., Kavunginpoothala Vellathil Kolappala*
12	Kairali	IR 36 / Jyothi	*Eravapandy, Thekkan, Thekkancheera*
13	Kanchana	IR 36 / Pavizham	*Eravapandy, Thekkan, Karivennel*
14	Aiswarya	Jyothi / BR 51	*Thekkancheera*
15	Bhadra	IR 8 / PTB 20	*Vadakkan chitteni*
16	Asha	IR 11-1-66 / Kochuvithu	*Kochuvithu*
17	Pavizham	IR 8/ Karivennel	*Karivennel*
18	Karthika	Triveni / IR 1539	*Thekkancheera, Kavunginpoothala*
19	Aruna	Jaya / PTB 33	*Arikkirai*
20	Remya	Jaya / PTB 33	*Arikkirai*
21	Kanakom	IR 1561 / PTB 33	*Arikkirai*
22	Aarathy	Jaya / PTB 33	*Arikkirai*
23	Renjini	Mo.5 / Improved Sona	*Kochuvithu*
24	Pavithra	Surekha / MO.5	*Kochuvithu*
25	Panchami	Pothana / Mo.5	*Kochuvithu*
26	Uma	Mo.6 / Pokkali 372	*Karivennel, Pokkali*
27	Revathy	Mo.6 / Cul 12814	*Karivennel, Chennellu*
28	Karishma	Mo.1 / Mo.6	*Chettivirippu, Karivennel*
29	Krishnanjana	Mo.1 / Mo.6	*Chettivirippu, Karivennel*
30	Lakshmi	Kottarakkara 1 / Poduvi	*Kutticheradi*
31	Onam	Kochuvithu / T(N)1 // Triveni	*Thekkancheera, Kavunginpoothala Kochuvithu*
32	Bhagya	Thadukan / TN 1	*Thadukan*
33	Dhanya	Jaya/ PTB 4	*Vellari*
34	Vyttila 4	Chettivirippu / IR4630-22-2-17	*Chettivirippu, Pokkali*
35	Ahalya	PTB 10/ TN 1 // TN 1	*Thekkencheera,*

Induced Mutagenesis of Traditional Tall *indicas*

Efforts to improve the traditional tall *indicas* of Kerala through induced mutagenesis were initiated in the Kerala Agricultural University during early seventies. The traditional tall varieties were subjected to irradiation at varying doses to change the plant type without disturbing other important characters viz., adaptability to varying agro-ecological conditions, resistance/tolerance to stresses and red kernel colour. Several early maturing, dwarf/semi-dwarf, non-lodging and awnless mutants with high yield and resistance to major biotic and abiotic stresses could be developed (Table 13.7). Desirable mutants so isolated were utilised directly for commercial cultivation and also as parents in cross breeding programmes. An awnless, early maturing (10-15 days early) mutant, isolated from the M_2 generation of the gamma (g) ray treated population of *Oorpandy*, the long awned saline resistant local variety of Ernakulam district in Kerala was released as Rasmi in 1985. The variety is resistant to gall midge, BPH and blast and is used as a donor for multiple resistance in the rice breeding programmes at the national level (Rao *et al.*, 1998). The dwarf mutant of MO1, a selection from *Chettivirippu*, the tall *indica* variety of the coastal region of Kerala, was released for general cultivation as Remanika in 1998 (Devika *et al.,* 2004). Induced mutagenesis of improved traditional rice varieties like PTB 1 (*Aryan*), PTB 7 (*Parambuvattan*), PTB 9 (*Thavalakkannan*) PTB 10 (*Thekkancheera*), PTB 20 (*Vadakkan, Chitteni*) and PTB 26 (*Chenkazhama*) etc. yielded mutants possessing high genetic variability for morphological and physiological attributes. Among these, two mutants of PTB 20 have been found to be excellent replacement for the parent variety which is the ruling second crop variety of the State. During 1990, a recombination breeding programme was initiated involving the mutants M 2 (Mutant of PTB 9), M 210 (mutant of PTB 10) and upland rice varieties like PTB 28 to develop high yielding and non-lodging varieties suitable for the rainfed uplands of Kerala. Two varieties were developed from this programme, Harsha released during 2000 and Varsha in 2002.

Table 13.7. Varieties developed through induced mutagenesis and inter-mutant hybridisation.

S. No	Variety	Parentage	Native germplasm involved
1	Rasmi	Mutant of Oorpandy	*Oorpandy*
2	Harsha	M. 210 / PTB 28	*Thavalakkannan, Thekkancheera., Kattamodan*
3	Varsha	M. 210 / PTB 28	*Thavalakkannan, Thekkancheera., Kattamodan*
4	Remanika	Mutant of Mo. 1	*Chettivirippu*

6	Gouri	Mo.4 / Cul. 25331	*Vadakkan chitteni,, Oorpandy*
7	Chingam	M2 / Cul. 25331-1	*Thavalakkannan, Oorpandy*
8	Dhanu	Mutant of PTB 20	*Vadakkan chitteni*

Use of Kerala Rice Germplasm in National and International Breeding Programmes

Many of the traditional rice varieties and the improved varieties developed from them through selection are internationally acclaimed for their performance and are accepted as international donors for biotic and abiotic stress breeding. The parentage of many national and international varieties can be traced back to these improved rice varieties and landraces of Kerala (DRR, 2001). As many as eighty varieties have been developed at the national level using Kerala rice varieties (Table 13.8) and about twenty rice varieties/cultures have been developed by IRRI using rice varieties from Kerala. The most frequently used parents are PTB 10, PTB 15, PTB 18, PTB 21, PTB 33 and Pokkali (Table 13.9).

Table 13.8. Varieties developed by various states in India using Kerala rice germplasm.

Sl. No.	State / Agency	Varieties released	Kerala rice germplasm involved
1	Orissa	Abhaya, Ananga, Bhanja, CR 55, CR 57 CR 93, CR 94, CR 157, CR 179, CR 190, Daya, Indravathi, Kharveli, Konark, Kshira, Lalat, Manika, Mehar, Neela, OBS 677, Pratap, Radhi, Sarasa, Shakti, Siva, Somasila, Sravani, Tara and Udaya	PTB 10, PTB 18, PTB 21, PTB 33, PTB 43 (Swarnaprabha)
2	Andhra Pradesh	Bharani, Penna, RP 10, RP 352, RP 825, RP 536, Rasmi, Sagar samba, Varsha and Vikramarya	PTB 2, PTB 18, PTB 21, Pokkali
3	Tamil Nadu	ADT 36, ADT 37, ADT 44, ASD 18, ASD 20, ASD 28 and CO. 47	PTB 33, Thekkancheera, Kavunginpoothala
4	Central release	Jawahar Rice 3-45 , PA 103 (6201), Pusa 834, Sakthiman, Suraksha and Swathi	MO 1, PTB 18, PTB 21
5	Gujarat	GR 3, GR 6, IR 22, IR 28, IR 579, SLR 52214	PTB 21, PTB 33, Tadukan,

6	West Bengal	Jamini, Mahananda, Sashi and Sebati	PTB 10, PTB 18, PTB 21, PTB 33
7	Bihar	Archana, Deepa and Rajendradhan 201	Tadukan
8	Himachal Pradesh	Himalaya 2216, HKR 120 and RP 2421	PTB 33
9	Karnataka	Mahaveera, Netravathi and Red Annapoorna	PTB 10
10	Maharashtra	Karjat 1, Karjat 3 and Ratnagiri	PTB 18, PTB 21
11	Uttar Pradesh	Narendra 80, Narendra 118 and Pantdhan 10	PTB 18, PTB 21
12	Haryana	HKR 120	PTB 33
13	Madhya Pradesh	Radhe	PTB 18, PTB 21
14	Pondicherry	Bharathidasan	PTB 18, PTB 21, PTB 33
15	Punjab	PR 108	PTB 21

Table 13.9. IRRI varieties having Kerala parentage.

Sl. No	Varieties / Breeding lines developed and released by IRRI	Germplasm involved
1	IR 22, IR 28, IR 32, IR 36, IR 38, IR 40, IR 42, IR 44, IR 48, IR 50, IR 52, IR 56, IR 58, IR 60, IR 62, IR 65.	PTB 18, PTB 21, PTB 33
2	IR 579-48-1-2, IR 1561, IR 2061, IR 4630	PTB 18, PTB 21, PTB 33

Future Thrust

The rice germplasm collection made so far in Kerala still fall short of the full spectrum necessary to meet future needs. As new genes or more loci of a similar genetic function are needed to meet the ever expanding needs of plant breeding, they need to be identified and used to cope with such situations. The following priorities need emphasis:

1) To assemble and inventorise all the existing germplasm now available with different institutions of the State and characterization of the same at multi-locations using standard descriptors (already developed by IRRI/NBPGR) including disease/pest resistance, quality attributes, etc. Thus, status of field genebank in the state can be assessed.

2) Accessioning of all existing germplasm with different partners.

3) Multiplication, seed increase and conservation of germplasm in the National Gene Bank at NBPGR with necessary passport and evaluation data.

4) To collect germplasm from unexplored/under-explored areas of germplasm availability including hot spot locations.

5) Collection and maintenance of wild germplasm with high priority, and

6) Registration of elite germplasm.

References:

Anonymous. 2000. *Farm Guide.* Farm Information Bureau, Government of Kerala 10:23.

Devika RN, Rema Bai, Leenakumary S. 2004 . Remanika: A short duration rice for Kuttanad, Kerala. *Journal of Tropical Agriculture* 42(1-2):16-18.

Directorate of Rice Research. 2001. *High Yielding Rice Varieties of India-2000. DRR Bulletin 2001-1.* Directorate of Rice Research, Hyderabad.

Heinrichs EA, Medrano FG, Prapusa HR. 1985. *Genetic Evaluation for Insect Resistance.* International Rice Research Institute, Manila, Philippines. 365 pp.

Ikeda R, Busto Jr GA, Ogawa T. 1990. Resistance of wild rices to Bacterial Blight (BB). *Intern Rice Res Newsl* 15(3):14.

IRRI (International Rice Research Institute). 1990. *Program Report for 1989.* Manila, Philippines.

Kalode MB, Krishna TS, Pophaly DJ, Lakshminarayana A. 1977. Note on new donors having multiple resistance to major insect pests of rice. *Indian J Agric Sc.* 47:626-627.

Khush GS, Coffman WR, Beachell HM. 2001. The history of rice breeding: IRRI's contribution. Pages 117-135 in Rockwood WG (ed) *Rice Research and Production in the 21st Century.* IRRI, Los Banos, Philippines.

Krishnaswamy N, Chandrasekharan P. 1958. A new species of Oryza L. *Madras Agric J* 45(12):471-472.

Leena Kumary S. 2004. Genetic diversity, characterisation and evaluation of medicinal rice in Kerala. Lead paper presented in the National Seminar on *Medicinal and Aromatic Rices of Kerala.* 19-21 August 2004, Regional Agricultural Research Station, Pattambi, Kerala.

Leena Kumary S, Tessy Joseph, Nayar NK, Rehmath Niza TJ. 1998. Field tolerance levels of *indica* rice varieties for major diseases of Kerala. *Crop Res* 15(1):112-114.

Leena Kumary S, Susamma P George, Dijee Bastin, Aipe KC, Rema Bai N. 2003. Aromatic rices of Kerala. Pages 317-326 in Singh RK and Singh US (eds) *A Treatise on the Scented Rices of India.* Kalyani Publisher, New Delhi.

Nayar NM. 1973. Origin and cytogenetics of rice. *Adv in Genetics* 17:153- 292.

Nair NR, Nayar NM, Sahadevan PC. 1964. On the occurrence of *Oryza* species in south-western India and its significance. *Curr Sci* 33:515-517.

Natarajan L, Elsy CR, Nayar N K. 1996. *Kerala rice as international donors for stress resistance*. Kerala Agricultural University, Vellanikkara, Kerala.

Rao NGP, Ambika Singh, Sivasubramanian V, Murty KS, Mukhopadhyay AN, Abraham CC. 1998. Rice research and production in India. In: *Present Status and Future Perspective: Quinquennial Review Committee Report 1988-96*. Directorate of Rice Research, Hyderabad. 110 pp.

Sahadevan PC (ed). 1966. *Rice in Kerala*. Agricultural Information Service, Deptt of Agriculture, Kerala.

Sujith Kumar CK. 1999. *Krishi Malayalam*. Akshara Samskrithi, Samskrithi Publications, Kannur, Kerala.

Tateoka T. 1963. Taxonomic status of *Oryza*. III. Key to the species and their enumeration. *Shakubutsugaku Zasshi* 76:165-173.

Vaughan DA. 1994. *The Wild Relatives of Rice: A Genetic Resources Handbook*. International Rice Research Institute, Manila, Philippines.

Vaughan DA, Muralidharan VK. 1989. Collection of wild relatives from Kerala State, India. *Plant Genet Resour Newsl* 42:2-6

Virmani SS, Shinjyo C. 1988. Current status of analysis and symbols for male sterile cytoplasms and fertility restoring genes. *Rice Genet Newsl* 5:9-15.

Genetic Resources of Rice in India : 431-464
Editor : S. D. Sharma
Today & Tomorrow's Printers and Publishers, New Delhi - 110 002 (India)

Chapter 14

Rice Germplasm of Chhattisgarh: Collection, Evaluation, Utilization and Conservation

G.R. Sahu, N.K. Motiramani, A.K. Sarawgi, D.J. Pophaly, R.L. Pandey and M.N. Shrivastava

The geographic tract, comprising of southeastern part of Chhattisgarh along with adjoining Koraput district of Orissa is considered to be a secondary centre of origin for rice *Oryza sativa*, encompassing within itself major part of biological diversity found within rice crop in India. Chhattisgarh, as part of erstwhile state of Central Provinces and Berar during British rule, happens to be among those states which took lead in rice germplasm collection and utilization.[1] In very early years of twentieth century, the British Government of Central Province and Berar established three seed production and distribution farms for rice, wheat and other arable crops respectively at Labhandi (Raipur), Powerkheda (Hoshangabad) and Adhartal (Jabalpur) in 1903, each under one Deputy Director of Agriculture. The objective of these farms was to multiply the seeds of popular cultivated strains of the mandated crops and distribute among farmers particularly during the years following drought or famine. Ensuring seed supply to farmers in such years became necessary as the country underwent devastating droughts in 1896 and 1899.

---

[1] During the British rule, Chhattisgarh was a part of the Central Provinces and Berar which included what is now Chhattisgarh, Vidarbha region of Maharashtra and Narmada valley of the present-day Madhya Pradesh with Nagpur as its capital. In 1956, when states were reorganized, Chhattisgarh became a part of the Madhya Pradesh state with Bhopal as its capital. Before the creation of Indira Gandhi Agricultural University in 1987, all agricultural research of Chhattisgarh was under the Jawaharlal Nehru Krishi Vishwa Vidyalaya, Jabalpur. (Editor)

Germplasm Collection

Early efforts

In order to ensure seed supply of desired strains to farmers, it was quite essential to first identify the cultivars popular with farmers. Further, a lot of natural variability, as expected, was existing within each landrace accumulated as a result of spontaneous mutations occurring over centuries, rather millenniums and forces of natural selection acting over them. It was therefore very natural for Dr. R.J.D. Graham, the first Deputy Director of Agriculture in-charge of Labhandi Seed Multiplication Farm to collect major variants of each major strain, identify the popular ones, purify and multiply the seed for distribution among farmers. This involved quite an extensive work.

Thus, the credit of initiating collection of local variability in rice crop in Central India goes to Dr. Graham. The lead was later followed by Dr. D.N. Mahata his successor. Several local types were collected. The details of this work are not available now. Nevertheless, those strains selected within and across each major type for seed increase and distribution to farmers are available. For the purpose, four selections in 1913, one each in 1916 and 1918 and three in 1923 were identified. The important landraces from which the selections were made were Nungi, Bhondu, Parewa, Gurmatia and Dilbuxa. The given designation numbers like No. 143 for Parewa, No. 35 for Dilbuxa and No. 11 for Bhondu tend to indicate that at least that many number of pure line selections were made within each of the landrace to identify the better ones for seed multiplication (Shrivastava and Kumar, 2004) (Table 14.1). In fact, these selections identified for seed increase, to a good extent, can be considered as varieties released since there was no formal committee constituted to identify and release new varieties in those years.

Dr. Graham's and Dr. Mahata's work was followed by Dr. B.K. Dave from 1930 onwards. He collected about 700 germplasm samples from the erstwhile state of Central Provinces and Berar. However, no catalogue was published. The importance of characterization of natural variability collected for the purpose of varietal improvement was not realized to that extent in those early years.

Table 14.1. The improved types isolated from landraces for seed multiplication during the period 1913-23.

S.N.	Selection No.	Selected from	Year
1.	EB-17	Nungi	1913
2.	Parewa No. 143	Parewa	1913
3.	Bhondu No. 11	Bhondu	1913

4.	G 13	-	1913
5.	Bhata Gurmatia	Gurmatia	1916
6.	Bhondu No. 10	Bhondu	1918
7.	Parewa No. 22	Parewa	1923
8.	Gurmatia No. 11	Gurmatia	1923
9.	Dilbuxa No. 35	Dilbuxa	1923

Source: Shrivastava and Kumar (2004)

Chhattisgarh rice germplasm

The Rice germplasm collection of Chhattisgarh presently maintained by Indira Gandhi Agricultural University, Raipur and numbering to 21,357 accessions owes almost exclusively to the efforts made by late Dr. Radheylal Hariram Richharia between the years 1971 and 1978. Dr. Richharia, after retiring as Director of Central Rice Research Institute, Cuttack in 1966 was appointed as Agricultural Advisor in 1971 to the Government of undivided Madhya Pradesh of which the current state of Chhattisgarh was a part. The collections were made through village level extension workers of Department of Agriculture as well as through trips undertaken by Dr. Richharia and his associates deputed to help him by the government. This work further got strengthened with coming into existence of 'Madhya Pradesh Rice Research Institute' (MPRRI), an autonomous body, in 1976 funded by the state government with Dr. R. H. Richharia as its Director.

The Institute maintained details of each survey undertaken by the staff members. The 'Survey Register' included expedition number, dates of survey, names of team members, area covered and total number of accessions collected. The individual samples were then entered in 'Block Registers' i.e. the geographic area from where they were collected, so as to get an idea about how exhaustively a block has been surveyed thereby to indicate the need of any further surveys. The individual samples were then cross entered in 'Alphabetic Accession Register', where the entry was made in alphabetic order as per the name of the variety as informed by the farmer. This register includes the passport traits such as name of the variety, maturity, grain type, village and farmers' name along with any other trait of importance as told by the farmer. The serial number of this register is the number for the purpose of maintenance. For example, accession number 'A:670' refers to the 670th accession of the collection whose name begins with letter 'A'. This very number is also used for recording observations, storage and seed supply; in essence it is the identity number of the accession. The 'Field Registers' and the 'Observation Registers' are yearly registers to keep details of planting in the field and for recording observations, respectively.

The collections made or received by the Institute until 1978 from different districts of Madhya Pradesh are given in Table 14.2.

Table 14.2. Number of samples collected from different districts of Madhya Pradesh until 1978.

S. N.	District	No. of Samples	S. N.	District	No. of Samples
1.	Bastar	5628	22.	Khandwa	74
2.	Raipur	4509	23.	Morena	60
3.	Rajnandgaon	1306	24.	Chhatarpur	50
4.	Raigarh	1239	25.	Dhar	38
5.	Bilaspur	1121	26.	Narsinaghpur	42
6.	Balaghat	744	27.	Bhind	47
7.	Surguja	651	28.	Chhindwara	31
8.	Durg	544	29.	Jhabua	50
9.	Shahdol	442	30.	Sagar	30
10.	Seoni	469	31.	Raigarh	29
11.	Mandla	357	32.	Mandsaur	26
12.	Jabalpur	344	33.	Shajapur	15
13.	Hoshangabad	192	34.	Bhopal	14
14.	Panna	191	35.	Vidisha	10
15.	Sidhi	158	36.	Dewas	9
16.	Damoh	131	37.	Guna	9
17.	Betul	123	38.	Shivpuri	8
18.	Satna	102	39.	Ujjain	8
19.	Tikamgarh	102	40.	Indore	3
20.	Rewa	78	41.	Sehore	1
21.	Gwalior	75		Total	**19,055**

Subsequent collections

The material (19,095 accessions) collected by Dr. R.H. Richharia and his associates constitutes the major source of current genetic wealth maintained by Indira Gandhi Agriculture University (IGAU), Raipur. However, considerable material has been added subsequently including working germplasm of Rice

Breeding Unit of the University. Table 14.3 gives details of the addition made later to the lot collected by Dr. Richharia and his associates.

Table 14.3. Subsequent additions made by the Rice Breeding Unit of IGAU to R.H. Richharia's collection.

S.N.	Year of addition	Number of accession	Source details of the accessions
1.	1979	1097	Working germplasm maintained by Raipur centre of State Agricultural University
2.	1993	210	Wild rice samples collected by Research Scholars
3.	2000	705	Collected under IRRI-NBPGR-IGAU "Collaborative On-farm Germplasm Conservation Project"
4.	2001	1052	Collected under NATP-funded "Plant Biodiversity Project"
	Total	**1967**	

The first addition by the Rice Breeding Unit of the Department of Plant Breeding was to transfer a collection of 1,097 lines of their working germplasm basically donors for various traits including that for gall midge, bacterial leaf blight, leaf blast, etc. These had been collected from different rice research centers of the country including Directorate of Rice Research, Hyderabad and were maintained for future use. This raised the total collection to 20,192 accessions (Sharma *et al.*, 1987). The total collection, currently maintained is given maturity group-wise in Table 14.4.

Table 14.4. The total collection currently maintained by the IGAU.

Group	Number of Accessions
A. Indigenous lines	
a. Old collections:	
(i) Extra early (up to 95 days)	480
(ii) Very early (96 to 110 days)	1197
(iii) Early (111 to 125 days)	3880
(iv) Medium (126 to 140 days)	5069
(v) Late (Above 140 days)	7915

b. New collections:	
1. Collected through on-farm conservation project (IRRI, IGAU, NBPGR)	705
2. Collected through NATP on Plant Biodiversity Project	1052
B. Breeding lines	849
Grand total	**21,357**

Evaluation and Characterization

Subsequent to collection, evaluation and utilization of material are two follow-up steps as otherwise maintaining material without utilization would be just like what is said frequently, 'keeping books safe in almirah without being read by anyone.'

Characterization of so huge material is clearly a gigantic job which continues not only for years but perhaps for decades and beyond. It is for more than a quarter of a century that the material is being maintained by Indira Gandhi Agricultural University (IGAU), Raipur. The efforts made by erstwhile Madhya Pradesh Rice Research Institute (MPRRI) under the guidance of Dr. Richharia and thereafter by IGAU scientists are summarized below.

Evaluation of the material started right from 1972 by Dr. Richharia and his associates in MP Rice Research Institute.

Evaluation of material during MPRRI period

Variety purification

Up to 1970s, major rice area in the state was under tall varieties; the Institute's main emphasis was therfore to exploit the genetic variation within each of the landrace, thereby recommending to the farmers purified and higher yielding version of that very cultivar (which they were cultivating for centuries) but in a purified form along with appropriate agronomy *viz.* moderate doses of fertilizers and minimum plant protection techniques as were appropriate for tall cultivars. With the above objective in view, the Institute developed as many as 1623 purified versions of existing landraces and called them 'BD selections' (BD stood for Baronda, a village where the Institute's research farm was located (Richharia *et al.*, 1978).

Classification of accessions

The material collected was classified and maintained in five maturity groups. These were (i) extra early (up to 95 days), (ii) very early (96 -110 days), (iii) early (111-125 days), (iv) medium (126 to 140 days) and (v) late (more than 140 days) (Richharia *et al.,* 1974). These maturity groups so defined are still followed and all accessions are maintained in the field and laboratory under these very groups.

Identifying major groups of land varieties in the material.

The Institute, in an effort to analyse the patterns of variability existing in the state of Madhya Pradesh, identified prominent groups of varieties. These groups, though not based on any numerical data analysis, represent some common traits. Prominence of the group was indicated by number of variant names given to the type. Table 14.5 below gives some of the major groups so identified along with number of variants collected (Richharia *et al.,* 1978). In turn, most of the variants are often represented by multiple samples collected from different places.

Table 14.5. Major groups identified in the germplasm collection.

S.N.	Group	No. of variants collected	Major trait(s)
1.	Gurmatia	94	Early maturity, bold grain
2.	Luchai	78	Medium maturity, medium bold grains
3.	Safri	44	Mid maturity, medium slender grains
4.	Chudi	42	Purple auricled, mid-late
5.	Phool	66	Early, drought tolerant
6.	Bhog	40	Aromatic, mid late

Semi-dwarf types

The Institute collected six semi-dwarfs (Table 14.6) from farmers' fields. These were believed to be grown by farmers for decades and were suspected to be existing locally prior to introduction of Taiching (Native)-1.

Table 14.6. Semi-dwarfs collected from farmers' fields.

S.N.	Parent Source	Given BD number
1.	Badalphool	BD-21
2.	Dhour	BD-23
3.	Huda	BD-16
4.	Assamchudi	BD-831

5.	Ajan	BD-1292
6.	Nunga	BD-830
7.	JS-5	BD-49

When the first three of these semi-dwarfs were crossed with Dee-geo-woo-gen-derived lines, no segregation occurred in F_2 generation indicating that these have the same gene for semi-dwarfness.

Rices having medicinal value

During collection trips, the farmers told the Institute scientists about medicinal properties of some of the rice cultivars that were being cultivated by them. The list of such cultivars is provided in Table 14.7. These are maintained in the germplasm collection. The verification of their claims could not be undertaken as it involves controlled medical tests.

Table 14.7. Some of the rice cultivars reported to have medicinal values against the health problem mentioned.

S.N.	Cultivar	Reported to be effective in ailments
1.	Alcha	Skin ailments
2.	Gathuwan	Arthritis
3.	Karhani	Strokes
4.	Maharaji	Weakness in women after delivering a child
5.	Baisur	Headaches, epilepsy
6.	Lohbhaddar	Gas
7.	Baisur	Migraine
8.	Resari	Cough

Characterization

Serious efforts were initiated by Dr. Richharia in 1978 to bring out a catalogue (entitled 'Encyclopedia of Rice Varieties of Madhya Pradesh') which continued even after merger of MP Rice Research Institute with the Agricultural University but no publication could come out. The Institute, however, published several booklets and circulars to describe the promising selections made to popularize them.

A publication entitled *Our Strategies on the Rice Production Front in Madhya Pradesh* basically gives recommendations to be followed for getting higher yields (Richharia *et al.*, 1974). However, it also describes the variability available for various traits in the germplasm collection. For example, it mentions that the accession 'Dokra Dokri' has the longest grain (over 14 mm), 'Bhimsen' has longest grain width, 'Gathuwan' has resistance to green plant hopper and 'Udanpakheru' has high protein content. Studies made until 1978 also indicated maturity duration in the germplasm collection to be varying from 55 to 170 days, of these 7.88% belong to extra early group and 62.37% belong to medium and late groups (126 days and above). Six strains were found to have resistance to gall midge. Besides these, the publication also describes 260 purified versions of cultivars for major (passport) traits i.e. place of collection, maturity in days at the locality collected from, grain type, apiculus and kernel colours.

The second publication *A Strategy for Rice Production to Ensure Sustained Growth* was published in 1977 (Richharia, 1977). Apart from giving revised agronomical practices for higher production, it also recommended additional purified versions of cultivars district-wise. Based on number of samples of a landrace collected from different blocks, a popularity index called 'wider adaptability index' was calculated for some of the major cultivars grown in the state.

Another publication *Dhan Sampada* (in Hindi) enlists and describes cultivars for their adaptability to low-lying areas, summer cultivation, submerged areas, drought prone areas, wild rice eradication and for their supposed medicinal use (Richharia, 1978). *Early Maturing Rice Varieties of Madhya Pradesh* is another publication which characterizes 402 improved versions of local cultivars for six traits and another 194 for four traits (Richharia *et al.*, 1978).

Evaluation and characterization - Recent efforts

The germplasm material comprising of 19,095 accessions maintained by Madhya Pradesh Rice Research Institute became nucleus of all varietal improvement activities of Rice Breeding Unit of Raipur campus of the University. Upon merger of the Institute with Jawaharlal Nehru Agricultural University, efforts were made to further systematise the maintenance work and to take up characterization of the material.

Systematising maintenance, initiating characterization and sorting-out duplicates

All accessions which were being maintained block-wise during MPRRI period were rearranged in alphabetic order within each maturity group. The five maturity groups identified by Richharia continued to be the basis of grouping of the accessions and these were (i) extra early (up to 95 days), (ii) very early (96 to 115 days), early (116 to 125 days), medium (126 to 140 days) and late (above 140 days). The Indian Council of Agricultural Research financed germplasm maintenance work under a cess fund aided *ad hoc* project "Strengthening of Madhya Pradesh Rice Germplasm" which was operative during 1989-1993. One of the objectives of this project was to bring out a catalogue of the material, which despite considerable efforts, could not be brought out. The project, however, helped in systematizing maintenance and creating a short term storage facility. Good progress was made in recording 19 traits of 11,000 accessions. This information was used to publish the catalogue discussed later in this section.

Evaluation of rice germplasm accessions for biotic stresses under a network project funded by ICAR and publication of a catalogue.

With regard to publication of the catalogue, a breakthrough was made with sanctioning of a network project by the Indian Council of Agricultural Research on "Evaluation of Rice Germplasm for Biotic Stresses" which remained operative during 1993-1998. This project had in all 13 centres at which 19 morphological and seven resistance traits were recorded on 13,620 accessions (of which 10,620 were from Raipur). The project was conceived by Dr. E.A. Siddiq, the then Project Director, Directorate of Rice Resarch, Hyderabad and included Dr. M.N. Shrivastava, Senior Rice Breeder of Raipur centre and Dr. R.S. Rana, Director, National Bureau of Plant Genetic Resources as Co-Principal Investigators. Although all the three Principal Investigators could not complete the work under their supervision (as they accepted other assignments), the project was successfully completed by their successors, namely, Dr. A.P.K. Reddy of Directorate of Rice Research, Hyderabad and Dr. N.K. Motiramani of Raipur centre along with cooperators of 11 other centres.

The results of practical utility obtained under this network project will be discussed in subsequent sections of this chapter. A very significant achievement of this project, however, had been the publication of a catalogue describing 13,620 accessions for 27 traits through pooled data collected by 13 centres brought out by the Directorate of Rice Research, Rajendra Nagar, Hyderabad (DRR, 1999).

The Catalogue (Figure 14.1) gives in its first column INRC (Indian Rice Collection Number) which is a serial number of Directorate of Rice Research, Hyderabad assigned by them to all samples entering into their collection from any source. This is followed by the name of landrace/cultivar/variety followed by the source of material and the accession number as given by original agency supplying them the seed. Columns 5 to 19 are for morphological traits recorded in code e.g. plant height has 3 codes 1 for tallness, 2 for semi-dwarfness and 3 for dwarf stature and so on. The 15 traits recorded in this catalogue are (i) plant height, (ii) sheath pigmentation, (iii) foliage colour, (iv) culm thickness, (v) days to 50% flowering, (vi) tillering habit, (vii) panicle type, (viii) panicle exertion, (ix) panicle length, (x) apiculus pigmentation, (xi) awning, (xii) shattering habit, (xiii) hull colour, (xiv) kernel colour, and (xv) grain shape. The resistance traits included in the catalogue are resistance or tolerance to brown plant hopper, gall midge biotype 1 and 4, stem borer, leaf blast, bacterial leaf blight and sheath blight. This is perhaps the first document of its kind published in the country recording resistances to biotic factors in an exhaustive manner.

Based on natural occurrence of diseases and insects at respective places, eleven centres participated in the evaluation programme. Most of these centres were "hot spots" for one or more of the insects and/or diseases. The Indira Gandhi Agricultural University acted not only as a source of rice germplasm but also as a lead centre for seed increase and also as a major evaluation centre for bacterial leaf blight and gall midge. Jagdalpur, another research centre of Chhattisgarh, was included as a 'hot spot' for blast. Table 14.8 gives details of the centres that participated in the project and the stresses they screened the material for.

Table 14.8. Details of research stations that participated in the network project on evaluation of rice germplasm for biotic stresses.

SN.	Research centre	Organization	Participated as hot spot for
1	Raipur	IGAU, Raipur	GM, BLB
2	Jagdalpur	IGAU, Raipur	LBL
3	Aduthurai	TNAU, Coimbatore	Blast, BLB
4	Hazaribagh	ICAR, Almora	Blast
5	Kapurthala	PAU, Ludhiana	SB, BLB
6	Mandya	UAS, Banglore	SB, BPH
7	Maruteru	ANGRAU, Hyderabad	BPH, BLB, Sh. Bl.
8	Moncompu	KAU, Trichur	Sh. Bl.
9	Pondicherry	KVK, Pandicherry	BPH
10	Ponnampet	UAS, Banglore	LBL
11	Ragolu	ANGRAU, Hyderabad	GM
12	Rajendranagar	DRR, Hyderabad	SB, BPH, GM, Blast
13	Sambalpur	OUAT, Bhubaneshwar	GM

GM: Gall midge, BLB: Bacterial Leaf Blight, LBL: Leaf blast, BPH: Brown Plant Hopper, Sh. Bl.: Sheath Blight, SB: Stem Borer

Evaluation of the accessions for morphological traits was carried out as per standard procedure (IRRI, 1976) and screening against leaf blast, bacterial leaf blight, sheath blight, brown plant hopper and gall midge was done through following procedures of laying out Uniform Blast Nurseries, leaf clipping, tiller inoculation, resurgence and late planting respectively. The Raipur material as expected proved a very rich source for almost all stresses it was evaluated for.

The collection particularly proved a 'gold mine' for resistance to brown plant hopper and gall midge. The names of a few cultivars of which one or more accessions were identified as source of resistance for any given stress are presented in Tables 14.9 and 14.10.

Table 14.9. Names of some of the cultivars from Chhattisgarh Rice Germplasm identified as source of resistance in multilocation testing for various insects under the network project "Evaluation of Rice Germplasm for Biotic stresses".

S.N.	BPH	GM1	GM4	SB (WE)
1	Safri	Ajan	Agyasal	Anjan
2	Pyagi	Jalpa	Makdo	Bargi
3	Sarsariya	Atmashital	Bacheli	Bhedakawar
4	Chudi	Bhatapyagi	Sakri	Bansgathi
5	Sonasari	Luchai	Kadamphool	Asamchudi
6	Ajan	Dihula	Sindursenga	Chudi
7	Badshabhog	Bhejari	Jodara	Burso
8	Gadhakhunta	Khatiyapati	Vishnubhog	Badi Barik
9	Sathiya	Chingi	Jalpa	Rajbanko
10	Ajan	Nainkajal	Gurmatiya	Chiraitishi

Bph = Brown plant hopper, GM1 = Gall midge biotype 1, GM 4= Gall midge biotye 4, SB (WE) = Stem borer white ears

Since the duplicates have not been sorted out, each major landrace (cultivar) have several accessions as resistant. Of course, not all the accessions were resistant within such cultivars as is expected. For brown plant hopper, several accession of Safri, Chudi and Pyagi were found resistant.

For diseases, quite a few accessions of Atmashital, Badshahhog, Gangabaru and Jhiniparas for leaf blast and Chudi, Agyasal and Lali Ajan for bacterial leaf blight were found to be resistant (Table 14.10).

Table 14.10. Some of the principal cultivars of which one or more accessions were found resistant for leaf blast, bacterial leaf blight and sheath blight in the multilocation testing under the network project on 'Evaluation of Rice Germplasm for biotic stresses'.

S.N.	Disease		
	Leaf Blast	Bacterial leaf Blight	Sheath Blight
1	Badshahbhog	Chudi	Buthabadhi
2	Gangabaru	Agyasal	Ajan
3	Atmashital	Lali Ajan	Lali ajan
4	Jhiniparas	Berag	Thakurbhog
5	Chinnor	Badshahbhog	Madrasichudi
6	Ratanchudi	Assamchudi	Jilo
7	Dubraj	Dauna Prasad	Chudi
8	Bangla	Bilayatchudi	-
9	Asamchudi	Budha Budhi	-
10	Balkeshar	Kotte	-

Very few accessions have been reported to be resistant or tolerant to sheath blight. Further details about the resistant cultivars as to their other morphological traits can be obtained from final report of the project (DRR, 1999).

Of the total 10,650 accessions from Chhattisgarh rice germplasm evaluated in the project, the number of resistant or moderately resistant cultivars for brown plant hopper, stem borer (white ears), gall midge biotype 4, leaf blast, bacterial leaf blight and sheath blight were 267, 249, 380, 782, 200, 28 and 15 respectively. Thus, this project led to identification of a good number of resistant/tolerant donors for almost all diseases against which the material was screened for with the exception of sheath blight.

Evaluation of germplasm by Indira Gandhi Agricultural University

Another major advancement in the field of characterization of Chhattisgarh Rice Germplasm occurred in the year 2007 when Indira Gandhi Agricultural University published first of its catalogue of rice germplasm describing 5,102 accessions for as many as 24 descriptors (IGAU, 2007). One significant feature of this characterization endeavour was to measure, record and report several traits in quantitative terms instead of reporting in scores as done earlier. The Catalogue has 36 columns in all, of which 1 to 11 give passport characters, viz, (i) collection number, (ii) national identity (IC) number, (iii) vernacular name, (iv) village, (v) block, (vi) district, (vii) province name from where the sample was collected, (viii) latitude, (xiv) longitude (x) altitudes of the place, and (xi) date of collection. Besides, the Catalogue provides data on twelve qualitative traits measured in standard scores as per Standard Evaluation System (IRRI, 1976). These include colours of (xii) basal leaf sheath (xiii) leaf blade, (xiv) collar, (xv) ligule (xvi) auricles, (xvii) leaf angle followed again by colours of (xviii) internode, (xix) stigma, and (xx) apiculus, further followed by presence of awns and colours of (xxi) lemma, palea and (xxii) seed coat.

The quantitative traits included in the study were (xxiii) days to flowering, (xxiv) effective tiller numbers, (xxv) plant height, (xxvi) panicle length, (xxvii) number of filled grains, (xxviii) number of unfilled grains, (xxix) total number of grains per panicle (n), (xxx) yield per plant (g), (xxxi) test weight (of 100 grains) (g), (xxxii) grain length (mm), (xxxiii) grain width (mm), and (xxxiv) grain length:width ratio. All these traits were recorded under standard agronomical management as spelled out in the Catalogue. Efforts have also been made to report overall variation for each of the quantitative traits and the relationship among them by computing correlation coefficients (based on as high as 5,102 accessions) which give quite a comprehensive view of the material. Tables 14.11 and 14.12 are reproduced from the Catalogue.

The Indira Gandhi Agricultural University continued its efforts and brought out second volume of the Catalogue titled as *Indigenous Rice germplasm of Chhattisgarh and Madhya Pradesh, Volume II* (IGAU, 2008) on the same lines including the same traits as in the first volume. Total number of accessions characterized in this volume is 4,946. Thus, the data are now available on 10,048 accessions for 12 qualitative traits recorded in standard scores and 11 traits recorded in a quantitative manner along with the passport information on 11 aspects. The third volume of the series is in the press and is likely to come out soon.

Multilocation evaluation of rice germplasm

Another network endeavour to characterize Chhattisgarh germplasm was undertaken by National Bureau of Plant Genetic Resources (NBPGR) under the leadership of Dr. N.K. Gautam in the year 2004. The other centres included in the network project were Directorate of Rice Research (DRR), Hyderabad, Central Rice Research Institute (CRRI) Cuttack; Central Upland Rainfed Rice Research Station, Hazarihagh; Punjab Agricultural University, Ludhiana; Govind Ballabh Pant University of Agriculture and Technology, Pantnagar; Tamilnadu Agricultural University, Coimbatore; Indira Gandhi Agricultural University, Raipur; Assam Agricultural University, Titabar; Assam and Agricultural Research Stations of Nellore, Malan, Ponnempet and Tirur.

Table 14.11. Descriptive statistics for 12 quantitative traits recorded on 5,102 accessions of the Rice Germplasm of Chhattisgarh and Madhya Pradesh.

Parameters	DF	ET	Ht.(cm)	PL (cm)	FG/Pl	UFG/Pl	TG/Pl	Yield/Pt (g)	TW(g)	GL(mm)	GW (mm)	L/W
SE	0.13	0.03	0.24	0.03	0.40	0.15	0.46	0.08	0.01	0.01	0.00	0.01
Median	97.0	8.00	121.0	23.66	94.33	15.33	111.66	13.66	2.40	8.90	2.80	3.04
Mode	100.0	7.00	118.33	22.66	82.00	10.00	119.66	15.00	2.40	8.50	2.80	3.00
SD	9.51	2.11	17.23	2.49	28.32	10.63	32.74	5.77	0.37	0.72	0.30	0.44
Sample variance	90.38	4.46	296.77	6.18	802.19	113.04	1072.10	33.33	0.14	0.52	0.09	0.20
Kurtosis	-0.42	1.97	0.67	0.22	2.13	8.75	1.65	0.51	0.61	1.57	0.46	1.98
Skewness	-0.33	0.94	-0.43	0.05	0.94	1.78	0.86	0.69	-0.66	-0.34	-0.46	0.90
Range min.	56.00	3.00	48.16	12.50	31.66	2.00	39.99	1.50	0.94	5.40	1.70	1.92
Rang max.	119.00	21.33	176.06	33.66	294.66	102.33	334.66	42.33	3.47	11.20	3.80	5.37
Confidence level (95%)	0.26	0.06	0.47	0.07	0.78	0.29	0.90	0.16	0.01	0.02	0.01	0.01

Reproduced from Sharma et al. (2007): Catalogue on Indigenous Rice Germplasm of Chhattisgarh and Madhya Pradesh, Vol. I, p. vii

SE: Standard Error, SD: Standard deviation, DF: Days to 50% flowering, ET: Effective tillers/plant, Ht: Plant height (cm), PL: Panicle length in cm, FG/Pn: Number of fertile grains per panicle, UPG/Pn: Number of unfertile grains per panicle, TG/Pl: Total grains per panicle, Yield/Pt: Yield in gram per plant, TW: test 100 grain weight (gm), GL: Grain length in mm, GW: grain width in mm, and L/W: ratio of grain length and grain width

Table 14.12. Correlation coefficients among the quantitative traits.

Traits	ET	Ht	PL	FG/Pn	UFG/Pn	TG/Pn	Yield/Pl	TW(Gm)	GL(mm)	GW(mm)
DF	-0.121	0.322	0.127	0.266	0.153	0.280	0.156	0.176	0.002	0.257
ET	1.000	0.013	-0.013	-0.042	-0.060	-0.056	0.517	0.091	-0.021	-0.103
Ht.		1.000	0.609	0.294	0.066	0.275	0.385	0.226	0.227	0.199
PL			1.000	0.208	0.072	0.203	0.257	0.139	0.207	0.044
FG/Pn				1.000	0.261	0.95	0.336	-0.198	-0.129	0.008
UFG/Pn					1.000	0.055	0.015	-0.131	0.007	-0.024
TG/Pn						1.000	0.296	-0.214	-0.109	-0.018
Yield/Pl							1.000	0.164	0.064	0.125
TW (g)								1.000	0.494	0.512
GL									1.000	0.035

Reproduced from IGAU (2007): Catalogue on Indigenous Rice Germplasm of Chhattisgarh and Madhya Pradesh, Vol. I. p. vii

DF: Days to 50% flowering, ET: Effective tillers per plant, Ht: Plant height in centimetre, PL: Panicle length in centimetre, FG/Pn: Number of fertile grains per panicle, UFG/Pn: Number of unfertile grains per panicle, TG/Pn: Total number of grains per panicle, Yield/Pl: grain yield per plant in grams, TW: Test weight (of 100 grains) in gm, GL: grain length (mm) GW: grain width (mm)

Under the project 1,083 accessions all from Chhattisgarh rice germplasm were evaluated (A) for 17 qualitative traits *viz*, (i) vigour, (ii) leaf sheath colour, (iii) leaf colour, (iv) leaf pubescence, (v) panicle exsertion, (vi) stigma colour, (vii) panicle type, (viii) maturity, (ix) threshability, (x) ligule structure, (xi) ligule colour, (xii) apiculus colour, (xiii) seed colour, (xiv) hull colour, (xv) coleoptile colour, (xvi) aroma and (xvii) collar colour, (B) for 10 quantitative traits which included (i) leaf length, (ii) leaf width, (iii) number of effective tillers, (iv) plant height, (v) grain yield/plant, (vi) ligule length, (vii) days to 50% flowering, (viii) days to maturity, (xix) panicle length, (x) grains/panicle, and (C) for 5 insect resistance traits including (i) brown plant hopper, (ii) white backed plant hopper, (iii) gall midge, (iv) stem borer and (v) leaf folder.

Although this project operated only for one year, the results were of considerable practical utility because of multilocation testing which involved several 'hot spot' locations. The final project report summarized the achievements of the project in an explicit manner by identifying resistant/tolerant accessions for all quantitative characters having mean values better than best standard check for the trait concerned. Number of accessions found resistant/tolerant and acceptable at a level of being used as donor for the four diseases and five insects have been highlighted and listed in Table 14.13 and 14.14.

Table 14.13. A few promising accessions identified as resistant to different biotic stresses studied.

S.N.	Stress	Accession found resistant (IC Numbers)				
1.	BLB	114340	252142	114335	114546	114653
2.	LB	115032	113991	114725	115474	114510
3.	BS	114315	114780	252028	114970	114972
4.	PH	114455	114458	114786	114835	-
5.	SB	114040	115633	115921	252180	115632
6.	LF	115876	115873	115737	116094	114189

IC: Accession number of indigenous collection of rice maintained at National Bureau of Plant Genetic Resources, New Delhi; BLB: Bacterial leaf blight; LB: Leaf blast; BS: Brown spot; PH: Plant hopper; SB: Stem borer, LF: Leaf folder.

Table 14.14. Germplasm accession with multiple resistance.

S.N.	Acc. No.	IC No.	Diseases		
			Leaf blast	Bacterial blight	Tungro
226	392	114335	x	x	
352	560	114507	x	x	
432	677	114653		x	x
493	768	114787		x	x

x: Resistant

Evaluation of aromatic accessions of Chhattisgarh rice germplasm

Chhattisgarh germplasm collection has a good number of aromatic rices all of which have almost exclusively medium or short slender grains. Studies have been made with regard to variation in different physical and physio-chemical traits of the grains of these aromatic rices. Patil (2001) studied 132 cultivated aromatic rice cultivars. The range recorded for different traits varied from 3.90 to 7.10 mm for kernel length, 1.25 to 1.83 mm for kernel width, 9.17 to 26.02 for amylose content, from 5.0 to 7.0 for alkali spreading value, 43.0 to 74.17 percent for head rice recovery and 4.32 to 18.28 gm for grain yield per plant.

Soni (1991) studied variation in 128 randomly selected accessions from germplasm and reported variation in grain length from 6.4 to 12.3 mm, in grain width from 2.3 to 3.7 mm, in harvest index from 22.66 to 63.96 percent and in grain yield from 3.65 to 13.57 gm per plant. The variation in grain quality traits was recorded for milled grain length and width from 4.3 to 7.6 and 1.9 to 2.8 mm respectively, in elongation from 29.7 to 92.6 percent, in water absorption from 9.3 to 24.5 percent, in gel length from 3.4 to 10.30 and in alkali value from 2.00 to 4.85.

Ekka (2007) studied 90 germplasm accessions from Bastar division (along with 6 standard checks). In her studies, grain length varied from 6.00 to 10.00 mm, breadth from 2.15 to 3.55 mm, kernel length from 4.00 to 7.40 mm, milling and head rice recovery varied from 65.77 to 75.10 and 24.70 to 66.14 percent respectively.

Under an ICAR cess fund financed project entitled "Genetic Enhancement of Short Grain Rices for Higher Productivity and Export", as many as 517 aromatic accessions of Chhattisgarh rice germplasm were evaluated during the years 2002-2005 at Raipur for agro-morphological, quality and pest tolerance traits. The objectives of the project were also to study yield potential, effect of environment on quality traits and to develop package of practices for short grain tall aromatic cultivars. Dr. A.K. Sarawgi, Senior Rice Breeder was Principal Investigator with Dr. V.S. Trimurty (Pathologist) and Dr. S.K. Sarawgi (Agronomist) as Associated Investigators.

A total number of 517 aromatic accessions of Chhattisgarh Rice Germplasm including all well known principal aromatic scented groups in germplasm viz. Dubraj, Badshahbhog, Chinnor, Banspatri, Vishnubhog, Kubri mohar, etc. were studied.

Of the 517 total accessions, 415 were recorded as aromatic, the rest were either non-aromatic or had very mild aroma. Of these, 265 were screened for leaf blight, brown spot, leaf blast, sheath blight and sheath rot using standard screening techniques. Twenty two morphological descriptors were recorded and tabulated. The most promising cultures based on quality traits and yield potential were Shukla phool, Acharmati, Ganjeikali, Dhuriawa, Kalimunchh and Dubraj. Other accessions identified as promising were accession numbers K:1019, R:399, C:399, D:1008, D:6661, D:137, B:122, D:939, D:1090, D:959, etc. In another study, Laxmibhog, Tarunbhog, Kalanamak, P 176-91-1 and Chini Kamini were found to have good quality characteristics. Highest yield from tall short slender aromatic types was obtained at fertility level of 60+50+40 kg N, P_2O_5 and K_2O kg/ha respectively blended with farm yard manure.

Evaluation of germplasm for nutritional traits

Chandel *et al.* (2006) initiated a comprehensive study to screen 8,680 accessions of Chhattisgarh rice germplasm for protein, lysine, iron and zinc content along with molecular characterization of elite accessions using SSR based molecular markers with generous funds received from Department of Biotechnology, Government of India.

The seed samples for analysis were procured from single plant harvests of *kharf*-2003 to avoid environmental and soil related effects. The rice grains were manually dehulled using hand dehuller-poly urethane rollers. They were polished with rubber rollers at commercial level of whiteness which was recorded by reflectance meter prior to grinding of grains for chemical analyses. Protein, lysine, iron and zinc were estimated following standard procedure using modified Micro-Kjeldahl method for protein, spectrophotometer for lysine and wet-digestion method for iron and zinc.

As expected, high variability was observed in all four parameters of nutritional quality traits of rice grains. It is interesting to note that wild rice strains in general had higher protein content but with lysine on lower side indicating possibility of using wild rices to improve protein content of cultivated rices. The aromatic group had narrower variation both for protein as well as lysine content though mean values for these traits were higher. Tables 14.15, 14.16 and 14.17 represent variation in respect of four nutritional traits in a section (n = 8,680) of germplasm studied at commercial level of whiteness.

Table 14.15. Variation in protein and lysine content in milled rice grains of a part of Chhattisgarh rice germplasm collection maintained at IGAU, Raipur.

S.N.	Category of accessions	Number of accessions analysed	Total protein (%)		Lysine content(g/16 g N)	
			Range	Mean	Range	Mean
1.	Early	1781	4.25-12.08	6.57	2.02-7.30	4.81
2.	Medium	2146	3.49-10.45	6.98	1.83-6.53	4.89
3.	Late	2412	3.95-10.14	6.74	2.10-6.21	4.49
4.	Aromatic	265	6.72-7.31	7.66	3.31-6.77	5.05
5.	Wild rices	261	6.33-14.05	10.00	2.09-7.00	5.05
6.	Semi-dwarfs	92	5.00-10.20	6.92	-	4.05
7.	Uplands	214	4.96-13.64	7.00	2.09-6.35	4.32
8.	Others	1509	4.50-10.29	6.51	2.00-7.74	4.89
	Total	**8680**				

Table 14.16. Variation in iron and zinc content in milled rice grains of a section of Chhattisgarh Rice Germplasm Collection maintained at IGAU Raipur.

S.N.	Category of accessions	Number of accessions analysed	Iron (ppm)		Zinc content (ppm)	
			Range	Mean	Range	Mean
1.	Early	1781	2.00-23.00	10.36	4.50-36.50	19.23
2.	Medium	2146	2.00-23.50	10.00	4.50-36.00	19.16
3.	Late	2412	2.00-19.50	10.08	6.50-31.50	18.38
4.	Aromatic	265	3.36-17.29	9.87	6.50-35.00	18.92
5.	Wild rices	261	1.50-24.50	10.36	4.50-33.00	21.21
6.	Semi-dwarfs	92	-	-	-	-
7.	Uplands	214	2.00-18.00	10.08	6.50-34.00	20.48
8.	Others	1509	-	-	-	-
	Total	**8680**				

Table 14.17. Some accessions identified to have higher level of each of four quality traits in a section of germplasm.

S.N.	Trait	Accessions possessing higher level	
		Name	Accession number
1.	Protein	Harad gundi	H:261
		Kakdo	K:1843
		*	USR:67
		Sathia	S:743
		Baggar	B:2909
		Kohikari	Bd:1625
2.	Lysine	Fingeshwar	H:117
		Chhotesarkar	C:774
		Bhata mathura	B:1468
		Jhingi	MS:34
3.	Zinc	Gangaprasad	AR:1008
		Nagpuri Gurmatia	N:198
		Barangi	B:680
		Dudhiya potia	D:711
		Laloo-14	Bd:2
4.	Iron	*	MS:36
		*	D:22
		Anjan	A:460
		Bhaya	B:1616
		Ajvin	A:345

*: Wild rice accessions with no name.

Some of the other findings of the project were

a. Protein content in the grain did not correlate with grain colour, aroma, grain shape and size.

b. In general, most of the released rice varieties have good level of protein compared to the mean of traditional tall cultivars.

c. The iron level in grains gets reduced by 70% with commercial level of polishing. The concerning values for zinc were 25-30%, for total protein 10-15% and for lysine around 10%.

d. DNA profile generated for the elite rice accessions using SSR analysis and rice lines were grouped based on SSR data using unpaired group arithmetic average (UPGMA) and similarity matrices computed according to Jaccards coefficient.

e. Protein profiles of the high and low lysine lines have been developed to analyse different fractions of rice protein.

f. Total protein and lysine in all groups of germplasm were in general found to be negatively correlated; the correlation co-efficients varied from -0.014 to -0.500, whereas iron and zinc had positive correlation ranging between +0.120 to +0.660.

Screening of germplasm for resistance to bacterial leaf blight

Bacterial leaf blight is an important disease of rice in Chhattisgarh. Screening of germplasm accessions was initiated by Dr. K.C. Agrawal, Plant Pathologist in 1980 and continued for around 15 years; over 1,000 accessions were screened every year through artificial inoculation using leaf clipping method. Mixed culture was used for the purpose. A large number of varieties were identified to have resistance/tolerance to this disease.

Screening of germplasm for resistance to brown plant hopper

Entomology Department of IGAU undertook screening of germplasm for brown plant hopper with financial help from Cess funds of ICAR. The objectives were to identify resistant/tolerant accessions of both cultivated and wild rice for use in breeding work as well as for evaluating the effectiveness of various genes identified at the International Rice Research Institute, Philippines to Indian biotype of the insect under glass house conditions. The mass-screening technology described by Kalode and Krishna (1979) was used (Pophaly, 1996, 1998). The project was sanctioned for the period 1992 to 1995 and later extended up to July 1997.

In all 5,877 accessions of cultivated and wild rice germplasm of Indira Gandhi Agricultural University were used in screening in two phases of the project. The extensive study led to identification of 172 resistant accessions in the first phase with damage score less than 3.00 (scored on 0-9 scale, where 0 is immune and 9 is highly susceptible). Of these, 21 accessions were classified as highly resistant having plant damage score of < 1. Thus, one in every 20 accessions of germplasm can be said to have resistance to the extent of being used as donor (Table 14.18).

Table 14.18. Germplasm accessions showing average plant damage below score 1 for brown plant hopper under glass house conditions.

S.N.	Name of variety	Accession No.	Damage score	District collected from
1.	Agma Chachhi	A:522	0.36	Raigarh
2.	Anjaniya II	A:14 J II	0.56	Rajnandgaon
3.	Bansangi	B:2682	0.58	Bastar
4.	Wasmati	W:91	0.60	Gwalior
5.	Ama Koysli	A:719	0.60	Rajnandgaon
6.	Jaybayrang	J:432	0.04	Raipur
7.	Farsa Phool	F:16	0.09	Bastar
8.	Hathi Panjar	H:28	0.18	Raipur
9.	Gadur sela	G:699	0.60	Rajnandgaon
10.	Hinga	H:435	0.65	Bastar
11.	Dhouri	D:1043	0.69	Raigarh
12.	Agma Chachhi	A:552	0.36	Raigarh
13.	KABRi	K:2381	0.50	Bastar
14.	UraiKana	U:51	0.14	-
15.	Badli	B:1974	0.40	-
16.	Badsatka	B:2472	0.52	-
17.	Sukulmunoi	S:1517	0.60	-
18.	badli	B:1960	0.64	-
19.	badshah bhog	B:667	0.79	-
20.	newari	N:434 II	0.90	-
21.	lungi	L:47 II	0.97	-
22.	T (N)-1	-	9.00	Susceptible check

The insect feeding tests were conducted through quantifying honey dew secreted. It was found that plant damage did not always correlate with honey dew secretion indicating that plant genotypes have variable capacity to show less damage even though feeding values may be high.

The IRRI identified resistant donors Mudgo (*Bph 1*), ASD-7 (*bph 2*) and Babawee (*bph 4*) proved susceptible. Rathu Heenati (*Bph 3*) and Balamawee

(*Bph 9*) were moderately resistant and ARC 10150 (*bph 5*) and Swarnalata (*Bph 6*) as resistant. Thus, Raipur biotype appears to be more virulent than those prevailing in the Philippines.

In addition to cultivated accessions, a few wild rice accessions were also screened (Table 14.19). Some of the accessions, particularly those collected from Kondagaon of Bastar division belonging to *Oryza nivara* species showed resistance (with plant damage score of <3). This work need to be further taken up particularly to know which of the resistance gene they possess, as these might prove to be entirely new source of resistance.

Table 14.19. Wild rice accessions found resistant (with average plant damage score <3) to brown plant hopper.

S.N.	Collection No.	Species	Place collected from	Ave. plant damage score
1	USR-23	*O. spontanea*	Kondagaon (Bastar)	1.20
2	MS-47	*O. nivara*	Basna (Raigarh)	1.93
3	USR-14	*O. nivara*	Kondagaon (Bastar)	1.95
4	USR-34	-	-	2.02
5	USR-8	*O. nivara*	Kondagaon (Bastar)	2.21

Variability in wild rice accessions

Variation with regard to plant height, days to 50% flowering , number of tillers per plant, growing habit, resistance to gall midge and anthocyanin pigmentation on 12 plant parts were studied during *kharif* 1992 in 112 accessions of wild rice belonging to *O. nivara* and *O. rufipogon*. These were collected a year before only from Raipur district of the state.

Plant height in this study was found to vary from 46 cm to 87.5 cm, days to flowering from 74 to 94 days and number of tillers per plant from 9 to 43 (Kumar *et al.*, 1997). Surprisingly, all accessions were found to be susceptible to gall midge, a prominent pest of rice in the state.

Of 112 accessions, 3 were erect type, 26 semi-erect, 49 semi-spreading and the rest 34 were spreading type. All accessions had pigmentation on one or the other plant parts. Five accessions had pigmentation on all the 12 plant parts.

Utilization

The sole purpose of collecting and maintaining germplasm is its direct or indirect use in increasing productivity and quality of the crop produce for the benefit of humanity. The study of biological variation also helps scientists to understand forces of evaluation and thereby integrating whole diversity towards sustainable development as well as preserving ecological balances.

Through the process of domestication, the human race has utilized the variation created by nature and has brought landraces into cultivation to meet the requirements of food, fodder and fuel. In fact, it is said that culture started with agriculture i.e. ever since human beings learned to cultivate crops. According to Ramiah and Ghose (1951), the southern districts of Orissa and Chhattisgarh form a secondary centre of origin of cultivated rices. According to Chang (1976), rice was domesticated in the sub-Himalayan belt stretching from India to south China and Vietnam. According to Sharma *et al.* (2000), the entire area encompassing Chhattisgarh, Jharkhand and western belt of Orissa is the primary centre of origin of *aus* ecotype of rice and middle and lower Gangetic valley that of *aman* rices. In any case, Chhattisgarh, represents an impressive diversity of cultivated rices that are grown in diverse ecosystems since millennia.

However, in terms of efforts specifically diverted towards selecting, improving and utilizing landraces through research and development agencies began in the first decade of the last century as already discussed in the first section of this chapter. The Deputy Director of Agriculture responsible for seed production farm of Labhandi (Raipur) identified eight landraces for seed multiplication and distribution to farmers in 1913 (Table 14.1). This can be said to be the first endeavour to collect and utilize native rice germplasm of Chhattisgarh. The efforts which began in 1913 are still in vogue. The variety 'Jaldubi' a selection from local germplasm (AR 1023) of Surguja district has been released and recommended as late as in 2007 for commercial cultivation in lowland (*Behra)* areas of north hill zone and similar *Gabhar* lands of southern plateau of Bastar division. This line is resistant to blast and gall-midge and is adapted under medium lowlands up to 50 cm water-stagnant areas (Sastri *et al.,* 2008).

A good number of varieties have been released by Raipur research centre by pure line selection which involves just purification of raw germplasm lines collected from farmers' field and little testing for yield potential and tolerance traits as may be necessary without involving lengthy procedures of hybridization and selection for generations. As many as 29 commercial varieties have been developed through the processes of pure line selection (Shrivastava and Kumar, 2004) (Table 14.20).

Table 14.20. Varieties developed and released for commercial cultivation by selection from germplasm through 'pure line' method of breeding.

S.N.	Variety	Selected from land race	Year of
1.	EB 17	Nungi	1913
2.	Parewa No. 143	Parewa	1913
3.	Bhondu No. 11	Bhondu	1913
4.	Bhata Gurmatia	Gurmatia	1916
5.	Bhondu No. 110	Bhondu	1918
6.	Parewa No. 22	Parewa	1923
7.	Gurmatia No. 11	Gurmatia	1923
8.	Dilbuxa No. 35	Dilbuxa	1923
9.	R 1	Nungi	1942
10.	R 2	Sultugurmatia	1942
11.	R 3	Surmatia	1942
12.	R 4	Ludko	1942
13.	R 5	Budbhiabako	1942
14.	R 7	Ajan	1942
15.	R 8	Benisar	1942
16.	R8 A	Luchai	1942
17.	R10	Chhatri	1942
18.	R 11	Dubraj	1942
19.	R 12	Banspatri	1942
20.	R 13	Kubrimohar	1942
21.	R 14	Badshahbhog	1942
22.	R 15	Chinnor	1942
23.	Laloo-14	Laloo	1964
24.	Safri-17	Safri	1964
25.	Kalimoonch-64	Kalimoonch	1964
26.	Pandhri luchai-16	Luchai	1966
27.	Safedhan	Safedhan	1966
28.	Assamchudi-49	Assamchudi	1966
29.	Jaldubi	Jaldubi (AR 1023)	2007

Table 14.21. Semi-dwarf varieties released in Madhya Pradesh and Chhattisgarh through hybridization with an accession of Chhattisgarh Rice Germplasm or its derivative line being used as one of the parents.

S.N.	Variety released	Parentage	Year of release
1.	Pragati	T (N)-1 x Langi 70	1976
2.	Kranti	Cross 116 x IR 8	1976
3.	Garima	Cross 4 x T (N)-1	1977
4.	Abha	Saket 4 x JR 2- 331	1980
5.	Tripti	Saket 4 x JR 2- 331	1980
6.	Madhuri	Jaya x Dubraj	1980
7.	Poorva	Saket 4 x JR 2- 331	1981
8.	Shyamla	R 60-2712 x R 2389	1993
9.	Mahamaya	Asha x Kranti	1994
10.	Poornima	Poorva x IR 8608	1996
11.	Danteshwari	Samridhi x IR 8608	2001

In addition to germplasm accessions being directly released as commercial varieties (through pure line selection), much more important, however, is their use in hybridization particularly in context of advent of semi-dwarfs as now they are grown almost exclusively by farmers in most of the states. Prior to introduction of semi-dwarfs, the germplasm accessions were used in hybridization among themselves to recombine desirable traits. Shrivastava and Kumar (2004) have listed 14 improved tall varieties evolved until 1966 in undivided Madhya Pradesh.

Identification of genes for resistance

Identification of individual genes for resistance is important to ensure durability of resistance in context of changing pathogen. The Indira Gandhi Agricultural University scientists identified nine independent genes for resistance to gall midge in germplasm accessions. The genes have been identified through conventional genetic studies; for example, the first two genes – *Gm 1* and *Gm 2* were detected by screening F_2 population of the cross made between Samridhi (a released semi-dwarf variety deriving its resistance imparting gene from W1263) and Surekha (another semi-dwarf gall-midge resistant variety owing its resistance to one of its parent Siam-29, a Thai variety) (Chaudhary *et al.*, 1985). Siam-29 and most of the other resistance-possessing donors are not from Chhattisgarh germplasm collection but are maintained in the collection as donors. The details of genes so identified are given in Table 14.22.

Table 14.22. Genes for gall midge resistance identified in germplasm maintained at Raipur.

S.N.	Gene designation	Parent	INGR No.	National Identity	Source of gene
1.	*Gm 1*	Samridhi	04003	IC 296614	W 1263
2.	*Gm 2*	Surekha	04004	IC 296615	Siam 29
3.	*gm 3*	RP 2068-18-3	04005	IC 296616	Velluthacheera
4.	*Gm 4*	Abhaya	04006	IC 296617	Ptb 21
5.	*Gm 5*	ARC 5984	04007	IC 296618	ARC 5984
6.	*Gm 7*	RP 2333-156	04008	IC 296619	ARC 10659
7.	*Gm 8*	Jhitpiti	04009	IC 296620	Jhitpiti
8.	*Gm 9*	Line 9	04010	IC 296621	Dubraj
9.	*Gm 10*	BG 380-2	-	-	Ob 677

It is to be noted that all genes are dominant except *gm* 3 which is recessive. All these genes are identified by IGAU scientists except *Gm* 6 which was identified in China and is susceptible to biotype 1 of gall midge found in Chhattisgarh. All genes, except *Gm* 6 and *Gm* 10 have been registered with National Bureau of Plant Genetic Resources.

Utilization of germplasm accessions in hybridization at Raipur

Landraces are the unselected strains cultivated by the farmers and are the primary source of all resistances to biotic or abiotic stresses. Though most of the breeding programmes currently use derived lines in hybridization to avoid unnecessary segregation for tallness and other undesirable traits (such as awning, husk colour, etc.), yet, during the last five years, out of 525 crosses made by Rice Breeding Unit of IGAU, Raipur, a tall parent from germplasm was involved in as many as 86 crosses. This highlights the need of preserving indigenous germplasm.

Evolving purple leaf varieties

In areas where direct seeding is the primary method of crop establishment in Asia, Europe or America, the volunteer red rice plants are a perpetual problem leading to considerable losses in yield and quality of produce. In Chhattisgarh and areas around, the problem is further aggravated as this tract is considered to be a secondary centre of origin of rice crop and the wild relatives of rice, particularly *Oryza nivara* and *O. rufipogon* that grow in nearby ponds and ditches abundantly and cross freely with the cultivated varieties, give rise to a wide spectrum of hybrid populations which are weedy and infectious in nature. These hybrid plants are difficult to identify until flowering and cross with plants of cultivated varieties. As a result, even if these hybrids are rouged out, the seeds harvested from plants of the grown variety give out hybrid weedy plants. Such plants have very lax panicles with fewer spikelets, possess long awns and shatter grains resulting in tremendous yields loss.

The noble approach initiated at Raipur Research Station, first time in the country, was to evolve purple leaf colour varieties with good yield levels and acceptable grain quality. Such varieties are recommended to be cultivated in problem areas consecutively for two to three seasons. The purple leaf colour of cultivated variety enables farmers to rogue out normal green leaf volunteer plants of previously grown variety along with the hybrid plants making the fields 'clean' within three years. In subsequent years the normal green leaf varieties (which usually have higher yield potential than the purple leaf one) can then be cultivated as usual.

Following this approach, five such varieties were released between 1940 and 1950 using Nagkesar, a purple leaf tall variety, as one of the donor parent in hybridization. Table 10.23 gives details of these varieties.

Table 14.23. Purple leaf varieties released from Raipur research station in pre-HYV era.

S. N.	Variety	Parentage	Duration (days)	Mean grain yield (kg/ha.)
1	Cross 1	R2 x Nagkesar	131	1112
2	Cross 52	(Nungi x Nagkesar) x Surmatia	147	1097
3	Cross 34	(Gurmatia x Nagkesar) x Ajan	153	1645
4	Cross 5/18	(Luchai x Nagkesar) x Luchai	164	1605
5	Cross B11/2	(Luchai x Nagkesar) x (Luchai x Burma18)	167	1900

These varieties considerably reduced yield losses in farmers' fields. However, the problem again shot up when semi-dwarf varieties became available and area under traditional tall varieties progressively got reduced. The state of Chhattisgarh was again the first in the country to release a semi-dwarf purple leaf variety named Shyamla (meaning 'having colouration') in 1997 through hybridization (R60-2712 x R2389). This variety matures in 130 days, has long slender grains and a very high milling out turn. While travelling across Chhattisgarh in cropping season, one can easily see fields with complete purple foliage, presenting a scenic view.

Supply of Material to Scientists and Institutions

In addition to material being used by breeders of the IGAU, a good number of samples used to be supplied to other scientists of different disciplines as well as to other institutions of the country and abroad. Details of utilization by such scientists are not available; however, Table 14.24 gives the number of samples supplied from the centre.

Table 14.24. Number of germplasm samples supplied to scientists and institutions.

S.N.	Period	Number of samples supplied to	
		Within IGAU	Outside IGAU
1.	1981-85	5525	4319
2.	1986-90	8537	842
3.	1991-95	10467	10556
4.	1996-00	10511	3690
5.	2001-05	10100	Nil
6.	2006-onwards	473	Nil
	Total	**45613**	**49,407**

It is apparent that a large number of samples have been supplied to the institutions outside the host university. This supply, however, has been discontinued since 2001 in view of various regulations and restrictions introduced pertaining to germplasm movement.

Conservation

With fast spread of only a few high yielding varieties, the genetic diversity in the field is shrinking rapidly in all commercially cultivated crops. It is, therefore to be ensured that the field variability is brought to laboratories and cold storages for safe-keeping to meet future requirements.

Field maintenance

Since cold storage facilities were not available in early years, the whole material had to be grown every year in the field. The representative individual panicles used to be harvested and threshed in two paper packets, one bigger and second a smaller one and were sun-dried for 3 days. The two sets were stored separately. The seeds of smaller packet were for the purpose of next year's sowing and those of bigger packet for supply to indenters during the following year. Seeds in this packet were also used for second sowing in the field in the coming year, in case the first sowing failed for any reason.

The field sowing is done in raised beds of one metre width and a suitable length to accommodate one hundred accessions in one bed. Sowing 100 entries in one bed is just for easiness in handling. Each accession is grown in two rows. Three weeks later, the plots are thinned, weeded and fertilized. Recommended doses of fertilizers are applied (normally 60, 40, 30 kg of N, P_2O_5 and K_20/ha, respectively)

Maintaining a duplicate set

One of the thumb-rules related with germplasm maintenance and conservation is that the whole material should be either grown or stored at two places to guarantee safety of the material against any unforeseen calamity like drought, flood, epidemic, fire or earthquake occurring at one place, howsoever well protected it may be. Realising this, the whole of Chhattisgarh germplasm was being grown every year at Baronda farm near Raipur and Government Farm, Fanda near Bhopal during 1975-1979. With merger of Madhya Pradesh Rice Research Institute with Jawaharlal Nehru Agricultural University, the system got disrupted for some years but soon an understanding was developed with National Bureau of Plant Genetic Resources (NBPGR), New Delhi wherein almost the whole of the material was transferred there systematically and kept in medium and long term cold storage modules maintained by them. The material transferred to NBPGR is shown in Table 14.25.

Table 14.25. Transfer of duplicate samples of germplasm accessions to National Bureau of Plant Genetic Resources, New Delhi between 1991 and 2004.

S.N.	Particulars	Number of accession sent
1	Early accessions	5557
2	Medium duration accessions	5069
3	Late duration accessions	7915
4	Collected under on-farm conservation project	705
5	Collected under NATP funded project on Plant Biodiversity	1003
6	'mproved versions of tall varieties (BD numbers)	938
7	Wild rice accessions	210
8	Donors and breeding lines	1408
	Total	**20,805**

Medium-term cold storage

The Indian Council of Agricultural Research allotted one medium-term cold storage module (that was received from Winrock International) for Raipur centre. This cold storage module was installed and made operative in 1997. However, the whole material continued to be grown every year in the field until 2006; thereafter, every year only half of the material is field grown the other half was stored in the medium term cold storage module. The material stored in the module is considered safe for around five years; therefore, only one-fifth of the material need to be grown in the field every year. However, by growing half of the material, extra-precaution has been taken to avoid possibility of losing viability of the material in view of frequent failure of power supply from public distribution system and/ or of generator attached to the module.

References:

Chandel G. 2006. Final Report of the Project *Evaluation and Characterization of Rice Accessions Maintained at IGAU, Raipur for Nutritional Quality Traits*. Department of Biotechnology, Indira Gandhi Krishi Vishwavidyalaya, Raipur, India. 24 pp.

Chang TT. 1976. The origin, evolution, cultivation, dissemination and diversification of Asian and African rices. *Euphytica* 25:411-425.

Chaudhary BP, Srivastava PS, Shrivastava MN, Khush GS. 1985. Inheritance of resistsance to gall midge in some cultivars of rice. Pages 523-528 in *Rice Genetics*. International Rice Research Institute, Manila.

Chaudhary M. 1999. *Genetic Variability and Inheritance of Important Morphological and Quality Traits in Rice*. Thesis submitted to Indira Gandhi Agricultural University, Raipur, India. 130 pp.

DRR. 1999. *Evaluation of Rice Germplasm for Biotic Stresses*. Directorate of Rice Research, Rajendranagar, Hyderabad, India. 624 pp.

Ekka RE. 2007. *Genetic Evaluation of Traditional Accessions for Morphological and Quality Characters in Rice* (*Oryza sativa* L.). Thesis submitted to Indira Gandhi Krishi Vishwavidyalaya, Raipur, India. 113 pp.

Gautam NK, Mishra SK, Raiger HL, Mahajan RK, Sharma SK. 2007. *Multilocation Evaluation of Rice Germplasm (Vol. 1)*. National Bureau of Plant Genetic Resources, Pusa, New Delhi-110 012. v+492 pp.

IGAU. 2007. *Catalogue on Indigenous Rice Germplasm of Chhattisgarh and Madhya Pradesh. Part I.* Indira Gandhi Agricultural University, Raipur, India. 207 pp.

IGAU. 2008. *Catalogue on Indigenous Rice Germplasm of Chhattisgarh and Madhya Pradesh. Part II.* Indira Gandhi Agricultural University, Raipur, India. 307 pp.

IRRI (International Rice Research Institute). 1976. *Standard Evaluation System for Rice*. International Rice Research Institute, Los Banos, Laguna, Philippines.

Kalode MB, Krishna TS. 1979. Varietal resistance to brown plant hopper in India. Pages 183-199 in *Brown Planthopper: Threat to Rice Production in Asia*. International Rice Research Institute, Los Banos, Laguna, Manila, Philippines.

Kumar Dharmendra, Shrivastava MN, Sahu RK, Sahu VN. 1997. Morphological variability in wild rices of Raipur. *Oryza* 34(4):288-291.

Patil PV. 2001. *Genetic Evaluation of Traditional Aromatic Accessions of Morphological and Quality Characters in Rice* (*Oryza sativa* L). Thesis submitted to Indira Gandhi Agricultural University, Raipur, India. 133 pp.

Patil PV, Sarawgi AK, Shrivastava MN. 2003. Genetic analysis of yield and quality traits in traditional aromatic accessions of rice. *J Maharashtra Agric Univ* 28(3):255-258.

Pophaly DJ. 1996. Final Report of the Project *Evaluation of rice germplasm for resistance to Nilaparvata lugens* Stal. Indira Gandhi Agricultural University, Raipur, India. 15 pp.

Pophaly DJ. 1998. Final Report of the Project *Evaluation of rice germplasm against Brown Planthopper (BPH), Nilaparvata lugens* Stal. Indira Gandhi Agricultural University, Raipur, India. 11 pp.

Ramiah K, Ghose RLM. 1951. Origin and distribution of cultivated plants of south Asia - Rice. *Indian Journal of Genetics & Plant Breed* 11:7-13.

Richharia RH, Sahu KR, Sharma S, Sharma RP, Sahu KR, Tiwari NK, Brahmankar DR, Sahu GR, Khan RA. 1978. *Early Maturing Rice Varieties of Madhya Pradesh*. MP Rice Research Institute, Raipur. 84 pp.

Richharia RH. 1974(a). *Our Strategy on the Rice Production Front in Madhya Pradesh*. Adaptive Rice Research Center, Baronda, Raipur. 125 pp.

Richharia RH. 1974(b). A super long paddy grain. *J Cytol Gen* 9-10:97.

Richharia RH. 1977. *A Strategy for Rice Production to Ensure Sustained Growth in Madhya Pradesh*. Madhya Pradesh Rice Research Institute, Raipur. 51 pp.

Richharia RH. 1978. *Dhan Sampada* (Hindi). MP Rice Research Institute, Raipur. 42 pp.

Richharia RH. 1979. An aspect of genetic diversity in rice. *Oryza* 16:1-31.

Sarawgi AK, Trimurty VS, Sarawgi SK. 2005. Final Report of Research Scheme *Genetic Enhancement of Aromatic Short Grain Rices for Higher Productivity and Export*. Indira Gandhi Agricultural University, Raipur, India. 85 pp.

Sastri ASRAS, Rao SS, Sharma RN (eds). 2008. *Catalogue on Indigenous Rice Germplasm of Chhattisgarh and Madhya Pradesh, Part-II*. Indira Gandhi Agricultural University, Raipur, India. 307pp.

Sharma SD, Tripathy S, Biswal J. 2000. Origin of *O. sativa* and its ecotypes. Pages 349-369 in Nanda JS (ed) *Rice Breeding and Genetics: Research Priorities and Challenges*. Science Publishers, Enfield, USA.

Sharma DK, Sahu GR, Shrivastava MN. 1987. Present status of rice germplasm in Madhya Pradesh (India). *Oryza* 24:127-131.

Sharma RB, Pandey RL, Rao SS, Sharma RN (eds). 2007. *Catalogue on Indigenous Rice Germplasm of Chhattisgarh and Madhya Pradesh, Part-I*. Indira Gandhi Agricultural University, Raipur,

492006, India, 207 pp.

Shrivastava MN, Kumar A. 2004. Genetic improvement of rice varieties in Chhattisgarh and Madhya Pradesh. Pages 435-475 in Sharma SD and Prasada Rao U (eds) *Genetic Improvement of Rice Varieties of India (Part-I)*. Today and Tomorrow's Printers and Publishers, New Delhi.

Soni DK. 1991. *Variability Study of Morphological and Quality Characters in Rice Genetic Stock.* Thesis submitted to Indira Gandhi Agricultural University, Raipur, India. 152 pp.

Genetic Resources of Rice in India : 465-491
Editor : S. D. Sharma
Today & Tomorrow's Printers and Publishers, New Delhi - 110 002 (India)

Chapter 15

Genetic Resources of Rice in Uttar Pradesh and Uttarakhand: A Critical Analysis

J.L. Dwivedi

Introduction

Rice is one of the important cereal crops of Uttar Pradesh and Uttarakhand and is grown under different types of favourable and unfavourable ecological conditions. At present, it is cultivated in approximately 5.9 m/ha in Uttar Pradesh covering all the ecological conditions viz. irrigated, rainfed upland, rainfed lowland and flood-prone including deepwater, floating rices areas and *boro* rice and problem soils especially *usar* conditions. In Uttrakhand, rice area as per 1999-2000 estimates is around 3,98,433 ha covering plains and hills. Due to varying micro eco-cultural conditions in both the states, farmers grow different rice varieties suiting to their local conditions. These areas, therefore, harbour large varietal diversity. Eastern Uttar Pradesh is rich in rice genetic resources with hundreds of local varieties covering upland to rainfed lowland and flood-prone ecologies while hilly areas located in Uttarakhand are the real source of rice biodiversity for cold tolerance and other abiotic stresses. Major factors including adaptation to edaphic and topographic conditions and varying water regime contribute to genetic diversification and ecological specialization.

Uttar Pradesh is no doubt rich and diverse in rice genetic wealth but, unfortunately, no systematic work has been done to collect, evaluate and catalogue these valuable rice landraces conserved by the local farmers. In the present era of semi-dwarf rices where high yield has been the main focus, the traditional rice varieties have been ignored. Exploration, collection, evaluation and conservation of landraces and traditional varieties grown in different eco-cultural conditions have been done in piecemeal and sporadic way. Therefore, the potential of genetic biodiversity in the state which is called "rice bowl" has never been fully realised. However, research work done with regard to rice germplasm management, utilization, varietal development, genetic erosion, etc. have been discussed in the following pages.

Historical Background

Uttar Pradesh

In the United Provinces (now Uttar Pradesh and Uttarakhand), research work on rice improvement was started in 1924 at Kanpur. More than 1,300 rice varieties were collected from all over the state and studied for variability in different traits. In 1932, a post of Research Assistant was created for this specific purpose and the headquarters was shifted from Kanpur to Nagina in Bijnor district as Kanpur was not suitable for growing rice. During this year, Indian Council of Agricultural Research provided financial support for expanding the research work on rice and a post of Assistant Paddy Botanist was created who worked under the over all supervision of Economic Botanist (Sugarcane and Cotton). Nagina played a crucial role in breeding aromatic rice varieties. Some of the well known aromatic rice varieties bred from this centre included Type 3, Type 9, N (10)B and N 12. In 1938, a research sub-station was established in Basti district to study the problems of rice in the eastern districts of UP; this substation was later shifted to Gorakhpur. In 1942, rice research work was separated and entrusted to Economic Botanist (Rabi Cereals). Up to 1945, rice research was the responsibility of provincial government; the centre was providing only financial assistance and advice. Keeping in view the importance of the crop, the Central Rice Research Institute was established at Cuttack in 1946 by Government of India.

In 1948, Government of India accorded sanction for establishment of two additional research stations in the eastern districts of the state to study the problems of late paddy. During 1949, work on late paddy started at Panchperwa (Dist Gonda) and Tissuhi (Dist Mirzapur). The Assistant Paddy Botanist was given an independent status and was made responsible for paddy research in the state. The Indian Council of Agricultural Research sanctioned a scheme on deep water rice in 1957-58 which started functioning at Bansdih (Dist Ballia) where the famous lake Surha Tal with an area of 18 sq miles is located. During 1959, a separate research station was established at Ghaghraghat (Dist Bahraich) to take up the work on agricultural problems related to flood-prone areas of the eastern UP. The major mandate of this research station was to evolve flood resistant rice varieties and develop suitable agro-technology for deep water rice. Subsequently, a hill rice scheme was started at Majhera (Dist Nainital, now in Uttarakhand) to work on rice husbandry relevant to hill areas of Utter Pradesh; simultaneously another scheme to work on rainfed upland paddy was started at Balchandpur (Dist Bahraich).

The research work on rice was geared up in 1958 with creation of a post of Economic Botanist with full financial and administrative power. Subsequently a new Rice Research Farm was established at Masodha, Faizabad. The Economic Botanist (Rice) started working from 1964 with Faizabad as his headquarters. The rice research in the state was further strengthened and closely linked up with the national programme in 1968 and 1969 when zonal and regional research centres of All India Coordinated Rice Improvement Project were set up at Masodha in (Faizabad) and Majhera (Nainital) respectively. Further, the Government of Uttar Pradesh sanctioned a new Rice Research Station at Kanpur during 1971-72 to develop suitable rice technology(s) for saline-alkaline soils of the state and to develop post-harvest technology of rice.

Rice research work of Uttar Pradesh can be broadly grouped in three phases: the first phase (1924 -1955) was the pure line selection phase. During this early phase, a large number of landraces were purified and developed as varieties through pure line selection. Altogether 1,300 samples were collected and evaluated at Kanpur and Nagina centres. The significant contribution of this phase included the development of 21 rice varieties viz. $T_{1,}$ $T_{3,}$ $T_{9,}$ $T_{17,}$ $T_{21,}$ $T_{22A,}$ $T_{23,}$ $T_{26,}$ $T_{36,}$ $T_{38,}$ $T_{43,}$ $T_{96,}$ $T_{100,}$ $T_{136,}$ $T_{138,}$ $T_{139,}$ $N_{22,}$ $N_{27,}$ $N_{32,}$ N_{10B} and Sudha through pure line selection. In the second phase, *japonica-indica* hybridization programme was launched at Nagina. This phase resulted in selection of five inter-subspecific cultivars viz. NSJ 200, NSJ 98, Dhaneswar, Varuna and Nagina for commercial cultivation. Three *indica* hybrids developed by crosses with Russian strains viz. Kashi, Rohila and Bundela were also released for cultivation during this phase. Development of semi-dwarf varieties forms the third phase of rice improvement in which semi-dwarf rice varieties of Taiwan and International Rice Research Institute (IRRI), Manila have played a significant role.

Uttarakhand

In 2000, a separate state was carved out of the hilly districts of the erstwhile Uttar Pradesh. Rice cultivars those are adapted to Himalayan hills, valleys and plains in the foothills are grown in the state. Traditionally, two types of rice cultivation are practised in the hills: irrigated transplanted and direct sown rainfed upland. Rice in the plains is mainly grown in irrigated transplanted ecosystem in well managed fields.

The agricultural research work in Uttarakhand hills was initiated in 1936 by Prof. Boshi Sen, the Founder Director of Vivekanand Laboratory now known as Vivekanand Pravatiya Krishi Anusandhan Sansthan (VPKAS) at Almora. Prof. Sen initiated work on a number of food and fiber crops in the hills of Kumaon and Garhwal by arranging funds through private donations and small grants. Later in 1959, this laboratory was taken over by the Government of Uttar Pradesh to work mainly on various aspects of rice. Efforts resulted in the development of VL Dhan 8 through selection from Kaohsiung 22 in 1969. In 1958, Government of Uttar Pradesh established a rice research centre at Majhera (Nainital) to work on hill rices. Later in 1969, ICAR further strengthened it with financial support to work on specific regional and local problems (Dikshit, 1972). Majhera 3, a variety for rainfed spring sown condition was selected from one of native cultivars.

G.B. Pant University of Agriculture & Technology (GBPUAT), formerly known as UP Agriculture University, was established in 1961 at Pantnagar, Nainital — the only agriculture university on land grant pattern in the country with its aim to give new direction to agriculture research, education and extension. Rice research work was initiated in 1963 and the most effective control for devastating *khaira* disease was worked out in 1966. The research work on varietal improvement began in 1969 by adopting multidisciplinary approach to rice improvement (Mani *et al.,* 2001a). Work was focused on development of early and medium maturing varieties for irrigated ecosystem of plains and early maturing, medium to tall statured, cold tolerant varieties for the hills (Kumaon and Garwal divisions).

The Indian Council of Agriculture Research (ICAR) took over the Vivekanand Laboratory, Almora from Government of Uttar Pradesh in 1974 and renamed it as Vivekanand Pravatiya Krishi Anusandhan Santhan (VPKAS). GBPUAT with its main centre at Pantnager and two sub-centers located at Majhera (Kumaon) and Ranichauri (Garhwal) and Vivekanand Pravatiya Krishi Anusandhan Sansthan (VPKAS), Almora (an ICAR institute) are the main research establishments responsible for developing rice varieties for the hills. While VPKAS, Almora is responsible for developing suitable varieties for diverse production systems of the state, GBPUAT initially shouldered the responsibility for irrigated ecology only but now has diverted its focus on two major abiotic stresses i.e. drought and cold to enhance rice productivity in the hills.

Landrace Diversity in Relation to their Adaptation

There are four main agro-ecosystems of rice in Uttar Pradesh. Out of around 5.9 mha under rice cultivation, 62% of the area is under irrigated, 23% under rainfed lowland, 11% under upland and 4% under flood-prone rices. Uttar

Pradesh harbours high genetic diversity of rice with its diverse eco-geographical conditions. The range of maturity is wide, varieties taking 60 days to 250 days to mature. Dr. N.N. Dixit and Dr. D.M. Maurya collected a large number of traditional rices from eastern Uttar Pradesh. Enormous genetic variability for grain characteristics like length, breadth, aroma, cooking, etc also exists. Based on maturity duration and grain type, traditional varieties/landraces have been catalogued in Table 15.1. Substantial genetic diversity for these traits existed among the listed varieties. Majority of the traditional varieties were long and short bold followed by medium slender and long and short slender. Now most of these traditional rices have disappeared from farmers' fields. Early varieties were adapted to flat land while long duration varieties were cultivated in low-lying areas (Dwivedi *et al.*, 2005). Besides, varieties adapted to highly salt-affected areas (around 1.20 mha) and *boro* rice grown in winter (4000 ha) also exist in the state.

Table. 15.1. Genetic variability for grain types and maturity duration among traditional rice varieties of Uttar Pradesh.

Grade-1 (Long & short slender)	Grade-2 (Medium slender)	Grade-3 (Long & short bold)
	EARLY DURATION VARIETIES ADAPTED TO FLAT & UPLAND	
Dudhiya, Senia, Bhabhaya, Shekhu, Shyamjeera	*Thakurprasad, Durihya*	*Kachni, Soken, Ghojhari, Bansphool, Didai, Dal-Badal, Dhaniya, Pesaari, Bagari, Bagari Safed, Bhanali, Moongphali, Raghuri, Sarya, Sajani, Sathi-Bend, Sathi-Kali, Kudiya, Usha, Karhani, Kashi, Garer, Ghaghori, Champa, Agatehwa, Jhona, Tin Pakhiya, Dal Kachari, Dehula, Nanhjyava, Paas Benj, Bakki, Bajarbang, Bagari Kali, Bhadai, Moti Pandy, Ram Vilas, Raj Bhog, Ram Kajal, Ruhela, Lal Kawa, Lalki Bhadar, Sanwali, Shyamghata, Safedwa, Samokhan, Sathi, Pragati, Pahuna Prasad, Parhari, Mutmuri*
	MEDIUM DURATION VARIETIES ADAPTED TO MEDIUM LAND & PLAINS	
Aknai, Hansraj, Kachanyee, Sekhuiya, Lakra, Sendoor *Rath,*	*Badshah Pasand, Reghuri*	*Karanga Kajri, Gajraj, Ghee Bhat, Cheena-4, Chaul, Manshara, Motmarda, Morphankhi, Reghuri, Sedhani, Kashi Prasad, Karangi, Jognia, Dhedai, Nandibore Mirzapur, Gharbharan, Benibhog, Mangla, Badli, Sare.*
	LATE DURATION VARIETIES ADAPTED TO LOWLYING AREAS	
Jarhan, Dunipat,	*Kalanamak, Gauria, Bhainsalot, Bhata Phool, Belar, Mahachinnour,*	

Kala Sukhdas, Maskan, Lohti, Latera, Kalamdan, Dehradun, Gondha, Vishnuparag, Raimunia, Anjanee	*Chinnhur, Juhi, Bengal, Mahasuri, Tundul, Adamchini*	*Manohar Sali, Rajal, Reshmi, Sengar, Sonkharcha, Aknai, Laduiyya, Chainphool, Tudanwa, Dhalhi, Jeerbatti, Bela, Mirchpooti, Madhukar, Kambhog, Lehra, Sehdeyya, Sugga Pankhi, Hazardana.*

Himalayan *tarai* of Uttarakhand is the place of origin of world famous Basmati rice (Khush, 2000). Local small and medium grained aromatic rices that are grown in this area include Dehradoon Basmati, Tapovan Basmati, Hansraj, Tilakchandan, Bindli, etc (Table 15.2).

During 1970-1973, Dr. J.S. Nanda and his team collected many accessions of *boro* rice from eastern districts of Uttar Pradesh. During eighties, based at Crop Research Station, Ghaghraghat (Bahraich Dist), a scientific team led by Dr. J.L. Dwivedi, surveyed low-lying areas of eastern Uttar Pradesh and collected 258 accessions from unexplored areas. During 2002–2005, under an *ad hoc* project entitled "Genetic enhancement of aromatic short grain rice for higher productivity and export", efforts were made to survey the areas in the eastern Uttar Pradesh where traditional short grain aromatic rices are grown. Altogether 50 traditional varieties/landraces were collected by a team led by Dr. Dwivedi and subsequently purified and characterized. These were highly variable for morphological and agronomic traits and quality parameters. Further, a total of 306 indigenous rice accessions shared by cooperating centres viz. DRR (Hyderabad), IGKV (Raipur), NDUAT (Faizabad), RAU (Pusa) and OUAT (Bhubaneshwar) with Dubraj, Vasumati and Kalanamak as local checks were also characterized for 22 morphological and agronomic traits (Dwivedi *et al.,* 2005). Large variation was observed for these traits among the accessions (Table 15.2). This diversity could be used to improve short grain aromatic rices across the states. An *ad hoc* project entitled "Networking project on collection, evaluation, cataloguing and improvement of traditional scented rices of eastern Uttar Pradesh" was sponsored by UP Council of Agricultural Research with Dr. J.L. Dwivedi as its Nodal Scientist based at NDUAT, Faizabad. Initially the project was sanctioned for three years (2003-2006) but later it was extended for two more years i.e. up to 2008. This project was implemented jointly by NDUAT, Kumarganj, Faizabad, Banaras Hindu University, Varanasi, and Participatory Rural Development Foundation (PRDF), Gorakhpur. One of the major outputs of this project was cataloguing 323 accessions cultivated by the farmers' of the eastern Uttar Pradesh for their morphological, agronomic and quality parameters (Table 15.3) in addition to their long term storage at NBPGR, New Delhi. Besides, Narendra Lalmati was developed, released and notified vide Gazette no. 2187 date 27th August, 2009.

Table. 15.2. Adaptability of aromatic rices to different zones/regions.

Zones/Region	Area of Adaptation	Land Races/Varieties
North Eastern Plain	Basti, Siddharthnagar, Maharajaganj, Gorakhpur, Gonda and Bahraich	*Kalanamak, Ramjawain, Dhaniya, Kanakjeera, Sakkarchini, Duniyapet, Benibhog.*
Eastern Plain	Faizabad, Barabanki, Sultanpur, Pratapgarh, Jaunpur, Azamgarh, Ballia, Ghazipur, Varansi	*Lamati, Sakkarchini, Vishnu Parag, Badshah Pasand, Selhi, Dulhaniya, Bhantaphool, Moongphali, Tulsi, Manjari, Tulsi pasand*
Vindhyan	Mirzapur, Sonbhadra, Part of Varansi	*Adamchini, Kesar Laungchoor, Phool chameli, Jeerabatti, Dubraj, Sonachoor.*
Western Plain	Dehradoon, Saharanpur & Tarai areas	*Basmati 370, Hansraj, Tilakchandan, Taroari Basmati, Type 3, Sarbati, Basmati 385*

Table: 15.3. Adaptation of ASG rices to different zones as influenced by maturity and topography.

Zones	Maturity group/ Plant stature		
	Early (115d) /MT	Medium (130d)/MT	Late (>145d)/tall
Topography	Flat and well drained lands	Medium and well drained lands	Low land
NEP	*Benibhog, Latsar*	*Sakkarchini, Ramjawain*	*Kalanamak, Dhania, Kanakjeera*
EP	*Lalmati, Dulhanika, Lalkhawa,*	*Visun parag, Moongphali Ram dhuni*	*Tulsi manjari, Badshah Pasand, Bhantaphool*
Vindhyan	*Adamchini*	*Dubraj, Kesar, Jeerabatti, Thakurbhog*	*Sonachoor, Laungchoor*

ASG = Aromatic Short Grain,
Source: Dwivedi *et al.* (2005)

Current Status of Working Germplasm

Though no systematic collection of traditional rices/landraces has been done in Uttar Pradesh and Uttarakhand to conserve the rice genetic wealth of the region, surveys of different rice growing areas especially of eastern districts of Uttar

Pradesh (which is also called 'rice bowl' of the region) and hilly areas of Uttarakhand have been done in piecemeal. A good number of accessions have been lost due to improper management, loss of viability of the accessions and lack of proper storage facilities. However, current status of working germplasm accessions maintained at different Agriculture Universities/Institutes of Uttar Pradesh and Uttarakhand is provided in Table 15.4.

Table 15.4. Current status of working germplasm in Uttar Pradesh and Uttarakhand.

S. No.	University/Institute	No. accessions (traditional + exotic)
Uttar Pradesh		
1.	NDUAT, Kumarganj, Faizabad	
	Masodha	1100
	Ghagharaghat	525
2.	BHU, Varanasi	200
3.	Sardar Vallabh Bhai Patel UAT, Meerut	300
4.	CSAUAT, Kanpur	200
Uttarakhand		
5.	GBPUAT, Pantnagar	1112
6.	VPKAS, Almora	1577
	Total	**5014**

Of the total working collection of 5014 maintained at various agriculture universities/institutes located in Uttar Pradesh and Uttarakhand, majority of these accessions have been deposited at the National Bureau of Plant Genetic Resources, New Delhi for long term storage. A part of the accessions have also been sent to CRRI, Cuttack for rejuvenation and seed multiplication. These germplasm accessions include landraces, exotic varieties and promising lines/cultures developed for different ecologies and areas. Due to variation in flood patterns, cold, drought spells and heterogeneous soil coupled with an array of maturity time, varieties being grown are adapted to different environmental situations. A full range of varieties from primitive to improved types are under cultivation in different ecosystems including favourable, rainfed upland, lowland, flood-prone, problem soils and hilly areas.

Owing to extremely diversified agro-ecological system, it is obviously not feasible for a single genotype to perform consistently well in different environments. Accordingly, certain competing traits such as submergence tolerance in flash flood, elongation ability, nodal tillering and photoperiod sensitivity in deep stagnant flooding, drought tolerance and weed competitive ability in drought-prone upland, cold tolerance in *boro* rice and cold tolerance coupled with drought tolerance in hilly areas are sought for. Many traditional varieties have been identified as donors for specific traits.

Singh and Srivastava (1997) collected and purified 19 aromatic rice accessions at Varansi. They reported a large intra-varietal variation. Awasthi *et al.* (1997) reported histological variation among the aromatic rices. Large to medium size metaxylum and phloem were observed in mild to medium scented and non-scented rice varieties whereas strongly scented varieties like Basmati 370, Kalanamak, Phoolchandi and Pusa Basmati-1 had smaller size of metaxylum and phloem.

Utilisation of Local Varieties

At present, altogether five Agriculture Universities/Agricultural Institutes in addition to Regional Agriculture Testing and Demonstration Station (RATDS) and Department of Agriculture are responsible for rice research and development in the state. Narendra Dev University of Agriculture & Technology has two major research centres, namely, at Masodha (Faizabad) and Ghaghraghat (Bahraich). Besides, the University has also established four environment specific zonal research centres at Tissuhi (Mirzapur), Ghazipur, Basuli (Maharajganj) and Ghaghraghat (Bahraich). Of these, Masodha shoulders the responsibility of upland, rainfed lowland, shallow water ecology, irrigated value-added and problem soils while Ghagharaghat is the only research centre for flood prone rices in the state. C.S. Azad University of Agriculture & Technology, Kanpur is responsible for problem soil areas and irrigated ecology of the central Uttar Pradesh. Sardar Ballabh Bhai Patel University of Agriculture & Technology, Meerut caters to the need of irrigated plains of western Utter Pradesh with special reference to basmati rices. Sam Higginbottom Institute of Agriculture, Technology & Sciences (formerly Allahabad Agriculture Institute, now deemed to be a University), Allahabad, BHU, Varanasi and a few NGOs including PRDF, Gorakhpur have recently initiated work on rice research.

Majority of the germplasm has been screened for biotic and abiotic stresses. This information has been documented time to time in various journals and other literature. Initially rice improvement programme was confined to pure line selections; later hybridization followed by various breeding methods was adapted. Sood *et al.* (1983) found that Bindli and Type 3 possess high kernel elongation and show better combining ability to be used as donors in breeding programme. During 1921-23, altogether 1,265 accessions including Basmati were maintained at Doiwale, Dehradoon (established in 1921). Type 3 was one of the most popular varieties for trade due to its better eating and cooking quality.

Uttar Pradesh is also rich in genetic diversity of wild Oryza species particularly of *O. nivara* and *O. rufipogon*. *O. nivara* is frequent and abundant especially in the eastern part of Uttar Pradesh. These are good sources for pest and disease resistance. A collection of *O. nivara* from Basti district of Uttar Pradesh proved to be the only source of resistance to grassy stunt virus at IRRI. *O. rufipogon* is perennial and photoperiod sensitive with thick and long culm and well exserted and spreading panicles; it possesses the ability to elongate fast under abrupt flooding.

The *indica* x *japonica* crosses not only show higher magnitude of heterosis for grain yield but also hold a great promise to improve the cold tolerance ability. The wide compatibility (WC) genes discovered by Ikehashi and Araki (1984) in varieties like Dular, N 22 and PBI 76 is of immense value to overcome the problem of sterility usually encountered in *indica* x *japonica* hybrids.

Donors identified and utilized for various abiotic stresses from among the traditional varieties include N 22, Kalkari (drought and weed competitiveness), Jalmagna, Jaisuria, Goanth (elongation ability), Madhukar, Sugapankhi, FR 13A (submergence tolerance), etc. N 22 has been used in developing 14 upland rice varieties in the country. Kalkari is used as donor for its drought tolerance in developing Shushk Samrat, a variety released for commercial cultivation in Uttar Pradesh, Orissa and Bihar under upland condition. The list of high yielding varieties developed by Uttar Pradesh is furnished in Appendix 15.1.

The main role of GBPUAT, Pantnagar and Sardar Vallabh Bhai Patel University of Agriculture & Technology, Meerut is to develop high yielding, disease and pest resistant aromatic rice varieties. The Basmati export zone comprises of four states i.e. Uttarakhand, Uttar Pradesh, Punjab and Haryana. These include 14 districts (namely Bareilly, Sahajahanpur, Pilibhit, Rampur, Badaun, Bijnore, Mooradabad, J.P. Nagar, Saharanpur, Muzafarnagar, Meerut, Bulandshar, Ghaziabad and Baghpat) of Uttar Pradesh and 4 districts (Uddham Singh Nagar, Nainital, Haridwar and Dehradoon) of Uttarakhand. A list of high yielding varieties developed by the state is furnished ecosystem/eco-condition-wise in Appendix 15.2.

Extensive evaluation and screening work done both at VPKAS, Amora and GBPUAT Crop Resarch Station, Majhera resulted in release of three semi-tall varieties viz. Majhera 3, VL Dhan 206 and Majhera 7 through pure line selection. These varieties possess high degree of resistance to blast.

VPKAS, Almora, using well adapted native germplasm accession L 12-2A and cold tolerant *japonica* variety Stajree 45 has developed VR 431 for spring season. Later VL Dhan 221 was released for cultivation in the hills of Uttarakhand and Himachal Pradesh. Efforts made at GBPUAT, Pantnagar resulted into the development and release of another variety Govind for low altitude (up to 900 msl) and plains especially for rainfed conditions.

A vibrant and dynamic hybridization programme initiated at NDUAT, Kumarganj, Faizabad, BHU, Varanasi, GBPUAT, Pantnagar and VPKAS, Almora has led to the development of a large number of varieties with different maturity groups for different ecologies and areas prevailing in Uttar Pradesh and Uttarakhand (Table 15.5).

Table 15.5. High yielding rice varieties developed in Uttar Pradesh and Uttarakhand for different eco-systems.

Uttar Pradesh	
Ecosystem	**Name of the varieties**
Rainfed upland	Baranideep, Jawahar Dhan 75, Nagina 22, Narendra Dhan 118, Narendra Dhan 80, Narendra Dhan 1, Renu, Sudha, Shushk Samrat, Narendra Lalmati
Irrigated early	Malviya Dhan 2, Narendra Dhan 2026, Saket 4, Ashwani, NDR 2064
Irrigated Mid early	Malviya Dhan 36, Narendra Sankar Dhan 2, Narendra 3112-1
Irrigated medium	Sarjoo 52, NDR 359
Rainfed shallow low land	Cross 116, Type 22A, NDR 8002, Jal Lahari
Semi Deep water	Chakia 59, Jalpriya, Madhukar
Deep Water	Jal Nidhi, Jalmagna
Intermittent Flooding	Madhukar (T), Barah Avarodhi, Swarna Sub-1 (Improved Swarna)
Sodic/Saline	Narendra Usar Dhan 2, Narendra Usar Dhan3, Narendra 2, Narendra Usar Sankar Dhan 3, Type 100, Usar 1, Narendra Usar Dhan 2008
Uttarakhand	
Ecosystem	**Name of the varieties**
Rainfed upland	Majhera 7, VL Dhan 97, VL Dhan 207, VL Dhan 208, VL Dhan 209, Vivek Dhan 154, Pant Dhan 16
Irrigated early	Manhar, Pant Dhan 6, Pant Dhan 12, Pant Sankar Dhan 3, Prasad
Irrigated Mid early	IR 24

Irrigated medium	Pant Dhan 4, Pant Dhan 10, Pant Dhan 18, Pant Sankar Dhan 1,
Hills upland	VL Dhan 163, VL Dhan 206, Majhera 7
Hills irrigated	Majhera 3 (T), Pant Dhan 11, VL Dhan 65, VL Dhan 85, VLK Dhan 39
Scented	Pant Sugandh Dhan 15, Pant Sugandh Dhan 17

Flood Prone Ecology

The genetic wealth of cultivated and wild rices in the flood-prone ecosystem is rich and diverse in Uttar Pradesh. Great varietal diversity among sub-ecosystem types within an ecology may be ascribed to variation in flooding pattern, heterogeneous soil, maturity requirement, plant stature, specific demand of the local ecology and farmers' needs. Mutually inclusive choices in the articulation of breeding objectives must include a deepwater survival strategy and maturity during October to December. Jalpriya was a crossbred developed from a cross between IET 4060 and Jalmagna. Elongation ability was transferred from Jalmagna to this culture. Barh Avarodhi, an improved crossbred variety for intermittent/flash flood situation possesses submergence tolerance trait from Madhukar which can survive up to 12 days of complete submergence. Jalnidhi was a pure line selection from a local variety 'Goant' which possesses excellent elongation ability.

Boro Rice

Boro rice is cultivated in the deeply flooded areas during dry season. Typical *boro* rices possess cold tolerance especially at their vegetative stage specifically in the nursery to early basic vegetative phase coinciding with the cooler months i.e. November to February. It is cultivated in eastern UP covering Gorakhpur, Basti, Deoria, Ballia, Ghazipur, Mirzapur and Varanasi districts. Total area under *boro* rice in eastern UP is around 4000 ha. Traditional *boro* rice varieties such as China 4, Sarraiya, Jharanga and Kalaboro are still cultivated only in certain pockets of these districts. No systematic research work on collection, evaluation and utilization of *boro* rices has been done in Uttar Pradesh except in the early seventies at G.B. Pant University of Agriculture & Technology, Pantnagar (Uttarakhand) and later at NDUAT, Kumarganj, Faizabad and PRDF, Gorakhpur. During seventies, 128 accessions were collected, evaluated and maintained at Pantnagar. However, in course of time, this entire germplasm got lost due to problems of maintenance/regeneration at Pantnagar though this has been conserved at NBPGR, New Delhi and also maintained at IRRI, Manila, Philippines.

Later, 558 germplasm accessions were collected, evaluated and categorized by PRDF, Gorakhpur (Chaudhary, 2003). Similarly, 72 accessions were also collected and evaluated by the scientists of BHU, Varanasi. Efforts were made in 2001 and 2003 by the Crop Research Station, Ghaghraghat, Bahraich to identify *boro* rice growing areas in Uttar Pradesh and evaluate the improved rice germplasm developed in other states through scientists-farmers' participatory approach. Accordingly, NDR 97, Baranideep, Gautam, Prabhat, Richhariya, Saroj, Dhanlaxmi, Joymati, Krishna Hamsa and Pant Dhan 12 were found suitable for *boro* rice areas of Uttar Pradesh.

Efforts need be concentrated on survey of *boro* rice growing areas, collection of traditional *boro* rice varieties and their proper evaluation and cataloguing. However, research work carried out at G.B. Pant University of Agriculture and Technology, Pantnagar revealed that besides the cold tolerance, *boro* rices are a good source of resistance for various diseases including leaf blast (UPRB 7, UPRB 8, UPRB 31, UPRB 32, UPRB 33, UPRB 37) (Nanda *et al.*, 1976) and bacterial leaf blight (UPRB 30, UPRB 31) (Nanda *et al.*, 1977). Chaudhary *et al.* (1984) developed a cytoplasmic male sterile line called 'Patna CMS 1' by using a *boro* line UPRB 31.

Hybrid Rice

The research work on the development and seed production technology of hybrid rice was initiated in 1989 at NDUAT, Faizabad and GBPAUT, Pantnagar through a research network financed by the ICAR/UNDP. This research work has led to the development and release of three-line hybrids Pant Shankar Dhan 1 in 1997 and Pant Shankar Dhan 3 in 2004 for irrigated medium and irrigated early areas of plains by GBPUAT, Pantnagar and Narendra Shankar Dhan 2 in 1998 by NDUAT for irrigated areas. Later, NDUAT developed Narendra Usar Shankar Dhan 3 in 2004, the first hybrid (yet to be notified) for problem soil (*usar*) areas.

Loss of Rice Biodiversity

The conservation of biodiversity has been a hot topic for decades but the destruction of ecosystems and genetic erosion of species and varieties have continued unabated. Dr. Peter H. Raven, Director of Missouri Botanical Gardens in St. Louis while emphasizing the importance of conservation of species, mentioned "there is no gift more precious to your children and grand-children and their grand-children than to be able to preserve the biological diversity including the genetic diversity of species". Once lost, the biological diversity cannot be regained. Rice biodiversity refers to the existing rice varieties on the farmers' field. In the era of high yielding rice varieties, traditional varieties are disappearing from their native places due to their poor yield and susceptibility to prevailing pests and diseases. The rate of loss of bio-diversity has been comparatively slower in the rainfed rices than in the irrigated ones. Among rainfed ecosystems, there is little genetic erosion in the flood-prone rice ecosystem in eastern Uttar Pradesh owing to the release of only a few high-yielding varieties for this ecosystem. The rate of replacement of landraces is slow because of the poor seed distribution system and farmers' preference for specific traits. The leading flood-prone rice cultivars grown in eastern Uttar Pradesh clearly indicate that traditional varieties still have a substantial share in the area. A recent survey carried out in flood-prone areas indicated that a good number of traditional varieties such as Malhi, Suapankhi, Amgaud, Kajari, Lalha, Venaga, Agahani, Gauria, and Dhanesar are still cultivated in flash flood; Barogar, Cross 116, and Sainagar in semi-deep stagnant water and Jaisuria, Sugapankhi, Bainslot, Goanth, Ghoghari and Tairaki in floating/deep rice areas besides the improved types.

An estimate of varieties lost in previous years may be 70-80% in favourable rainfed lowland and 60-70% in upland. Further, 15-20% more may be lost in the next 20 years (Singh *et al.,* 2004). However, genetic erosion is not very alarming in the flood-prone ecosystem (Dwivedi, 1996) (Table 15.6).

Table 15.6. An estimate of loss of varieties in different ecosystems.

Ecosystem	% Varieties lost
Upland	60-70
Rainfed lowland favourable	70-80
Rainfed lowland unfavourable	20-30
Flood prone	<5

Out of about six hundred aromatic rice germplasm in the world, two hundred are native to India. Due to weak stem, susceptibility to many diseases especially panicle blast, pests and poor and unstable grain yield, these valuable local scented varieties are disappearing and a large number of small and medium grained aromatic varieties have gone out of cultivation (Singh and Singh, 2005; Dwivedi *et al.*, 2005). Study revealed that some 35 odd number of these short grain aromatic rices were grown in eastern UP covering 18 districts two decade back but at present hardly 5-6 varieties viz. Kalanamak, Lalmati, Vishun Parag, Adamchini, Badshahbhog, etc. are in cultivation in certain pockets in small acreage only.

Conclusion

The great diversity in ecosystems coupled with limited available resources for research makes research more challenging for scientists. Suitability of rice varieties will depend on land situation, hydrology and other ecological conditions, tolerance to submergence, flowering time, cold and drought in different ecological conditions. Because of the problem soils and unpredictable combination of flood and drought, rice yields are still low. By and large, rice is the only option for these two states during wet season in problematic areas. Therefore, collection of existing landraces/traditional varieties and their evaluation and conservation should be a never ending process and need emphasis in future rice research programmes. Unfortunately, none of the varieties developed so far could cover full range of prevailing diversity of ecosystems. Therefore, farmers' excellent empirical understanding of adaptation of different rice types and their involvement in breeding programme need to be encouraged. In view of increased requirement of rice in future and drastic changes in weather especially due to global warming, the world must accept the challenge and commit resources to achieve higher and economical production from rainfed ecologies as rice will remain staple food for the majority of Asians in future covering 2/3 of the population of the world. The efforts made by rice breeders working in different agricultural universities and institutes has resulted in development of a large number of rice varieties suitable for different ecosystems but only a limited number of improved high yielding varieties is in cultivation in any particular ecosystem. This indicates lack of local adaptation of these varieties to the prevailing ecosystems. Therefore, there is an urgent need to utilise the existing genetic wealth possessing different desirable traits for development of high yielding varieties for different ecologies and areas.

Keeping in view the prevailing diversity in the ecosystem, development of more number of improved varieties with requisite traits and adaptability to localty/ field will be the appropriate strategy. This will enable us to meet the requirement of farmers with regard to enhancing the productivity and production and also to conserve bio-diversity in the fragile ecology including upland, flood-prone and hilly conditions on the farmers' fields.

References:

Awasthi LP, Sharma AK, Maurya DM. 1997. Histological variations among aromatic rices. *Intern Rice Res Notes* 22(2):334.

Chaudhary, N. 2003. Collection, conservation and morphological evaluation of Boro rice (*Oryza sativa* L) with reference to incidence of diseases and pests from eastern Uttar Pradesh, India. PhD Thesis, University of Gorakhpur, India.

Chaudhary RC, Sahai VN, Saran S. 1984. A possible new male sterile line of rice with cytoplasm from Boro rice of Uttar Pradesh. *Intern Rice Res Newl* 1:104

Dikshit NN. 1972. Varietal improvement of rice in U.P. Page 11-20 in *Rice in Uttar Pradesh.* UP Institute of Agricultural Sciences, Kanpur-2.

Dwivedi, JL. 1996. Conserving genetic resources and using diversity in flood prone ecosystem in eastern India. Page 230-237 in Sperling L, Loevinsohn M (eds) Proceeding of the workshop on *Using Diversity* held at IDRC, New Delhi.

Dwivedi JL, Singh BB, Tripathi AK, Singh RK. 2005. Diversity in relation to traditional aromatic rice adaptation in eastern Uttar Pradesh: A critical analysis. Pages 22-32 in Singh RK, Singh US (eds) *Scented Rices of Uttar Pradesh and Uttarakhand.* Kalyani Publisher, New Delhi.

Dwivedi JL, Tiwari Triveni, Chaudhary RC, Pandey Alok, Singh Ratnesh, Singh RK. 2005. Short grain aromatic rices: Strategies for improvement and future opportunities. Pages 13-21 in Singh RK, Singh US (eds) *Scented Rices of Uttar Pradesh and Uttarakhand.* Kalyani Publisher, New Delhi.

Ikehashi H, Araki II. 1984. Varietal screening of compatibility types revealed F_1 fertility of distant crosses of rice. *Japan J Breed* 34:304-313.

Khush GS. 2000. Taxonomy and origin of rice. Pages 5-14 in Singh RK, Singh US and Khush GS (eds). *Aromatic Rices.* Science Publishers Inc, USA.

Mani SC, Mishra CH, Shukla SK. 2001. A history of rice breeding in UP and Uttarakhand. Pages 10-13 in *Towards Better Rice Production in Uttarakhand.* (Souvenir). GB Pant Univ of Agri and Tech, Pantnagar.

Nanda JS, Chaudhary RC, Rana OS. 1976. Blast resistance in cold tolerant *boro* rice collections of U.P. *Intern Rice Res Newl* 1:10-11

Nanda JS, Chaudhary RC, Singh H, Singh BN, Singh JP. 1977. Two possible new source of resistance to bacterial leaf blight in rice. *Intern Rice Res Newl* 2(1):2

Singh R, Srivastava VK. 1997. Reports on *Basmati and Scented Fine-grained Types at Varanasi Centre*. BHU, Varanasi, India. 29 pp.

Singh RK, Singh US. 2005. *Scented Rices of Uttar Pradesh and Uttarakhand.* Kalyani Publisher, New Delhi. 169 pp.

Singh RK, Dwivedi JL, Thakur R, Mallik S, Ahmad T. 2004. Rice biodiversity and genetic wealth of flood-prone environments in eastern India. Pages 115-128 in Bhuriyan SI, Abedin MZ, Singh VP, Hardy B (eds) *Rice Research and Development in the Flood-prone Ecosystem.* Proceedings of the international workshop on *Flood-prone Rice System,* 9-11 January, 2001, Gazipur, Bangladesh. International Rice Research Institute (IRRI), Manila. 283 pp.

Sood BC, Siddiq EA, Zaman FU. 1983. Genetic analysis of kernel elongation in rice. *Indian J Genet* 43(9):40-43.

Appendix 15.1. High yielding rice varieties developed for different ecosystems in Uttar Pradesh.

Variety name / IET No	Cross combination	Year of release	Eco-system	Maturity (d)	Yield q/ha	Grain type	Released by	State/ area recommended	Notification details	Special features
Upland and Drought-prone Ecology										
Nagina 22	Sel from Rajbhoj	1978	Rainfed Upland	85-90	20-25	SB	SVRC	UP	13(E) dated 19.12.1978	Drought tolerant at veg & reproductive stage; droopy leaves; wide adaptability
Narendra Dhan 1 IET 2233	Belle Patna/IR8	1981	Rainfed Upland	105	35(DS) 40 (T)	MB	SVRC	UP	13(E) dated 19.12.1978	Resistant (R) to blast (Bl) and RTV; tolerant to saline alkaline soils
Narendra Dhan 80 IET 7626	N22/IR 36	1986	Rainfed Upland	105	55	MS	SVRC	UP	165(E) dated 6.3.1987	Resistant (R) to blast (Bl); non-lodging type
Narendra Dhan 118 IET 8681	IR36/ Hansraj A	1987	Rainfed Upland	95	35(DS) 45(T)	MS	SVRC	UP	471(E) dated 5.5.1988	Tolerant to drought, having good recovery after drought spell; resistant to Helminthosporium, leaf spot; moderately resistant (MR) to ShR
Narendra 97 IET 9210	N22/ Ratna	1992	Rainfed Upland	100	40	LS	CVRC	UP, WB &Orissa	860(E) dated 25.11.1992	Drought and lodging resistant; MR to BLB and sheath rot

Baranideep IET 13194	C1064-5 /IR 9129 -320-3-3 -3//IR 54	2001	Rainfed Upland	100	35	LS	SVRC	UP	1572(E) dated 20.9.2006	Resistant to BS, BLB, ShR; MR to ShB; tolerant to leaf folder
Shusk Samrat IET 17458	C 1064-5/ Kalkari// IR 54	2006	Rainfed Upland	110	35-40	LB	CVRC	UP, Orissa & Bihar	122(E) dated 6.2.2007	MR to drought; resistant to BL
Narendra Lalmati IET 21051	Selection from local Lalmati	2009	Rainfed Upland	105	30-35	SS	SVRC	UP	2187 (E) dated 27.8.09	Mild aroma; light red kernel colour; MR to drought; suitable for double cropping system especially in potato growing areas.
				Irrigated Ecology						
Early (115-120 d)										
Saket 4 IET 1410	TKM 6/ IR 8	1971	Irrigated mid early	115	40-45	LS	SVRC	UP	598(E) dated 8.10.1974	MR to BLB, GH and ShB
Narendra Dhan 2026 IET 14998	Sipi 632063/ Chianing Sen Yu47 //Taichung Sen 12	2005	Irrigated mid early	115	40-45	MS	SVRC	UP	-	Suitable for double cropping system; MR to SBWE; resistant ShR, BLB and FS
Narendra 2064 IET 17475	Pant Dhan 4/Saket 4 //NDR 2017	2009	Irrigated mid early	120-125	50-55	LB	SVRC	UP	2187 (E) 27.08.09	MR to BLB, ShR and FS diseases, MR to SB, LF; good cooking quality'

Malaviya Dhan 2	IR36/HR 137	2004	IRE	100-105		LS	SVRC	UP	1177(E) 25.8.2005	-
Medium (120-130d)										
Sarjoo 52	TN1/ Kashi	1980	Irrigated medium	135	60-65	LB	SVRC	UP	19(E) dated 14.1.1982	Photoperiod insensitive; good threshability; MR to BLB
IR 24	IR8/IR 127-2-2	1972	IRME	120-125		LS	SVRC	UT	598(E) 8.10.1974	Medium resistance to blast, bacterial leaf blight and green leafhopper
Madhuri	Sabarmati/ Ratna	1983		128		-	SVRC	UP	-	-
Narendra Dhan 359 IET 11005	BG 90-2-4/OB677	1993	Irrigated medium	130	60-65	SB	CVRC	UP, Bihar, Assam, Orirsa	636(E) dated 2.9.1994	Wide adaptability; resistant to lodging, leaf blast, BLB, BS
Narendra Sankar Dhan 2	IR58025 A/NDR 3026-3-1-R	1998	Irrigated medium early	128	65-70	LS	SVRC	UP	425(E) dated 8.6.1999	Most stable hybrid with high yield potentiai
Malviya Dhan 36	Mutant of Mahsuri	2002	IRME	125-130		SB	SVRC	UP	647(E) 9.9.1997	Resistance to bacterial leaf blight; medium resistance to stem borer insect

Malviya Dhan 1	HBR 92/ Pusa Basmati/ Kasturi	2005	SCR	128		LS	SVRC	UP	1177(E) 25.8.2005	-
Narendra Dhan 3112-1 IET 19335	NDR 313-1/NDR 359	2009	Irrigated medium	125-130	50-55	LB	SVRC	UP	2187 (E) 27.08.09	Suitable for double cropping system
				Water-logged and Flood-prone Ecology						
Rainfed Lowland (30-50 cm water depth)										
Cross 116	Bhondu/ Parewa	1978	Rainfed Shallow low land	150	40-45	SB	SVRC	UP	13(E) dated 19.12.1978	Tolerant to 40-50 cm water stagnation for a limited period; resistant to blast
Semi Deep Water (50-100 cm water depth)										
Chakia 59	Sel from local	1969	Semi deep water	155-160	30-35	SB	SVRC	UP	13(E) dated 19.12.1978	Useful donor for deep water and flood prone
Jalpriya IET 10530	IET 4060 /Jalmagna	1993	Semi deep water	160-170	35-40	LS	SVRC	UP	408(E) dated 4.5.1995	Suitable for stagnant flooding; moderate elongation ability; MR to stem borer and brown spot; quality grain
Jallahiri IET 13534	Pankaj/ Mahsuri// TKM 6	1993	Semi deep water	170	45-50	MS	SVRC	UP	636(E) dated 2.9.1994	MR to ShB, ShR; resistant to lodging; tolerant to 10-12 days submergence

Intermittent/Flash flooding										
Madhukar	Sel from Gonda	1969	Flash flooding	150-155	30-35	LB	SVRC	UP	13(E) dated 19.12.1978	Tolerant to submergence up to 10 days; tall and weakly photoperiod sensitive
Barh Avarodhi IET 11295	Madhukar /Sona	1995	Intermit-tent/Flash flooding	155	30-35	LB	SVRC	UP	647(E) dated 9.9.1997	Survives up to 12 days complete submergence and better regeneration ability; MR to stem borer; attractive grain
Deep water (> 100cm water depth)/ Floating rice										
Jalmagna	Sel from Badhon	1969	Deep water	180-200	30-35	SB	SVRC	UP	13(E) dated 19.12.1978	Photoperiod sensitive (PS); possessing good elongation ability; MR to stem borer
Jalnidhi IET 10235	Sel from Goanth	1993	Deep water	180-200	35-40	MB	SVRC	UP	636(E) dated 2.9.1994	Possesses excellent elongation ability, kneeing ability & nodal tillering; can be cultivated up to 2 m water depth; photoperiod sensitive; suitable for deep water/ floating rice

										areas; MR to stem borer
Problem Soils / Inland Salinity & Alkalinity										
Type 100	Sel from Bhainslot	1978	Irrigated saline & alkaline	150	25-30	SB	SVRC	UP	13(E) dated 19.12.1978	Tall and tolerant to sodicity
Narendra Dhan 2 IET 4555	IR8/ Tadukan //TKM6 /TN1// IR8xIR24	1982	Irrigated saline & alkaline	115	45	LS	SVRC	UP	165(E) dated 6.3.1987	Tolerant to salinity; R to BL, BLB & RT
Narendra Usar Dhan 2 IET 13556	Sel. From IRRI line	1998	Irrigated saline & alkaline	125	35-40	LB	SVRC	UP	401E) dated 15.5.1998	Tolerance to salt, lodging and shattering, no major disease or insect pest affect the crop
Narendra Usar Dhan 3 IET 14659	Lung YAI 148/IR 9125-209-2-2-2-1// IR 1872 -27-3-1	2000	Irrigated saline & alkaline	135	40-45	LS	SVRC	UP	92(E) dated 2.2.2001	Tolerance to salt; no major disease & insect affect the crop
Narendra Usar Dhan 2008 IET 18699	TCCP 266-249 -B-B-3/ IR 262-43-8-1	2009	Irrigated inland salinity & coastal salinity	120-125	45-50	LB	CVRC	Orissa, A.P. & West Bengal	2187 (E) 27.08.09	Suitable for coastal salinity and inland salinity; can be cultivated up pH 9.5

Appendix 15.2. High yielding rice varieties developed in Uttarakhand.

Variety name / IET No	Cross combination	Year of release	Eco-system	Maturity (d)	Yield q/ha	Grain type	Released by	State/ area recommended	Notification details	Special features
IR 24	IR8/IR1 27-2-2	1972	IRME	120-125	50-55	LS	SVRC	UT	598(E) 8.10.1974	Medium resistance to blast, bacterial leaf blight and green leafhopper
Majhera 3(T)	Selection from Jelia	1968	HRIR	160-165	18-20	LS	SVRC	UT	-	Susceptible to blast
Majhera 7	Selection from a local landrace	-	HRUR	160-165	21-23	MB	SVRC	UT	-	-
Prasad	IR 747B-26-3/IR 57948	1978	IRE	115-120	50-55	MS	SVRC	UT	13(E) 19.12.1978	Resistance to blast
Manhar	IR20/ Cauvery	1985	IRE	110-115	50-55	LS	SVRC	UT	832(E) 18.11.1985	Medium resistance to bacterial leaf blight & white backed planthopper
Pant Dhan 4	IR 262/ Remadia	1983	IRM	125-130	55-60	LS	SVRC	UT	-	Resistance to blight and brown spot; medium resistance to bacterial leaf blight

Pant Dhan 6	IR8608-298-3-1/IR10179-23	1986	IRE	110-115	40-42	MS	SVRC	UT	834(E) 18-9-1987	Resistance to blast, bacterial leaf blight, brown spot; medium resistance to stem borer
Pant Dhan 10	IR 32/Mahsuri//IR 28	1992	IRM	115-120	58-60	LS	SVRC	UT	615(E) 17.8.1993	Medium resistance to blast , bacterial leaf blight, sheath blight, stem borer, white backed planthopper, leaffolder
Pant Dhan 11	VL 206/Dagi	1992	HRIR	125-130	42-48	LB	SVRC	UT	615(E) 17.8.1993	Medium resistance to blast, bacterial leaf blight, brown planthopper
Pant Dhan 12	Govind/UPR201-1-1	1994	IRE	120-125	55-58	LS	SVRC	UT	1(E) 1.1.1996	Medium resistance to bacterial leaf blight & brown spot
Pant Dhan 18	SPR 85163-5-1-2-4	2002	IRM	130-135	50-55	LS	SVRC	UT	1178(E) 20.7.2007	Medium resistance to sheath blight
Pant Shankar Dhan 1	UPR 195-178A/UPR192-	1997	IRM	130-135	65-70	LS	SVRC	UT	425(E) 8.6.1999	Medium resistance to blast, bacterial

	133 R									leaf blight, rice tungro virus, brown spot, stem borer, brown planthopper & white backed planthopper
Pant Shankar Dhan 3	UPRI 95-18A /UPRI 93-287R	2004	IRE	120-122	55-60	LS	SVRC	UT	599(E) 25.4.2006	Medium resistance to blast, bacterial leaf blight, rice tungro virus, brown spot, stem borer, brown plant hopper & white backed plant hopper
Pant Sugandh Dhan 15	Basmati 370/Sadri //Bahral/ Muskan 41	2002	SCR	140-144	50-65	LS	SVRC	UT	1178(E) 25.4.2006	Medium resistance to blast
Pant Sugandha Dhan 17	Pusa Basmati 1/UPRI 95-154	2004	SCR	130-135	45-50	LS	SVRC	UT	1178(E) 25.4.2006	Medium resistance to blast & stem borer
Vivek Dhan 154	VL Dhan 221/VL24	2003	RUP	105-110	40-45	-	SVRC	UT	599 (E) 25.4.2006	Resistance to blast & brown spot; medium resistance to leaf folder insect

VL Dhan 65	Himalaya 2/VL Dhan 16	2006	HRIR	130-133	40-45	LS	SVRC	UT	122(E) 6.2.2007	Resistance to blast & brown spot
VL Dhan 85	HPU 799/VL 221	2005	HRIR	110-115	40-45	LS	SVRC	UT	599(E) 25.4.2006	Resistance to blast & brown spot
VL Dhan 97	N 22/ Ratna	1991	RUP	110-115		LS	SVRC	UT	-	-
VL Dhan 163	IR 747/ KN 16-36-1//IR 2053-521 -1-1-1	1987	HRUR	105-110	20-25	MS	SVRC	UT	471(E) 5.5.1998	Resistance to blast; medium resistance to stem borer and leaf folder insect
VL Dhan 206	Selection from Bamani	1983	HRUR	140-145	20-22	MS	SVRC	UT	540(E) 24.7.1998	Medium resistance to blast; medium resistance to stem borer and leaffolder
VL Dhan 207	VL Dhan 206/ Annada	2005	RUP	145-150	40-45	MS	SVRC	UT	599(E) 25.4.2006	Resistance to blast & brown spot; medium resistance to stem borer and leaffolder
VL Dhan 208	VR 410-19/VR 212	2005	RUP	150-155	40-45	MS	SVRC	UT	599(E) 25.4.2006	Resistance to blast & brown spot; medium resistance to stem borer and leaf folder

VL Dhan 209	Him dhan/K 39//VL dhan221	2006	RUP	150-155	35-40	SB	SVRC	UT	122(E) 6.2.2007	Resistance to blast; medium resistance to stem borer and leaf folder

NB: HRIR = Hill rice, irrigated; HRUR = Hill rice, upland rice; IRE = Irrigated, early; IRM = Irrigated, medium; IRME = irrigated, medium early; RUP = Rice, upland; SCR = scented rice.

UP = Uttar Pradesh, UT = Uttarakhand

Genetic Resources of Rice in India : 493-556
Editor : S. D. Sharma
Today & Tomorrow's Printers and Publishers, New Delhi - 110 002 (India)

Chapter 16

Genetic Resources of Rice in North East India – Role of Assam and Assam Agricultural University

T. Ahmed, G.R. Das and S.K. Chetia

Northeast India has been highlighted as the repository of diverse rice varieties and valuable gene systems (Sharma *et al.*, 1971; Vairavan *et al.,* 1973). The entire region falls within the area of diffused origin of rice. According to Watabe *et al.* (1980), the place of origin of cultivated rice is to be sought in a relatively high latitude zone from Yunnan province in China to Assam state in India. Eco-geographical diversity and ethnic diversity are supposed to be instrumental in the creation and perpetuation of diverse varieties and ecotypes of rice in the region.

Eco-geographical and Ethnic Diversity

Northeast India comprises of seven sister states, viz., Assam, Arunachal Pradesh, Nagaland, Manipur, Mizoram, Tripura and Meghalaya. The region is bordered by China and Bhutan on the north, Burma on the east and Bangladesh on the south and west; it is connected with the rest of India by a small corridor in the northwest joining Assam with West Bengal (Fig. 16.1). It is predominantly a mountainous hilly region with stretches of flat valley lands and scattered pockets of marshy lowland particularly in Assam. The climate is sub-tropical and tropical in the low altitude plains but, in high altitude mountainous areas, it is temperate. Both humidity and rainfall are high in the region. In fact, the world's wettest place, Cherrapunji – Mawsynram area of Meghalaya, with an annual rainfall of about 12 thousand mm – lies in this region. In the plains of Assam, however, the average annual rainfall varies from about 1200–1500 mm in the rainshadow areas of Karbi Anglong to 3500-4000 mm in Karimganj of Barak Valley. Being the staple food of the people of the region, rice is grown as the principal crop in all the seven states over widely different eco-geographical situations. To match with, there are groups of varieties with specific adaptations (Fig. 16.2).

Fig. 16.1. Map of Northeast India showing state boundaries.

The region is also unique in ethnic diversity. Many ethnic tribes and communities with distinct linguistic and cultural identities have been living in different parts of the region for a long time. They mainly belong to Tibeto-Burman, Tibeto-Himalayan, Austro-Asian and also Indo-Aryan groups. In Arunachal Pradesh alone, the number of main tribes is 20 and each tribe has a few sub-tribes. Ethnic conflicts act as a sort of isolating mechanism causing varietal differentiation between ethnic territories while ethnic preferences help in the maintenance of diverse rice ecotypes. Ethnic migration and immigration in the state over many years have also resulted in the introduction of different types of rice cultivars into the region leading to further diversification of rice genetic stock.

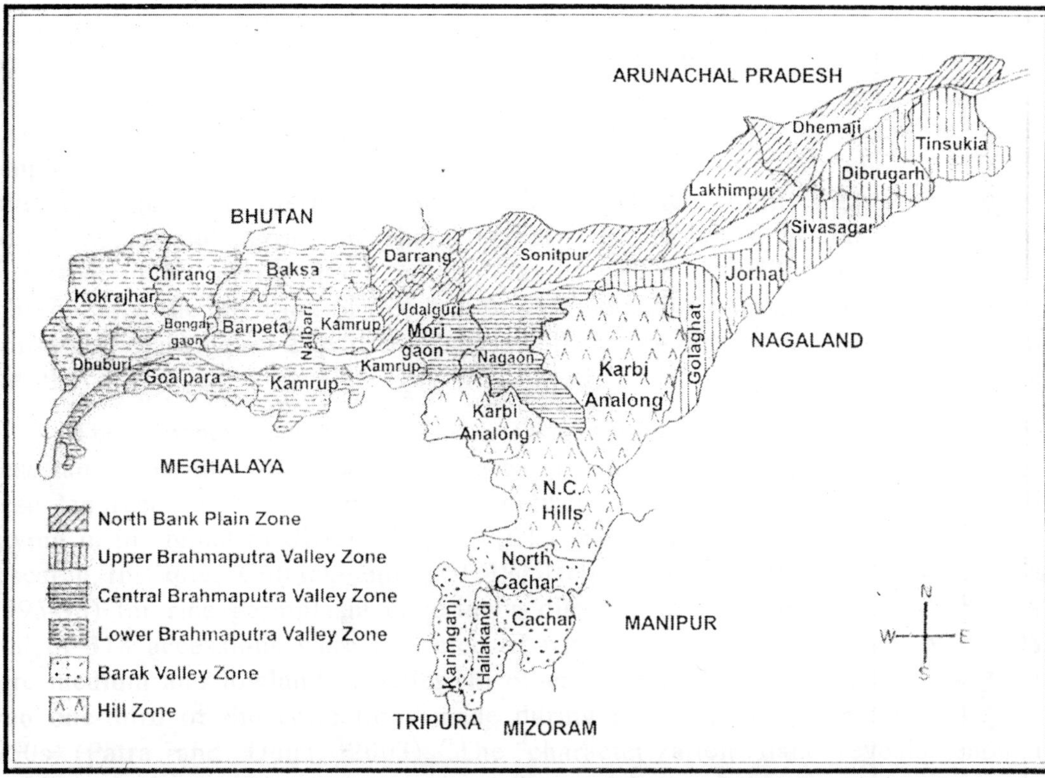

Fig. 16.2. Map of Northeast India showing agroclimatic zones.

Rice Growing Situation

Depending on eco-geographical variation in the region, rice is grown over a wide range of situations varying from hill slopes to an altitude of 2000 m and above, through drought affected upland and rainfed lowland to deep water situations where standing water may be 2–4 m at peak season. The factors defining the rice growing situations are altitude, climate i.e. tropical, subtropical or temperate, land topography such as hill slopes or plains, field water condition viz. drought affected areas, rainfed lowland semideep or deepwater condition and growing season *viz. sali* (winter rice), *boro* (summer rice) or *ahu* (autumn ric). Rice growing situations dictate the type of cultivation system and nature of rice varieties (Table 16.1, Fig. 16.3). The rice crop in this region has to overcome all such physical and physiological stresses like drought, flood, low light intensity, low temperature and biological problems like pests and diseases. Continuous growing of rice varieties for centuries through all such situations thereby exposing the crop to adaptive selection either through nature or farmers discretion has resulted in the development of multitude of rice varieties with specific adaptations.

Table 16.1. Rice growing situations, cultivation systems and varietal groups.

Altitude and climate	Rice growing situations			Cultivation system	Varietal groups
	Land situation	Field water condition	Season		
A. Low altitude tropical or sub-tropical climate	1. Upland plain	No standing water	Feb to Jun	Dry broad cast	*Ahu*
	2. Rainfed lowland	Water level 0-25cm, higher in flood prone areas	Mar-Apr to Jul–Aug	Wet broadcast and transplanting	*Ahu*
	3. Rainfed lowland	Water level 0-25cm, higher in flood prone areas	Jun-Jul to Nov-Dec	Transplanting	*Sali, bora* (glutinous) *chokuwa* (semi glutinous) and *joha* (aromatic)
	4. Semi deep water situation	Water level up to 60cm	Apr-Dec	Wet broadcast, rarely transplanting, sometimes mixed with *ahu*	*Asra*
	5. Deep water situation	Water level up to 1m	Mar-Apr to Dec	Dry broadcast sometimes wet broadcast	*Bao* group with submergence tolerance
	6. Very deep water situation	Water level above 1 m up to 4m	Mar to Dec	Dry broadcast, sometimes wet broadcast	*Bao* group with floating habit
	7. Beels, haors and marshy lands	Water stagnates with premonsoon	Nov to Apr-May	Transplanting	*Boro* (cold tolerance)

		showers, becoming virtually dry in winter			
B. High altitude temperate climate	8. Hill slopes	No standing water	Apr to Oct	Jhuming	Hill rice (cold tolerant)
	9. Plain strips	No standing water	Mar–Apr to Sep-Oct	Dry broadcast	Hill rice (cold tolerant)
	10. Plain terrace	Water level 0-15cm	Apr-Oct	Wet broadcast & transplanting	Hill rice (cold tolerant)

* *Ahu, boro* and hill rice are photoperiod insensitive, all others are sensitive.

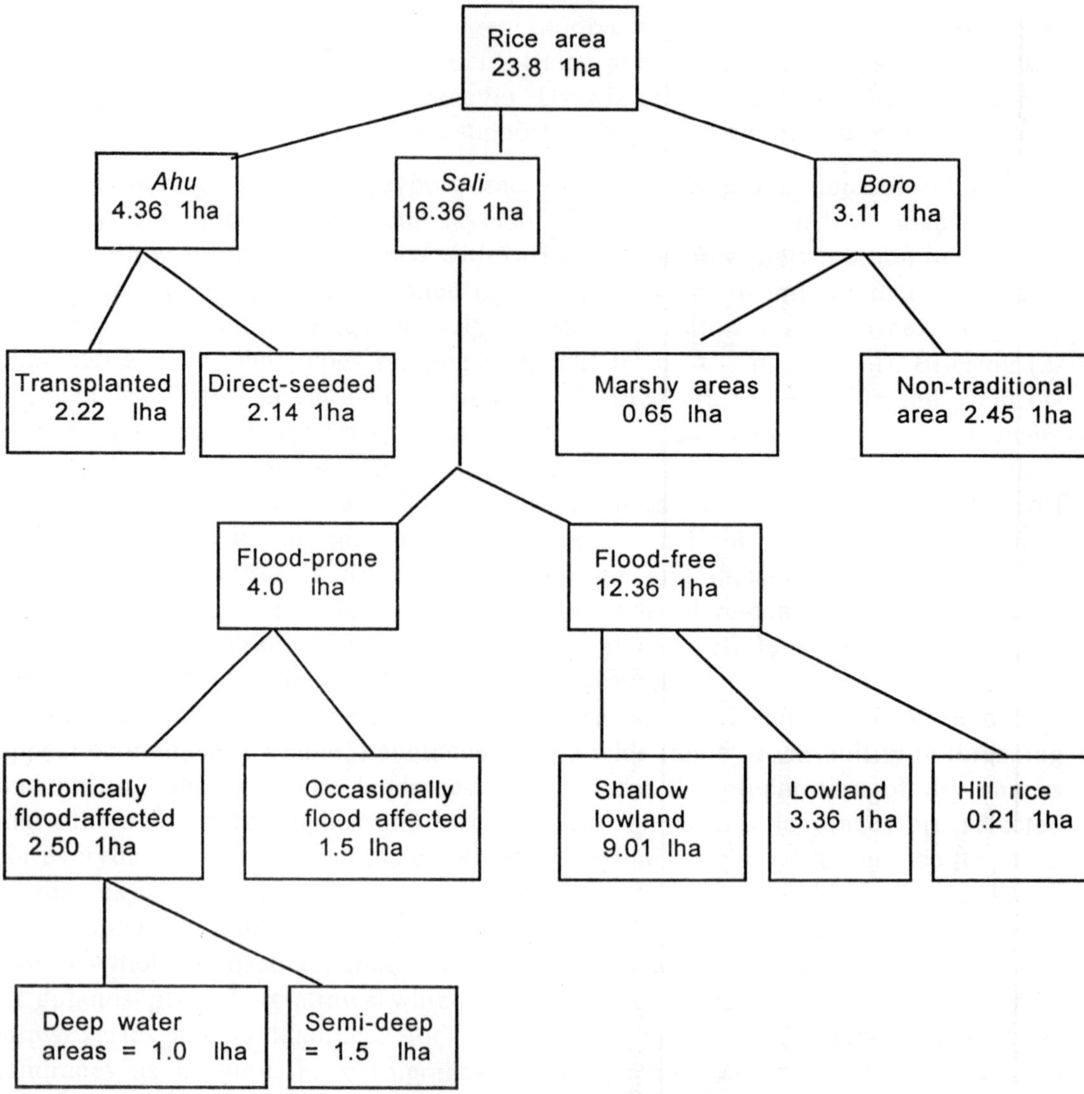

Fig. 16.3. Approximate area (2004-05) under various rice growing cultures/ ecosystems (lha = lakh hectares).

Wild Rice in the Region

Wild rice is found in scattered lowlying rice fields and swampy areas, sometimes mixed with other members of Poaceae such as *Leersia* spp. Katayama (1980) isolated 17 different strains of wild rice belonging to *O. rufipogon* Griff. (identified as *Oryza perennis* Moench by him) and *O. sativa* var. *spontanea* Roschev. from road side swamps, edges of ponds, rice fields and canals during his short trip from Guwahati to Sivasagar of Brahmaputra Valley of Assam. Bor (1960) reported availability of *Oryza officinalis* (identified as *O. minuta* by him) and *O. rufipogon* in the swampy habitats of Assam. Considering the fact that northeast India lies within the core area of incipient domestication of rice, it is highly probable that had there been an extensive survey in all the states of the region, greater diversity in wild and weedy forms of rice could have been obtained.

Traditional Classes of Rice of Northeast India

Assam has a locally accepted system of classification of indigenous rice according to season of cultivation, crop duration, field water condition, grain type and endosperm nature (Mitra and Ganguli, 1932). But no such established traditional system of classification is known for hill rices of northeastern hill states of the region. Under the circumstances, adding hill rice as a separate category and making slight modification in the earlier scheme for Assam rice, the entire variability in rice germplasm of the region is classified in the manner given in Table 16.2.

Diversity at Taxonomic Level

It has been observed that the indigenous rice germplasm of northeast India represents a wide range of phenotypes resembling various grades from *indica* to *japonica* subspecies of *O. sativa* (Sharma *et al.,* 1971; Vairavan *et al.,* 1973).

Glazmann (1986) classified Asian domestic rice varieties into six enzymatic groups on the basis of isozyme polymorphism at twenty one loci and indicated that *sali* (=*aman* of Bangladesh) belonged to *indica* subspecies while upland rice and hill rice along with *bulu* of Indonesia and *keng* of China belonged to *japonica*. Further, *ahu* (=*aus*) and deep water rice (*rayada* group) belonged to two independent groups separated from both *indica* and *japonica*. *Ahu* rice was, however, identified with *javanica* subspecies for its intermediate characteristics between *indica* and *japonica* (Chang, 1985; Oka, 1988). Interestingly, it has also been reported that some rice varieties from northeast India resemble *O. glaberrima* in certain features such as glabrous plant parts, short ligules, red kernels, short panicles rarely having secondary branches and shy tillering habit, for example, ARC 6186, 6187, 6189, 6191, 6192 and 6200 (Annual Report, CRRI 1974).

Sharma *et al.* (1971) made an extensive collection of indigenous cultivars numbering 6630 accessions from all over NE India except the hilly state of Mizoram and the Barak Valley of Assam and reported tremendous variability in the germplasm collection in respect of quantitative and qualitative characters and the

Table 16.2. Traditional classification of rice.

Photoperiod reaction	Agricultural class	Sub class	Growing season	Duration (days)	Chief characteristics
Sensitive (winter rice category)	*Sali* (non glutinous)	*Sali*	Jun-Jul to Nov-Dec	150-180	Coarse grain, generally white kernel, tall stature, thick culm called *sail* in Barak Valley
		Lahi	Jun-Jul to Nov-Dec	150-180	Medium grain, generally white kernel, tall stature, thick culm
		Joha	Jun-Jul to Nov-Dec	150-180-	Fine grain, white kernel generally scented, tall, thin culm
	Bora (glutinous rice)	-	Jun-Jul to Nov-Dec	150-180	Glutinous or waxy rice, negligible amylose fraction but high amylopectin generally pigmented in one or more plant parts, tall stature, variable grain size.
	*Chokuw*a (semi glutinous/ soft rice)	-	Jun-Jul to Nov-Dec	150-180	Coarse grain, generally white kernel, low amylose, very soft but slightly sticky rice, tall stature, thick culm.
	Asra (medium deep water rice)	-	Apr-May to Dec-Jan	240-270	Coarse grain, white or red kernel, tall, thick culm, broadcast or transplanted in low lying areas, can endure water depth <100cm. Sown at the time of *ahu* rice and harvested at the time when *sali* rice is harvested.
	Bao (deep water rice)	-	Mar-Apr to Dec-Jan	270-300	Deep water or floating rice, normally grown broadcast, can endure water depth >100 cm (up to 400 cm). Sown at the time of *ahu* rice and harvested at the time when *sali* is harvested.

					Coarse grain, white or red kernel, thick culm, very tall depending on water level, elongation ability, kneeing ability, nodal rooting and branching ability, some with submergence tolerance.
Insensitive	*Ahu* (autumn rice)	*Dumai*	Mar-Apr to Jun-Jul	85-90	Medium grain, red kernel, usually grown broadcast, intermediate height, weak straw, usually prevalent in Barak Valley.
		Murali and Chengri	Mar-Apr to Jun-Jul	91-100	Medium grain, red kernel, usually grown broadcast, intermediate height, weak straw, usually prevalent in Barak Valley.
		Ahu	Mar-Apr to Jun-Jul	>100	Generally medium grain, mostly red kernel, some with white kernel, medium to tall stature, weak straw, generally grown transplanted, called *aus* in Barak Valley.
	Boro (Summer rice)		Nov to may-Jun	180-200	Medium grain, mostly red kernel, some with white kernel, generally awned, medium to tall stature, weak straw, cold tolerant at the vegetative stage ,grown transplanted traditionally in beel and marshy land situation with minimum or no tillage.
	Hill rice (Jhum)		Mar-Apr to Sep-Oct	210-250	Variable grain size, finer type and *japonica* type grain mostly prevalent, semidwarf to intermediate stature, grown in hills slopes as direct seeded, adaptable to temperate climate.

presence of many economic characters of applied value. These germplsam collections, designated as Assam Rice Collections (ARC) have since been variously investigated both nationally and internationally. To study the nature of genetic divergence in ARC, Varivan *et al.* (1973) studied a representative sample of 190 ARC varieties along with 3 *indica* and 1 *japonica* standards in respect of 10 characters and classified the varieties by canonical analysis into 42 preliminary groups which were further classified by using D^2 statistic into 9 divergent clusteres. Height followed by flag leaf area was mainly responsible for primary differentiation and 100-grain weight followed by amylose content was important for secondary differentiation. The characterization of the clusters may be done by cluster means for these 4 charcters as given in Table 16.3.

Rice germplasm collections of northeast India have also been classified in a different manner on the basis of specific pattern of character constellation associated with a key character (Seetharaman *et al.,* 1974; Srivastava, 1977, 1978). This is outlined after inserting probable locally-recognized category of rice in Table 16.4. It should be evident that classification of the entire rice germplasm into well differentiated groups, as shown in Table 16.3 and 16.4, gives an insight into the nature of diversity and its taxonomic implications. Nevertheless, traditional agricultural classes, mentioned in Table 16.2, have practical utility from the point of agro-ecological adaptation and have correspondence to taxonomic categories.

Table 16.3. Characterization of varietal clusters.

Cluster No.	Height (cm)	Flag leaf area (%)	100-grain wt (g)	Amylose content (cm^2)
I	Tall stature (155.9)	Large Leaf (43.9)	Medium grain (1.9)	High amylose (27.1)
II	Tall stature (168.9)	Large leaf (53.0)	Medium grain (2.2)	Low amylose (14.4)
III	Semi-tall stature (134.8)	Large leaf (43.8)	Fine grain (1.6)	High amylose (26.8)
IV	Semi-tall stature (133.9)	Large leaf (46.8)	Coarse grain (2.6)	Medium amylose (19.8)
V	Semi-dwarf (109.0)	Medium leaf (38.6)	Medium grain (2.0)	High amylose (26.3)
VI	Medium stature (118.6)	Medium leaf (37.7)	Coarse grain (2.8)	Very low amylose (4.8)
VII	Semi-dwarf (109.3)	Small leaf (33.0)	Fine grain (1.3)	High amylose (25.3)
VIII	Semi-dwarf (101.3)	Small leaf (34.7)	Coarse grain (2.6)	Medium amylose (19.3)
IX	Dwarf (79.9)	Small leaf (34.2)	Coarse grain (2.4)	High amylose (27.6)

NB: Adapted from Vairavan *et al.*, 1973

Table 16.4. Classification of rice germplasm of NE India.

Group (key character)	Character constellation of rice	Probable local category
Semidwarf (*indica* grain)	Photoperiod insensitive; moderately high to high tillering; relatively longer panicle than semidwarf japonica type; varying grain number and grain size; thin, narrow-angled flag leaves with early senescence; non lodging.	Mostly hill rice and some *ahu* rice
Japonica grain	Semidwarf to intermediate stature; photoperiod insensitive, shy tillering habit; relatively short panicles; closely arranged grains; thick, wide – angled flag leaves with late senescence; non lodging.	Mostly hill rice
Thick culm	Tall stature; partially to fully lodging; photoperiod sensitive; relatively shy tillering; long heavy panicle with coarse grain.	*Sali* and *asra* rice
Thin culm	Tall stature; partially to fully lodging; photoperiod sensitive; moderate tillering; long panicle; fine to superfine grains, generally scented.	*joha* group of *sali* rice
Beaded grain	Tall stature; partially to fully lodging; photoperiod sensitive; moderate tillering; long panicle; short bold to medium bold grains with moniliform arrangement on panicle branches	*Sali* rice

NB: As conceived by Seetharaman *et al.*, 1974; Srivastava, 1977 & 1978

Role of Assam Agricultural University in Germplasm Collection, Evaluation and Utilization

The collection of indigenous rice germplasm in Assam was initiated in the second decade of the 20th century with the establishment of the first Rice Experiment Station at Karimganj in 1913. The works, however, started in 1914 in the form of collection, maintenance and evaluation of Surama valley rice in accordance with the plan and programme chalked by G.P. Hector, Economic Botanist of Bengal; A.G. Birt, Deputy Director of Agriculture, Bengal and Rai Bahadur B.C. Basu, Special Officer of Agriculture, Bengal. Even in those early days, the importance of rice germplsm collection was realized and pure line selections were made at the initial step followed by yield evaluation of pure lines. These early efforts were rewarded by the identification and recommendation of Indrasail (1918), George sail (1918) and Latisali (1920) in the initial years. Systematic research works on the crop were, however, taken up by the state in the year 1921 with the appointment of Dr. S.K. Mitra as the first Economic Botanist of Assam along with botanical staff.

The experimental station at Karimganj dealt with *ahu*, *sali* and *asra* rice to address the rice problems of Surama valley of erstwhile Assam. A second Rice Research Station (now RARS) was established at Titabar in 1923 to deal with *ahu* and *sali* rice of Upper Assam Valley (Upper Brahmaputra Valley Zone). A third Rice Research Station was established at Habiganj in 1934 to deal with *aman* (deep water rice) and *boro* rice of lowlying areas but this station went to East Pakistan (now Bangladesh) after partition of India. Therefore, a new station was eventually established at Raha in Nagaon district in 1948 for *aman* and *boro* rice. Presently, Raha station has been diverted to fishery research under Assam Agricultural University (AAU) and the works on deep water rice is being looked after by RARS, North Lakhimpur.

Historically, the greatest contribution to rice research in Assam was by Mr. P.M. Ganguli, Botanical Assistant of Karimganj Station who under the guidance of Dr. S.K. Mitra, performed invaluable works on rice starting from collection, preservation and classification of Surama Valley rice to inheritance studies and breeding of higher yielding varieties and also some amount of basic agronomical studies. During 1928, as many as 161 rice germplasm collections of Rice Research Station, Titabar were categorized into early (74 varieties with 51-70 days flowering duration), medium (82 varieties with 71-90 days flowering duration) and late (5 varieties with duration more than 90 days). Upto 1931, about 759 *sali* germplasm including *sali, lahi, joha, bora* and exotic were collected and evaluated for yield. However, 927 germplasm consisting of *sali, lahi* and *ahu* were conserved during 1937 at Rice Research Station, Titabar (unpublished report). By 1946, as many as 3080 landraces (including some synonyms and homonyms) were kept in the research stations (Choudhury and Ganguli, 1948). Initial collections of Karimganj came from Cachar and Syhlet districts (now in Bangladesh, formerly part of Assam) and were collectively designated as Surama Valley rice while, at Titabar, the collections were from the Upper Brahmaputra Valley zone. In subsequent years, landraces from other parts of the state including hilly states enriched the collection at both the stations. During the later years, rice varieties received for research purpose from other states of India and a few foreign countries were also included in the germplasm collections maintained at Karimganj and Titabar. Before independence, germplasm collection trips were frequent but in the recent years such trips have been very few. However, with the launching of nationwide variety and technology testing network under the All India Co-Ordinated Rice Improvement Project and the IRRI sponsored International Network for the Genetic Evaluation of Rice (INGER), numerous varieties were screened and the more promising ones were added to the collection.

Besides early efforts made by Karimganj and Titabar stations for collection of indigenous rice, the first national level venture to collect rice germplsam from NE India was initiated by Dr. R.H. Richharia in 1965 and, consequently, 874 collections were made from Manipur state (Krishnamurthy and Sharma, 1970). A PL-480 project for collection of rice geremplasm for NE India also became operative during 1967-1972 with Dr. M.S. Swaminathan and Dr. S.V.S. Shastry as the Principal Investigators. The programme was, however, executed by Dr. S.D. Sharma and 6630 collections were made from Arunachal Pradesh, Nagaland, Manipur, Meghalaya, Tripura and from North Lakhimpur, Kamrup and Goalpara districts of Assam (Table 16.5). Evaluation of these collections helped in identifying a number of accessions having resistance to specific pests and diseases (Sharma *et al.*, 1971; Sharma, 1982). A collaborative collection programme was undertaken in 1978-1980, involving State Agricultural Universities, State Departments of Agriculture, Indian Council of Agricultural Research (ICAR) centres and the National Bureau of Plant Genetic Resources (Roy *et al.*, 1985) and covered almost all the rice growing states. Under this programme 412 varieties were collected from Assam. Rice germplasm maintained at the Central Rice Research Institute (CRRI), Cuttack contains 5052 accessions from Assam out of 12,256 Indian Rice Collections (Paroda and Sharma, 1986) (Table 16.6).

Table 16.5. Number of landraces of rice collected from different districts of NE India.

State	District	Number of samples	Total
Arunachal Pradesh	Kameng	138	1527
	Subansiri	383	
	Siang	330	

	Luhit	374	
	Tirap	302	
Nagaland	Tuensang	270	849
	Mokokchang	349	
	Kohima	230	
Manipur	Manipur North	70	898
	Manipur East	172	
	Manipur South	109	
	Manipur West	61	
	Manipur Central	486	
Assam (hills)	Mikir hills	518	593
	North Cachar hills	75	
Meghalaya	Garo hills	808	1399
	Khasi & Jayantia hills	548	
	Research station, Shillong	43	
Tripura	Tripura North	109	246
	Tripura South	137	
Assam (Plains)	North Lakhimpur	658	1118
	Sivasagar	20	
	Kamrup	231	
	Goalpara	27	
	Res. Station, Titabar	182	
	Total		**6630**

Source: Sharma *et al.,* 1971.

Table 16.6. Present stock of rice germplasm of NE India maintained at CRRI, Cuttack.

States	No. of accessions
Assam	2054
Arunachal Pradesh	757
Nagaland	354
Manipur	1043

Mizoram	9
Tripura	187
Meghalaya	565
Unspecified	83
Total	**5052**

The current stock of rice germplasm in the rice research stations of the Assam Agricultural University consists of around 4050 accessions, including local varieties, improved strains and introductions (Table 16.7). The actual stock may, however, be smaller, if one takes into account the possibility of duplication of materials (Ganguli 1950; Sharma *et al.*, 1971; IRRI, 1976; Richharia, 1979; Srivastava *et al.*, 1984). In different field trips for the collection of germplasm, the same region is likely to be visited more than ones, giving rise to duplication of the samples in the collection. Studies on glutinous rice germplasm have revealed two types of anomalous collection (i) the same variety bearing different names in different localities (synonymous types) and (ii) different varieties bearing the same name (homonymous types). By considering some important characteristics, such as pigmentation, grain dimension, grain weight, plant height, flowering behaviour and panicle characteristics, in addition to an iodine test for confirmation of glutinous nature, 157 indigenous glutinous rice accessions were reduced to 85 potentially different varieties through elimination of the duplicates (Ahmed, 1986). Such systematic efforts are considered essential for other classes of rice, especially to reduce recurring heavy expenditure involved in maintaining the germplasm in the field every year. Elimination of duplicates is, however, an extremely difficult task because minor genetic differences among the entries may not express as phenotypic difference to be detected by the naked eye. Obviously, unless extreme caution and proper techniques are employed, there are chances of losing useful genotypes as duplicates. Molecular markers may be utilized for detecting the differences between those varieties which bear very close resemblance to one another in order to eliminate duplicates.

The stock of germplasm maintained in the research stations of Assam Agricultural University is far from being full representation of the available diversity of landraces of NE India. It is, therefore, of utmost importance to collect and conserve local germplsm in view of the serious threat of extinction of some of the land races.

Table 16.7. Number of different classes of rice varieties preserved at Assam Agril. University.

Class	Karimganj			Titabar		North Lakhimpur	Diphu	
Ahu			954		219*		24	
	Dumai	37						
	Murali and Chengri	38						
	Ahu	879		219				
Sali			1282**		749			
	Sali			555				
	Lahi			151				
	Joha			43				
Bora			153		198			
	Chokuwa				18			
	Asra		263***		24			
Bao					62	90		
Boro			14		21			
Total			**2666**		**1270**	**90**	**24**	**4050**

* Classification of *ahu* is not prevalent in the Brahmaputra valley
**The number includes *sali, lahi* and *joha*
****Asra* collection at Karimganj includes *Bao* varieties also.

Risk of Genetic Erosion

Reasons which may lead to depeletion of genetic variability in a crop species are of two broad categories - natural calamities and human interventions. Nature, which is regarded as the most potent agent in creating genetic diversity can also cause depletion of it when weather conditions are extreme and such natural phenomena are instrumental in wiping out valuable landraces.

Similarly, human interventions such as war and short sightedness in agricultural approach may also lead to genetic erosion of crop species. In fact, it is apprehended that genetic variability in crop plants has been disappearing at an alarming rate (Frankel and Hawkes, 1975) with sophistication of agriculture. In rice, genetic variability of the region is threatened by persistent propaganda for growing modern high yielding varieties which, having related parentage and being the product of directional selection, have less genetic differences among themselves. Such apprehension has already been reported by Das *et al.* (1981). The danger of losing valuable landraces through replacement by modern varieties is, however, more in easily accessible plain areas of the region. In hilly areas, indigenous varieties are better preserved, to a large extent, due to inaccessibility of the region.

Diversity at Intra-class Level

It has been observed from our studies with indigenous rice collections of Assam maintained at Karimganj and Titabar stations of Assam Agricultural University that variability in any agricultural class of rice, such as *ahu, sali,* etc. is substantial. Further, unlike inter-class diversity which shows discontinuity in certain key characteristics leading to class delineation, the variation in metric characters at intra-class level is continuous. During the early period of rice research in Assam, evaluation of rice germplasm was based on the classification into agricultural classes and their agro- botanical characters and was also classified on the basis of season, duration, land and water requirement, seed size and shape, aroma and endosperm characteristics (Mitra and Ganguli, 1932; Mitra *et al.,* 1942; Choudhury and Ganguli, 1948; Ganguli, 1950). However, overlapping in the range of variation observed during subsequent studies indicated that the basis of original classification had not been rigidly maintained, despite the fact that the names were still prevalent among farmers (Das *et al.*, 1981).

During the 1950s and 1960s, there was a decrease in the effort to collect, evaluate and utilize Assam rice germplasm. However, more recently, several workers from both inside and outside the state studied the germplasm and reported wide diversity and the presence of a valuable gene system in the collections (Sharma *et al.*, 1971; Das and Borthakur, 1973; Srivastava, 1977; Das *et al.,* 1981). The semidwarf varieties were reportd to be present in the Assam rice germplasm and are similar in nature to Taichung Native 1 with regard to their genetic constitution. Das *et al.* (1981) reported two important findings of the studies: availability of semidwarf varieties in traditional *ahu* and *sali* germplasm and availability of varieties with high yield potential in the germplasm collection. It was however observed that indigenous semidwarf *ahu* were earlier in maturity but lower in yield compared to modern semidwarf varieties. Nevertheless, such semidwarf indigenous rice, which owed semidwarfism presumably due to accumulation and fixation of negative polygenes controlling height, may serve as an utilizable source for transmitting this character in varietal development.

Ahu class

Existence of genetic diversity in *ahu* rice was shown by Das and Borthakur

Table 16.8. Variability in three sub-classes of traditional *ahu* of Karimganj.

Class	Sub classes of *ahu*	Mean values of characters						Frequency of varieties under different yield status				No. of accession
		Days to flowering	Plant height (cm)	Panicle number hills^{-1}	Panicle weight (g)	250-grain weight (g)	Yield hill^{-1} (g)	Low (L)	Medium (M)	High (H)	Population of H (%)	
1	2	3	4	5	6	7	8	9	10	11	12	13
Ahu	*Dumai*	75.1 (67-86)	129.8 (104-159)	11.98 (4.7-16.2)	0.79 (0.35 -1.55)	5.00 (4.0-6.0)	9.02 (0.4-13.4)	4	30	7	17.07	41
	Murali-Chengri	80.8 (71-95)	141.5 (116-163)	9.77 (5.4-15.2)	0.94 (0.45-2.14)	4.94 (3.0-5.9)	8.87 (4.5-21.2)	13	76	13	12.75	102
	Ahu or *aus*	87.0 (70-126)	141.60 (88-192)	8.60 (3.7-19.2)	1.106 (0.24-3.03)	5.07 (3.10-8.3)	9.19 (1.8-22.6)	68	364	111	20.44	543
	Tall *ahu*	85.20	144.30	9.02	1.069	4.98	9.24	7	435	124	19.62	632
	Dwarf ahu	87.70	100.70	8.47	0.987	5.75	7.91	12	35	7	12.96	54
	Overall	85.40	140.90	8.98	1.062	5.04	9.13	85	470	131	19.10	686
	Modern Varieties	95.20 (68-112)	101.40 (86-113)	8.64 (4.9-12.9)	1.20 (0.35-2.29)	5.79 (3.2-7.9)	10.13 (3.2-5.4)	10	68	29	27.10	107
Overall for *ahu* germplasm		86.70	135.60	8.93	1.081	5.14	9.27	98	538	160	-	793

Within parentheses are the ranges
Source: Das *et al.*, 1981

(1973) by using D^2 statistic. Later, large scale evaluation was initiated for the first time by Das *et al.* (1981) under transplanted condition taking 793 photoperiod insensitive varieties of Karimganj collection comprising of 686 traditional *ahu* and 107 modern semidwarf varieties on the basis of ten quantitative characters. The varieties were characterized on the basis of three broad category levels (high, medium and low) for each of five key characters and grouped into 96 groups according to common characterization. The mean and range of the key characters and yield show the nature of variation in the three subclasses of traditional *ahu* (Table 8). *Dumai* subclass was clearly distinguished from *ahu* subclass by having shorter duration, shoter height, less panicle weight and relatively more panicle number while *murali-chengri* subclass was generally intermediate for those characters.

Some promising *ahu* varieties identified from the evaluation are presented in Table 16.9. Baruah (1990) made an extensive study of 144 *ahu* varieties of Titabar station with respect to 21 quantitative characters and 24 qualitative characters like pubescence, awnedness, leaf angle, pigmentation, etc. and found significant variation in the characters (Tables 16.10 and 16.11). Using D^2 analysis, Baruah *et al.* (1994) grouped the varieties into eight divergent clusters (Table 16.12). A distinction between *ahu* rice of Barak Valley and that of Brahmaputra Valley was also observed in certain selective crosses with IR8 showing very high sterility when *dumai* type of Barak Valley was used and low sterility when *ahu* rice (cv. Garem) of Brahmaputra Valley was used (unpublished data of Titabar). This, however, needs to be confirmed by extending the observations over several crosses.

Table 16.9. Description of some *ahu* varieties showing excellence in one or more characters.

Accession numbers	Local classification	Characters						Remarks
		DF	HT	PN	PW	TW	YD	
D 208-2	DM	67	113	12.3	0.79	5.3	9.7	Earliness
M 37-2	MC	76	137	9.9	2.14	5.4	21.2	Earliness, yield
AS 417	MV	68	102	10.4	1.47	5.5	15.4	Earliness, yield
D 51-1	DM	74	152	8.7	1.55	6.0	13.4	Earliness, yield
M 135-3	MC	74	143	12.8	1.03	5.3	12.8	Earliness, yield
AS 13-6	AS	74	130	8.7	1.40	6.3	12.1	Earliness, yield
AS 189	AS	82	165	19.2	0.57	4.4	11.0	Tillering
AS 363	AS	81	135	16.0	1.04	5.2	16.7	Tillering, yield

AS 789	AS	116	117	9.2	1.69	7.9	15.5	Heavy grain, yield
AS 24–1	AS	81	148	8.4	1.33	3.4	11.3	Fine grain
ASC 1520-85	AS	90	143	10.8	1.82	3.7	19.6	Fine grain, yield
AS 786	AS	81	150	7.6	2.96	5.7	22.6	Heavy panicle, yield
AS 12-2	AS	82	166	6.8	3.03	4.9	20.5	Heavy panicle, yield
AS 374	MV	80	99	9.6	1.78	5.5	17.1	Yield
AS 234	AS	80	147	11.1	1.50	5.1	16.7	Yield

DF: Days to 50% flowering (days) HT: Plant height (cm)
PN: Panicle number plant $^{-1}$ PW: Grain weight plant^{-1} (g)
TW: 250 grain weight (g) YD: Yield plant^{-1}(g)
D: *Dumai*; M: *Murali* C: *Chengri*; AS : *Ahu* or *Aus*
MV: Modern variety ASC: *Ahu* cross.
Source: Das *et al.*, 1981

Table 16.10. Mean, range, co-efficient of variation of *ahu* germplasm of Titabar collection.

Characters	Mean	Range	Co-efficient of variation (%)
Days to flowering (days)	90.33	66-117	8.85
Days to maturity (days)	118.84	110-144	6.47
Plant height (cm)	126.33	89-156	9.26
Panicle number plant $^{-1}$	9.96	5.9-19.2	26.17
Panicle length (cm)	22.95	17.7-28.5	8.29
Spikelet number	113.56	55-241	32.63
Panicle weight (g)	1.71	0.7-3.3	31.42
Spikelet density	4.91	2.6-9.8	28.39
Fertility percent (%)	79.04	48.0-97.9	11.03
Flag leaf length (cm)	31.28	20.7-43.6	11.41
Flag leaf breadth (cm)	1.24	0.7-2.3	20.73
Flag leaf weight (mg/cm^2)	6.90	3.2-16.0	29.43
Grain length (mm)	7.71	6.0-8.7	6.32
Grain breadth (mm)	3.01	1.9-3.7	8.93
Grain length: Grain breadth	2.60	2.0-4.5	12.83
Kernel length (mm)	5.59	4.2-6.4	7.24

Kernel breadth (mm)	2.53	1.6-3.0	8.21
Kernel length: Kernel Breadth	2.25	1.6-4.0	13.23
250- grain weight (g)	5.21	3.3-7.10	11.36
Yield plant $^{-1}$(g)	8.97	2.0-18.0	33.68

Source: Baruah, 1990

Table 16.11. Variation in qualitative characters (discrete variables) in *ahu* germplasm of Titabar station.

Characters	Proportion of different categories
Pubescene	0 pc glabrous, 26 pc intermediate, 74 pc pubescent
Leaf blade colour	1 pc pale green, 51 pc green, 40 pc dark green, 2 pc purple margin, 3 pc purple blotch, 3 pc purple
Basal leaf sheath colour	90 pc green, 6 pc light purple, 4 pc purple
Flag leaf angle	3 pc erect, 74 pc intermediate, 23 pc horizontal
Ligule colour	99 pc white, 1 pc purple lines
Ligule length	84 pc short (below 19 mm), 16 pc medium (between 19 to 34 mm)
Ligule structure	100 pc thick, 2 cleft
Collar colour	99 pc pale green, 1 pc purple
Auricle colour	99 pc pale green, 1 pc purple
Auricle structure	42 pc small, 52 pc large
Culm angle	5 pc erect, 12 pc intermediate, 27 pc open, 56 pc spreading
Internode colour	93 pc green, 2 pc purple lines, 5 pc purple
Panicle type	19 pc compact, 29 pc intermediate, 52 pc open
Panicle branching	99 pc heavy, 1 pc clustering
Panicle axis	98 pc straight, 2 pc droopy
Awnedness	63 pc awnless, 26 pc short partially awned, 7 pc long partially awned, 4 pc long fully awned
Awn colour	63 pc colourless, 31 pc straw, 1 pc golden, 1 pc brown, 1 pc purple, 3 pc black
Apiculus colour	7 pc white, 64 pc straw, 19 pc purple, 10 pc purple tip
Stigma colour	84 pc white, 8 pc light purple, 8 pc purple
Lemma palea colour	30 pc straw, 10 pc gold furrows, 10 pc brown spots, 20 pc brown furrows, 8 pc brown, 1 pc light purple, 1 pc purple spots, 9 pc

	purple furrows, 1 pc purple, 10 pc black
Sterile lemma colour	89 pc straw, 2 pc red, 9 pc purple
Septum colour	6 pc white , 2 pc pink, 5 pc light green, 4 pc green, 39 pc light yellow, 15 pc yellow, 3 pc red, 3 pc light purple, 21 pc purple, 2 pc black
Kernel colour	17 pc white, 20 pc light red, 52 pc red, 11 pc deep red
Grain pubescence	100 pc long velvety hairs

pc = per cent
Source: Baruah, 1990

Table 16.12. Cluster means for 7 quantitative characters of *ahu* germplasm of Titabar collections.

Cluster	Days to flowering	Plant height (cm)	Panicle number	Spikelet density	Panicle weight (g)	250 grain weight (g)	Yield hill^{-1}(g)
I	90.5	126.5	8.0	5.47	1.89	4.57	8.11
II	90.6	**140.0**	7.0	5.71	**2.36**	5.27	8.83
III	**80.9**	116.1	9.1	4.72	1.79	5.23	8.90
IV	88.0	117.5	9.0	**3.22**	1.48	**6.16**	9.32
V	101.3	134.6	**5.7**	4.37	**1.25**	**4.11**	**2.60**
VI	106.0	112.3	8.7	**6.66**	2.02	5.89	9.61
VII	**107.0**	**94.2**	**11.3**	5.26	1.33	4.85	**11.82**
VIII	89.7	106.2	**11.3**	3.66	1.34	4.74	9.69

*Bold figures are highest and lowest mean values for a particular character

Source: Baruah *et al.,* 1994

Sali class

Traditionally *sali* rice is grouped into three subclasses, viz., *sali, lahi* and *joha*, according mainly to fineness or coarseness of grains. Srivastava (1978), however, classified *sali* rice into three groups, viz., thick culm, thin culm and beaded grain groups. In a large scale evaluation, Das *et al.* (1983) characterized 978 *sali* rice strains of Karimganj station into 82 groups following the same methodology that they used for *ahu* rice and found that the variation was the maximum for panicle weight and least for duration and plant height (Table 16.13). The varieties were, by and large, photoperiod sensitive and flowered between October 5 and November 16 (114 to 156 days from sowing). However, second fortnight of October appeared to be optimum flowering time for this class of rice. The yield potential of *sali* rice was much higher on an average compared to *ahu* rice. Of 978 *sali* strains evaluated, 246 strains (25.15%) recorded yield of 4-6 t/ha and 35 strains (3.58%) registered yield exceeding 6 t/ha. Some promising strains showing excellance in one or more characters are given in Table 16.14. In an extensive investigation with 547 *sali* rice collections of Titabar station, Sarmah (1990) reported considerable variation among the varieties in respect of 15 quantitative and 11 qualitative characters like culm angle, leaf angle, awnedness and pigmentation in different plant parts (Tables 16.15 and 16.16).

Investigations revealed that apart from possessing photoperiod sensitiveness and long duration, the *sali* class differs from *ahu* class in having greater variability in grain size and shape, greater proportion of varieties with erect and intermediate culm angle and flag leaf angle, higher proportion of awnless varieties and lower proportion of pigmented varieties and also in having aromatic strains. Moreover, the presence of semidwarf varieties in both *ahu* and *sali* classes is of great practical significance.

Table 16.13. Mean and variability of six quantitative characters including yield of *sali* rice germplsam of Karimganj collection.

Charcters	Mean	Range	Variance	CV %
Days to flower (days)	137	114-156	9.2438	2.22
Plant height (cm)	145	90-187	25.8589	3.51
Panicle No. hill $^{-1}$	8.01	2.9-15.0	5.2779	28.68
Panicle weight (g)	1.679	0.174-3.98	1.3957	70.36
250 grain weight (g)	6.32	2.0-9.6	4.3571	33.03
Yield hill $^{-1}$ (g)	13.237	1.11-45.00	10.1356	24.05

Source: Das *et al.*, 1983

Table 16.14. Character manifestation of first 15 best yielding *sali* varieties.

Designations	Days to 50% flowering	Plant height (cm)	Panicle number	Panicle weight (g)	250 grain weight (g)	Yield plant^{-1} (g)
Sc 1834-14	135	155	15.8	2.00	6.7	30.00
Sc 1834-16	139	149	12.6	2.43	7.9	30.50
Sc 1834-82	138	155	10.8	2.93	8.1	31.68
Sc 1848-20	136	157	12.4	2.42	7.5	30.00
Sc 1848-24	137	162	12.1	2.75	7.4	33.33
Sc 1853-28	136	150	12.2	2.46	8.0	30.00
Sc 1860-21	137	151	10.7	2.80	8.2	30.00
Sc 1861-10	135	161	10.9	2.75	7.3	30.00
Sc 1863-10	139	156	11.9	2.85	7.2	33.89
Sc 1863-11	139	150	11.8	2.78	7.0	32.70
Sc 1863-12	138	176	11.3	3.98	7.3	45.00
Sc 1863-31	135	157	12.9	2.33	8.1	30.00
Sc 1863-38	140	151	9.4	3.31	7.5	31.11
Sc 1863-39	137	153	10.0	3.17	7.7	31.67
Sc 1863-44	140	162	9.9	3.26	7.7	32.22

Source: Das *et al*, 1983

Table 16.15. Mean, range, co-efficient of variation (cv) of *sali* germplasm of Titabar collection.

Characters	Mean	Range	CV (%)
Days to flowering (days)	126	107-142	4.6
Plant height (cm)	147.88	111-195	8.4
Panicle number hill^{-1}	7.52	2.4-15.4	30.1
Panicle length (cm)	26.84	21.5-34.1	9.3
Spikelet No. panicle^{-1}	172.15	81-384	27.6
Panicle weight (g)	1.85	0.43-7.08	42.1
Spikelet density	6.42	3.4-15.3	26.6
Fertility percent (%)	79.06	27.9-96.7	15.2

Flag leaf length (cm)	35.23	19.2-55.8	15.2
Flag leaf breadth (cm)	1.35	0.7-2.9	20.7
Kernel length (mm)	6.06	3.8-8.0	12.5
Kernel breadth (mm)	2.57	1.7-3.3	12.6
Kernel length : kernel breadth	2.59	1.4-4.3	17.9
250 grain weight (g)	5.84	2.0-9.4	19.8
Yield hill $^{-1}$ (g)	12.85	3.3-25.2	30.8

Source: Sarmah, 1990

Table 16.16. Variation in qualitative characters (discrete variables) in *sali* germplasm of Titabar station.

Characters	Proportion of different categories
Culm angle	22 pc erect, 47 pc intermediate, 30 pc open, 1 pc spreading
Flag leaf angle	21pc erect, 67 pc intermediate, 12 pc horizontal
Awnedness	84 pc awnless, 8 pc short partially awned, 1 pc short fully awned, 4 pc long partially awned, 3 pc long fully awned
Basal leaf sheath	87 pc green, 12 pc light purple, 1 pc purple
Septum colour	67 pc white, 21 pc yellow, 11 pc pink, 1 pc purple
Stigma colour	91 pc white, 9 pc purple
Apiculus colour	65 pc straw, 1 pc brown, 28 pc purple, 6 pc purple tip
Lemma palea colour	60 pc straw, 8 pc golden yellow and golden furrows, 1 pc brown spots, 2 pc brown furrows, 10 pc brown, 2 pc reddish to light purple, 9 pc purple spots, 2 pc purple furrows, 6 pc purple
Sterile lemma colour	77 pc straw, 1 pc golden, 22 pc purple
Kernel colour	93 pc white, 7 pc red
Aroma	8 pc scented, 92 pc non scented

pc: per cent

Source: Sarmah, 1990

Joha class

Joha is a special class of *sali* rice in Assam characterized by aromatic endosperm. *Joha* varieties are mostly short, medium grained and have very low yield potential. However, this class of aromatic rice differs from Basmati rice grown elsewhere in the country in terms of grain type and elongation after cooking. But *joha* rices possess strong aroma, hence there is possibility of exploring the market in the Middle East countries where aromatic rices are preferred. Forty two indigenous *joha* rices were studied for their important quantitative and qualitative characters by Sarmah (1990) and substantial variability was observed (Table 16.17). Apart from their photoperiod sensitivity and longer growth duration, wide variation in grain size is a special feature of this class of rice.

Table 16.17. Mean and range of *joha* germplasm of Titabar.

Characters	Mean	Range
Days to flowering (days)	126	121-135
Plant height (cm)	151	131-165
Panicle number hill^{-1}	10.2	3.4-13.2
Panicle length (cm)	24.4	21.5-31.1
Spikelet number panicle $^{-1}$	153	92-243
Panicle weight (g)	1.0	0.51-3.2
Spikelet density	6.26	3.9-10.2
Fertility percent (%)	83.96	64.5-96.1
Flag leaf length (cm)	28.48	19.2-39-5
Flag leaf breadth (cm)	1.17	0.7-2.1
Kernel length (mm)	5.48	4.0-7.1
Kernel breadth (mm)	2.07	1.7-2.7
Kernel length : kernel breadth	2.66	1.6-3.4
250 grain weight (g)	3.59	2.0-6.7
Yield hill $^{-1}$ (g)	9.83	5.0-13.6

Bora and Chokuwa class

Bora and *chokuwa* are two special categories within *sali* rice in Assam, differeing in their endosperm characteristics. *Bora* or glutinous rice has high amylopectin which confers opaqueness, stickiness and waxiness to the endosperm. Mitra and Ganguli (1932) separated glutinous rice from the entire indigenous rice collection of Assam on the basis of opaqueness of kernel. However, glutinous rice varieties of Assam as well as advanced glutinous lines were found to have opaque kernels without exception, the reverse was not necessarily true as revealed by iodine test (Ahmed and Das, 1990). Kawahara and Motogi (1951) reported variation from translucent to opaque kernels in 'Tohoku' glutinous rice. Anthocyanin pigmentation in one or more plant parts is most common in *bora* rice. The most frequent combination of pigmented parts invariably included purple apiculus and inner side of leaf sheath and least frequent ones included stigma and ligule (Das and Ahmed, 1982). Ahmed (1986) reported that the waxy locus is linked with pigmentation in 10 different plant parts with varying crossover percentages.

On the basis of pigment distribution, 85 glutinous rice germplasm was grouped into 47 groups (Ahmed and Das, 1994). These varieties were broadly characterized into 52 groups based on high and low status of eight quantitative characters including yield and presence or absence of kernel colour (Table 16.18). Only 25 varieties in this collection could show high yield status based on either grain weight per panicle or panicle number per plant but never both. Existence of genetic diversity of *bora* rice was studied by Ahmed and Bora (1999) by using D^2 statistic. The varieties were grouped into 12 clusters based on 13 agronomic characters. Whereas a maximum of 10 genotypes each were placed in Cluster X and XII, a minimum of three genotypes were confined in Cluster I. Tiller number, panicles per hill, grains per panicle, grain fertility and grain yield accounted for major portion of divergence (Tables 16.19 and 16.20). The extent of variation in quantitative characters in *bora* and *chokuwa*, however, appears to be somewhat less than that of *sali* rice (Ahmed, 1986; Baruah, 1990). Moreover, specialized morphological characteristics such as clustered grain, long sterile lemma are also found in certain *bora* varieties.

Compared with *bora* rice, amylose content of *chokuwa* rice is higher whereas amylopectin content is lower. Most of the *chokuwa* varieties do not become sticky on cooking but become soft in consistency. Dutta and Barua (1980) showed that starch percent of dehusked rice (oven dried) varied from 75.5 to 80.7 for *sali* class, 77.8 to 78.7 for *chokuwa* and 73.3 to 73.5 for *bora* class; amylose/100 g of starch ranged for 28.15 to 31.97 for *sali,* 17.14 to 19.0 for *chokuwa* and trace for *bora;* and crude fibre percent ranged from 0.74 to 0.78 for *sali* and 1.18 to 1.22 for *bora* and chokuwa.

Table 16.18. Mean and variability of important characters of glutinous rice.

Characters	Mean	Range	CV%	Heritability (h^2)
Yield (kg/ha^{-1})	1741	517-3133	1.24	0.66
Days to flowering (days)	115	107-129	5.0	0.99
Days to maturity (days)	143	133-156	4.22	0.99
Plant height (cm)	133	119-156	6.0	0.83

Tiller number hill^{-1}	6.65	3.80-12.20	31.0	0.95
Panicle number hill^{-1}	5.73	2.80-11.8	27.0	0.62
Panicle weight (g)	1.27	0.55-2.84	36.0	0.91
Panicle length (cm)	22.97	19.40-28.30	8.00	0.74
No. of grains panicle^{-1}	70	33-106	22.0	0.64
Sterility percentage (%)	25.93	6.70-56.85	37.82	0.59
250-grain weight (g)	6.19	3.60-8.20	10.00	0.98
Spikelet density	4.15	2.21-8.98	22.73	0.82
Grain breadth (cm)	0.31	0.25-0.37	10.0	0.90
Grain length (cm)	0.89	0.78-1.10	19.0	0.88
Length : Breadth	2.89	2.07-3.79	18.62	0.99
Flag leaf length (cm)	37.03	26.80-50.60	13.0	0.76
Flag leaf breadth (cm)	1.58	1.10-2.30	13.0	0.67
Penultimate leaf length (cm)	57.61	45.0-47.60	9.0	0.66
Penultimate leaf breadth (cm)	1.31	0.90-2.0	17.0	0.83

Source: Ahmed and Das, 1994.

Table 16.19 . Composition of clusters among 85 genotypes of glutinous rice.

Cluster	No. of genotypes	Genotypes included
I	3	Rajloxmi bora, Tita Phulia bora, Poita bora 2
II	9	Chepa bora, Tepra bora, Jhonki bora, Uba bora, Hati bora, Baruital bora, Nodi bora, Khaldhar bora, Mokli bora
III	6	Mow bora 1, Kacha bora, Moglai bora, B 87, Uki bora, Tangan bora
IV	9	Hatipar bora, Dang bora, Shona bora, Malagkit, Haldi bora, Shankha bora, Helochi bora,Ahu bora, Tipera bora
V	5	Kala bora, Jengoni bora, Pakhori bora, Mow bora 2, Kharika bora
VI	7	Gandhi bora, Madhu bora, Khaki bora, Mow bora 4, Kharika bora, Tipera bora, Mokli bora
VII	7	Sauduung bora, Bakul bora, Bor bora, Hupori bora, Kakua bora, Santiboka bora, Til bora
VIII	8	Agisjal bora, Pakhura bora, Hupani bora, Bhat bora, Rupohi bora 1, Pakhia bora, Gela bora, Noga bora

IX	4	Chakua bora, Bormalbhog bora, Rupohi bora 2, Joha bora
X	10	Puti bora, Jangon bora, B 92, Tulashi bora, Rajat bora, B 113, Poita bora 1, Narkuljapa bora, Naringatu bora, Thepa bora
XI	7	Gandhi bora 2, Khagra bora, Mow bora 3, Charainekhia bora, Kajoli bora, Ghew bora, Boga bora
XII	10	Sylhet bora, Puspa bora, Lal bora 1, Lal bora 2, Kachulatu bora, Agarsali bora, Bct 21/C-69, Jeori bora, Misiri bora, Batra bora

Source: Ahmed and Bora, 1999

Asra and Bao classes

Asra may be considered as an ecotypic sub-group of *sali* rice with adaptation to lowland and semi-deep water condition. Without exception the varieties of this group are tall, thick culmed, leafy, coarse-grained and non-scented but variability exists for other characters including adaptability to different field water levels and tolerance to submergence. *Bao* varieties are adapted to deep water situation and have tall stature and non-scented coarse grains but the varieties show variation in submergence tolerance, elongation ability, kneeing ability and rooting ability. Evaluating 60 medium to lowland rice genotypes including 34 *sali*, 21 *bao* (deep water) and five modern varieties for 15 morpho-physiological characteristics, N.K. Sarma (1989) reported significant variability for all the characters studied including seedling submergence tolerance and kneeing ability. 'Asra', 'Bao 55' and 'Kanti' were identified promising for seedling vigour, submergence tolerance and kneeing ability with average stability in yield per plant.

Boro class

This class is supposed to be an ecotypic variant of *ahu* class with adaptation to winter temperature in subtropical region. The varieties are almost invariably awned, mostly red kernelled and reasonably tall to resist water stagnation due to pre-monsoon showers. Glazmann (1986) grouped at least some of them with *ahu* rice under the same enzymatic category.

Hill rice

It represents a group of rice varieties adapted to high altitude conditions. The varieties are cold tolerant, mostly semidwarf with *japonica* grain type. Wide variation is, however, observed in grain size and shape and for amylose content. The group is also a rich source of economically important genes and gene systems.

Table 16.20. Cluster means for thirteen characters in 85 glutinous rice germplasm.

Characters	Characters means in different clusters											
	I	II	III	IV	V	VI	VII	VIII	IX	X	XI	XII
Days to flowering (days)	112.7	**125.9**	116.1	114.4	110.7	119.0	**109.0**	112.7	111.2	113.3	114.6	115.1
Plant height (cm)	125.0	130.0	**140.5**	135.0	128.6	123.2	139.0	131.1	132.5	125.5	137.9	**114.0**
Tiller number	5.49	5.23	**4.66**	5.00	6.77	**10.60**	7.71	8.06	6.74	5.81	5.27	7.51
Panicle hill^{-1}	**3.63**	5.09	4.22	4.43	4.27	9.72	6.72	**7.03**	5.33	5.12	4.61	6.84
Panicle length (cm)	23.49	**20.37**	24.24	22.02	23.83	21.57	23.78	23.38	**26.63**	22.82	24.57	22.62
Grain panicle^{-1}	45.33	**95.11**	91.33	69.04	59.27	**43.67**	67.76	53.88	81.33	73.57	82.24	64.93
Grain sterility (%)	**54.84**	**89.70**	72.39	72.39	56.06	75.12	81.06	72.59	62.10	76.44	75.37	73.25
250- grain weight (g)	6.56	6.18	**6.94**	6.75	5.55	6.51	6.87	5.79	5.78	6.56	5.89	5.37
Panicle weight (g)	**0.69**	**2.13**	1.88	1.42	0.79	0.82	1.32	0.77	1.24	1.18	1.44	1.05
Grain length (mm)	**1.01**	**0.82**	0.85	0.95	0.85	0.97	0.93	0.85	**0.82**	0.87	0.96	0.88
Grain breadth (mm)	0.30	0.32	**0.34**	0.31	0.32	**0.27**	0.32	0.31	0.33	0.33	**0.27**	**0.27**
Length : Breadth ratio	3.31	2.54	**2.48**	3.10	2.66	**3.65**	2.88	2.72	2.51	2.63	3.51	3.29
Grain yield hill^{-1}(g)	**2.60**	**10.76**	8.02	6.16	3.40	8.02	8.83	5.45	6.62	5.99	6.64	7.23

Bold figures are highest and lowest mean value for a particular character
Source: Ahmed and Bora, 1999

Special Morpho-physiological Characteristics

Existence of varieties with diverse morpho-physiological peculiarities and extreme expression of certain traits (Table 16.21) illustrate further the type of genetic variousness and potential available in the rice germplasm of the region. Apart from rare or less common traits like double grained rice, clustered spikelets, beaked lemma, long sterile lemma and glutinous rice with coarse, scented grains, there are also varieties with economically important characteristics such as high protein content, high lysine content, high hulling recovery, high spikelet density and semidwarf stature. Das *et al.* (1981, 1983) reported that some semidwarf accessions from *ahu* and *sali* groups were derivatives of crosses between indigenous tall varieties and semidwarfism in those accessions were presumed to be due to the accumulation of negative polygenes. A new recessive gene for semi-dwarfism was discovered in Ngoba (a semidwarf variety from Meghelaya) and it was found to be complementary non-allelic to the semidwarf gene in Jaya (Ghosh *et al.*, 1981). It was reported by Seshu *et al.* (1974) that two good yielding Assam dwarfs, Khagaribiroin and Haldiram, owed their dwarfism to a new gene system comprising a hypostatic tall gene and a dominant inhibitor gene.

Table 16.21. Rice varieties of NE Region with special morpho-physiological characteristics.

Characters	Varieties	Reference
Clustered grains	Chidon, Charimohia ahu 2	RARS, Titabar
Double grain	S- 555	RARS, Karimganj
Beaked lemma	Sorainakhia Chokuwa	Baruah, 1975
Long sterile lemma	Akiab biroin, Gorur Pakhi	RARS, Karimganj
	Necklace bora	RARS, Titabar
Fine slender grain in ahu (L=83 mm, B=1.9 mm)	Basanta Bahar	RARS, Karimganj
Very small grain (1000-grain wt: 8-8.8g)	Arab joha, Maniki madhuri joha	Sarmah, 1990
Very heavy grain (1000-grain wt. above 40 g)	Malbhog bora	Baruah, 1975
Stout pointed awns	Sorokphola	RARS, Titabar
High hulling recovery (above 80 pc)	Badshabhog	CRRI, 1971
High spikelet density (10-15 grains per cm)	Bordhan, Hatibondha sali, Nepali Tulsi, Arab Joha, Badshabhog	Sarmah, 1990

High protein content (11.5 pc to 14.0 pc)	Garem, Hati sali, Khalaru, Khurupi sali, Khali sali, Malbhog, Bogamanohar	Govindaswamy *et al.*,1973 Kondoli, 1992
High lysine content	ARC 10525	IRRI, 1974
High aromatic oil content	Kola joha 2, Krishnabhog	Ahmed *et al.*, 1995
Glutinous type with coarse scented grains	Mnp 53, Mnp 90, Mnp 93, Mnp 94, Mnp 99	CRRI, 1971,
	Gandhi bora, Joha bora	Ahmed, 1986
Very hard rice	Dolkuchi, Zengoni	RARS, Titabar
Promising fertility restorer	ARC 11353	Ramalingam *et al.*, 1994
Semidwarf	Sabit, Theki, K. Chikari-34, Khagori biroin, Maizumgajaw, Haldiram, Bukongsei, Tinfooti, Sencher, Batdong, Ngoba,	Seshu *et. al.*, 1974
	Khawehaku, Muzulo, Akughu Khoghi	Seshu *et al.*, 1974
	ARC 6160	CRRI, 1978
	ARC 5901, ARC 12422	Srivastava, 1978
	Baki sali, Beti sali, Kulki sali	Sarmah, 1990

Resistance to Biotic and Abiotic Stress

Rice varieties of the region provide a wealth of diverse genes for resistance to pests and diseases, flood, drought and various other stresses (Tables 16.22, 16.23 and 16.24). New sources of resistance to pests and diseases including blast have been identified in a large number of ARC collections and the resistant types have been reported to be in greater frequency in ARC than in other stocks of germplasm (Shastry *et al.,* 1971; Krishnamurty *et al.,* 1975; Devdath and Rao, 1976; Chatterjee *et al.,* 1977) (Table 16.25). Non-allelic genes for resistant to blast were reported by Mathur and Srivastava (1981). It has also been reported that some of the Assam dwarfs have the ability to withstand moisture stress and some of them have also multiple resistance to various pests and diseases (Seshu *et al.,* 1974). For example, semi-dwarf varieties such as Khagaribiroin and Haldiram have sufficient tolerance to bacterial leaf blight, tungro virus, stem borer, gall midge and green leafhopper.

Table 16.22. Possible sources for some important traits.

Traits	Sources
Resistance to different pests and diseases	All agricultural classes of rice germplasm in the state
Drought tolerance and recovery at vegetative stage	Dry seeded upland *ahu* germplasm and the varieties grown in the hill districts in *jhum* cultivation
Cold tolerance at vegetative phases	*Boro* germplasm
Low light stress tolerance	All agricultural classes of rice germplasm in the state
Adaptation to flash flood submergence and waterlogged situations	*Sali* and *asra* germplasm
Adaptation to deepwater situations	*Bao* germplasm
Tolerance of phosphatic deficiency and aluminium toxicity	Direct seeded upland *ahu* rice germplasm
Tolerance of iron toxicity	Transplanted *ahu* and *sali* germplasm
Earliness	*Dumai* group of *ahu* germplasm
Non aromatic high quality grain	Fine grain *sali* germplasm (*lahi* rice)
Aromatic quality grain	Scented germplasm (*joha* rice)

Table 16.23. Number of rice geremplasm samples resistant to different pests and diseases in Assam.

Pest/diseases	Hills			Plains		
	T	R	%	T	R	%
Stem borer	120	10	8	167	51	30
Gall midge	120	4	3	163	10	6
Green leaf hopper	110	9	8	224	60	27
Blast	132	5	4	110	4	4
Bacterial leaf blight	232	0	0	311	12	4
Rice tungro virus	61	24	40	7	1	14

T = Total number of collection
R = Total number of resistant types
% = percentage of resistance
Source: Sharma, 1987

Table 16.24. Rice varieties of NE India with resistance to biotic and abiotic stresses.

Stress condition	Variety/ARC number	Reference
1	2	3
1. Bacterial blight (BLB)	ARC 5767, ARC 6063 (Zengoni bora), ARC 6175 (Laodhumra), ARC 6605 (Boradhan), ARC 10156 (Manila joha), ARC 10204 (joha), ARC 10266 (joha), ACR 10435 (Bordhan), ARC 10539 (Mangsang Malbok) ARC 10969 (Soru Mekuri), ARC 11257 (Jangira), ARC 11264 (Malsira)	AICRIP, 1971
	ARC 6003, ARC 6009, ARC 7260, ARC 7327, ARC 7328, ARC 10516	IRRI ,1974
	ARC 5758, ARC 5793, ARC 6068, ARC 6109, ARC 6230, ARC 7090, ARC 7303, ARC 7347, ARC 10520, ARC 10945, ARC 10959	CRRI, 1976
	Kataribhog	CRRI, 1978
	ARC 5756, Kolongi bao	Sidhu *et al.,* 1978
	ARC 5827, ARC 6023, ARC 6135, ARC 7330, ARC 10215, ARC 10272, ARC 10424, ARC 10610, ARC 10908, ARC 10981, ARC 11300, ARC 11323, ARC 12057, ARC 12615, ARC 12637, ARC 14377, ARC 14387, ARC 14396, ARC 14723, ARC 14926, ARC 15399, ARC C18146, ARC 18562	Devadath and Rao, 1978
	ARC 7325, ARC 11094 (resistant to 5 races)	IRRI, 1984
	Manohar sali	AAU, 1983
	Dholamula, Moina sali, Kartik kolma, Japori sali, Jati sali, Gajep sali	Rathaiah *et al.,* 1986
	Begunbichi-2, Ronga ahu, Rongagutia ahu, Rikhojoi-2, Kolimekuri, Solpona, Sailbadal, Ahu şuri, AS 330, Kola ahu	Bora, 1992
	Kuntalan, Khachamota, Gogyeku	Ali *et al.*, 1995
2. Bacterial leaf streak	ARC 5827, ARC 6023, ARC 6135, ARC 7327, ARC 7328, ARC 7330, ARC 10272, ARC 12637, ARC 14377, ARC 14387, ARC 14396, ARC 14723, ARC 14926, ARC 15399, ARC 18146, ARC 18562	Devadath and Rao, 1978
3. Blast	ARC 10622	AICRIP, 1972
	ARC 13840	CRRI, 1978
	Kalamdani, Sial sali, Terabali, Mugi sali, Madri sali, China 63, Solpona, Hybrid 14	Rathaiah and Das, 1987

	ARC 10372	DRR, 1987
	AS 36, AS 152, AS 313, Ronga ahu, Rongagutia ahu, Dusri ahu, Padumani ahu	Borah *et al.*, 1994(a)
	Kalamdani 3, Agis sali	Borah *et al.*, 1994(b)
	Kolimekuri, Andrew sali, Manohar sali, Gajep sali	Bhattacharjee *et al.*, 1995
	Yumuk, Korpu, Balilife, VL – 8	Ali *et al.*, 1995
	ARC 11353	McWilliam *et al.*, 1995
4. Neck blast	Ikora sali, Yabor, Kesa plu, Ame	AAU, 1983; Ali *et al.*, 1995
	Banki sali, Kolimekuri, Andrew sali, Manohar sali	Bhattacharjee *et al.*, 1995
5. Sheath blight	ARC 15762, ARC 18119, ARC 18275, ARC 18545	CRRI, 1977
	Manohar sali, Terabali-1, Joporasali, Dholamula, Supkheru	AAU, 1983
	ARC 10315, ARC 10378, ARC 10600, ARC 11214, ARC 11220, ARC 11262, ARC 11274, ARC 11275, ARC 11283, ARC 11284, ARC 11292, ARC 11329	DRR, 1997
	Behari sali	Borah *et al.*, 1994 (b)
6. Sheath rot	ARC 14727, Negheri bao	CRRI, 1974
	Andrew sali	Ali *et al*, 1995
7. Leaf scald	ARC 6563, ARC 10654	IRRI, 1979
8. Rice Tungro virus (RTV)	ARC 5981 A	Shastry *et al.*, 1971
	ARC 7054	CRRI, 1973
	ARC 6202, ARC 7318, ARC 10342, ARC 10531, ARC 10980, ARC 13515, ARC 13560, ARC 13677, ARC 13754, ARC 13771, ARC 13804, ARC 13820, ARC 13829, ARC 13901, ARC 13925, ARC 13943, ARC13959, ARC 14787, Kataribhog	CRRI, 1975
	Lati sali, ARC 7125, ARC 7140	CRRI, 1976
	Manohar sali	CRRI, 1977
	ARC 11554	IRRI, 1984
9. Ragged stunt virus (RSV)	ARC 11206, ARC 11280, ARC 11921, ARC 12629, ARC 12632, ARC 13363	IRRI, 1987

10. White backed plant hopper (WBPH)	ARC 5803, ARC 6064, ARC 7138, ARC 7318, ARC 10340	CRRI ,1973
	ARC 14539B, ARC 14766A	CRRI, 1976
	ARC 6003	IRRI, 1977
	ARC 6564, ARC 6650, ARC 13349, ARC 14342A,	IRRI, 1978
	ARC 10239	IRRI, 1979
	ARC 6632, ARC 15382, ARC 18024, ARC 18037	CRRI,1979
	ARC 5752	IRRI, 1984
	Andrew sali	DRR, 1997
11. Stem borer (SB)	ARC 5929, ARC 5981A, ARC 11229	AICRIP, 1971
	ARC 10217, ARC 10443	CRRI, 1972
	ARC 6158, ARC 10379, ARC 10386	CRRI, 1973
	ARC 10451, ARC 10979, ARC 11150, ARC 15221	CRRI, 1974
	ARC 5984	DRR, 1987
	Kataribhog	Baruah and Das, 1995
12. Yellow stem borer (YSB)	ARC 5500, Manohar sali	DRR, 1997
13. Gall midge (GM)	ARC 12619, ARC 12627, ARC 12774	CRRI, 1973
	ARC 5959, ARC 5987, ARC 6632, ARC 13500, ARC 13516, ARC 13564, ARC 13902, ARC 13929, ARC 14148, ARC 14378, ARC 14421, ARC 14549, ARC 14725, ARC 14748, ARC 14787, ARC 15067, ARC 15151, ARC 15159, ARC 18596, ARC 18601, ARC 15905, ARC 17789, ARC 18595	CRRI, 1974
	ARC 5929	AICRIP, 1976
	ARC 6157, ARC 6221	Shobharani *et al.*, 1992
	ARC 5981, ARC 5984, ARC 10654	Rana *et. el.*, 1994
	ARC 6605, ARC 10660	DRR, 1997
14. Brown plant hopper (BPH)	ARC 5772, ARC 7229	CRRI, 1973
	ARC 5918, ARC 5984, ARC 6079, ARC 6650, ARC 7289, ARC 7313, ARC 7328, ARC 11255, ARC 14342, ARC 14529, ARC 14729, ARC 14768, ARC 15223, ARC 15284, ARC 15378, ARC 15821, ARC 18095, ARC 18529, Manohar sali	CRRI, 1975

	MNP 76, ARC 10176, ARC 12627, ARC 14524, ARC 14736, ARC 14766A, ARC 14768, ARC 14810, ARC 15831	CRRI, 1976
	ARC 13506, ARC 13530, ARC 13536, ARC 13556, ARC 13872, ARC 13909	CRRI, 1978
	ARC 6632, ARC 15382, ARC 18024, ARC 18037	CRRI, 1979
	ARC 1583A, ARC 6564, ARC 10550, ARC 10572C, ARC 10945, ARC 11570C, ARC 11857, ARC 14394, ARC 14539B, ARC 15694, ARC 15694A	Rao *et al.,* 1981
	ARC 6858, ARC 10659, ARC 10660, ARC 10666, ARC 11321, Andrew sali	DRR, 1987
	Balam	Pathak *et al.*, 1995
	ARC 7080	DRR, 1997
15. Green leaf hopper (GLH)	ARC 5929, ARC 5981A, ARC 6030, ARC 6049, ARC 6176, ARC 10376, ARC 11226, ARC 11229	AICRIP, 1971
	ARC 12619, ARC 12627, ARC 12774	CRRI, 1973
	ARC 6602,	IRRI 1976
	ARC 13554,	IRRI, 1984
	Badal, ARC 10313	Velusamy *et al.,* 1994
	ARC 6606	DRR, 1997
16. Rice hispa	ARC 7080	DRR, 1987
	AS 36, AS 36/14, AS 36/20, AS 48, AS 75, AS 93/1, Bizor 3, Begunbichi, Garem	Dutta and Hazarika, 1994
17. Leaf folder	ARC 10932	IRRI, 1978
	ARC 6650, ARC 10650	IRRI, 1984
	ARC 11128, Balam	DRR, 1987
	ARC 11281	DRR, 1997
18. Whorl maggot	ARC 5760, ARC 5811, ARC 5811B, ARC 5906, ARC 6954, ARC 7217, ARC 7251, ARC 7292, ARC 7308D, ARC 7321A	AICRIP, 1972
19. Root knot-nematode	Garem, Dumai	Roy, 1973
	Ikora sali	Kalita, 1991
20. Submergence tolerance	ARC 12749, ARC 12751, ARC 12753, ARC 12765, ARC 12773, ARC 12797, ARC 10508, ARC 11198, ARC 12175, ARC 12559	CRRI, 1974

	Negheri bao, ARC 5756, ARC 5764, ARC 5938, ARC 5939, ARC 5955, ARC 5984, ARC 6066, ARC 6077, ARC 6133, ARC 6205, ARC 6239, ARC 6572, ARC 6605, ARC 6626,	IRRI, 1975
	ARC 6629	
	ARC 6586	CRRI, 1976
	ARC 12037, ARC 12038, ARC 12059, ARC 12172, ARC 12503, ARC 12606	IRRI, 1977
	ARC 614-25 b	IRRI, 1978
	Sali badal, Dhola badal, Solpona, Kola sali, Bogabordhan, Ronga sali, S 22, S 24, S 25, S 28	Pathak *et al.*, 1995
21. Drought tolerance	ARC 6565, ARC 7060, ARC 7102, ARC 7327, ARC 11964	IRRI, 1976
	ARC 7001, ARC 7046, ARC 7303	IRRI, 1977
	ARC 10372, ARC 10952, ARC 10959, ARC 11836, ARC 11940, ARC 12851	IRRI, 1978
	AS 47, AS 313/11, AS 180, Hansa Kumra, Koijapori, Bairing, Ahu joha, Dagaranga, Noga ahu, Soraituni, Maibee, Lakhi	Kalita *et al.*, 1990
22. Deep water tolerance	Kekowa bao, Negheri bao	Paroda and Malik, 1990
23. Kneeing ability	ARC 5955	IRRI, 1977
24. Cold tolerance	ARC 6000	IRRI, 1979
	Khonorullo, Abor red B, Ryllo red	Majumder and Borthakur, 1991
25. Low light intensity tolerance	Sorokphola	Baruah, 1993
26. Tolerance to iron toxicity	Badal, Sial sali, Phoolpakhari	AAU, 1997
27. Tolerance to Aluminium toxicity	Azucena, Ate local, Thalanchuah, Chao, Nikhunkati	IRRI, 1986
28. Tolerance pre	Maibela, Soraituni, Padumoni, Chaklaicha	AAU, 1997; Kalita *et*

harvest sprouting	Maichu, Chakpirine, Cheni ahu, Goalbhog, Nilazee, Banglami, Guni ahu, Rongadaria 4, AS 93-1, AS 229	*al.*, 1994
29. For late planting situation	Manohar sali, Andrew sali, Prafulla, Gitesh	RARS, Titabar

Table 16.25. Number of accessions resistant to diseases and pests in AC, JBS and ARC collection.

Diseases/pests	AC	JBS	ARC
Blast	12	19	250
Helminthosporium	10	-	231
Bacterial leaf blight	4	2	25
Stem borer	6	-	4
Gall midge	9	12	23
Brown plant hopper	1	2	167
Green leaf hopper	-	-	52
Tungro virus	1	2	9
Total collection	**5619**	**1568**	**6630**

AC: Accessions already listed in CRRI, Cuttack
JBS: Jeypur Botanical Survey (Collection made by CRRI from south Orissa)
ARC: Assam Rice Collections made by IARI, New Delhi
Source: Krishnamurthy *et al.*, 1975

Germplasm Utilization

In the early period of rice research, variety development was initiated following the pureline selections of the promising germplasm. Following this breeding method, 49 rice varieties (16 *ahu,* 16 *sali,* 12 *asra/bao* and 5 *boro*) were evolved and were recommended for cultivation under different situations (Table 16.26). As already mentioned, Assam experienced a diverse rice growing ecosystem (Fig. 16.2). The diverse ecosystem demands basketful of rice varieties for different rice growing situations. Therefore, in the process of varietal development, due attention was given to select need based, situation specific varieties to meet the need of the farmers. For instance, several varieties viz., Kmj D 138-6, Kmj D 204-1, TTB AS 86 and TTB AS 35 that mature within 90 days were recommended for flood affected areas to be grown as pre- and post-flood rice. In selecting the varieties, emphasis was given on growth duration, early vigour, tillering ability, non-lodging habit, easy thereshability, uniformity in flowering and maturity, preferred grain quality and drought tolerance. Hybridization programme was initiated as early as 1921 (Assam State Govt, 1923) with the dual purpose of studying the inheritance pattern of different characteristics and evolving varieties with high yield and quality. During this period, considerable genetic studies were made on the inheritance of various characters, especially, pigmentation in rice plant and many valuable contributions were made in this line. However, the hybridization programme carried out until the 1960s resulted in little improvement in yield over the pureline selections. This could be mainly due to the utilization of narrow genetic base involving closely related tall *indica* parents.

Besides, in the absence of a modern plant type concept, parentage with traditional types, which failed to show high yield potential, may have been selected. However, these programmes succeeded in incorporating some desirable traits such as preferred grain quality, easy threshability, better culm strength, etc. From this programme, 14 varieties were developed (Table 16.27). It is worth mentioning that varieties Andrewsali and Manoharsali which were developed during sixty's are still popular in Assam because of their wide adaptability.

As a part of the ICAR sponsored *japonica-indica* hybridization scheme, attempts were made in 1954 to develop non-lodging, fertilizer responsive, high yielding varieties. As a result, a sizeable number of croses were made, involving *japonica* and several promising tall indigenous *indica* varieties. Selection in the segregating generations of *J x I* crosses, varietal trials with promising hybrid derivatives and cultural-manurial trials on them were conducted. Varietal improvement through JxI hybrid scheme was, however, far below expectation and several crosses showed high sterility due to the possible reason of cryptic structural hybridity. The scheme was rather pre-maturely abandoned within a period of 10 years before any tangible results could be obtained.

With the advent of major breakthrough in the modern plant type concept and the introduction of high yielding semidwarf varieties, viz., TN (1) in 1965 and IR 8 in 1966 in the country, the rice improvement programme entered into a new era and the impact of such a concept became visible in Assam also.

New era of rice research began in eighties with systematic approach of research based on need and priority of the farmers. As a result of the breeding programme launched with the new concept, 42 rice varieties have been developed (Table 16.28) over the last three decades through the utilization of either improved semidwarf only or both semidwarf and indigenous tall parents. It is worth mentioning that the indigenous rice germplasm have played a major role in evolving high yielding varieties especially for unfavourable situations. Varieties like Prafulla and Gitesh for staggered planting and TTB 303-2-23 and TTB 303-18-3 for water-logged situations are some of the examples of utilization of this diverse germplasm in evolving varieties for specific situations. Some of the germplasm from ARC collection have been directly introduced in different states of the country (Table 16.29). In addition to these, two improved deep water rice varieties were developed in the late 1980s as part of the collaborative programme between CRRI, Cuttack and Assam Agricultural University from a three-way cross involving two semidwarfs and one local deep water variety, Negheribao. Thus the rice germplasm of Assam has been successfully utilized for evolving rice varieties of diverse adaptability for the farmers. However, the rice breeding in the state still has a long way to go in developing rice varieties, particularly those that may be suitable for growing under stress situations such as dry seeded uplands, flood-prone areas, low lying deep water areas, areas affected by low temperature during *boro* season and with multiple resistance to major insects pests and diseases. Many of the rice germplasm accessions of Assam have also been utilized in specific breeding programmes outside the state (Table 16.30).

Conclusion

Vast genetic diversity in the rice germplasm of NE India is a proven fact but the collection made and characterization, evaluation and documentation done so far is still not complete. This is particularly true for certain hilly regions which are relatively inaccessible for various reasons including ethnic conflicts and difficult terrain. On the other hand, indigenous varieties are constantly being replaced by higher yielding modern varieties at an unprecedented rate. This genetic depletion is more in *ahu* class and fine grained *joha* group. For example *dumai* group of *ahu* is almost wiped out from cultivation, so also varieties like Aus Jaria, Bherapoa, Rata and many others which were once popular with the farmers. Obviously, the highest importance should be attached to the collection of rice germplasm of the region before losing any more traditional varieties.

Evaluation of rice germplasm needs to be intensified to the molecular level. Identification of valuable nuclear genes and plasma genes and assessment of their interactions under different external environments would be of great theoretical and practical significance. Utilization of genetic potential for developing situation specific varieties will come as a natural follow up.

Table 16.26. Rice varieties evolved for different agro-climatic situations in Assam through pureline selection during pre-AAU period.

Sl.No.	Name	Originating station	Year of recomm-endation	Duration (days)	Adaptation	Varietal characteristics
1	2	3	4	5	6	7
	Ahu					
1.	Kmj D 138-6 (Selection from Kala Dumai)	Karimganj	1920	80	Direct seeded *ahu*	Grown broadcast in March/April and harvested in May/June, also grown in Aug/Sept after recession of flood owing to its early maturity. Grains are mottled black in colour, medium in size and red kernelled
2.	Kmj D 204-1 (Selection from Lal Dumai)	-do-	1940	80	Direct seeded *ahu*	Plant features same as that of Kala Dumai, grains are blackish brown in colour, medium in size, red kernelled.
3	Kmj M 36-20 (Selection from Baruash Murali)	-do-	1920	95	Direct seeded *ahu*	Belongs to Murali sub class, non lodging. Grains are straw coloured with traces of awns, medium in size, red kernelled.
4	Kmj M 142 (Selection from Koimurali)	-do-	1925	95	Direct seeded *ahu*	Belongs to Murali sub class, grains are straw coloured with traces of awns, coarse grain, red kernelled
5	Kmj M 175-1 (Selection from Dhola Jali)	-do-	1938	95	Direct seeded *ahu*	Belongs to Murali sub class, grains are straw coloured with traces of awns, coarse grain, red kernelled
6.	Kmj C 148-1 (Selection from Chengri)	-do-	1932	90	Direct seeded *ahu*	Belongs to Chengri sub class, grains are blackish brown, medium in size, red kernelled.

7	Kmj C 203-3 (Selection from Chengri	-do-	1940	95	Direct seeded *ahu*	Belongs to Chengri sub class, grains are blackish brown, medium in size, red kernelled.
8	AS 3 (Selection from Basmati)	Karim-ganj	1923	122	Transplanted *ahu*	Grains are straw coloured, white kernel, scented, good eating quality
9	Kmj AS 24-1 (Selection from Basanta Bahar)	-do-	1924	105	Transplanted *ahu*	Grains are long and very fine ,scented with white kernel, good cooking quality
10	Kmj AS 2 (Selection from Kasalath)	-do-	1927	110	Transplanted *ahu*	Grains are long and very fine ,scented with white kernel, good cooking quality
11	Kmj AS 46 (Selection from Aus Jaria)	-do-	1927	120	Transplanted *ahu*	Stiff strawed, non lodging, grains are brown in colour, medium in size and red kernelled
12	AS 86 (Selection from Rangadaria)	Titabar	1927	95	Direct seeded *ahu*	Non lodging, high tillering ability, grains are brown in colour, red kernelled
13	AS 25 (Selection from Fapori)	-do-	1930	95	Direct seeded *ahu*	Stiff strawed with good tillering ability, grains are blackish brown, red kernelled
14	TTB AS 35 (Selection from Farma)	-do-	1934	100	Direct seeded *ahu*	Strong culm and good tillering ability. Grains straw coloured, medium in size, red kernelled
15	AS 20-1 (Selection from Garem)	-do-	1938	100	Direct seeded *ahu*	Grain are straw coloured, medium in size, red kernelled
1 6	TTB AS 48 (Selection from Dubaichenga)	-do-	1938	105	Direct seeded *ahu*	Grain are straw coloured, medium in size, red kernelled
	Sali					
17	Kmj S 22 (Selection from Lati sali)	Karimganj	1920	165	Transplanted *sali*	Wide adaptability, grains are coarse and straw coloured, course, white kernel, high yielding

18.	Kmj S 156 (Selection from Nagra sali)	-do-	1923	160	Transplanted *sali*	Not suited to all types of soil. Yielding ability as good as that of Lati sali. Grains are straw coloured, medium fine with white kernel.
19.	Kmj S 61 (Selection from Prasadbhog)	Karimganj	1926	150	Transplanted *sali*	Grains are straw coloured with medium slender, white kernel with good milling and cooking quality
20.	Kmj S 155 (Selection from Badsabhog)	-do-	1927	165	Transplanted *sali*	Grains are straw coloured with very fine, scented white kernels. It fetches better price in the market for its quality.
21.	Kmj S 161 (Selection from Dhepi sali)	-do-	1929	155	Transplanted *sali*	Stiff straw, grains are coarse with white kernel.
22.	TTB SL 70 (Selection from Ahom sali)	Titabar	1929	160	Transplanted *sali*	Erect habit, good tillering ability. Grains are straw coloured with medium, white kernel
23.	TTB S 114 (Selection from Fulgaj)	-do-	1931	155	Transplanted *sali*	Stiff straw, grains are straw coloured with whith kernels.
24.	TTB S 115 (Selection from Latamaguri)	-do-	1931	160	Transplanted *sali*	Erect and possesses drought tolerance, straw coloured grains with white kernels and good milling quality
25.	TTB Sj 226 (Selection from Bengali joha)	-do-	1931	165	Transplanted *sali*	Grains are fine, scented, white kernels with good quality
26.	TTB SB 279 (Selection from Gomiri bora)	-do-	1931	160	Transplanted *sali*	Grains are brown, medium with white glutinous kernels.
27.	TTB S 126 (Selection from Laudumra)	-do-	1933	160	Transplanted *sali*	Non lodging, can withstand drought, grains are straw coloured, coarse and white kernelled with good milling quality.

28.	TTB SL 240 (Selection from Johori)	-do-	1933	160	Transplanted *sali*	Grains are yellow, medium and white kernelled, suitable for making *chira, muri* and *khai*
29.	TTB SL 533 (Selection from Kanaimuluk)	-do-	1933	160	Transplanted *sali*	Grows well in high and medium paddy land. Grains are yellow coloured, medium, white kernelled with good milling quality
30.	TTB Sj 311 (Selection from Lakshmi joha)	-do-	1935	155	Transplanted *sali*	Grains are fine, scented with white kernels of good quality
31.	TTB S 36 (Selection from Hati sali)	-do-	1935	165	Transplanted *sali*	Suitable for growing late and in relatively low lying areas. Grains are coarse, white kernelled with good milling quality
32.	TTB SL 202 (Selection from Kalamdani)	Titabar	1945	165	Transplanted *sali*	Can be grown under shallow and medium deep water situation like *asra* paddy variety. Grains are deep yellow with white kernel
	Asra* and *Bao					
33.	Kmj Ar 1 (Selection from Sail Badal)	Karimganj	1921	200	*Asra* (semideep water situation, generally direct seeded)	Withstand water depth up to 90cm, Grains are coarse, straw coloured with white kernel
34.	Ar. 2 (Selection from Laki)	-do-	1921	200	*Asra* (semideep water situation, generally direct seeded)	Withstand water depth up to 90cm, Grains are coarse, straw coloured with white kernel
35.	Ar 28 (Selection from Birpak)	-do-	1926	190	*Asra* (semideep water situation, generally direct seeded)	Withstand water depth up to 90cm. submergence tolerant, white kernelled

36.	Kmj Ar.108-1 (Selection from Dhola badal)	-do-	1935	205	*Asra* (semideep water situation, generally direct seeded)	Vigorous in growth, grains are coarse, straw coloured with white kernel
37.	Hbj Amon 1A 13-20 (Selection from Bagdar)	Habiganj	1938	270	Deepwater situation, generally direct seeded	It grows well in 90-120 cm of water depth and is grown on the upper part of the depressed situations. Grains are awn less, medium in size, straw coloured with faint brown furrow and are amber kernelled.
38. 180	Hbj Aman II A 192 (Selection from Goda Laki)	-do-	1938	270	Deepwater situation, generally direct seeded	It grows well up to water depth of cm. grains are awnless, medium, straw coloured with amber kernels.
39.	Hbj Aman II1 A38-13 (Selection from Gowai)	-do-	1938	270	Deepwater (*Bao*/ floating rice) situation, direct seeding	It grows well in 2.4-3.6 m of water. Grains are awned, coarse, with dull straw coloured husk and red kernel.
40.	Hbj Aman IV A 185 (Selection from Gudh Laki)	Habiganj	1943	275	Deepwater (*Bao*/ floating rice) situation, direct seeding	Grows well in 1.5 to 2.0 m of water depth. Grains are straw coloured, awnless with amber kernel of better quality character than that of aman II
41.	Hbj Aman V A 51-1 (Selection from Dhola Aman)	-do-	1944	275	Deepwater (*Bao*/ floating rice) situation, direct seeding	It also grows well in 2.0 to 3.6 m of water depth. Grains are straw coloured, awnless, coarse and red kernelled.
42.	E B 1 (Selection from Negheri bao	Roha	1959	285	Deepwater (*Bao*/floating rice) situation,	Thrives well up to 4.5 m of water depth.

					direct seeding	
43.	E B2 (Selection from Kolonghi bao)	-do-	1959	280	Deepwater (*Bao*/ floating rice) situation, direct seeding	Thrives well up to 2.4 to 3.0 m of water depth. Grains are brown with yellow ribs, long awned, coarse and red kernelled.
44.	E B 3 (Selection from Kekoa bao)	-do-	1959	280	Deepwater (*Bao*/ floating rice) situation, direct seeding	Thrives well up to 3.6 m of water depth. Grains are yellow with brown furrows, awned and red kernelled.
	Boro					
45.	Hbj B III-17 (Selection from Boro 1)	Habiganj	1937	160	Transplanted *boro*	Tall, early maturing variety. Grains are straw coloured with prominent blackish brown furrows, awned and red kernelled.
46.	(Hbj B 43 III-17 (Selection from Boro 2)	-do-	1938	165	Transplanted *boro*	Grains are straw coloured, short awned with white scented kernels. It exhibits good elongation on cooking.
47.	(Hbj B 78- III-4 (Selection from Sali boro)	-do-	1938	165	Transplanted *boro*	It responds remarkably well to irrigation. Grains are straw coloured, awnless, medium in size with white scented kernels of good cooking quality.
48.	Hbj B IV- 16-3	-do-	1946	155	Transplanted *boro*	The variety is similar to *Boro* 1 except that it is about 5 days earlier in duration than the later.
49.	Hbj B V	-do-	-	165	Transplanted *boro*	Grains are medium yellow in colour with reddish tinge and white, scented kernels.

- the *ahu* and *boro* rice varieties are photoperiod insensitive
- the *sali* and *asra* rice varieties are photoperiod sensitive and tall

Source: Pathak, 2001

Table 16.27. Rice varieties evolved through hybridization during pre-AAU period.

Sl. No.	Name	Parentage	Originating station	Year of recommendation	Duration (days)	Adaptation	Salient features
	Ahu						
1.	Aus Hybrid I (Kmj ASC 536-143)	Kasalath/ Basmati	Karimganj	1938	105	Transplanted *ahu*	First ahu variety developed in the state through hybridization programme. Developed with the objective of combining earliness of Kasalath with grain quality of Basmati. Grains are straw coloured, medium in size and white kernelled.
2.	Aus cross (TTB ASC 313-11)	Rangadaria/ Dacca No.6	Titabar	1944	100	Direct seeded *ahu*	First variety developed at Titabar Station through hybridization programme. Inspite of its poor grain yield, the variety gained popularity among farmers because of its white kernel. Grains are brown, medium in size and white kernelled.
	Sali						
3.	Vijoy sali (Kmj S.C. 54-60)	Lati sali/ George sali	Karimganj	1929	150	Transplanted *sali*	Non lodging type, white kernel, with threshings problem
4.	Kerr sali (Kmj S. C. 94-47)	Nagra sali/ Mayna sali	Karimganj	1929	165	Transplanted *sali*	Grains are medium with white kernelled
5.	Sali hybrid 3 (Kmj S.C. 412-56)	Nagra sali/ Dhepi sali	Karimganj	1935	160	Transplanted *sali*	Stiff strawed, easy threshability. Grains are elongated with white kernels.

6.	Swarna sali (Kmj S.C. 412-56	Lati sali/ Guachari	Karimganj	1941	165	Transplanted *sali*	Adaptable to adverse situation. It has combined general good performance and uniform ripening characters of lati sali. Grains are straw coloured, coarse and white kernelled.
7.	Sali hybrid 5 (Kmj S.C. 1177-36	Vijoy sali /Kerr sali	Karimganj	1945	155	Transplanted *sali*	White kernelled
8.	Handique sali (Kmj S.C. 1177 – 6)	Vijoy sali/ Kerr sali	Karimganj	1946	165	Transplanted *sali*	Non lodging and uniform ripening habit. Grains are straw coloured, medium and white kernelled.
9.	Manohar sali (Kmj S.C. 412-125)	Lati sali/ Gauchari	Karimganj	1967	155	Transplanted normal and late *sali*	Wide adaptable variety, suitable for planting late in *sali* season. Grains are straw coloured, white kernelled.
10.	Andrew sali (TTB S.C. 308-51)	Lata maguri/ Lakshmi joha	Titabar	1940	153	Transplanted normal and late *sali*	Suitable for plantingh late in the *sali* season. Grains are straw coloured, medium and white kernelled.
11.	Sali hybrid (TTB S.C. 308-372)	Lata maguri/ Lakshmi joha	Titabar	1946	157	Transplanted *sali*	Sister line of Andrew sali. Better than Andrew sali in respect of spikelet fertility and milling recovery. Grains are deep yellow, elongated and white kernelled.
12.	TTB S.C. 406-93		Titabar	1962	155	Transplanted *sali*	Grains are straw coloured with medium, white kernel and good milling recovery.
	Asra						
13.	Asra hybrid – 1 (Kmj Ar.C. 353-148)	Sail badal/ Lati sali	Karimganj	1935	185	*Asra*	It does well in 60-120 cm depth of water. It carries many of the good characters of Lati sali eg.

							Earliness, uniform maturity etc., and has the typical *asra* varietal characteristics of Sali badal.
							Grains are coarse, straw coloured and white kernel.
14.	Asra hybrid – 2 (Kmj Ar. C 614-256)	Sailbadal/ Birpak	Karimganj	1944	200	*Asra*	It thrives well in 60-180 cm waterdepth. Grains are blackish brown, coarse and white kernelled.

Table 16.28. Rice varieties developed under AAU programme.

Sl. No.	Name	Parentage	Originating station	Year of recommendation	Duration (days)	Adaptation	Salient features
1	2	3	4	5	6	7	8
1.	Madhab (TTB 4/7)	IR 8/ Ch 63	Titabar	1977	125	Transplanted *ahu* and *sali*	Semidearf in plant stature. Recommended to be grown as transplanted *sali* also. Grains are medium bold.
2.	Kmj 1-19-1	IR 8/ Manohar sali	Karimganj	1978	145	Late *sali*/ Delayed planting	Semidwarf, photoperiod sensitive variety, suitable for growing late in the *sali* season as such fit well in the flood prone situation, white kernel.
3.	Kmj 1-17-2	IR 8/ Manoha sali	Karimganj	1978	145	Late *sali*/ delayed planting	Semidwarf, photoperiod sensitive variety, suitable for growing late in the *sali* season as such fit well in the flood prone situation, white kernel.
4.	Rongdoi (TTB 2-6-1-1)	Prasadbhog /IR8	Titabar	1981	125	Transplanted *ahu* and *sali*	Semidwarf, recommended for transplanted *sali* also. Medium slender grain.
5.	Lachit (TTB 14-1)	CRM13-3241/ Kalinga 2	Titabar	1987	120	Transplanted *ahu*	Semidwarf in plant stature, medium bold grain, tolerant to BLB, BPH and WBPH. Notified by CVRC in 1992.
6.	Chilarai (TTB 15-1)	IR 24/CR 44-118-1	Titabar	1987	125	Transplanted *ahu*	Semidwarf in plant stature, medium bold grains. Notified by CVRC in 1992

7.	Padmapani	Pureline selection	N. Lakhimpur	1988	185	*Bao* (Deep water/ Floating rice, direct seeded)	Tall photoperiod sensitive variety with relatively short crop growth period can stand upto 2.0 m water depth, and escapes ufra disease.
8.	Panindra	Pankaj/ Jagannath// Negheri bao	N. Lakhimpur	1990	240	*Bao* (Deep water rice, direct seeded)	High yielding deep water rice that can withstand up to 1.5m water depth
9.	Padmanath	Pankaj/ Jagannath// Negheri bao	N. Lakhimpur	1990	240	*Bao* (Deep water rice, direct seeded)	High yielding deep water rice that can withstand up to 1.5m water depth
10.	Ranjit	Pankaj/ Mahsuri	Titabar	1992	150	Transplanted *sali*	Developed for replacement of Mahsuri, which was the most popular variety for *sali* season, during 1980s. The variety with higher yield, resistance/ tolerance to several insect pests and diseases, lodging and shattering and grain quality similar to Mahsuri is recommended for growing in *sali* season as beter alternative to Mahsuri.Notified by CVRC in 1994
11.	Bahadur	Pankaj/ Mahsuri	Titabar	1992	150	Transplanted *sali*	Developed for replacement of Mahsuri, which was the most popular variety for *sali* season, during 1980s. The variety with higher yield, resistance/ tolerance to several insect pests and diseases, lodging and shattering and grain quality similar to

							Mahsuri is recommended for growing in *sali* season as beter alternative to Mahsuri.Notified by CVRC in 1994
12.	Kushal	Pankaj/ Mahsuri	Titabar	1992	150	Transplanted *sali*	Developed for replacement of Mahsuri, which was the most popular variety for *sali* season, during 1980s. The variety with higher yield, resistance/ tolerance to several insect pests and diseases, lodging and shattering and grain quality similar to Mahsuri is recommended for growing in *sali* season as beter alternative to Mahsuri. Notified by CVRC in 1994
13.	Maniram	Pankaj/ Mahsuri	Titabar	1992	155	Transplanted *sali*	Developed for replacement of Mahsuri, which was the most popular variety for *sali* season, during 1980s. The variety with higher yield, resistance/ tolerance to several insect pests and diseases, lodging and shattering and grain quality similar to Mahsuri is recommended for growing in *sali* season as beter alternative to Mahsuri
14.	Piolee	Pankaj/ Mahsuri	Titabar	1992	155	Transplanted *sali*	Developed for replacement of Mahsuri, which was the most popular variety for *sali* season, during 1980s. The variety with higher yield, resistance/ tolerance to several insect pests and diseases, lodging and shattering

							and grain quality similar to Mahsuri is recommended for growing in *sali* season as beter alternative to Mahsuri
15.	Bhogali	Kmj 1 -52 -2/Ghew bora	Karimganj	1992	155	Transplanted *sali*	Photoperiod sensitive, semidearf with pigmented plant parts, high yielding glutinous (bora) variety recommended for growing in *sali* season.Notified by CVRC in 1994.
16.	Rongilee	Kmj 1-52-2/ Ghew bora	Karimganj	1992	155	Transplanted *sali*	Photoperiod sensitive tall, stiff strawed, high yielding glutinous rice. Notified by CVRC in 1994.
17.	Aghoni	Kmj1-52-2/ Gandhi bora	Karimganj	1992	155	Transplanted *sali*	Photoperiod sensitive, semidwarf, high yielding glutinous rice.
18.	Luit (TTB 127-209-2)	Heera/ Annada	Titabar	1993	90-100	Direct seeded or transplanted in early *ahu* and late *sali*	Suitable for chronically flood affected areas. a very short duration variety with reasonable high yield. Can be grown early in the *ahu* season and harvested before occurrence of flood, can also be grown late in the *sali*
19.	Kapilee	Heera/ Annada	Titabar	1993	90-100	Direct seeded or transplanted in early *ahu* and late *sali*	Suitable for chronically flood affected areas, a very short duration variety with reasonable high yield. Can be grown early in the *ahu* season and harvested before occurrence of flood, can also be grown late in the *sali* season after recession of flood and harvested without being affected by cold.

20.	Barak	Pankaj/ Mahsuri	Titabar	1995	155	Transplanted *sali*	Developed for replacement of Mahsuri, which was the most popular variety for *sali* season, during 1980s. The variety with higher yield, resistance/ tolerance to several insect pests and diseases, lodging and shattering and grain quality similar to Mahsuri is recommended for growing in *sali* season as beter alternative to Mahsuri
21.	Hybrid 14	Indica/ Japonica (Names are not available)	Karimganj	1995	150	Transplanted *sali*	Tall, stiff strawed variety, photoperiod sensitive, recommended for growing in *sali* season. Derived from Indica/ Japonica hybridization programme initiated in the fifties.
22.	Longai	Pusa 2-21/ China 63	-do-	1995	130	Transplanted *sali* in multiple cropping areas	Semidwarf, recommended to be grown as *sali* rice under multiple cropping sequence
23.	Paresh	Kmj1-52-2/ Gandhi bora	Karimganj	1995	155	Transplanted *sali*	Photoperiod sensitive, semidwarf, high yielding glutinous rice.
24.	Ketekijoha	Savitri/ Badsabhog	Titabar	1995	155	Transplanted *sali*	Medium tall, photoperiod sensitive, fine grained, high yielding aromatic variety (joha)
25.	Bokul joha	Savitri/ Badsabhog	Titabar	-	155	Transplanted *sali*	Medium tall, photoperiod sensitive, fine grained, high yielding aromatic variety (joha)

26.	Satyaranjan	IET 9711/ IET 11162	Titabar	1996	130	Transplanted *sali* in multiple cropping areas	Photopereiod insensitive, multiple resistant, fine grained variety, suitable for growing as a *sali* rice in areas where *sali* rice is followed by a rabi crops.
27.	Basundhara	IET 9711x IET 11161	-do-	1996	130	Transplanted *sali* in multiple cropping areas	Photopereiod insensitive, multiple resistant, fine grained variety, suitable for growing as a *sali* rice in areas where *sali* rice is followed by a rabi crops.
28.	Jaymati	Jaya/Mahsuri	Titabar	1996	170	Transplanted *boro*	A fine grained, high yielding variety recommended for growing in the *boro* season, Mahsuri type grain.
29.	Jyotiprasad	K343-29-1-1/ Suweon 334	Titabar	1997	165	Transplanted *boro*	Developed specifically for the irrigated *boro* growing situation.
30.	Bishnuprasad	K 343-29-1/ Suweon 334	Titabar	1997	165	Transplanted *boro*	Developed specifically for the irrigated *boro* growing situation.
31.	Gopinath	Pusa 2-21/ IR 36	Titabar	1997	115	Transplanted *ahu*	Early *ahu* variety with semidwarf plant stature with good grain type. season after recession of flood and harvested without being affected by cold.Notified by CVRC in 1998
32.	Jalashree	Pankaj/FR 13A	Titabar	2000	150	Transplanted *sali* in flash flood area	Tall, high yielding photoperiod sensitive variety having submergence tolerance up to 15 days

33.	Jalkunwari	Pankaj/FR 13A	Titabar	-	150	Transplanted *sali* in flash flood area	Tall, high yielding photoperiod sensitive variety having submergence tolerance up to 15 days
34.	Disang	Chilarai/ Kalinga 3	Titabar	2005	90-100	Direct seeded or transplanted in early *ahu* and late *sali*	Suitable for chronically flood affected areas, a very short duration variety with reasonable high yield. Can be grown early in the *ahu* season and harvested before occurrence of flood, can also be grown late in the *sali* season after recession of flood and harvested without being affected by cold.
35.	Kolong	Chilarai/ Kalinga 3	Titabar	-	90-100	Direct seeded or transplanted in early *ahu* and late *sali*	Suitable for chronically flood affected areas, a very short duration variety with reasonable high yield. Can be grown early in the *ahu* season and harvested before occurrence of flood, can also be grown late in the *sali* season after recession of flood and harvested without being affected by cold.
36.	Dikhow	Chilarai/ Kalinga 3	Titabar	-	100	Direct seeded or transplanted in early *ahu* and late *sali*	Suitable for chronically flood affected areas, a very short duration variety with reasonable high yield. Can be grown early in the *ahu* season and harvested before occurrence of flood, can also be grown late in the *sali* season after recession of flood and harvested without being affected by cold.

37.	Inglongkiri (AAUDR 9301-14-1)	Maibee/ CRM 53	Diphu	2009	115	Hill zone for direct seeded *ahu*	Recommended for growing in hill zone of Assam. Tolerant to drought.
38.	Rongkhang (AAUDR 9312-13-1)	Maibee/ CRM 53	Diphu	2009	120	Hill zone for direct seeded *ahu*	Recommended for hill zone in *ahu* season and suitable for jhum areas. Good cooking quality with white kernel colour.
39.	Dehangi (AAUDR 9313-14-3)	Maibee/CRM 49//Maibee	Diphu	2009	120	Hill zone for direct seeded *ahu*	Recommended for hill zone in *ahu* season and suitable for jhum areas. Good cooking quality with white kernel colour.
40.	Prafulla	Aki sali/ Kushal	Titabar	2009	150	Transplanted *sali*/ Delayed planting	Photoperiod sensitive variety with stiff strawed and high yield. Suitable for delayed planting with aged seedlings (up to 90 days old seedling) up to 2nd week of September
41.	Gitesh	Aki sali/ Kushal	Titabar	2009	150	Transplanted *sali*/ Delayed planting	High yielding semidwarf variety suitable for delayed planting with aged seedlings (up to 60 days old seedlings), medium slender grains.
42.	Kanaklata	-do-	Titabar	2009	170	Transplanted *boro*	A fine grained medium tall variety, with stiff straw, recommended for growing in the *boro* season, high yield, grains are straw coloured

Table 16.29. Varieties developed/improved outside Assam from the ARC germplasm.

State	Name/IET No	Parentage	Year of release	Duration	Resistance
Karnataka	IET 7575	Sona/Manohar sali	1988	165	Blast
	IET 8116	Vikram/Andrew sali	1998	130	-
Kerela	IET 9268 (Makan)	ARC 6650/Jaya	1989	110	Blast
	Suvarna madan	Pureline selection of ARC 11775	1976	105	Blast
Orissa	Vanaprabha	ARC 12422/ARC 12751	1988	65	RTV
	Keteki joha	Badsabhog/Savitri	2005	155	-

Table 16.30. Varieties/entries developed by using ARC collections.

IET No	Cross combination	Average plant damage score	
		BLB	BPH
12767	Ratna/ARC 5981	5.0	0.5
12770	Ratna/ARC 10654	6.0	0.7
12793	Ratna/ARC 10654	5.0	1.1
12206	Ratna/ARC 10654	5.0	1.1
11481	Ratna/ARC 19666	3.0	1.1
2030	IET 6315/ARC 5984	6.0	1.2
12805	Ratna/ARC 5981	2.3	1.3
12785	Nagarjuna/ARC 5984	3.0	1.6
12204	Phalguna/Andrew sali	3.0	1.7
Resistant check	-	1.5	1.3
Susceptible check		9.0	9.0

Source: Rana *et al.*, 1994.

References:

AAU (Assam Agricultural University). 1997. Proceedings, *Crop and Horticulture Seminar and AAU-DOA Interface*. Assam Agril Univ, Jorhat.

AAU. 1983. *Research Highlight.* Assam Agril Univ, Jorhat (Unpublished)

Ahmed SA, Barua I, Sarkar CR. 1995. Volatile component (2 Acetyl-1-pyroline) in scented rice.

Pages 55-57 in Proceedings of seminar on *Problems and Prospects of Agricultural Research and Development in NE India.* Assam Agril Univ, Jorhat.

Ahmed T, Bora P. 1999. Genetic diversity in glutinous rice germplasm of Assam. *Oryza* 36(1):74-75.

Ahmed T, Das GR. 1990. Inheritance of anthocyanin pigmentation and endosperm character of glutinous rice. *Oryza* 27:220-223.

Ahmed T and Das GR. 1994. Evaluation and characterization of glutinous rice germplasm of Assam *Oryza* 31:77-83.

Ahmed T. 1986. *Diversity, genetic associations and breeding potentiality of glutinous rice.* PhD Thesis, Assam Agril Univ, Jorhat.

AICRIP. 1971. *Annual Progress Report. Vol 3, 1971.* All India Coordinated Rice Improvement Project, Hyderabad.

AICRIP. 1972. *Annual Progress Report. Vol 3, 1972.* All India Coordinated Rice Improvement Project, Hyderabad.

AICRIP. 1976. *Annual Progress Report. Vol 3, 1976.* All India Coordinated Rice Improvement Project, Hyderabad.

Ali MS, Bhagabati KN, Baruah DK, Pathak AK. 1995. Non-chemical methods of rice disease management in Northeast India. Pages 284-289 in Proceeding of seminar on *Problems and Prospects of Agricusltural Research and Development in NE India.* Assam Agril Univ, Jorhat.

Assam State Government. 1923. *Annual Report.* India Department of Agriculture.

Barthakur BC. 1983. *Summary of Research Results, 1980-81 to 1982-83.* Directorate of Research, Assam Agril. Univ., Jorhat.

Baruah AM. 1993. *Studies on Nitrate Reductase, Seed Storage Protein and Starch in Rice Varieties Differing in Grain Yield and Grain Protein Content.* MSc Agri Thesis, Assam Agril Univ, Jorhat.

Baruah RKSM. 1975. *Variability and Interrelationship of Characters in* Chokuwa *and* Bora *Varieties of Assam Rice.* MSc Agri Thesis, Assam Agril Univ, Jorhat.

Baruah RKSM. 1990. *Evaluation and Genetic Analysis of* Ahu *Rice of Assam.* PhD. Thesis, Assam Agril Univ, Jorhat.

Baruah RKSM, Das GR. 1995. Utilization of indigenous land races in developing rice varieties of Assam. Pages 21-36 in Proceedings seminar on *Problems and Prospects of Agricultural Research Development in NE India.* Assam Agril. Univ, Jorhat.

Baruah RKSM, Das GR, Hazarika MH. 1994. Diversity of *ahu* rice of Assam and its characterization. *Oryza* 31:163-168.

Bhattacharjee A, Saikia UN, Puzari KC and Rathaiah Y. 1995. Reaction of some rice varieties of Assam to blast. *JASS* 8:226-227.

Bhattacharya KR. 1989. Need for widening the scope of rice breeding in India for the purpose of table rice. *Oryza* 26:117-122.

Bor NL. 1960. *The Grasses of Burma, Ceylone, India and Pakistan (excluding Bambuseae).* Pergamon

Press, Oxford.

Borah P, Baruah DK, Ali MS, Kalita UC. 1994b. Evaluation of indigenous *sali* rice germplasm for resistance to leaf blast, bacterial blight and sheath blight diseases. *JASS* 7:224-226.

Borah P, Baruah DK, Ali MS, Kalita UC, Hazarika MH. 1994a. Evaluation of indigenous *ahu* rice germplasm for resistance to leaf blast, bacterial blight and sheath blight diseases. *JASS* 7:111-113.

Borah P. 1992. *Evaluation of Indigenous Rice Germplasm for Resistance to Major Diseases.* MSc Agri Thesis, Assam Agril Univ, Jorhat.

Chang TT. 1985. Field collection of rice germplasm and dispersal of upland and *javanica* rice in SE Asia. *East Asian Tertiary/Quarternary Newsletter* 2:52.

Chatterjee SM, Prasad K, Misra BC, Rajamani R. 1977. Identification of gall midge (*Orseolia oryza* Wood-Mason) resistant rice germplasm and their utilization in breeding. *J Entomol Res.* 1:111-113.

Choudhury SD, Ganguli PM. 1948. *Experiments on Rice Having Practical Application in Agriculture Conducted in Assam during the Period from 1916 To 1946.* Departmental Bulletin, Dept Agric, Assam.

CRRI. 1971. *Annual Report 1971.* Central Rice Res Institute, Cuttack.

CRRI. 1972. *Annual Report 1972.* Central Rice Res Institute, Cuttack.

CRRI. 1973. *Annual Report 1973.* Central Rice Res Institute, Cuttack.

CRRI. 1974. *Annual Report 1974.* Central Rice Res Institute, Cuttack.

CRRI. 1975. *Annual Report 1975.* Central Rice Res Institute, Cuttack.

CRRI. 1976. *Annual Report 1976.* Central Rice Res Institute, Cuttack.

CRRI. 1977. *Annual Report 1977.* Central Rice Res Institute, Cuttack.

CRRI. 1978. *Annual Report 1978.* Central Rice Res Institute, Cuttack.

CRRI. 1979. *Annual Report 1979.* Central Rice Res Institute, Cuttack.

Das GR, Ahmed T, Bhattacharyya HC and Barthakur BC.1981. Evaluation of rice germplasm. I. Characterization and evaluation of *ahu* rice for important quantitative characters in relation to yield. *J Res Assam Agril Univ* 2(2):156-164.

Das GR, Ahmed T. 1982. Distribution of pigments in glutinous rice. *Oryza* 19:114-116.

Das GR, Ahmed T. 1995. Conservation of rice genetic resources of NE India. Pages 10-20 in Proceedings of seminar on *Problems and Prospects of Agricultural Research and Development in N E India.* Assam Agril Univ, Jorhat.

Das GR, Barthakur DN. 1973. Genetic divergence in rice. *Indian J Genet* 33 (3):436-443.

Das GR, Ahmed T, Bhattarcharyya HC, Barthakur B C. 1983. Evaluation of rice germplasm. II. Characterization and evaluation of traditional *sali* rice. *J. Res. Assam Agril Univ* 4(2):92-97.

Devadath S, Rao CS. 1978. Identification of sources of resistance to bacterial leaf blight and bacterial leaf streak of rice. *Oryza* 15:93-95.

Directorate of Rice Research. 1988. *Research Highlights 1987*. Directorate of Rice Research, Hyderabad. 61 pp

Directorate of Rice Research. 1997. *Vision 2020: DRR Perspective Plan*. Directorate of Rice Research, Hyderabad. 74 pp

Dutta L, Barua JN. 1980. Effect of processing on nutritional quality of rice. Part I. *J Res Assam Agril Univ* 1(2):143-147.

Dutta MC, Hazarika LK. 1994. Screening of some rice cultivars for resistance to rice hispa. *JASS* 7:31-34.

Frankel OH, Hawkes JG. 1975. *Crop Genetic Resources for Todays and Tomorrow*. Cambridge Univ Press, Cambridge, UK. 492 pp.

Ganguli PM. 1950. *Rice in Assam, Paddy Series*. Bulletin No 6. Dept of Agriculture, Assam.

Ghosh AK, Bhattacharya PK, Asthana AN. 1981. Genetic variability of indigenous rice varieties of Meghalaya. *Indian J Agric Sci* 51(5):281-283.

Glaszmann JC. 1986. A varietal classification of Asian Cultivated rice (Oryza *sativa* L) based on isozyme polymorphism. Pages 83-90 in *Rice Genetics*. Proceedings of the Intern Rice Genetics Symp, 27-31 May, 1985. IRRI, Manila, Philippines.

Govindaswamy S, Ghosh AK, Sinha MK, De RN, Das AB. 1973. Genetic variability of protein content in rice. *Indian J Agric Sci* 43:805-809.

IRRI. 1974. *Annual Report for 1973*. Int. Rice Res Institute, Los Banos, Philippines.

IRRI. 1975. *Annual Report for 1974*. Int. Rice Res. Institute, Los Banos, Philippines

IRRI. 1976. *Annual Report for 1975*. Int. Rice Res Institute, Los Banos, Philippines.

IRRI. 1977. *Annual Report for 1976*. Int. Rice Res Institute, Los Banos, Philippines.

IRRI. 1978. *Annual Report for 1977*. Int. Rice Res Institute, Los Banos, Philippines.

IRRI. 1979. *Annual Report for 1978*. Int. Rice Res Institute, Los Banos, Philippines.

IRRI. 1984. *Annual Report for 1983*. Int. Rice Res Institute, Los Banos, Philippines.

IRRI. 1986. *Annual Report for 1985*. Int. Rice Res Institute, Los Banos, Philippines.

IRRI. 1987. *Annual Report for 1986*. Int. Rice Res Institute, Los Banos, Philippines.

Kalita M. 1991. *Management of Root-knot Nematode (*M. graminicola*) on Rice*. MSc Agri Thesis, Assam Agril Univ, Jorhat.

Kalita UK, Barkakaty R, Baruah DK. 1990. Evaluation of rice germplasm for drought tolerance in *ahu*. Pages 9-10 in Proc Int Symp on Rice, New Frontiers.

Kalita UC, Baruah DK, Upadhaya LP. 1994. Seed dormancy in germplasm collection of rice (*Oryza sativa*) insensitive to photoperiod. *Indian J Agric Sci* 64(13):160-164

Kandali R. 1992. *Studies on Seed Protein and Leaf Nitrate Reductase Activity in Some Local Rice (Oryza sativa* L*) Cultivars of Assam*. MSc Agri Thesis, Assam Agril Univ, Jorhat.

Katayama TC. 1980. Distribution and some morphological characters of the wild rice in the Northeastern India. Pages 34-43 in Tadayo Watabe (ed) *Some Aspects of Historical Changes of Rice*. Report of the Kyoto University Scientific Survey to India and Burma, 1978/79. Kyoto University, Japan.

Kawahara E, Motogi T. 1951. Studies on the quality of upland rice. I. The degree of transparency of the hulled grain of glutinous rices and the relation between transparency and various other characters of the hulled grain. *Proceeding Crop Sci Japan* 20:205-214.

Krishnamurthy A, Sharma AC. 1970. Manipur rice in rice germplasm. *Oryza* 7(1):45-50.

Majumder ND, Barthakur DN. 1991. Stability for yield in cold tolerant rice. *Oryza* 28:23-26.

Mathur SC, Srivastava DP. 1981. Blast reaction in certain groups of rice cultivars for Northeast India and allelic relationship for resistance in two resistant genotypes. *Oryza* 18:204-207.

McWilliam JR, Ikehashi H, Singha SK 1995. Progress in hybrid rice in India. *IRCN* 44:80-87.

Mitra SK, Ganguli PM 1932. Studies on Surma Valley rices and their classifications. *Indian J Agric Sci* 2:571-606.

Mitra SK, Nandi HK, Ganguli PM. 1942. *Monograph of Paddy (*aus, sali *and* asra*) in the Surma Valley, India.* Dept of Agriculture, State Govt of Assam.

Oka HI. 1988. *Origin of Cultivated Rice*. Elsevier Science Publishers. 254 pp

Paroda RS, Malik SS. 1990. Rice genetic resources, its conservation and use in India. *Oryza* 27:361-369.

Paroda RS, Sharma SD. 1986. Collection of rice germplasm in India. Paper presented in Annual Rice Workshop for 1986. NDUAT, Faizabad.

Pathak PK, Pathak AK, Ahmed T, Roy AK. 1995. Status of rice germplasm in Assam, India. *IRC Newsl.* 44:69-74.

Pathak PK. 2001. Major cereal crops of Assam. Pages 153-74 in *Agriculture in Assam*. Assam Agricultural Univrsity, Jorhat.

Ramalingam J, Nadarajan N, Vaunirajan C, Rangaswamy P. 1994. Natural outcrossing potential in cytoplasmic male sterile lines of rice. *IRRN* 19(3):5.

Rana DK, Pophaly DJ, Kotasthane AS, Kaushik UK. 1994. Reaction of advanced IET rice varieties to major pests in Raipur, India. *IRRN* 19(3):10-11.

Rao AV, Prasad ASR, Krishna TS, Seshu DV, Srinivasan TE. 1981. Genetic divergence among some brown plant hopper resistant rice varieties. *Indian J Genet* 41:179-185.

Rathaiah Y, Das GR. 1987. Manohar sali, neck blast resistant variety. *IRRN* 12:13.

Rathaiah Y, Das, GR, Sarmah BN. 1986. *Bacterial Leaf Blight of Rice*. Research Handout, AAU Rice Res Station, Titabar, Assam. 11 pp

Richaria RH. 1979. An aspect of genetic diversity in rice. *Oryza* 16(1):1-31.

Roy AK. 1973. Reaction of some rice cultivars to the attack of *Meloidogyme graminicola. Indian J Nematology* 3:72-73.

Sarma NK 1989. *Evaluation of* Sali *and* Boro *Rice Cultivars under Different Situations and Their*

Sensitivity to Environmental Variations. MSc Agri Thesis, Assam Agril Univ, Jorhat, India.

Sarmah KC. 1990. *Evaluation of Indigenous* Sali *Rice Germplasm of Assam for Agrobotanical Characters and Adaptation*. PhD Thesis, Assam Agril Univ, Jorhat.

Seetharaman R, Srivastava DP, Ghorai DP. 1974. Preliminary studiers in rice cultivars from northeast India. *Indian J Genet* 34(2):43-149.

Seshu DV, Rao VP, Srinivasan TE, Shastry SVS. 1974. New source of dwarfism in the Assam Rice Collections. *Indian J Genet* 34:390-394.

Sharma S D. 1982. Collection and evaluation of rice from NE India. *Plant Genetic Resources News letter* 50:62-69.

Sharma SD, Vellanki MR, Hakim KL, Singh RK. 1971. Primitive and current cultivars of rice in Assam: A rich source of valuable genes. *Curr Sci* 40:126-128.

Shastry SVS, Sharma SD, John VT, Krishnaiah K. 1971.New sources of resistance to pests and diseases in the Assam Rice Collection. *IRC Newsl* 22(3):1-10.

Shobharani N, Shastry MVS, Kalode MB, Bentur JS. 1992. Inheritance of gall midge resistance in rice. *Oryza* 29:7-9.

Siddiq EA. 1991. Genes and rice improvement. *Oryza* 28:1-17.

Sidhu GS, Khush GS, Mew TW 1978. Genetic analysis of bacterial blight resistance in seventy four cultivars of rice, *Oryza sativa* L. *Theor Appl Genet* 53:105-111.

Srivastava D P. 1977. Gene-ecological variations from northeast India and their importance in rice breeding programme. *Proceedings of the 64th Indian Sc Congr. Part II,* p 77. (Abstract)

Srivastava DP. 1978. Diversity in the late duration rice cultivars from northest India. *Oryza* 15:26-33.

Srivastava MN, Sahu RK, Tiwari NK, Sharma RP, Srivastava PS. 1984. Rice germplasm conservation at JN Agricultural University Campus, Raipur, India, *IRRI Newsletter* 9:7-8.

Vairavan S, Siddiq E A, Arunachalam V, Swaminathan M S. 1973. A study on the nature of genetic divergence in rice from Assam and Northeast Himalayas. *Theor Appl Genet* 43:213-221.

Velusamy R, Kumar MG, Edward YSJT, Babu PCS. 1994. Resistance to green leaf hopper in rice varieties with different resistant genes. *IRRN* 19(3):15-16.

Watabe T, Tanaka K, Torigoo Y, Biswas S. 1980. A note on agriculture and rice cultivation in Assam. Pages 44-52 in Tadayo Watabe (ed) *Some Aspects of Historical Changes of Rice, Report of the Kyoto University Scientific Survey to India and Burma 1978/79*. Kyoto University, Japan.

Genetic Resources of Rice in India : 557-585
Editor : S. D. Sharma
Today & Tomorrow's Printers and Publishers, New Delhi - 110 002 (India)

Chapter 17

Landraces of Rice: Traditional and Participatory Conservation Systems in Jeypore Tract of Orissa

Smita Mishra and Susanta Sekhar Chaudhury

Introduction

Landraces have a certain genetic integrity. They are recognizable morphologically; farmers have names for them and different landraces are understood to differ in adaptation to soil type, time of seeding, date of maturity, height, nutritive value, use and other properties. Most important, they are genetically diverse. Such balanced populations – variable, in equilibrium with both environment and pathogens, and genetically dynamic – are our heritage from past generations of cultivators. They are the result of millennia of natural and artificial selections and are the basic resources upon which future plant breeding must depend.

—— Harlan (1975)

A landrace identified with a unique feature and selected for uniformity over a period of time for maintenance of the characteristic features of the population can evolve into a farmers' variety or even a modern cultivar as, for example, Maruti in the case of pigeon pea (Saxena and Singh, 2006). Conversely, a modern cultivar grown over a time by the farmers and not maintained as per the principles of breeding can 'evolve' into a landrace (Ramanandan, 1997).

According to Ramiah and Ghose (1951), Jeypore tract of Orissa is a secondary centre of origin of Asian cultivated rice (*Oryza sativa* L.) and, according to Sharma, Tripathy and Biswal (2000), it is the primary centre of origin of *aus* ecotype of rice. Before the introduction of high yielding rice varieties in mid-sixties, a large number of landraces were being cultivated in the Jeypore tract but, with the advent of high yielding varieties (HYV) and intensive campaigning by the Department of Agriculture, Government of Orissa to introduce and popularise the high yielding rice varieties, the tribal farm families have progressively given up cultivation of their traditional varieties and have gone for the modern rice varieties. This has resulted in the loss of genetic diversity of rice in one of the centres of origin and genetic diversity of cultivated rices. This chapter describes the situation with regard to landraces of rice in this Jeypore tract of Orissa and efforts of the M.S. Swaminathan Research Foundation, Chennai to conserve the traditional rice germplasm in farmers' fields with a participatory approach, to enhance the value of their landraces and help them to claim Farmers' Rights for these landraces.

Background Information

The southern part of Orissa, now comprising of Malkangiri, Nabarangpur, Koraput and Rayagada districts and its adjacent areas is often referred to as Jeypore tract in rice literature after its main commercial town Jeypore. This area was known as Jeypore Agency Area during British rule as it comprised of princely states and was looked after by a Political Agent of the British Government. Before the creation of a separate Orissa province in 1936, this area formed a part of Madras Province. After Independence and merger of the princely states with Orissa province in 1948, this area formed a single district called Koraput. In 1992, Koraput was split into four districts as mentioned already. The whole tract is covered with hills and forests and inhabited by tribal people who belong to Austro-Asiatic stock. Their main crops are rice and *ragi* (*Eleusine coracana*). Earlier, the tribal people used to follow slash-and-burn method of cultivation in the hill slopes but, due to discouragement by the government agencies, they are gradually giving up this practice. In the plains, they sow rice seeds by broadcasting. They transplant rice in lowlands and valleys of streams and brooks and now in irrigated lands if available.

Dr K. Ramiah (who had presumably visited this area when he was the Paddy Specialist of Madras Province during 1924-37) was highly impressed with the genetic variability of the cultivated rices of this area. He opined in his monograph *Rice Breeding and Genetics* (completed in 1944 but published in 1953) that "although no detailed studies have been undertaken, the wealth of forms present in the Agency tracts of Madras and Orissa is probably an indication that this area might form another centre of origin." The ICAR approved a project for collection and study of the traditional rice varieties of this tract. This project popularly known as Jeypore Botanical Survey (JBS) and executed by the Central Rice Research Institute (CRRI) during 1955-59 collected about 1745 traditional rice varieties and 150 wild rice samples mostly of *rufipogon-nivara-spontanea* complex (Govindaswami and Krishnamurty, 1959). These are known as JBS collections in rice literature. At that time, CRRI could study only some morphological and agronomic characters of these varieties, as the rice breeders were too much preoccupied with *japonica-indica* hybridization programme. Subsequently, the attention of rice breeders of the country was diverted to breeding semi-dwarf plant types. The JBS collections therefore did not receive due attention that they deserved.

The Japanese scientist, Dr H.I. Oka studied the population genetics of these landraces, the wild rices of the area and their natural hybrids and regarded these landraces as forms intermediate between cultivated and wild types "still staying in the midst of differentiation" (Oka and Chang, 1962). The Central Rice Research Institute, Cuttack and NBPGR, New Delhi jointly explored the area (1995-96) for rice germplasm covering a distance of over 3080 km by road. A total of 318 accessions were collected of which 120 were upland varieties, 198 were medium and lowland varieties. In other words, they could collect less than even one-third of the collections made during the Jeypore Botanical Survey in 1950s (Patra and Dhua, 2003). The characterization data reveals enough morphological as well as agronomic variation among the landraces.

During 1995-96, the M.S. Swaminathan Research Foundation, Chennai formulated a project to document and characterize the rice germplasm of Jeypore tract of Orissa to safeguard the intellectual property rights (IPR) of the different tribal and rural communities who have been responsible for the conservation and genetic enhancement of the rice genetic resources with the financial assistance made available by the Ramon Magsaysay Award Foundation, Philippines. It has also taken a leading role in enabling the tribal and rural farm families to derive economic benefit from their past and present contributions in conservation of plant genetic resources and has shaped the operational concept of the Farmers' Right developed by the FAO. In this project, the four districts, namely Koraput, Nabarangpur, Malkangiri and Rayagada of Orissa State (Fig. 17.1) were explored; landraces of rice were collected and the following information was recorded: (a) the names of the women and men who own/conserve/cultivate these landraces, (b) the photographs of the farmer-conservers as a mark of their identity, (c) the audio-recording of the farmer-conservers describing the traits of the landraces conserved by them (as an additional claim of their ownership), and (d) the morphological characters of the seed recorded in a data-sheet and signed by district Agricultural Officers as a still additional proof of ownership of the landraces.

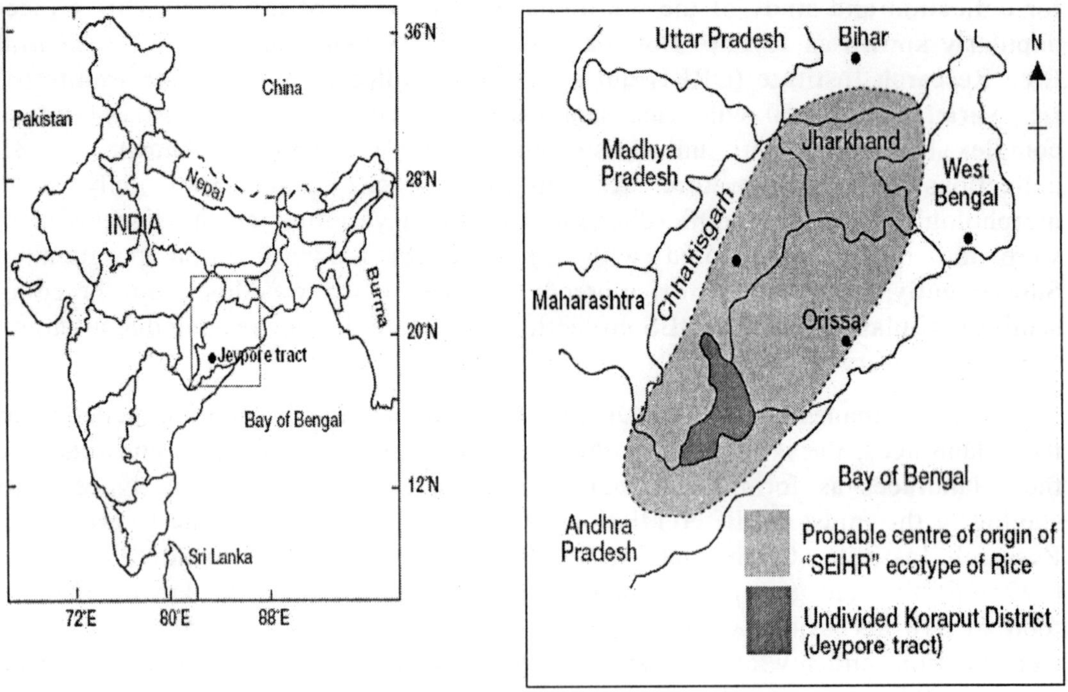

Fig 17.1. Postulated centre of origin of *aus* ecotype in southeast India. (seihr = southeast Indian hill rices)

During the period July 1995–March 1996, the MSSRF team made six tours of the area. Altogether, 256 landraces of rice were collected from 67 villages. Out of these, 58 were short duration, 64 were medium duration and 134 were long duration. These landraces show many primitive features such as presence of awn, black/brown husk and red/light red kernel. A number of landraces were documented and collected with tolerance to biotic and abiotic stress conditions and with special characters like aroma. The tribal communities who had shared their traditional knowledge on landraces and indigenous technical knowledge on rice cultivation and storage were: Paroja, Bhumia, Bhatara, Halva, Gadaba, Kandha, Pentia, Amanatya, Koya, Dora, Kutia Kandha, Bonda, Gond, Kolar, Langia Soura. A list of selected landraces with multiple characters is given in Appendix-17.1 to highlight the value of rice germplasm.

The genetic diversity of the landraces of rice in the Jeypore tract of Orissa has been recorded by many workers (Ramiah and Ghose, 1951; Govindaswami and Krishnamurty, 1959; Oka 1974; Patra and Dhua 2003; Tripathy *et al.*, 2005). These landraces are claimed to contain sources of useful genes for tolerance to physiological and ecological stress, disease and pest resistance and good quality traits (Sharma *et al.,* 1996; Tripathy *et al.,* 2005). Besides the richness in genetic resources of landraces, the area is also rich in the variability of the perennial and annual wild rices, namely, *O. rufipogon* and *O. nivara* respectively. Moreover, the natural hybrids between these landraces and the wild species (known as *O. sativa* f. *spontanea*) are also found in cultivated rice fields.

A wide range of agro-ecological and socio-cultural settings in Jeypore tract has given rise to impressive diversity in landraces of rice, which are highly adapted to local conditions. They have low yielding ability and normally do not respond to high input management. Generally, they are tall with long leaves and many unproductive tillers. They possess grains with better cooking, nutritive and medicinal qualities, besides they are tolerant / resistant to biotic and abiotic stresses of the region.

After the green revolution, government extension agencies started introducing HYVs giving priority to tribal areas for improving production and productivity. In the process the local landraces were replaced gradually with high yielding rice varieties and genetic diversity of rice was lost slowly.

Fifty years back, in 1958-59, landraces were cultivated in 3,19,708.3 hectares in Jeypore tract whereas, in 2009, landraces are grown in only 28,770 hectares which accounts for 9% of the total rice area. Small and marginal farmers who constitute 70% of the farming community and stay in the interior areas cultivate these landraces primarily for their food habit and yield stability under low input agricultural systems with varying biotic and abiotic stress. Swaminathan and Castillo (2000) point out that "local landraces are still being maintained largely by the rural and tribal poor" mainly by default because of their limited access to modern varieties (MVs) and inputs. Landraces may persist because of their good agronomic performance in specific fields or because they are highly valued by farmers for their post-harvest end use qualities (Hugo *et al*., 2003). It has been shown that diversity in an agro-ecosystem is more productive than monocultures (Trenbath, 1974).

Genetic Diversity

The genetic diversity existing in the landraces of rice at Jeypore tract is fairly notable when compared with many other crop plants growing in the region. Multiple landraces are maintained on farm in different areas to address different needs (Bellon, 1996).

Landraces are grown at 60 to 1200 m above the sea level. They show wide range of variation for morphological characters as well as for ecological and physiological adaptability. Crop genetic diversity can be measured by the level of crop varieties, agro-morphological traits and genetic markers (Brown and Hodgkin, 2007). In his *Origin of Species*, Darwin (1971) states a general principle: "The greatest amount of life can be supported by great diversification of structure".

Landraces of rice in Jeypore tract often exhibit tremendous morphological diversity. The colour of the outer layer of the grain can range from black/ purple to red, yellow, brownish and straw with varied anthocyanin pigmentation in different parts of the pericarp. The presence of awn, which is a primitive feature, is also seen in many landraces exclusively in short duration upland rice. They are characterized by plant types ranging in height from (70 cm-147 cm), tillering (2-7), panicle length (16 cm-27 cm), number of filled grains (35-170) and yield (1-3 t/ha). The grain shape varies from short to long and course to fine with many intermediate combinations. The grain quality shows a wide range for palatability, texture and nutritional value. Kernel colour varies from red and light red to off white and white. In the case of specific landraces, light yellow colour has also been noticed. Kernel is diverse in its colour and shows opaque, white and translucent nature externally illustrating primitive to advanced characters. The most highly valued landraces are reserved for social and religious ceremonies and at times offered to distinguished guests. Limited numbers of landraces have a high market value for their superior quality, compared to the normal HYVs and thus appear economically viable. Landraces show a wide range of variability in threshing and milling quality. Some landraces are used for the preparation of dry snacks and others for sweet dishes and sweet snacks. Numerous landraces are selected for everyday food looking at the cooking quality and individual preference for taste. Landraces give reasonably more straw than HYVs for cattle feed and house thatching. Long duration landraces growing in lowlands give comparatively long straws which are used for thatching houses and short duration landraces growing in uplands give short straws which are mostly used as animal feed. Quality of straw varies among landraces for both the purposes. On the other hand, these landraces as a whole have tolerance to biotic and abiotic stresses, aroma and other desirable traits, which may be useful in breeding better varieties.

There are landraces that grow in uplands where water does not accumulate to areas where depth of water is up to three meter. There are short duration photo-insensitive (60-90 days) as well as medium and long duration photosensitive (100-280 days) landraces. Short duration rice is normally known as *chhoto dhan* and long duration rice is known as *bada dhan.* In hilly areas and valleys, specific landraces with short and long duration are grown together in the terraced fields, which are watered by perennial streams locally known as *jhola dhan*. The majority of these landraces are susceptible to lodging at maturity.

Landraces are grown in three different seasons depending on their maturity period. Short duration landraces are grown in uplands; these are sown in May-June and harvested in September-October. They are popularly known as *donger dhan* (autumn rice). Medium and lowland rice is sown in June-July and harvested in November-December and are locally known as *bhata dhan* and *khala dhan* respectively (winter rice). Summer rice (*khara dhan*) is sown in December-January and harvested in April-May. Autumn and winder rice are together known as *kharif* rice.

Community Conservation: Earlier Efforts

The conservation ethos of the tribal communities has enabled to maintain the landrace genetic diversity with associated indigenous technical knowledge. The landrace diversity is an essential component of sustainable crop production, secured food and nutrition in addition to household income for many of the subsistence and marginal tribal and rural farm families. Farmers' decisions on choice of varieties are primarily governed by their intended uses to support household livelihood and food security (Campilan, 2002). Besides being a staple food, rice is used in making processed snacks, alcohols, medicines and has an ornamental value as various decorative items are made out of rice panicles and grains. Several landraces of rice have diverse and multiple uses, which in turn have cultural significance that have co-evolved over a period of time with the genetic diversity. The landraces' names in tribal and local languages often reflect the morphological appearance like husk colour, presence of anthocyanin rings in the nodal region and distribution of anthocyanin pigments in the pericarp/husk. Some of the examples like *Pandakagura* (white and coarse rice), *Muktabali* (smooth and glossy rice like pearl), *Kalajeera* (black, small and scented rice). Landraces ending with the name *chudi (Umuriachudi, Baringchudi, Haldichudi, Dongerchudi, etc)* mean the presence of anthocyanin rings at the leaf sheath or nodes. Similarly, *kavuri (Bodikavuri, Bhatakavuri, etc)* means distribution of anthocyanin pigments in the pericarp/ husk. The names ending with *bhog (Badshabhog, Bastabhog, Deulobhog)* indicate scented rice highly preferred by royal and high society people like kings. Tribes have emotional attachments with traditional rice varieties since they think that these are the only legacy they have received from their forefathers and they respect it. In spite of the easy availability of the high yielding varieties, many farm families cultivate their ancestral landraces for generations. Some of the very popular ancestral landraces still cultivated are *Para, Mati* and *Lalu* in uplands, *Bodikavuri, Gathia, Sapuri* in the medium land and *Umuriachudi, Pathangada* and *Machhakanta* in the lowland, *Sapru, Kerandi* and *Louchei* in the *jhola* and *Kandulkathi, Gurumukhi* and *Baikhana* in the flood plains. They also have a feeling that traditional varieties are God's gift for them whereas HYVs are man-made. They obtain more social recognition from their friends and relatives if they provide traditional landraces/varieties as a gift on specific occasions. This long established belief has encouraged their conservation traditions.

Earlier Studies

Evaluation of landraces and release of improved varieties

Improvement of rice through research has been a continuous process in India and dates back to early 1920s. Pure line selection in the indigenous tall landraces was the most important approach till 1950 with little attention for hybridization. Most of these landraces were tall and evolved by the farmers mainly on the basis of their adaptation to local conditions. These landraces in general are poor yielder but most of them have many other desirable traits. The second period (1950-65) emphasized on inter-racial hybridization between *japonica* and *indica*. The *japonica-indica* hybridization programme taken up by the ICAR and FAO aimed at transferring high yielding ability and nitrogen responsiveness of *japonica* ecotypes into the local *indica* ecotypes which were well-adapted to local conditions and had tolerance to the prevalent disease and pests of the region. As a whole, the programme had very limited success and virtually failed.

The Rice Research Station at Jeypore undertook research on rice in 1937 through a scheme sponsored jointly by the Indian Council of Agricultural Research and the then Maharaja of Jeypore and continued up to 1946 when the State Government took it over. Its objective was to collect samples of local varieties of rice from Koraput and other adjoining districts (except Ganjam which already had a rice research station) and make selections from them with a view to identify high yielding strains, to test newly evolved types against the existing recommended strains of rice and to evolve new varieties of rice by hybridization followed by selection.

The Jeypore station released J1 (a selection from *Bobbilibhutta*) and J2 (a selection from *Mohulkunchi*) for the autumn season, J4 (a selection from *Sorumundabali*) and J5 (a selection from *Chudi*) for mid-winter season and J7 (a selection from *Kerandi*) for the late winter season for the Jeypore tract. Later two more varieties were released as the "Jeypore New Series" varieties, namely, JNS 941-8 (a selection from *Saprulu*) and JNS 601-8 (a selection from *Godakhunta*). Table 17.1 provides the list of rice varieties bred at the Rice Research Station at Jeypore during 1937 to 1942.

Table 17.1. Details of improved rice varieties of Jeypore tract released in 1940s.

Sl no	Landraces name	Duration (days) Land type	Leaf sheath colour	Husk colour	Kernel colour	Quality	Average yield (q/acre)
J1	Bobbli bhutta	105 UL	Purple	Straw	Translucent White	Superfine	4.5-5.4
J2	Mohulkunchi	110 UL	Purple	Brown	White	Medium fine	5.4-6.5
J4	Sorumundabali	145 ML	Green straw	Bright	Translucent White	Medium fine	9.97
J5	Chudi	150 ML	Light purple	Straw	White	Medium fine	10.88-11.79
J7	Kerandi	165 LL	Green	Straw	White	Course	10.88
JNS. 941-8	Saprulu	135ML	Green	Fawn	White	Superfine	NA
JNS. 601-8	Godakhunta	145 ML	Purple	Fawn	White	Fine	NA
J10	Barang Chudi	130 UL	Green	Fawn	White	Course	NA
J11	Bayagunda	150 LL	Green	Fawn	White	Medium	NA fine

Source: Chalam and Mohanty (1956)

Seed multiplication and distribution

By 1942, nine promising improved varieties were bred and recommended by the Rice Research Station, Jeypore. Subsequently, to multiply and distribute the seeds among the farmers, small seed farms were established in spite of the restricted condition of finance due to the Second World War. During 1968 to 2005, the Orissa University of Agriculture and Technology (OUAT) have used these landraces in the breeding programmes and developed new varieties, namely, *Padma, Jayanti, Vijaya, Pooja, CR 104* and *Ketaki Joha.*

Traditional Knowledge

The varietal diversity in the cultivated rice in Orissa as a whole is quite large and reflects the great diversity in the agro-ecological condition under which the crop has been cultivated in the state for thousands of years. In an exploration mission in early 1930s, nearly 3000 landraces were collected from six districts of Orissa including Koraput (Chalam and Behera, 1956). Botanical and agronomical studies revealed that there were repetitions of the same genotype grown under different names at different places without any morphological differences among them. Many of these indigenous collections are now lost due to lack of proper maintenance.

Food value

Rice is the staple food cultivated in diverse agro-ecosystems. Early maturing (60-100 days), photo-insensitive landraces which are grown in bunded as well as unbunded but fragmented uplands, mature in September/October and provide food in the scarcity period of the annual agricultural cycle. The tribal farm families have been cultivating these landraces exclusively for their own consumption and to celebrate *Nuakhia* i.e. the 'new rice eating ceremony'. Of all the landraces grown in the uplands, the tribal communities have been cultivating *Para* and *Mati* since time immemorial. Many of these landraces are course and red and hence suitable for making rice flakes. Broad, thick and red rice flakes are very popular in the region. A particular indigenous tribe, *Soura* has the expertise in making sweet balls using these rice flakes. In addition, some landraces are used for preparing specific dishes. Carpenter (2005), in a study of *in situ* conservation of rice genetic diversity in the Philippines, for example, found that traditional varieties of glutinous rice were primarily conserved for the production of local sweet delicacies. Sometimes tribal people barter upland rice for other essential commodities like edible oil, salt, etc. At times, this rice is also given as wages to labourers working in their rice fields. Early maturing landraces with different periods of maturity have been conserved as these reduce the burden of work in the field. At the same time, they continue to supply food until the medium duration rice is ready for harvest. Besides, they also insure against total crop failure, as these are tolerant to disease and pest. A few of these are drought and lodging resistant making the farm families assured of a reasonable harvest. Short duration landraces are suitable for mixed cropping practice, which provides a combination of cereals, pulses, millets and oilseeds from a common field and ensures food supply over a period of time. The straw is used as an animal feed (Mishra *et al.*, 2008).

Landraces, which grow in medium lands, mature between 100-130 days. The tribal farm families have conserved these landraces as these fulfil the food needs over a longer period of time due to their higher yield. In this category, there are course grained as well as fine grained landraces which are used for home consumption and marketing. Farm families prefer these landraces as they are suitable for broadcasting as well as transplanting and fit with the weather conditions and also with the economic condition of the tribal farmers. Most importantly, yield potential of these landraces is the same for irrigated and rainfed systems. Many landraces are suitable for making pop, puffed and flake rice for social functions, ceremonies and rituals. Course grained landraces are used for flake rice; white and fine grains are preferentially used for making pop rice and both red and white grain types are used for making puffed rice. Tribal community like *Soura* have made it as their livelihood and prepare sweet balls for marketing. Aromatic landraces with medium duration are used for making special dishes during specific ceremonies and functions. Landraces grown in this land category possess good milling quality and hence give them a higher market price. Tall and lodging resistant plants provide straw for thatching and also fodder for the livestock. Landraces with long and broad leaves have been preferred to suppress the weed growth. Besides their food need, the tribal communities have conserved many landraces with purple leaf sheath and, every year or once in two years, they alter such landraces with that of green leaf sheath to make weeding easy and less expensive. Contrary to the short duration landraces, white kernel landraces are offered in the functions and ceremonies.

Landraces, which take more than 130 days to mature, have been grown in lowland where water is retained either from perennial streams or from rain. Tribal communities fulfil their economic needs from these landraces as yield is more when compared with the other two types of land. Tribal farm families put all their efforts and invest money for these landraces as they get a good return and a superior market price. They have, therefore, conserved many more landraces suitable for this type of land. Landraces with various morphological and agronomic traits are available in this group. Panicles are large in size and grains are fine with good palatability. Other characters are well matched with the medium duration landraces. These landraces are used in social ceremonies like marriage.

Tribal communities living in the hilly regions have recognized and conserved a number of landraces suitable for their terraced fields in the valleys or in the beds of small streams. They mix both long and short duration landraces and broadcast them together long before the onset of rainy season. They harvest short duration rice in three months which provides them immediate food in the lean period; the long duration rice yields more and provides food for the rest of the year. These landraces have coarse grain, red kernel and awn, are tasty and give good yield. Long duration landraces also provide enough fodder for their livestock. These landraces are often tolerant to flood and submergence.

The wild relatives of rice are frequent and abundant in Jeypore tract. It is assumed that before its domestication, tribal communities were gathering these for their consumption along with other foods. Even today wild rice (*O. rufipogon*) is found in the ponds of Kotpad and Borigumma. The two ponds seen in the plains of Borigumma and Kotpad are most probably a part of the religious culture adhered to by Jains since the fifth and sixth century AD. They grew wild paddy in one of the ponds which still exists in the region. Small boats were used to harvest the paddy. The processed wild rice called *Titing chuda* (rice flakes) is still sold during festivals (Das, 2002). The annual wild rice (*O. nivara*) grows in seasonal pools in the hills of Banuaguda. Tribal communities continue to collect wild rices from seasonal ponds in and around the cultivated fields even in present time i.e. after a long period of its domestication. The annual wild species is gregarious and synchronous in flowering and seeds are abundant at the time of its maturity. It is "harvested' even today by primitive tribes for subsistence as it matures during food shortage period and by non-tribal people for consumption on days of fasting as it is considered as 'god-given' (*deobhat*) rice (Tripathy, 1994; Sharma *et al.,* 2000; Nanda, 2002).

During the month of November, tribal communities celebrate *Diwali* festival when they prepare parboiled rice from the wild rice and feed it to the cattle. On this day offerings are made to the forefathers, cattle are worshipped to show their gratitude to them and annual bonus is given to the cattle-boy. Many of the tribal communities have been following this tradition from the time of their ancestors. They collect the wild rice (*O. sativa* f. *spontanea*) locally known as *jhara* or *balunga* from the cultivated field several times before it sheds/ shatters seeds. In a day, they can collect nearly 5-6 kg seeds of wild rice which can provide them food for a day or two during the period of scarcity. Rice flakes of wild rice are sweet, tasty, bigger in size and yellow in colour. Sometimes, rice flakes are made which plays a role in conveying the marriage proposal for their children.

Nutritional value

Tribal communities believe that the nutritional value of their landraces is better than that of the modern high yielding rice varieties. They feel that their landraces provide all the energy needed by them to work. According to them, the rice cooked from their local varieties stays in their stomach for a longer period giving them the feeling of being satiated and thus enables them to work for a longer period. With a single meal of the local rice in the morning, they manage to work till evening without feeling tired. Rice prepared from their landraces is tastier than that from the modern varieties and tribal people hardly need any side dishes for that. It swells more on cooking giving them a larger volume to eat. Some of the landraces like *Umuriachudi, Kalachudi, Kalamalli, Machhakanta,*

Meyer, Sapuri, Pathangada, Bhatagunda, Haldichudi are highly preferred as they swell more on cooking as compared to other landraces. For example, 750 gm of the above mentioned rice when cooked would be enough for a family of four members as compared with one kg of other varieties. Grains of some landraces are heavy while that of others are light. Ramiah and Rao (1953) have reported that iron and zinc content of landraces with red kernel is 2-3 times more than that of white rices.

Tribal people process their rice manually thus preserving the bran layer which has a high nutritional value. Proteins, fats, vitamins and minerals are present in greater quantity in the outer layer of the grain than in the inner layer which is starchy endosperm. They parboil the paddy before pounding. In this process paddy is soaked in water after which it is steamed for a short period (Frei and Becker, 2004; Rahman *et al.,* 2006; Ahuja *et al.,* 2007).

Tribal people prefer to consume course grains which are often red rices. The gruel of the course rice is considered very useful which keeps them energetic and fresh. They have preserved a number of landraces having red kernel which when cooked provide a thick gruel and thus help them to work hard during agricultural season. Some of the well-known red rices are *Para* of short duration, *Kerandi* of *jhola*, *Bodikaveri* of medium duration and *Kandulkanti* of long duration. Course grain with red kernel gives them food along with thick gruel. These are also suitable for making pressed rice.

The tribal people believe that red rices are suitable for various agro-ecological conditions and adverse situations. They show resistance to drought, flood, submergence, pests and diseases. Red rices are comparatively more resistant to storage insects than normal rice varieties.

Cultural value

Rice is an integral part of tribal culture which is evident from the existence/ maintenance of large diversity in its landraces. Tribal communities have conserved such a large diversity for traditional usages on various occasions. The evolution of rice-related cultural diversity has influenced the conservation of rice genetic diversity. Studies by de Candolle, Vavilov, Harlan, Wellhausen and many others have shown a positive correlation between cultural diversity and genetic diversity demonstrating the impact of human communities on the conservation and use of biodiversity. Agrobiodiversity is largely the result of interaction between human and nature (Swaminathan, 2002). Rice landraces have been selected and developed to meet specific cultural needs, for example, early maturing landraces with red kernel have been conserved to celebrate *Nuakhia* in addition to yield stability.

Tribal farm families grow red rice to suit to various agro-ecological conditions and to yield under adverse circumstances.

Some landraces have cultural and religious "use value"; thus conservation of agricultural biodiversity is closely intertwined with the conservation of indigenous culture (Gonzales, 2000). Specific landraces with white kernel are meant to be used during the festivals and social and religious functions. These landraces have different maturity periods to match harvest with various occasions. Long, slender, scented and white grains are selectively grown to be given as bride price in tribal marriages which provide a higher social status in the community. Long duration landraces like *Sapuri, Machhakanta, Umuriachudi* and *Haldichudi* etc. have long been used for this purpose. Certain ancestral landraces have been maintained across different communities and are used in rice-based traditions like *Nuakhia* (eating of new rice) in September, *Dussera* (Durga / Shakti Puja) in October, *Diwali* (worshipping cattle, offering to forefathers) in November, *Push Parab* (end of cropping season, merry making) in December and *Chait Parab* (visiting friends and relatives) in April. Besides different landraces of rice are being used in rituals associated with rice sowing, planting, first harvest eating and harvesting rites which are common among different tribes. In addition to this, tribes make use of rice in different functions related to restoring the soil fertility, protection of crops from damage and human and livestock welfare which shows that rice plays a significant role in their cultural heritage. Brush and Meng (1998) and Gurung and Vaidya (1998) indicated that economic status and food culture or culinary preferences promote diversity on farm. Tribal farm families have managed to preserve the seeds of precious rice landraces inside the village/within the community through barter systems with a continuous series of seed saving and replanting leading to safe conservation. As a part of tribal culture, they have been cultivating landraces with purple leaf sheath after every two to three years to identify and weed out the wild rices from their fields. Pigmented grains have been protected as they help the farm families in rouging before seed selection and removing off types while cleaning the seeds. In each life cycle ritual, rice plays a prominent role. During naming ceremony, marriage and death ceremony, they use several kinds of rice currently available with them (red, light red, white) but for offering to god and in magico-religious practices, they strictly use white rice. Doors of houses are decorated with the ear heads of rice showing their deep touch with rice culture.

In tribal culture of Jeypore tract, magico-religious rites are performed throughout the annual agricultural cycle and food collection. The process starts in the month of February from clearing hill forest for shifting cultivation and ends in December/ January with the threshing of rice. A vital part of the tribal culture is the celebration of festivals before consuming new crops and fruits.

Ecological value

The ancestors of the present day tribes have carefully selected landraces which were best suited to their diverse agro-ecological conditions. There are landraces which grow in the rainfed waterlogged lowlands (where sometimes water height exceeds 150-180 cm), in terraced lands which are irrigated by perennial streams, in the irrigated or rainfed medium lands and in uplands without any standing water.

Specific Characters of Landraces as Documented by MSSRF

Twenty four aromatic landraces were documented from upland, medium land and lowland conditions. Majority of the aromatic landraces were collected from the northern part of Jeypore and were cultivated mostly by non-tribal farm families. These landraces are economically important and fetch a good market price compared to the other rices. Some of the farm families cultivate these for their own consumption during special functions and ceremonies while many others grow for commercial purpose in addition to their own consumption. Small, oval and white grains with both mild and strong scent are preferred for making *payas* (a sweet dish) on special days.

Twenty seven landraces grown in the flood plains were documented mostly from the northern part of Jeypore which the farm families claim as flood resistant. These landraces survive for 7-8 days under 2 m water. Many of them are tall and yield long straw for thatching and animal feed. They are late maturing and have bold grains and good yield potential. Several farm families have been cultivating these landraces for the past 50-60 years.

The tribal farm families have recognized two landraces which can survive for a month in submerged condition even under more than 4 m deep water. These landraces yield about 3.5 t/ha and deliver 2.5 m long straw which is useful for thatching their roofs.

The tribal farm families have preserved 15 landraces which are grown in the unbunded upland and are used for mixed cropping with *ragi*, *arhar* and *jowar*. These landraces are claimed to be drought resistant as they can grow under dry condition in soils with low fertility. A medium duration landrace was also reported as drought resistant which can tolerate a long spell of dry weather.

Two landraces were documented with profuse tillering habit. Individual plants of these landraces look like a bush with 16-18 tillers per plant under good cultural practices with average yield potential. Besides grains, these landraces provide plenty of straw for thatching and feeding their cattle.

Pigmentation

One landrace has purple leaves. This particular landrace is grown to identify all rogue (green leaf) plants and weed them out from the field at the vegetative stage. It is grown once in three years to eradicate the mixtures and weed rice completely from the rice field.

Red rices

The indigenous communities have conserved a number of landraces that have red or light red kernel. There are early and late maturing landraces of such types adapted to different agro-ecological zones. Most of the red rices are coarse-grained and have purple leaf sheath and awns. However, some short and long grained landraces are also available with aroma. On the whole, 125 landraces were documented as having red and light red kernel and accounted for 48.82% of the total collection. Duration-wise, 63.79% of short duration landraces (grown in uplands), 39.06% of medium duration landraces and 47% of the long duration landraces were of red and light red kernel.

Rice flakes

Red and light red course rice is often used for making pressed rice as it gives flat, thick and broad rice flakes which is considered superior in quality and is preferred across the rural and tribal communities. At times, they add more value to it by double frying and making sweet balls for house and market.

Pop rice

Tribal communities make use of white kernel rice to make pop rice as it has a high demand during marriages and in other social functions. White look gives a pleasant sense and is appreciated by all. Big and small sweet balls are also prepared for own consumptions, for functions and ceremonies and even to sell in the local markets.

Puffed rice

Both white and red rice are used to make puffed rice. Course rice is preferred to get big and heavy puffs which is profitable in addition to being sweet and tasty.

Participatory Conservation Programme (PCS)

The MSSRF has established the Scarascia Mugnozza Community Genetic Resource Centre and Gene Bank at Chennai whose mission is to revitalize conservation of agrobiodiversity and cultural ethos of tribal and rural farm families through social prestige and economic reward for their invaluable contributions to food and nutrition security. The Gene Bank is a medium term storage facility maintained at 4°c and 25% RH. A duplicate sample of each accession is also stored in the long-term storage at the National Gene Bank as an additional safeguard. They are accessible, subject to Indian laws, by any party with Prior Informed Consent (PIC) of the holder community. MSSRF facilitates such access through mutually agreed terms and Material Transfer Agreement (MTA).

All accessions have detailed digitized database called Farmers' Right Information System (FRIS) is devised to establish the intellectual property rights of farmers on their landraces. The database contains traditional knowledge associated with each accession, their passport data and nationally and internationally accepted scientific descriptors.

Since 1998 MSSRF has attempted to recognize community conservation and mainstream it into poverty reduction. MSSRF commenced work on a project mode in partnership with local tribal communities on revitalization of *in-situ* and on-farm conservation traditions and livelihood systems through conservation of agrobiodiversity.

In the first phase of the work, the project facilitated tribal communities to collect and conserve 98 landraces of rice on-farm with the technical guidance of MSSRF staff. In addition, a Participatory Plant Breeding initiative helped to impart and evolve techniques for quality seed production and scientific/improved methods of cultivation to increase crop productivity and to maintain soil fertility.

To ensure conservation of landraces of rice as well as food security, tribal communities established a Community Gene-Seed-Grain Bank; MSSRF staff facilitated this work at seven tribal dominated villages. These banks have been designed to work in an integrated manner to support the chain from conservation to cultivation and consumption. Landraces with distinct characteristics are conserved for long term, both by ensuring their continued cultivation *in situ* and on-farm through community participation and by depositing samples at the Community Gene Bank at MSSRF, Chennai. Loss of viability or shortage of seeds can be met by the Gene Bank samples. Excess and non-viable seeds can be diverted to the Grain Bank to be used as grains (Arunachalm *et al.,* 2006).

In 2000 under Field Gene Bank activity, 99 farmers in 114 plots multiplied 98 landraces of rice. Around 55 agronomic characters were collected in different stages of plant growth according to International Board for Plant Genetic Resources (IBPGR) data format. Some of the landraces gave very good yield and showed early maturity and hence farm families decided to multiply those landraces in large scale. The selected landraces were *Haladichudi, Bhaktichudi, Pathanagada, Umurichudi, Osagathiali, Assamchudi, Kuyerkuling, Machhakanta, Rajamuan, Tulashiganthi and Karloili* (MSSRF, 2006).

With the experience gained from field Gene Bank, farm families started realizing the economic/commercial value of their landraces and the urgent need to conserve them. They volunteered land and labour to conserve landraces having different maturity dates. MSSRF provided seeds of desired landraces as start up. The following year, after regeneration, farm families returned good quantity of seeds, which were distributed in the next cropping season leading to a self-sustaining cycle. The process has become a part of the cultivation practice and the number of purified landraces started increasing along with larger participation of farm families and area under cultivation. The process of Participatory Conservation System (PCS) has become institutionalized with the guidance of MSSRF. One of the chief objectives of PCS is to conserve and validate good quality rice landraces on farm. Preferred landraces were grown along the main crop as border crop and cultivated in the same plot. In this process each farm family conserved 10 landraces along both the borders in 3 to 5 lines each. Seven farm families have conserved 70 landraces in the border. This activity created an attitude among farm families towards conservation and promotion of other landraces for future utilization. Thus 45 landraces were revitalized in 2001, 64 in 2002, 64 in 2003 and 39 in 2004 through PCS. An analysis of 17 common landraces across the years based on six productivity parameters viz, plant height, number of tillers/plant, number of panicles/plant, panicle length, number of grains/ panicle and grain-fill index, has placed two landraces, Alasikiba and Baramasi, at the top, with the rest aligned in medium (13 LRs) and low (2 LRs) performing groups (MSSRF Annual Report, 2003-04).

This organisational structure of participatory conservation has provided a network to save and revive site-specific rice genetic resources and make valuable genes available for scientific research and provide secure and sustainable livelihood of farm families. Participatory conservation with a bottom-up approach empowering farmer-conservers through field gene banks has been acknowledged to have singular options for benefit sharing (Arunachalam, 2000).

Worede and Mekbib (1993) provide case histories and varied instances from Ethiopia in support of the participatory genetic resource conservation. A few important ones are re-emphasised below:

- *In-situ* on-farm conservation provides a mechanism through which the evolutionary systems responsible for the generation of variability are sustained.

- Field gene banks can be a source of seeds which are adapted to extreme conditions for post-drought planting or extreme climate events like floods and cyclone.
- Material in field gene banks can be used to restore cultivation in areas abandoned due to consistent crop failures.
- On-site maintenance would sustain distinctive traits, which were acquired over a long time under varying traditional, cultural and cropping practices.
- Field gene banks can provide valuable genetic material to formal interventions for breeding research opening new avenues of benefit accrual to farmers like participatory breeding and royalty income.
- Field gene banks reduce the load of regeneration of germplasm.
- Farming community-managed field gene banks can take adequate care of site-specific genetic resources and prevent genetic erosion.
- Field gene banks would provide an ideal framework to serve as genetic enhancement centres. Scientists and farmers can work together to expand the utility of landraces and local varieties by generating high yielding populations (pure lines, mixtures or open-pollinated varieties) carrying farmer-desired traits. They can even take up plant improvement based on specific molecular techniques. The successful case histories of *Hevea* and sugarcane employing 'portable (molecular) laboratories' provide encouraging evidence (Lenaud and Lebot, 1997).

MSSRF's partnership efforts with tribal communities in Jeypore tract of Orissa in conservation of rice genetic diversity *in situ* and on-farm for poverty reduction was recognized at the World Summit on Sustainable Development (WSSD) in Johannesburg in August 2002 when the communities received the *Equator Initiative's* Innovative Partnerships Award for Sustainable Development in Tropical Ecosystems. The *Equator Initiative*, a partnership of UNDP, UN's Global Development Network, IUCN, Government of Canada and few other agencies, is a global movement committed to identifying and supporting innovative partnerships that reduce poverty through the conservation and sustainable use of biodiversity. The programme is based on the recognition that, while biodiversity loss and poverty are increasing in most tropical countries, indigenous and other local communities are rising to meet these challenges in creative and effective ways.

Profitable Use of Landraces

A participatory work plan was prepared to carry farm families along the scientists of MSSRF to derive maximum benefits from the existing landraces of rice. In this process, farmers' indigenous technical knowledge (ITK) and ITK-derived agricultural practices were refined with scientific methods of crop improvement.

Around 26 landraces from upland, medium land and lowland were field-tested by farm families in about 9 villages using both their traditional practices and modified methods side by side. They realised striking advantages from modified practices like no increase in cost inputs, significant increase in yield and quality of grains with associated benefits of less time and labour for transplanting and weeding. Transfer of knowledge from farmer-to-farmer on concepts and practice of modified cultivation occurred rapidly across the farming communities resulting in willingness of voluntary participation in modified cultivation by more farm families from other villages.

MSSRF showcased the work of Jeypore farmers on *Kalajeera,* a black and highly scented rice by displaying around 100 kg of rice in the Orissa Government sponsored fair, "Parab" in 2002. Observing the attractive price of *Kalajeera* rice Rs 25 a kg tribal farmers came forward to go for commercial multiplication of *Kalajeera*. They preferred *Kalajeera* for its high aroma, taste, good cooking quality and expected higher market price. There were scopes for improving its productivity and production through participatory seed purification and other improved agronomic practices. Commercial cultivation of *Kalajeera* expanded among farm families spread over 10 hamlets around Jeypore during *kharif* 2004, for example. Steady expansion of farm families involved, acreage under cultivation and total production of *Kalajeera* reached around 25 hamlets, 112 acres and 100 tonnes respectively during 2006. *Kalajeera* got high consumer appeal not only within India but also outside for its good nutritional and cooking quality (Arunachalam *et al.,* 2007). The *Kalajeera* example showed that conservation and commercialization can become mutually reinforcing. This is Jeypore's unique contribution to the world of rice.

In recognition of their outstanding selfless services in conserving, improving and making available plant genetic resources for the development of new plant varieties, the Protection of Plant Varieties and Farmers' Rights Authority (PPVFRA) awarded the certificate of "Plant Genome Savior Community Recognition" for the year 2007 to the tribal communities of Jeypore tract.

The Koraput region of Orissa has been identified as agrobiodiversity hotspots and needs to be protected from genetic erosion. A National Gene Fund

Box 17.1

Conservation of Landraces of Rice

The rice landraces dwindled from 1745 to 325 by 1960 in the Jeypore tract. To mitigate this fast erosion, efforts were made to conserve and preserve these varieties *in-situ* on farm and also in cryogenic system after collecting passport data with the prior inform consent. These varieties possess desirable traits like resistance to drought, diseases, pests and flood. An integrated 4C Approach (Conservation, Cultivation, Consumption and Commercialization) is adopted for sustainable use and equitable sharing of benefits arising out of biodiversity.

Participatory Conservation System

In this scenario, conservation of plant genetic resources by communities is a priority area. The farming communities have conserved 106 landraces of rice through field gene-seed bank from which 37 are cultivated on large scale. Five popular varieties (Machhakanta, Kalajeera, Haladichudi, Gathia, Umeriachudi) are under consideration of PPVFR for farmers variety registration. Conservation efforts in the area of agrobiodiversity by tribal farm communities in Jeypore received international and national recognition: Equator Initiative Award – 2002, Krishi Bisharad Sanman – 2002, Genome Saviour Award – 2006, Folk Fair Award – 2008.

Participatory Plant Breeding (PPB)

PPB was chosen as the path to demonstrate and test cultivation processes in participation with tribal farmers. The major focus was on pure seed production from mother and primary tillers in a participatory approach so that the farm families could derive the benefit of pure seeds. A set of 19 modules of modified methods were prepared and implemented to enhance the yield potential of rice landraces through reduced seed rate and improved agronomic practices. The extent of yield improvement varied from 30-70% with no extra inputs and costs.

(Reproduced from a handout distributed by M.S. Swaminathan Research Foundation at State Level Consultation on "Community Management of Agro-biodiversity in an Era of Climate Change" held at Orissa University of Agriculture and Technology, Bhubaneswar on January 30, 2010.)

is constituted under Section 45 of the Plant Variety Protection and Farmers' Rights Act (PPV& FR Act) 2001 and is meant for supporting and rewarding farmers, community of farmers particularly tribal and rural communities engaged in conservation, improvement and preservation of genetic resources of economic plants and their wild relatives in areas identified as agrobiodiversity hotspots. It will also help for capacity building and supporting *in situ* conservation at the level of the local body.

The poor tribal farm families of Jeypore tract should receive Government support for conservation, cultivation, consumption and sustainable marketing of valuable but vanishing rice diversity in the centre of origin and diversity. The path laid over the past one decade using the principles of community conservation approach is a model path for secure and sustainable livelihood of tribal farm families for conservation and enhancement of rice genetic diversity.

Acknowledgements:

The authors wish to thank Swiss Agency for Development and Cooperation, New Delhi for financial support and tribal farm families of Jeypore tract for sharing their knowledge and time.

References:

Ahuja U, Ahuja SC, Chaudhary N, Thakrar R. 2007. Red Rices - Past, Present and Future. *Asian Agri-History* 11(4):291-304.

Arunachalam V. 2000. Participatory Conservation: A Means of Encouraging Community Biodiversity. *Plant Genetic Resources Newsletter* (IPGRI, Rome) 122:1-6.

Arunachalam V, Chaudhury SS, Ray T, Sarangi SK, Ray T, Mohanty BP, Nambi VA, Mishra S. 2006. *Rising on Rice: The Story of Jeypore.* A manual published by Ms Swaminathan Research Foundation. AMM Prints, Chennai. 39 pp.

Arunachalam V, Chaudhury SS, Ray T, Sarangi SK, Mohanty BP, Nambi VA. 2007. *Jeypore Kalajeera: Orissa's Gift to Quality Rice Lovers.* A manual published by Ms Swaminathan Research Foundation and Swiss Agency for Development and Cooperation. AMM Prints, Chennai. 16 pp.

Bellon MR. 1996. The dynamics of crop infra-specific diversity: A conceptual framework at the farmer level. *Economic Botany* 50:26–39.

Brown AHD, Hodgkin T. 2007. Measuring, managing and maintaining crop genetic diversity on farm. Pages 13-33 in Jarvis D, Padoch C and Cooper D (eds) *Managing Biodiversity in Agricultural Ecosystems.* Colombia University Press, New York.

Brush SB, Meng E. 1998. Farmers' valuation and conservation of crop genetic resources. *Genetic Resources and Crop Evolution* 45:139–150.

Campilan, D. 2002. *Farmer Management of Sweet Potato Diversity in a Changing Livelihood System: A Case from Central Luzon, Philippines.* A sourcebook produced by CIPUPWARD in partnership with GTZ GmbH, IDRC, IPGRI, and SEARICE.

Carpenter D. 2005. The *in situ* conservation of rice plant genetic diversity: A case study from a Philippine barangay. *Agriculture and Human Values* 22(4):421–434.

Chalam GV, Behera RN. 1956. Botanical Research on Rice. Pages 20-56 in *Rice in Orissa* published by Department of Agriculture, Government of Orissa. (Reprinted by the Home (Public Relation) Department, Government of Orissa, Bhubaneswar in 2009).

Chalam GV, Mohanty HK. 1956. Improved Varieties of Rice. Pages 88-102 in *Rice in Orissa.* published by Department of Agriculture, Government of Orissa. (Reprinted by the Home (Public Relation) Department, Government of Orissa, Bhubaneswar in 2009).

Darwin C. 1882. *The Origin of Species* (Sixth Edition). Reprint edition published in1972 by Dent JM and Sons Ltd, London. 488 pp.

Das Kornel. 2006. *Tribal Crop-Livestock Systems in South-East India.* Manohar Publishers & Distributors, New Delhi. 283 pp.

Frei M, Becker K. 2004. Agro-biodiversity in subsistence-oriented farming systems in a Philippine upland region: Nutritional considerations. *Biodiversity and Conservation* 13(8):1591–1610.

Gonzales T. 2000. The culture of the seed in the Peruvian Andes. Pages 193-216 in Brush SB (ed) *Genes in the Field: On-Farm Conservation of Crop Diversity.* Lewis Publishers, Boca Raton, FL, USA.

Govindaswami S, Krishnamurty A. 1959. Genetic variability among cultivated rices of Jeypore tract and its utility in rice breeding. *Rice News Teller* 7(3/4):12-15.

Gurung JB, Vaidya A.1998. The benefits to agrobiodiversity of the wild range of food culture in Nepal. Pages 55-62 in Tej Pratap and Sthapit B. (eds) *Managing Agrobiodiversity: Farmers' Changing Perspectives and Institutional Responses in the Hindu Kush Himalayan Region Kathmandu.* IPGRI/ICIMOD, Rome, Italy.

Harlan, JR. (1975). Our vanishing genetic resources. *Science* 188:617-621.

Hugo PR, Brush SB, Qualset CO. 2003. Landraces of maize in Central Mexico: An altitudinal transect. *Economic Botany* 57(1):7–20.

Lenaud C, Lebot V. 1997. Molecular techniques for increased use of genetic resources. Pages 92-97 in Ayad WG, Hodgkin T, Jaradat A and Rao VR (eds) *Molecular Genetic Techniques for Plant Genetic Resources.* IPGRI, Rome, Italy.

Mishra S, Chaudhury SS, Swain S, Ray T. 2008. Multiple Cropping System: A Way for Conservation and Sustainable Use in Jeypore tract of Orissa, India. *Asian Agri- History* 13(1): 39-51.

MS Swaminathan Research Foundation. 2003-04. *Fourteenth Annual Report.* Biodiversity and Biotechnology, p 47-48.

MS Swaminathan Research Foundation. 2006. *Biodiversity Conservation, Integrated Natural Resource Management and Poverty Reduction.* Phase II completion Report submitted to Swiss

Agency for Development and Cooperation (SDC*)*, New Delhi.

Nanda JS. 2002. Antiquity and spread of rice cultivation. Pages 331-346 in Nanda JS and Sharma SD (eds) *Rice Breeding and Genetics: Research Priorities and Challenges.* Science Publishers, Enfield and Oxford and IBH, New Delhi, p 349-369.

Oka HI. 1974. Experimental studies on the origin of cultivated rice. *Genetics* 78:475-486.

Oka HI, Chang WT. 1962. Rice varieties intermediate between wild and cultivated forms and the origin of the *japonica* type. *Bot Bull Acad Sinica* 3(1):109-131.

Patra BC, Dhua SR. 2003. Agro-morphological diversity scenario in upland rice germplasm of Jeypore tract. *Genetic Resources and Crop Evolution* 50(8):825-828.

Rahman S, Sharma MP, Sahai S. 2006. Nutritional and medicinal value of some indigenous rice varieties. *Indian Journal of Traditional Knowledge* 6:454–458.

Ramanandan P. 1997. Pigeonpea: Genetic Resources. Pages 89-116 in Nene, YL (ed) *The Pigeonpea.* CAB International, Wallingford, UK.

Ramiah K. (1953). *Rice Breeding and Genetics.* Scientific Monograph No. 19, Indian Council of Agricultural Research, New Delhi. 276 pp.

Ramiah K, Rao MVBN. 1953. *Rice Breeding and Genetics.* ICAR Scientific Monograph No- 19, Indian Council of Agricultural Research, New Delhi, India. 276 pp.

Ramiah K, Ghose RLM. 1951. Origin and distribution of cultivated plants of south Asia - Rice. *Indian Journal of Genetics* 11:7-13.

Saxena S, Singh AK. 2006. Revisit to definitions and need for inventorization or registration of landrace, folk, farmers' and traditional varieties. *Current Science* 91(11):1451- 1454.

Sharma SD, Tripathy S, Biswal J. 2000. Origin of *O. sativa* and its ecotypes. Pages 349-369 in Nanda JS (ed) *Rice Breeding and Genetics: Research Priorities and Challenges.* Science Publishers, Enfield and Oxford and IBH, New Delhi.

Sharma SD, Tripathy S, Gurung P. 1996. *Chronicling Intellectual Property Contributions of Tribal/ Rural Women and Men of Orissa with reference to Conservation and Enhancement of Rice Genetic Resources.* Unpublished report submitted by MS Swaminathan Research Foundation, Chennai to Ramon Magsaysay Award Foundation, Philippines.

Swaminathan MS, Castillo G. 2000. "Foreword." Pages xi–xii in Almekinders C and de Boef W (eds) *Encouraging Diversity: The Conservation and Development of Plant Genetic Resources,* Intermediate Technology Publications, London.

Swaminathan MS. 2002. The Past, Present and Future Contributions of Farmers to the Conservation and Development of Genetic Diversity. Pages 23-31 in Engles JMM, Rao VR, Brown AHD and Jackson MT (eds) *Managing Plant Genetic Diversity*, CABI Publishing, Wallingford, Oxon, UK.

Trenbath B. 1974. Biomass productivity of mixtures. *Advanced Agronomy* 26:177-210.

Tripathy S, Gurung P, Sharma SD. 2005. Intellectual property contributions with regard to rice genetic resources by tribes of south Orissa, India. *Plant Genetic Resources Newsletter* 141:70-73.

Tripathy S. 1994. *Ecogenetic Differentiation in Oryza sativa* L. Ph D Thesis, Central Rice Research Institute, Cuttack, Orissa, India. 238 pp.

Worede M, Mekbib H. 1993. Linking genetic resource conservation to farmers in Ethiopia. Pages 78-84 in Boef W de, Amanor K, Wellard K and Bebbington A (eds) *Cultivating Knowledge*, Intermediate Technology Publications, UK, p 78-84.

Appendix 17.1. Landraces of Jeypore Tract of Orissa with multiple use value.

Scented LRs:

Sl.no	Collection No.	Name of the Variety	Type of Land	Duration (Days)	Remarks
1	ST 215	*Loctimachhi Dhan*	LL	115-120	Scented
2	ST 217	*Basuabhog Dhan*	LL	140	Scented
3	ST 218	*Dudhmani Dhan*	LL	140	Scented
4	ST 219	*Tikichudi Dhan*	LL	110	Scented
5	ST 220	*Chatianaki Dhan*	LL	130	Scented and survives in flood water, helps in weeding as leaf sheath colour is purple
6	ST 226	*Samudrabali Dhan*	LL	180	Scented, not susceptible to disease and pest, tolerates more water in the field
7	ST 263	*Bastabhog Dhan*	LL	150	Scented, Deep water
8	ST 69	*Loctimachhi Dhan*	LL	135	Scented
9	ST 239	*Donger Haldichudi Dhan*	UL	95	Scented, tasty,
10	ST 82	*Benia Dhan*	Flood Plain	130-140	Scented
11	ST 83	*Badshabhog Dhan*	Flood Plain	130-140	Scented, 9 q per acre
12	ST 88	*Dubraj Dhan*	Flood Plain	140	Scented, survives 7-8 days in submerged condition, no diseases
13	ST 92	*Basuabhog Dhan*	LL	160	Scented
14	ST 252	*Chhota Badsabhog Dhan*	Flood Plain	120	Scented, not susceptible to diseases, 9 q/acre
15	ST 98	*Kalajeera Dhan*	LL	140	Scented
16	ST 101	*Benyakudi Dhan*	ML	120	Scented
17	ST 102	*Pandkaduda Dhan*	UL	100	Scented
18	ST 104	*Nadiarasa Dhan*	LL	140-150	Scented

Flood resistant LRs:

Sl.no	Collection No.	Name of the Variety	Type of Land	Duration (Days)	Remarks
1	ST 221	*Kalamangara Dhan*	LL	145	Survives in flood water
2	ST 222	*Dhoblichudi Dhan*	LL	145	Survives in flood water

3	ST 76	*Sopuri Dhan*	Flood Plain 140	Flood resistant
4	ST 77	*Gurmukhi Dhan*	Flood Plain 150	Flood resistant
5	ST 78	*Dhoabkhuji Dhan*	Flood Plain 150	Flood resistant
6	ST 84	*Nimchudi Dhan*	ML 120	Flood resistant
7	ST 86	*Basnimudi Dhan*	Flood Plain 160	Flood resistant, survives 7-8 days in submerged condition, no diseases, 3ft long straw
8	ST 87	*Haldigandhi Dhan*	Flood Plain 160	Flood resistant, survives 7-8 days in submerged condition, no diseases
9	ST 89	*Kandulkanti Dhan*	Flood Plain 160	Long glumes present, survives 7-8 days in submerged condition, no diseases
10	ST 89A	*Kandulkathi Dhan*	Flood Plain 160	Flood resistant, survives 7-8 days in submerged condition, no diseases
11	ST 206	*Kandulkathi Dhan*	Flood Plain 160	Flood resistant
12	ST 206A	*Haldiganthi Dhan*	Flood Plain 165	Flood resistant
13	ST 211	*Bagdichudi Dhan*	Flood Plain 135	Flood resistant
14	ST 212	*Bansaganthia Dhan*	Flood Plain 155	Flood resistant
15	ST 213	*Gurumukhi Dhan*	Flood Plain 155	Flood resistant
16	ST 214	*Neemchudi Dhan*	Flood Plain 125-130	Flood resistant
17	ST 246	*Baikhana Dhan*	Flood Plain 160	Flood resistant
18	ST 247	*Badamakada Dhan*	Flood Plain 180	Flood resistant, survives for 8-10 days under water, 3ft long straw
19	ST 249	*Kasamjopa Dhan*	Flood Plain 190	Flood resistant, tolerates 7-8 days of submergence, 3-4 ft long straw, yields 10-11 q per acre, not susceptible to diseases
20	ST 253	*Madraschudi Dhan*	Flood Plain 150	Flood resistant
21	ST 254	*Manchha Dhan*	Flood Plain 150	Flood resistant, survives 8 days under water, tasty
22	ST 255	*Kharkoili Dhan*	Flood Plain 160	Flood resistant, survives 8 days under water, tasty
23	ST 256	*Bilatichudi Dhan*	Flood Plain 150	Flood resistant, not

					susceptible to diseases, tasty, good market price, 8-9 q yield per acre
24	ST 257	*Kanakchudi Dhan*	Flood Plain	180	Flood resistant, survives 4-5 days in submerged condition, not susceptible to diseases, tasty, good market price, 8-9 q per acre, good for puffed rice and pressed rice
25	ST 259	*Kosamchudi Dhan*	Flood Plain	170	Flood resistant, survives 8-15 days in submerged condition, notsusceptible to diseases, suitable for puffed and pressed rice, tasty
26	ST 260	*Banktichudi Dhan*	LL	120	Flood resistant, survives 8-15 days under water, no disease, good for pressed and puffed rice, tasty and yield is good
Pressed, Puffed and Popped rice:					
1	ST 216	Pathangada Dhan	LL	140-145	Puffed rice
2	ST 227	Mallimakada Dhan	LL	180	Not susceptible to disease, Puffed and pressed rice
3	ST 261	Kaisari Dhan	LL	150	No disease and pest, puffed rice, pressed rice, tasty, 7.5 q per acre
4	ST 264	Pandkaguda	LL	120	Pressed rice, puffed rice, no disease and pest
5	ST 266	Kasikamala Dhan	LL	135	No disease, puffed and pressed rice, tasty
6	ST 267	Gathiachila Dhan	LL	150	No disease, puffed and pressed rice, tasty
7	ST 229	Khatra Dhan	UL	90	Tasty, pressed rice and puffed rice
8	ST 230	Manchhamalli Dhan	LL	135	Pressed rice and puffed rice
9	ST 233	Kalamatia Dhan	LL	135	Disease resistant, puffed rice, pressed rice, tasty
10	ST 234	Lodiari Dhan	LL	125-130	Not susceptible to any disease, pressed rice, tasty
11	ST 235	Docrikuji Dhan	LL	125-130	Not susceptible to any disease, puffed rice, tasty

12	ST 236	Kasikamana Dhan	LL	140	Puffed and pressed rice, no disease
13	ST 238	Sunaseri Dhan	LL	150	Pressed and puffed rice, tasty
Deepwater:					
1	ST 263	Bastabhog Dhan	LL	150	Small grain, scented and can survive under deep water
2	ST 258	Goindi Dhan	Flood Plain	180	Flood resistant and deep water i.e. it survives for a month under 14 ft water, 7-8 ft long straw, no disease, 15 q per acre, tasty
Profuse tillering:					
1	ST 242	Lakmipati Dhan	LL	160	Not susceptible to any disease, 15-17 tillers per plant, 7.5-9 q /acre, tasty,
Terraced Land:					
1	ST 244	Sapru Dhan	LL	195	Coarse grain, awned, tasty, good yield
2	ST 245	Jhara Dhan	LL	175	Red kernel
3	ST 105	Gobarluchei Dhan	LL	160	Light red kernel
4	ST 106	Kerandi Dhan	LL	180	Red kernel
5	ST 107	Chipti Dhan	UL	100	Red kernel
6	ST 110	Sonakhadi Dhan	LL	210	-
7	ST 186	Nini Dhan	LL	210	Scented, short oval grain
8	ST 187	Rangalochei Dhan	LL	180	Course grain
9	ST 188	Sapur Dhan	LL	210	Course grain
10	ST 189	Bayagunda Dhan	LL	180	Course grain
11	ST 190	Kerandi Dhan	UL	90	Red kernel
12	ST 197	Bada Dhan	LL	150	-
13	ST 198	Luchei Dhan	LL	150	Awned

Genetic Resources of Rice in India : 587-594
Editor : S. D. Sharma
Today & Tomorrow's Printers and Publishers, New Delhi - 110 002 (India)

Chapter 18

Collection of Unique Rice Germplasm from the Cradle of Rice *(Oryza sativa* L.) in Eastern Uttar Pradesh

R. C. Chaudhary and S. B. Mishra

Introduction

Eastern Uttar Pradesh (UP) in India borders Nepal in the north, Bihar in the east and Chhattisgarh and Madhya Pradesh in South (Fig. 18.1). Activities in germplasm collection, cataloguing and conservation have been handled by a civil society organization called Participatory Rural Development Foundation (PRDF) located in the district of Gorakhpur in eastern UP. Founded in 1997, PRDF has under its "Nature and Care" component, major activity of collecting, cataloguing and conserving germplasm. Collection is focused from eastern part of UP (Table 18.1). The area is typically called *Tarai*, which is formed by the soil brought by rivers originating in Himalaya Mountain. Himalaya was formed more than 20 million years ago when the Indian plate, while moving across Tethys Sea, collided with southern tip of fixed Asian Plate. Some 65 million years ago, the place of Himalaya was the floor of Tethys Sea on the northern part of India. After the collision this part rose to form Himalayan ranges; the principal uplift occurred around 12 million years ago.

Uttar Pradesh: the Cradle of Rice Cultivation

Uttar Pradesh (UP) that forms a synclinorial depression between peninsular India and Himalayan ranges, lies on the western rim of the geographic area where rice (*Oryza sativa* L.) is supposed to have been domesticated. The history of rice cultivation in this region dates back almost to 5,000 BC and some of the fossilised rice excavated are yet to be studied fully (Chaudhary *et al*. 2005a, 2007b) (Fig. 18.2). This sample was discovered at Piprahwa in Siddharth Nagar district

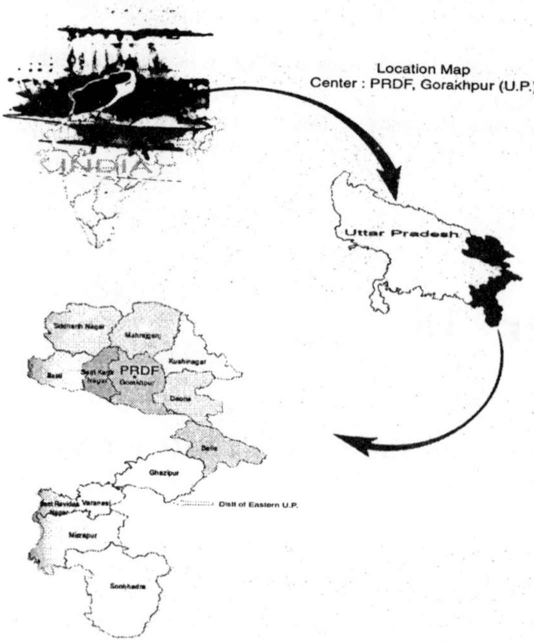

Fig. 18.1. Map of the area in India depicting the collection sites.

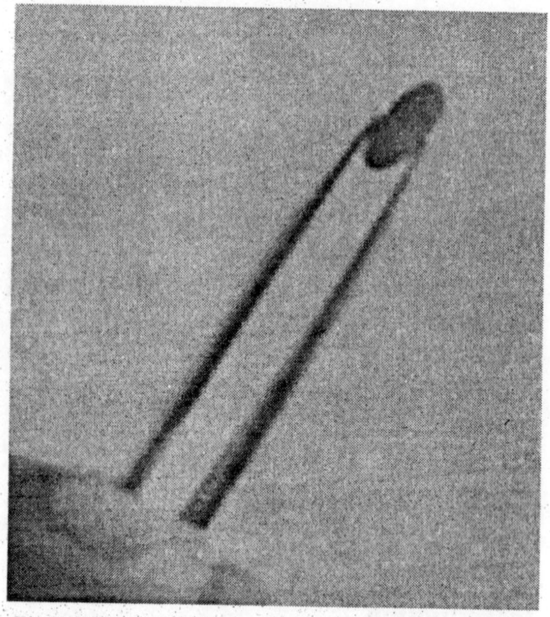

Fig. 18.2. Undated carbonized grain of rice collected from Siddharth Nagar, UP.

Fig. 18.3. Raising of *boro* crop during winter as per the traditional procedure.

Fig. 18.4. An accession of semi-dwarf Boro (left) as compare to normal *boro* (right).

of UP. Many such samples are available at the local universities of Gorakhpur and Allahabad discovered from excavations made in UP. Even the most primitive rice called *boro* is cultivated in this area (Chaudhary, 2003; Chaudhary and Chaudhary, 2004). *Boro* is supposed to be the link between *indica* and *japonica* forms of *Oryza sativa* L. (Chaudhary *et al.* 2007a, 2007b).

In addition to normal rice, there are several types of aromatic rices cultivated in this region (Chaudhary and Tran, 2001). There are most photosensitive and long duration varieties like Kalanamak, cultivated under Late Rainfed Lowland situation as long duration varieties. There are varieties of deepwater or floating rice which reach up to heights of five metres and can elongate up to 30 cm in one night. Besides, there are upland rices with short duration (60 days) and drought tolerance. This is in addition to normal irrigated rice situation. Of special mention is the *boro*, which grows during winter season tolerating 4° C, and matures in summer tolerating 46°C during reproductive stage. These types of rice ecosystems have developed their unique setting for rice cultivation. Rice is life in this region as during *kharif* (rainy season) rice is cultivated in over 98% of the cropped area.

No wonder the available genepool in eastern part of UP is highly diverse (Dwivedi *et al.*, 2005, Roy *et al.*, 1975, Chaudhary *et al.*, 2007b). Even the rice germplasm collections from lower hills of Himalaya are diverse and distinct (Nanda *et al.*, 1976, 1977a, 1997b, Chaudhary *et al.*, 1978). Thus it is adapted to various land forms and agro-ecosystems prevalent in the region. The genepool evolved under pressure of natural selection and also under domestication has resulted in the diversity in cultural practices vis-à-vis rice varieties. This germplasm has greatly contributed to the genetic diversity of rice varieties of other countries. Based on all these factors, Chaudhary (1999, 2000) proposed to protect the right of the farmers to use the germplasm.

Collection of Unique Rice Germplasm

Uttar Pradesh is quite rich in genetic diversity for rice and earlier explorations have yielded rich dividends (Roy *et al.*, 1975, Chaudhary *et al.*, 1978, 1984, Nanda *et al.*, 1976, 1977a, 1977b) be it common rice, hill rice, aromatic rice or *boro*. Collection of common rice germplasm from this state has been described in Chapter 15 of this book. Therefore, collection and conservation of only unique rice germplasm like *boro* rice and aromatic rice have been described in this chapter.

Collection of boro *rice*

Boro (*Oryza sativa* L.) is a distinct group of primitive rice which is grown sown during the winter season (October seeding) and harvested during summer season (May). Thus it is exposed to extremes of cold (4^{0}C) during its vegetative phase and extreme heat (46^{0}C) during reproductive phase. It is cultivated in abandoned and inundated areas around lakes, rivers and depressions which remain un-cropped during rainy season. As the water recedes, *boro* seedlings are raised during October and transplanted around February and harvested around May. Thus it takes 7 to 8 months to mature. Specific tribes of fishermen cultivate this *boro* and claim to receive very high yield. *Boro* rice grown in the eastern part of UP is very distinct from normal rice and even "*boro*" grown in Assam and West Bengal which connotes to a season. As will be clear, later *boro* can tolerate extreme cold of 4^0 C during vegetative phase and extreme heat of 46^0 C during reproductive phase. It can tolerate drought also. It has very weak stem and collapses at maturity. Farmers differentiate between *boro* and common rice. In their terminology, rice is *dhan* but *boro* is not *dhan* rather *boro* only. Morphologically, it is also different from rice. Besides crosses of *boro* with common *Oryza sativa* L. gives sterile hybrids and even cytoplasmic-genetic sterility was identified (Chaudhary *et al.*, 1984) from the earlier collection made from the same area (Nanda and Chaudhary, 1977).

In the second collection which is very exhaustive, a total 558 accessions were collected and identified after rejecting the duplicates. For this, collection teams collected directly from fields and some from threshing floors and barns in May–June months of year 2000 and 2001.

Collection of aromatic rice

The area is known also for aromatic rice of various grain types among different duration groups (Dwivedi *et al.*, 2005). No previous systematic collection was made from this area. Thus by sending teams of plant explorers at different times of the cropping season more than a hundred accessions were collected. After rejecting possible duplicates, 90 accessions were thoroughly evaluated and catalogued for 43 morpho-agronomic and grain quality characters.

Evaluation, Cataloguing and Conservation

The first collection of *boro* rice from eastern UP was made during 1974-75 (Table 18.1) which was evaluated and a catalogue entitled *Germplasm of Boro Paddy (Oryza sativa) from Uttar Pradesh, India* was published in 1977. The

Table 18.1. Rice germplasm collected, funded by GBPUAT Pantnagar (1974-75) and NATP of Government of India (1999-2004).

Crop / Group	1974-75	2000-01	2001-02	2002-03	2003-04	Total
1.*Oryza sativa* L. Boro	90	-	-	-	-	90
2.*Oryza sativa* L. Boro	-	-	558	-	-	558
3.*Oryza sativa* L. Traditional	-	11	60	-	42	113
4.*Oryza sativa* L. Scented	-	-	35	25	206	266
5.*Oryza sativa* L. Boro	-	-	-	-	37	37
TOTAL	**90**	**11**	**653**	**25**	**285**	**1,064**

collection was well studied and well utilized by the scientific community (Nanda and Chaudhary, 1977; Nanda *et al.* 1976; 1977a, 1977b; Roy *et al.*, 1975). Barring a few accessions stored with the prefix UPRB (UP Rice Boro) in the International Rice Gene Bank located at International Rice Research Institute (IRRI), the entire collection was lost during the civil disturbances at Pantnagar University during 1978. Second and the most comprehensive collection (Table 18.1) of *boro* (558 accessions) was evaluated for 43 morpho-agronomic and grain quality characters (Table 18.2) (Fig. 18.3) and a 147–page catalogue entitled *Catalogue of Boro Rice (Oryza sativa L.) Germplasm from Uttar Pradesh, India* was published. Third, and probably the last, as *boro* has almost disappeared from this area, collection of 37 *boro* accessions (Table 18.1) was made and sent for long-term conservation at National Gene Bank at NBPGR, New Delhi without further evaluation for cold tolerance.

Another catalogue on aromatic rice entitled *Catalogue of Aromatic Rices from Eastern UP* is under print (Chaudhary *et al.*, 2010) by the UP Council of Agricultural Research, Lucknow. Both the collections, *boro* rice and aromatic rice, are conserved under long-term storage at National Gene Bank located at the National Bureau of Plant Genetic Resources (NBPGR), New Delhi. IC numbers assigned for *boro* collections are IC 440918 – IC 141475 and for aromatic ice are IC 551797 – IC 551926. Another set has been conserved under medium term storage at Central Rice Research Institute (CRRI), Cuttack (Orissa), India. Thus these have been conserved for use by the present and future generations of the mankind.

Accessions of Special Significance

Boro rice collections represent a reservoir of genepool of rare and unique

Table 18.2. Evaluation of rice germplasm for 43 morpho-agronomic and grain quality characters.

Morpho-agronomic characters		Grain quality
1. Seedling vigour	16. Panicle exsertion	1. Brown rice length
2. Seedling height	17. Panicle length	2. Brown rice breadth
3. Cold tolerance	18. Apiculus colour	3. Length: Breadth ratio
4. Leaf length	19. Lemma & Palea colour	4. Grain class
5. Leaf width	20. Lemma & Palea pubescence	5. Brown rice colour
6. Leaf blade colour	21. Spikelet fertility	6. Head rice recovery
7. Basal leaf sheath colour	22. Heat tolerance	7. Amylose content
8. Leaf blade pubescence	23. Maturity	8. Grain chalkiness
9. Tillering ability	24. Husk colour	9. Aroma
10. Day to 50% flowering	25. Grain length	10. Hulling %
11. Awning	26. Grain width	11. Milling %
12. Awn colour	27. Resistance to pests	12. Water uptake
13. Culm length	28. Resistance to diseases	13. Volume expansion
14. Culm strength		14. Cooked kernel length
15. Plant height		15. Elongation ratio

germplasm. Good donors for extreme cold and heat tolerance (4° C to 46° C) can be identified and used in the breeding programme. PRDF-Boro 51 can be used as an alternate source (Chaudhary *et al.*, 2003) of semi-dwarfism (Fig. 18.4). Based on the traditional knowledge of the tribal people who grew *boro* since time immemorial, nutritional and medicinal value of *boro* is well established. Thus several of these will be good source of high protein, iron, Zn, etc. PRDF-Boro 271 has unique kernel colour not reported in any accession. This could be used as a marker character.

Among the aromatic rice, there are specific types having aroma that is different from that of Basmati. Once methods to specify aroma become known, it would help to differentiate different varieties which, at present, only consumers can differentiate. These aromatic rices represent photoperiod sensitive to non-sensitive types and early (90 days) to late types. Pureline selection from one lot was already released as Kalanamak KN3 by UP State Variety Release Committee in 2007 (Chaudhary *et al.*, 2008).

Advocacy on Farmers' Right and IPR

Importance of these unique collections in breeding and genetic studies can be hardly emphasised. Thus under the new Seed Act and IPR and Farmers' Right, right of the farmers has been advocated by several forums and PRDF has been strong and vocal about it (Chaudhary, 1999, 2000). As farmers are the custodian of these germplasm, any financial gains and recognitions must go to them. Thus in the catalogues and information available with the National Gene Bank, GPS readings and names and addresses of the farmers have already been provided.

Conclusion

After domestication, rice evolved in the farmers' fields under cultivation according to the preference of the farmers, vagaries of the weather and pressure of biotic and abiotic stresses. Thus farmer is the real custodian of this heritage of mankind. Anything lost can be recovered but not the germplasm. Thus we are thankful to the present and past generations of the farmers who were able to maintain this enormous and unique diversity in the rice germplasm. Local officials of the Department of Agriculture of UP and PRDF staff who participated in various explorations deserve our appreciation and thanks. The responsibility and credit for conserving this genetic diversity now rests with the new generation of germplasm conservers, namely, Gene Banks at National Bureau of Plant Generic Resources at New Delhi and Central Rice Research Institute, Cuttack (Orissa) who are maintaining this unique germplasm *ex-situ* for long and medium term storages respectively.

References:

Chaudhary N, Chaudhary RC, Chaturvedi ON, Verma IJ, Dubey DN. 2003. First semi-dwarf *boro* rice (*Oryza sativa* L.) germplasm collected in eastern UP, India breeds true. *Intern Rice Res Newsn* 28.1/2003. 31 p.

Chaudhary N. 2003. *Collection, Conservation and Morphological Evaluation of Boro Rice (Oryza sativa L) with Reference to Incidence of Diseases and Pests from Eastern Uttar Pradesh, India.* Ph D Thesis, University of Gorakhpur, Gorakhpur, India.

Chaudhary Nisha, Chaudhary RC, Proverbio L. 2007a. The most primitive rice under cultivation by tribals: Reality of *Boro* rice (*Oryza sativa* L.) of eastern Uttar Pradesh, India. Page 177–199 in *Geographical Researches on Rice: A Comparative Analysis of Rice Districts in the European Union and India.* Edizioni Mercurio, EU–ECCP, Edizioni Mecurio, Vercelli, Italy.

Chaudhary RC, Chaudhary N. 2004. *Catalogue of Boro Rice (Oryza sativa* L.) *Germplasm from Uttar Pradesh, India.* PRDF Gorakhpur, India and UPCAR Lucknow, India, 129 pp.

Chaudhary RC, Dwivedi JL, Singh J. 2010. *Catalogue of Aromatic Rices from Eastern UP.* UP Council of Agricultural Research, Lucknow, India. 150 pp (under print).

Chaudhary RC, Mishra SB, Dubey DN. 2008. Scented rice variety Kalanamak and its cultivation for better quality and high yield. *Rice India* August 2008: 21-26.

Chaudhary RC, Prajapati SC, Chaudhary N, Anita, Shukla AK. 2007b. Diversity in rice, its cultivation and its uses in the area of its origin: Eastern Uttar Pradesh, India. Pages 165-176 in Davide Papotti, Carlo Brusa (eds) *Geographical Researches on Rice: A Comparative Analysis of Rice Districts in the European Union and India.* EU-ECCP, Edizioni Mecurio, Vercelli, Italy.

Chaudhary RC, Sahai VN, Saran S. 1984. A possible new male sterile line of rice with cytoplasm from *Boro* rice of Uttar Pradesh. *Intern Rice Res Newl* 1:104.

Chaudhary RC, Singh H, Singh JP, Singh BN. 1978. Rice germplasm from higher hills of Himalayas in Uttar Pradesh. *Internatl Rice Res. Newsl* 3(3):12.

Chaudhary RC, Tran DV. 2001. *Speciality Rices of the World: Breeding, Production and Marketing.* FAO Rome, Italy. 358 pp.

Chaudhary RC. 1999. Ethics and mechanisms to internationalise elite germplasm while still honouring IPR and farmers' right. In Langrdige *et al.* (eds) *Proc 11th Australian Plant Breeding Conf* Vol 2. Adelaide, Australia. 119 pp.

Chaudhary RC. 2000. Exploitation of plant genetic resources: Many tools, many rules. Paper presented in *International Conference on Managing Natural Resources for Sustainable Agricultural Production in 21st Century.* New Delhi, India, February 2000. P 12.

Chaudhary RC. 2005. Collection and cataloguing of *Boro* rice (*Oryza sativa* L.) germplasm from eastern Uttar Pradesh. Pages 5-7 in *Salient Achievement from Shod Nidhi-funded Projects 1995-2004.* UPCAR, Lucknow, India.

Dwivedi JL, Tiwari T, Chaudhary RC, Pandey A, Singh R, Singh RK. 2005. Short grain aromatic rices: Strategies for improvement and future opportunities. Pages 13-21 in Singh RK, Singh US (eds) *Scented Rices of Uttar Pradesh and Uttarakhand.* Kalyani Publisher, New Delhi.

Nanda JS, Chaudhary RC, Rana OS. 1976. Blast resistance in cold tolerant *boro* rice collections of UP. *Intern Rice Res Newl* 1:10-11.

Nanda JS, Chaudhary RC, Rana OS. 1977a. Field resistance for blast in rice collections from Kumaon hills of UP. *Sci & Cult* 43:343-345.

Nanda JS, Chaudhary RC, Singh H, Singh BN, Singh JP. 1977b. Two possible new source of resistance to bacterial leaf blight in rice. *Intern Rice Res Newl* 2(1):2.

Nanda JS, Chaudhary RC. 1977. *Germplasm of Boro Paddy (*Oryza sativa*) from Uttar Pradesh, India.* (Tech. Bulletin No. 102). Directorate of Experiment Station, GBPUAT, Pantnagar, India. 11 pp.

Roy JK, Mehra KL, Borthakur DN, Chaudhary RC, Hamdani AR. 1975. Collection, maintenance and utilization of genetic resources in rice: A status report for India. Proc Symp *IBPGR-IRRI Workshop on Conservation of Rice Genetic Resources.* IRRI, Manila, Philippines, 12 -15 December 1977. p 54.

Genetic Resources of Rice in India : 595-604
Editor : S. D. Sharma
Today & Tomorrow's Printers and Publishers, New Delhi - 110 002 (India)

Chapter 19

Varietal Diversity of Rice as Recorded in Ancient Texts of India

K. Pavithran and S. Anitha

Introduction

Plants are the living factories of nature, manufacturing many medicinally important drugs. Since time immemorial, man has used these drugs to relieve pain and cure diseases. Black rice in India, Japan and also in China, perhaps, is known as "blood strengthening rice/drug rice/contributed rice" as it has body strengthening function and pharmaceutical value (Jian and Zhan, 1989) with low yield.

Rice has a greater quality index than other crop plants and consolidation of information on the quality index of rice (Table 19.1) showed that coloured and coarse grains have more protein, minerals, phosphorus and thicker bran layer; short duration rices are coarse and superior; parboiled rice has higher nutritive value; rice protein is superior to all cereal proteins and is richer in essential amino acids; coloured rices contain 2-3 times more iro;, rice oil is rich in vitamin E and is more nutritious; rice contains small amount of Vitamin A but is rich in vitamin B; whole grains (kernels) contain the same amount of protein as in wheat; low content of lysine is the most serious amino acid deficiency and wetland rices have more proteins and minerals than dryland rices (Table 18.1). Recent studies have shown that improvement of medicinal quality of rice grains/food grains would also help development of the concept of therapy through food by introducing medicinal and nutritive properties through genetic engineering. This has been recently recognised as an area of potential research with broad future (Jian and Zhan, 1989).

Table 19.1. Quality index of rice.

No.	Character/Particulars	Quality indications
1.	Coarse, coloured grains	More protein and minerals, richer in phosphorus
2.	Coarse grains	Thicker bran layer
3.	Short duration rices	Superior and coarse grained
4.	Parboiled rice	Higher nutritive value
5.	Rice protein	Superior to even wheat
6.	Rice protein	Richer in essential amino acids
7.	Coloured rice	2-3 times more iron
8.	Rice oil (bran oil)	Richer in vitamin E and nutrition
9.	Rice vitamins	Rich in B, poor in A, D
10.	Vitamin B complex	Thiamin = 4, Riboflavin = 0.6, Niacin = 0.50 microgram/gram
11.	Kernels	Nearly same amount of protein as in wheat
12.	Lysine	Low amount / most serious amino acid deficiency
13.	Split dose fertilizers	Protein content increases upto 20%
14.	Wetland rice	More minerals and proteins
15.	Polishing	12.5% loss
16.	Washing rice and loss of boiled rice water	Reduces nutrients of cooked rice

The present study relates to the preliminary search (a) to identify and consolidate the available information on ancient medicinal rices, (b) to bring forth information on their nomenclature, conservation and classification systems followed in ancient days and (c) to consolidate studies on biochemical characterisation of such rice genetic resources of high medicinal values.

The information recorded in the ancient literature such as *Astangasamgraham, Astangahridayam, Susrutasamhita, Charakasamhita, Amarakosam, Pharmacopoeia of India, Indian Medicinal Plants, Glossary of Indian Medicinal Plants, Medicinal Wealth of Andhra Pradesh, Krishigita,* etc. on ancient medicinal rices was scrutinised and their relevance in terms of nomenclature and classification systems and medicinal values was examined critically.

Nomenclature, Classification and Characterisation of Medicinal Rices in Ancient India

The ancient Ayurvedic literature such as (1) *Susrutasamhita* written by the great Susruta in 1500 BC, (2) *Charakasamhita* by Charaka in later period, (3) *Astangahridaya* and (4) *Astangasangraha* by Vagbhata in the 15th century (5) *Pharmacopea of India* by Warring in 1868, *Indian Materia Medica* by Dutt in 1900, and books published later in the 20th century have consistently referred to the use of rices for medicinal purposes, both for direct treatment and in the preparation of certain medicines. The ancient literature also has mentioned about classification, identification and certain characterisation of such rices, of course, in a broad sense (Tables 19.2 and 19.3).

Table 19.2. Ancient system of nomenclature of rice varieties.

Basis for nomenclature	Nomenclature/ varietal names	Meaning	Remarks
Grain colour	Pandu	White colour	refers to leucoderma
	Peethakam	Yellow/golden yellow	
	Padalam	light red colour	
	Raktam	deep red	blood colour
	Rakta	deep red	deep red
	Lohabalam	red (chennellu)	metalic strength/ iron red
	Kanchanam	golden colour	
	Kakalakam	black like crow	crow colour
	Kuravakam	black wild rice	
	Kedarakam	black navara	
	Chennellu	red paddy	
	Kanchanakam	golden colour	
	Perumchennellu	deep red/ big sized chennellu	
	Cheenam	yellow-chinese	
	Tapaneeyam	golden red	hot gold
	Gauram	white rice	
Grain/Kernel odour	Sugandhikam	sweet smell	

	Mukundakam	red onion smell	
	Ujwalakam/ Ujwalam	deep scented	
	Choornakam	sweet smell of talcum powder	
	Jeerakasala	coriander smell/ size or shape	
	Gandhakasala	scented rice	
	Saribam	smell like hemidesmus	
	Pramodam	pleasant smelling	
Comparison with animals	Langulam	horse tail like awned	
	Sakunahritam	bird-loving paddy	
	Sookaram/ Pannikuttan nellu	easily cultivable in wet lands	refers to pig/ panni
	Mahisham	buffallo	refers to body building
	Kukkutandakam/ Kozhikuttan nellu	shaped like egg/ chicken	
	Kakalakam	black navara	crow coloured
Seed qualities	Pramodam	causing pleasant experience by eating	
	Lodhrakom	disease resisting	
	Lohabalam	strengthening quality	
	Seetabheerukam	susceptible to cold	
	Saradam	dry season variety	
	Sukaram	easily cultivable	
	Saramukham	high quality rice	
	Mahat	high quality/ high yielding rice	
	Mahisham	body building rice	
Habit	Deerghasookam	Long awned chennellu	
	Patakam	shattering type	
	Langulam	horse tail awned	
	Kolambakam	throat clearing	
Seasonal	Hayanam	insensitive types	

Kalamam	late variety – wet land
Twaritakam	early flowering
Jangalam	dry land paddy

Table 19.3. Genetical control of certain biochemical characters in rice.

No.	Character	Symbol	Ratio	Reference
1.	Amylose content	*Am*	3:1	Hsieh and Wang, 1989
2.	Gelatinisation temperature	*Gt*	3:1	”
3.	Hardness of cooked rice	*H*	3:1	”
4.	Phenol reaction	*Ph*	3:1	Morinaga *et al.*, 1943, Pavithran, 1985
5.	Alkali degeneration	*alk*	3:1	Kudo, 1968; Oka, 1953a
6.	Dull endosperm	*du*	3:1	Satoh and Omura, 1981; Yano *et al.*, 1980
7.	Scented kernel	*Sk-a, b, c, d*	3:1, 9:7, 15:1, 37:37, 81:175	Dhulappanavar, 1976; Jodon, Kadam & Patankar, 1938; Ramiah and 1955; Rao, 1953; Reddy and Sathyanarayanaiah, 1980; Tripathi and Rao, 1979
8.	Sugary endosperm	*su*	3:1	Isono *et al.*, 1978; Satoh and Omura, 1981; Yano *et al.*, 1980.
9.	Waxy endosperm	*wx (am, gl)*	3:1	Chao, 1928; Nagao and Takahashi, 1963; Chang, 1964

Susrutasamhita referred to three types of rices – awned type (*sali vargam*), *navara* or early duration type and *vreehi* type or 60 days duration rices with their corresponding medicinal uses (Fig. 19.1).

According to *Charakasamhita* only two types – awned and early duration types – are described with little information on their medicinal uses (Fig. 18.2). However, in both classifications the types are recognised on the basis of awning habit and early duration. It is also noted that both Susruta and Charaka recognised the use of early duration rices like *Navara* (*Shastikam*) for "njavarakishi". Njavarakishi is an effective form of Ayurvedic treatment for acute rheumatic and allied complaints. It has been reported commonly that even completely paralytic patients have been cured of their troubles and could walk again and live long by this treatment. Similarly, *Astangasamgraha* also recognised early duration and awned rices to have medicinal properties. The later literature showed only repetitions of the ancient ideas rather than any new information.

Medicinal Values of Rices Mentioned in Ancient Literature

Awned rices are reported to have body building quality, anti-anaemic, rheumatic cure, urine inducing, ulcer curing qualities, and the *nivara* and early types are used to cure rheumatism and anaemia, for body building, to cure urinary diseases and for *navarakishi*, constipation, and to produce gastric enzymes (Tables 19.2 and 19.3). *Astangasamgraham* refers to the use of rice roots and flour of awned rices in medicine preparations such as *sukumaraghritam*, *ajamamsarasayanam*, *gatwaram kushampu*, etc. besides other uses (Fig. 19.3).

Early flowering types have medicinal properties for curing rheumatic and bronchial diseases, for body building, as anti-diabetic and to induce urine and heat in the body (Figs. 19.1, 19.2, 19.3).

Rice (Nellu)

Awned types (*Sali vargam*) Late duration No. described = 17	Nivara types (early duration) No. described= 10	*Vrihi vargam* (early 60 days) Nil
Medicinal uses:	*Medicinal uses:*	*Medicinal uses:*
1. body building	1. cure for rheumatism, anaemia	1. causes constipation
2. anti-anaemic & related diseases	2. body building & strengthening	2. chennellu is useful for bronchial anaemia
3. induce rheumatic & bronchial troubles	3. phlegm producing	3. induce foul gas formation & gastric enzymes
4. induce urine	4. shashtikam nivara is used for "Njavarakizhi"	
5. curing ulcers, fever, chronic rheumatic, anaemic & bronchial diseases	5. Navara of 3-4 months old is good for urinary diseases	

Fig. 19.1. Classification, characterisation and medicinal properties of rices (*Susrutasamhita*, 1500 BC).

Rice (Nellu)

Awned rice (*Sali vargam*) (late flowering) No. described = 10	Early duration rice (*Vrihi vargam*) No. described = 7
Medicinal uses:	*Medicinal uses:*
Nil	1. for urinary infection 2. Shashtikam is used for 'Njavarakizhi'

Fig. 19.2. Classification, characterisation and medicinal properties of rices (*Charakasamhita*).

Rice (Nellu)

Awned rice (*Sali vargam*) No. described = 29	Early flowering (*Vrihi vargam*) No. described = 16
Characterisation:	*Characterisation:*
long awned / awned/ white-rice / 4 – 4½ months duration	early duration of 60 days/red rice
Medicinal utility:	*Medicinal utility:*
Roots used in preparations like *Sukumaraghritam* & *trinapanchakralam*	
1. for rheumatic and bronchial diseases 2. to induce urine 3. for healthy eyes 4. for body building & body colour 5. remedy for thirst 6. against ulcers, fever. 7. for rheumatic & bronchial diseases	1. used against anaemia & related diseases 2. for body building 3. as anti-diabetic 4. induce urine and feacus 5. causes heat in body

Fig. 19.3. Classification, characterisation and medicinal properties of rices in *Astangasamgraham* (15th century).

Ancient System of Nomenclature of Rice Varieties

The ancient system of nomenclature of rice varieties differed from the most modern approach to naming the varieties with the names of 'stars' and 'superstars' in the galaxy of socio-political 'luminaries'.

The ancient system considered most salient characteristics such as grain colour, kernel odour, comparison with animals, seed qualities, habits and seasonal or agronomic characteristics, etc., while naming the varieties. The vernacular names of ancient varieties of rice and their meaning in English with necessary remarks are furnished in Table 19.2. Yet in another later literature like *Krishigeeta*, a Malayalam poetry written on rice varieties and their cultivation (these collections are known as Macqueen's collections, 1950s, in India) refers to over 100 varieties of rices with different names in terms of their qualities and agronomic characters. Such vernacular names of rice varieties have been reported from the tribal rice centre of Wynad, Attapadi of Kerala, Jeypore of Orissa, Assam and Bastar of Madhya Pradesh in India (Richharia, 1979). It may also be noted that such names are often forgotten or ignored in the new systems of nomenclature. Unfortunately such varieties that existed earlier are already lost in several centres of diversity or being eroded by modern introduction of high yielding varieties. The loss of such medicinally valuable rices and other quality rices including characters of breeding value such as resistance to pests and diseases are disappearing rapidly. The senior author collected in early 1970s from the Wynad hilly tracts over 400 tribal rice varieties with traditional nomenclature; for example *Thondi, Marathondi, Perumal, Mullan puncha, Kodaku, Veliyan, Punnadan Thondi, Peruvaya, Chetuveliyan, Chennellu, Navara, Kuruva, Gandhakasala, Jeerakasala, Thonnuran Thondi, Kalladiaryan, Adukkan,* etc. referring to place of cultivation/names of persons (probably the ones who selected them or so), colour and quality of grains etc. However, subsequent visits in the recent past in early 1980s indicated rapid disappearance of many of these varieties and a rough estimate revealed that over 50% of these varieties are virtually lost today.

Exploitation of the medicinal potentialities of the rice genetic resources, however, depends much on our knowledge of the genetical control of subtle biochemical and physiological characters of the rice plant. Earlier workers cited (Table 19.3) have contributed to the genetical understanding of amylose content, gelatinisation temperature, hardness of cooked rice, phenol reaction, alkali digestion, dull endosperm, scented kernel, sugary endosperm and waxy endosperm, etc. and reported mostly monogenic control of these characters except that scented kernel showed up to four complementary genes/ duplicate genes for its control (Table 19.3).

Conclusions

Biodiversity is the primary basis of sustainable agriculture and crop improvement. The rapid rate at which the genetic variations within the species are being eliminated is a disturbing situation which demands our immediate attention and concern. The genetic erosion in rice, which has the highest quality index among the crop plant species, has become more serious today. The ancient rice genetic resources existing in the remote corners are being rapidly eroded by the so called developmental advances and spread of HYVs. The ancient systems of nomenclature, characterisation and classification reveal that there are certain desirable factors for breeders to follow up further including the concept of therapeutic application of food for curing or preventing diseases. Study has shown that most of the biochemical characters in these rices are controlled monogenically, while studies are urgently needed to have a clear picture of the genetic architecture of the rices of medicinal significance in relation to the active ingredients in the context of transgenic transformation of rice genetic resources.

References:

Chang TT. 1964. *Present Knowledge of Rice Genetics and Cytogenetics. (IRRI Tech Bull No 1).* International Rice Research Institute (IRRI), Manila. 96 pp.

Chang TT. 1988. Taxonomic key for identifying the 22 species in the genus Oryza. *IRRN* 13:4-5.

Chao LF. 1928. Linkage studies in rice. *Genetics* 13:133-169.

Dhulappanavar CV. 1976. Inheritance of scent in rice. *Euphytica* 25:659-662.

Hsieh SC, Wang LM. 1989. Genetical studies on physiochemical properties of rice grains. Pages 325-328 in Iyama S and Takeda G (eds) *Proceedings of VI International SABRAO Congress, 1989.*

Isono Y, Satoh H, Omura T. 1978. Characteristics of carbohydrate synthesis mutants, sugary and shrunken in rice. *Jap J Breed* 28 (Suppl):130-31.

Jian LB, Zhan LL. 1989. The study on the breeding of "Black Superior Rice" by using biotechniques. Pages 289-291 in Iyama S and Takeda G (eds) *Proceedings of VI International SABRAO Congress, 1989.*

Jodon NE. 1955. Present status of rice genetics. *J Agric Assoc China* 10:5-21.

Kadam BS, Patnakar VK. 1938. Inheritance of aroma in rice. *Chron Bot* 4:496-7.

Kudo M. 1968. Genetical and breeding studies on physiological and ecological characters in hybrids between ecological groups of rice. *Bull Natl Inst Agric Sci, Ser D* 19:1-84.

Micke A. 1989. Mutagen induced genetic changes as genetic resources for plant improvement. Pages 93-101 in Iyama S and Takeda G (eds) *Proceedings of VI International SABRAO Congress, 1989.*

Morinaga T, Nagamatsu T, Kawahara E. 1943. New linkage relation in rice. *Japan J Genet* 19:206-208.

Nagao S, Takahashi M. 1963. Genetical studies on rice plant. XXVII. Trial construction of twelve linkage groups of Japanese rice. *J Fac Agric Hokkaido Univ* 53:72-130.

Nayar NM. 1973. Origin and cytogenetics of rice. *Adv Genet* 17:153-292.

Oka HI. 1953. Variation of various characters and character combinations among rice varieties. *Japan J Breed* 3:33-43.

Pavithran K. 1985. Interrelationships of rice genes: Linkage and pleiotropy. 1st International Rice Genetics Symposium, IRRI, Philippines.

Ramiah K, Rao MBVN. 1953. *Rice Breeding and Genetics. Science Monograph.* Indian Council for Agriculture Research, New Delhi.

Reddy PR, Sathyanarayaniah K. 1980. Inheritance of aroma in rice. *Indian J Genet & Plant Breed* 40:327-329.

Richharia RH. 1961. Recent research in rice breeding and genetics. *Span Suppl* 4:1-39.

Richharia RH. 1979. An aspect of genetic diversity in rice. *Oryza* 16:1-31.

Satoh H, Omura T. 1981. New endosperm mutations induced by chemical mutagens in rice *Oryza sativa* L. *Japan J Breed* 31:316-326.

Tripathi RS, Rao MJB. 1979. Inheritance and linkage relationship of scent in rice. *Euphytica* 28:319-323.

Yano M, Satoh H, Omura T. 1980. Gene analysis on the endosperm mutants induced by NMU treatment in rice. *Japan J Breed* 30 (Suppl 1):260-261.

Index